POST-AGB OBJECTS AS A PHASE OF STELLAR EVOLUTION

ASTROPHYSICS AND SPACE SCIENCE LIBRARY

VOLUME 265

POST-AGB OBJECTS AS A PHASE OF STELLAR EVOLUTION

Proceedings of the Toruń Workshop held July 5–7, 2000

Edited by

R. SZCZERBA

and

S.K. GÓRNY

Nicolaus Copernicus Astronomical Center,
Toruń, Poland

KLUWER ACADEMIC PUBLISHERS

DORDRECHT / BOSTON / LONDON

A C.I.P. Catalogue record for this book is available from the Library of Congress.

ISBN 0-7923-7145-3

Published by Kluwer Academic Publishers,
P.O. Box 17, 3300 AA Dordrecht, The Netherlands.

Sold and distributed in North, Central and South America
by Kluwer Academic Publishers,
101 Philip Drive, Norwell, MA 02061, U.S.A.

In all other countries, sold and distributed
by Kluwer Academic Publishers,
P.O. Box 322, 3300 AH Dordrecht, The Netherlands.

Printed on acid-free paper

Printed in the Netherlands.

Contents

Preface xi

Acknowledgements xiii

Organizing Committees xiv

List of Participants xv

Conference Photo xx

Introductory Review

Proto-planetary Nebulae as a Phase of Stellar Evolution 3
S. Kwok

Part I **Surveys and Observational Data**

Observational Database and Statistical Properties of Proto–planetary Nebulae 13
R. Szczerba, S.K. Górny and M. Zalfresso–Jundziłło

Optical Survey of Post-AGB Candidates 21
A. Manchado, O. Suárez, P. García-Lario, M. Manteiga and S.R. Pottasch

Hot Post-AGB Stars 29
M. Parthasarathy, G. Gauba, T. Fujii and Y. Nakada

Properties of Post-AGB Stars with IRAS Colors Typical of Planetary Nebulae 41
G.C. Van de Steene, P.A.M. van Hoof and P.R. Wood

$BVRIJHK$ Photometry of Post-AGB Candidates: Evolution of Post-AGB Central Stars 45
T. Fujii, Y. Nakada and M. Parthasarathy

Near IR Observations of the Arecibo Sample of OH/IR Stars 49
F. Jiménez-Esteban, D. Engels and P. García-Lario

HST Imaging of Proto-planetary Nebulae and Very Young Planetary Nebulae – Towards a New Understanding of Their Formation 53
R. Sahai

Morphological Properties of PPNs: Mid-IR and HST Imaging Surveys 65
T. Ueta and M. Meixner

MSX Sources in the LMC: Dust-enshrouded AGB Stars and Post-AGB Stars 71
P.R. Wood and M. Cohen

Post-AGB Stars in Globular Clusters and Galactic Halos 77
H.E. Bond and D.R. Alves

Part II **Stellar Evolution and Winds**

The Evolution of AGB Stars Towards Planetary Nebulae 85
D. Schönberner and M. Steffen

Mass Loss Characteristics of Dust Forming Long-Period Variable Stars 93
J.M. Winters and T. Le Bertre

Pulsations in Proto-Planetary Nebulae as seen From Light Curve Studies 101
B.J. Hrivnak et al.

Atmospheric Dynamics of Post-AGB Stars: New Results on the Nature of the Pulsation of HD 56126 105
A. Lèbre, A. Fokin, D. Barthès, D. Gillet and N. Mauron

Light Variability of Color Selected IRAS Sources: Variables in Late AGB and Early Post-AGB Phase 111
T. Fujii, Y. Nakada and M. Parthasarathy

Circumstellar Arcs in AGB and Post-AGB Stars 115
S. Kwok, K.Y.L. Su and J.A. Stoesz

Simulations of Quasi-Periodic Shells in Dust Forming AGB Winds 121
Y. Simis, V. Icke and C. Dominik

Polarimetry and Direct Imaging of Detached Shells Around Carbon Stars 127
D.G. Delgado and H. Olofsson

Evolution of Thin Gas Shells Along the AGB and Beyond 131
M. Steffen and D. Schönberner

Modelling of the Observed Post-AGB Evolution 139
R. Tylenda, K. Jeske, R. Szczerba and N. Siódmiak

The Effect of the Transition Times in the Formation of Planetary Nebulae 147
E. Villaver, A. Manchado and G. García-Segura

Are Microstructures in the Outer Shells of PNe Fossil Condensations of the AGB Wind? 153
D.R. Gonçalves, R.L.M. Corradi, E. Villaver, A. Mampaso and M. Perinotto

The Infrared [WC] Stars 157
A.A. Zijlstra

Analysis of 5 PN with [WC 8-11] Nuclei 163
K. Gesicki and A. Acker

Masses of Central Stars of Planetary Nebulae 167
K. Gesicki and A.A. Zijlstra

Part III **Binarity and Onset of Asymmetrical Structures**

AGB Evolution in Binary Systems 173
A. Frankowski

Planets and Axisymmetric Mass Loss 181
N. Soker

Bipolar Outflows from OH/IR Stars 189
A.A. Zijlstra

The Shape and Shaping of Proto-Planetary Nebulae 197
B.J. Hrivnak, S. Kwok and K.Y.L. Su

SPH Models of Proto-Planetary Nebulae 205
A.J. Gawryszczak, J. Mikołajewska and M. Rózyczka

Dynamics of the Collimated Outflows of the PPN Hen 3-1475 209
A. Riera, P. García-Lario, A. Manchado and M. Bobrowsky

Strong Asymmetries in the Neutral Envelope of Mira 215
E. Josselin, N. Mauron, P. Planesas and R. Bachiller

Mass Loss Geometry in Symbiotic Binaries 221
R.L.M. Corradi

Circumstellar Environment of RX Puppis 227
J. Mikołajewska, E. Brandi, L.G. García, O.E. Ferrer, P.A. Whitelock and F. Marang

The Sudden Loss of 0.01 $M_\odot$ from the VV Cep's Primary as a Characteristic of AGB Evolution in the Binaries 235
D. Graczyk and M. Mikołajewski

Part IV **Evolution, Spectroscopy and Chemical Composition of Central Stars**

AGB and Post-AGB Evolution: Structural and Chemical Changes 241
T. Blöcker, R. Osterbart, G. Weigelt, Y.Yu. Balega and A.B. Men'shchikov

Stellar Evolution and Nucleosynthesis of Post-AGB Stars 249
F. Herwig

What Can we Learn from Post-AGB Chemical Studies on the AGB 3rd Dredge-Up Phenomena? 257
H. Van Winckel and M. Reyniers

Optical Spectroscopy of Peculiar Supergiants 265
V.G. Klochkova, V.E. Panchuk and R. Szczerba

Spectroscopical Masses of Post–AGB Stars 271
M.R. Schmidt

BD+33°2642: Abundance Patterns in the Central Star of a Halo PN 277
R. Napiwotzki, M. Herrmann, U. Heber and M. Altmann

The Very Lithium Rich Post-AGB SB2 Binary HD 172481 283
M. Reyniers and H. Van Winckel

CIT 6: the Early Phase of PPN? 289
L. Začs, A. Alksnis and R. Spēlmanis

Spectroscopy of Post-AGB A-Supergiant HD 101584 (IRAS 11385-5517) 295
T. Sivarani, M. Parthasarathy, P. García-Lario, A. Manchado and S.R. Pottasch

Spectroscopy of Post-AGB Candidate HD 331319 (IRAS 19475+3119) 301
T. Sivarani, M. Parthasarathy, P. García-Lario and A. Manchado

Spectroscopy of Post-AGB Candidate IRAS 10215-5916 305
T. Sivarani, M. Parthasarathy, P. García-Lario and A. Manchado

Spectroscopy of HD 168625 (IRAS 18184-1623) 309
P. García-Lario, T. Sivarani, M. Parthasarathy and A. Manchado

The Evolutionary Status of HD 179821 313
F. Thévenin, G. Jasniewicz and M. Parthasarathy

IRAS 01005+7910, a High Galactic Latitude Post-AGB Star? 317
J. Hu

Part V **Dust around Post-AGB Sources**

ISO Results for Protoplanetary Nebulae 323
K. Volk, S. Kwok, B.J. Hrivnak and R. Szczerba

An Atlas of ISO SWS Spectra: The Transition from AGB Stars to PNe 329
P. García-Lario and J.V. Perea Calderøn

Big, Bumpy Dust Shells around Protoplanetary Nebulae 333
A.K. Speck, M. Meixner and T. Ueta

Spatial Distributions of Multiple Dust Components in the PPN/PN Dust Shells 339
T. Ueta, A.K. Speck, M. Meixner, A. Dayal, J.L. Hora, G. Fazio, L.K. Deutsch and W.F. Hoffmann

Structure and Properties of IRC+10216 343
A.B. Men'shchikov, Y.Yu. Balega, T. Blöcker, R. Osterbart and G. Weigelt

Spatially Resolved Spectra of the Red Rectangle in the N-band 351
T. Miyata, H. Kataza, Y. Okamoto, S. Sako, S. Takubo, T. Onaka and T. Yamashita

Part VI **Molecular and Atomic Gas around PPNe**

Molecules in AGB (and Post-AGB) Circumstellar Envelopes 357
A. Omont

Molecular Gas Images of Proto-planetary Nebulae 369
M. Meixner, D. Fong, E.C. Sutton and W.J. Welch

H_2 Spectroscopy of Protoplanetary Nebulae 377
D.A. Weintraub, J.H. Kastner and I. Gatley

Molecular Hydrogen Around Post−AGB Stars 383
D.A. García-Hernández, A. Manchado, P. García-Lario, C. Dominguez−Tagle, G. Conway and F. Prada

Morphology and Kinematics of the Egg Nebula: CO and H_2 387
P. Cox, R. Lucas, P.J. Huggins and J.-P. Maillard

O-bearing Molecules in Carbon-rich Proto-planetary Objects: Study of CRL 618 397
F. Herpin and J. Cernicharo

Low-excitation Atomic Gas Around Evolved Stars: ISO Observations of C-rich Nebulae 403
D. Fong, M. Meixner, E.C. Sutton, A. Castro-Carrizo, V. Bujarrabal, W.B. Latter, A.G.G.M. Tielens and D.M. Kelly

Low-excitation Atomic Gas in PPNe: ISO Observations of FIR Line Emission in O-rich Objects 409
A. Castro-Carrizo, V. Bujarrabal, D. Fong, M. Meixner, A.G.G.M. Tielens, W.B. Latter and M.J. Barlow

Probing the Post-AGB Nature of HD 179821 413
E. Josselin and A. Lèbre

Masers Diagnostics to Identify Stars Departing from the AGB 417
D. Engels

IRAS 18455+0448: The Birth and Death of OH/IR Stars 423
B.M. Lewis

OH Maser Mapping of Post-AGB Stars 429
M. Szymczak and A.M.S. Richards

The Time Variation in the Water Vapor Bands in Oxygen-rich Mira Variables 433
M. Matsuura, I. Yamamura, H. Murakami and T. Onaka

Further Detections of OH Masers in Carbon Stars with Silicate Features 439
M. Szymczak, R. Szczerba and P.S. Chen

Object Index 443

Author Index 447

Preface

The term proto–planetary nebulae (PPNe) in the context of the late stages of stellar evolution was created only slightly more than 20 years ago to express the belief that in the *near future* these objects will become planetary nebulae (PNe). The first proto-planetary nebulae (called also post–Asymptotic Giant Branch, or shortly post–AGB objects) AFGL 2688 and AFGL 618 were discovered in mid–seventies in course of the Air Force Sky Survey. Investigation of this phase of stellar evolution developed very rapidly in 1980's after the IRAS mission when it became clear that proto–planetary nebulae emit a significant part of their energy in the mid– and far–infrared. Hundreds of new candidates have been proposed but the recognition of the real proto–planetary nebulae is not a simple task and needs a substantial effort to exclude cases that represent different evolutionary stages. High resolution spectroscopy of stellar atmospheres is of much importance in this respect. Surprisingly, only a small group of central stars, the so called 21 μm emitters, show chemical signatures of the 3rd dredge–up process. Very recently, a more detailed studies of mid–infrared spectra from the Infrared Space Observatory (ISO) allow for a better understanding of chemical composition and evolution of circumstellar material around these stars.

A new impetus in the field of proto–planetary nebulae research was started in the 1990's with high spatial resolution imaging in mid–infrared and optical wavelength ranges. Similar to classical PPNe (AFGL 2688 and AFGL 618), evidence has been found that PPNe (at least their central parts) are quite asymmetric while shells of AGB stars display spherical shapes. There seems to be an agreement that the shaping of circumstellar envelopes occurs during end of the AGB (beginning of the post-AGB) evolution. The most intriguing challenge now is to understand which physical process(-es) is (are) dominant during formation and evolution of proto-planetary nebulae. Is it magnetic field, evolution in binary systems, planets, axisymmetric superwind mass loss and its further shaping by fast wind, collimated jets interacting with spherically symmetric AGB remnant or maybe an interplay between these different processes? The beautiful Hubble Space Telescope images show that the mass loss process near the end of the AGB evolution is modulated on timescales of a few hundred years. The timescales of the concentric arcs observed around AGB stars, in PPNe and PNe do not fit the pulsational nor nuclear (helium shell flash) timescales. Two images presented on the cover page of this proceedings show comparison between arcs in IRC+10216 and computations which lead to quasi–periodic mass loss with similar timescales.

In order to discuss the above problems, we have brought together specialists in different fields of research at this workshop. The main aim of the workshop was an attempt to formulate a consistent picture of the proto–planetary evolution by means of extensive comparison between observations and theoretical modeling. The meeting devoted to PPNe, which until the 1970's represented the missing link between AGB and PNe, was organized in July, 2000 in the 800 year–old Toruń (the home town of Nicolaus Copernicus) after the IAU Symposium 191 (Asymptotic Giant Branch Stars, held in Montpellier in 1998) and before the IAU Symposium 209 (Planetary Nebulae, to be held in Canberra in 2001). The large attendance of our meeting and especially the contents of the proceedings show that investigation of proto-planetary nebulae has become a very important field in stellar astrophysics. We hope that further meetings devoted to discussion of this fascinating class of stellar objects will take place on a more regular basis.

THE EDITORS

Acknowledgements

The organizers are very grateful to the Scientific Organizing Committee members for their advices during preparation of the scientific program and to the members of the Local Organizing Committee for their effort to make this workshop an important scientific and cultural event. We are also indebted to the Institute of Physics of the Nicolaus Copernicus University in Toruń for the audio–visual and organizing support provided. The organization of the workshop was possible partly due to the financial support from grant No.2.P03D.024.18p01 and No.2.P03D.002.13 of the Polish State Committee for Scientific Research.

The editors would like to thank for the permission to reproduce on the cover page the B+V image of IRC+10216 from Mauron & Huggins, 1999, A&A 349, 203 for which the copyright was granted by ESO, and for Drs. Y. Simis, V. Icke and C. Dominik for allowing to use the results of their computations for comparison with the observational data.

Scientific Organizing Committee

Local Organizing Committee

List of Participants

Thomas Blöcker
Max-Planck-Institut für Radioastronomie, Bonn, GERMANY
bloecker@mpifr-bonn.mpg.de

Howard E. Bond
Space Telescope Science Institute, Baltimore, USA
bond@stsci.edu

Romano Corradi
Isaac Newton Group of Telescopes, Santa Cruz de la Palma, SPAIN
rcorradi@ing.iac.es

Pierre Cox
Institut d'Astrophysique Spatiale, Orsay, FRANCE
Pierre.Cox@ias.u-psud.fr

David G. Delgado
Stockholm Observatory, Saltsjöbaden, SWEDEN
delgado@astro.su.se

Dieter Engels
Hamburger Sternwarte, Hamburg, GERMANY
dengels@hs.uni-hamburg.de

David Fong
University of Illinois at Urbana–Champaign, Urbana, USA
d-fong@astro.uiuc.edu

Adam Frankowski
N. Copernicus Astronomical Center, Toruń, POLAND
torque@ncac.torun.pl

Takahiro Fujii
Institute of Astronomy, University of Tokyo, Mitaka, JAPAN
fujii@ioa.s.u-tokyo.ac.jp

Domingo García-Hernández
Instituto de Astrofísica de Canarias, La Laguna, Tenerife, SPAIN
agarcia@ll.iac.es

Artur Gawryszczak
N. Copernicus Astronomical Center, Warszawa, POLAND
gawrysz@camk.edu.pl

Krzysztof Gęsicki
Centre for Astronomy, UMK, Toruń, POLAND
gesicki@astri.uni.torun.pl

Denise R. Gonçalves
Instituto de Astrofísica de Canarias, La Laguna, Tenerife, SPAIN
denise@ll.iac.es

Sławomir K. Górny
N. Copernicus Astronomical Center, Toruń, POLAND
skg@ncac.torun.pl

Dariusz Graczyk
Centre for Astronomy, UMK, Toruń, POLAND
weganin@astri.uni.torun.pl

Fabrice Herpin
Depto Física Molecular, I.E.M., Madrid, SPAIN
herpin@isis.iem.csic.es

Falk Herwig
Institut für Physik, Universität Potsdam, Potsdam, GERMANY
fherwig@uvastro.phys.uvic.ca

Sylwia Hołowacz
Centre for Astronomy, UMK, Toruń, POLAND
sylwiah@astri.uni.torun.pl

Sacha Hony
University of Amsterdam, THE NETHERLANDS
hony@astro.uva.nl

Bruce J. Hrivnak
Department of Physics and Astronomy, Valparaiso University, USA
bruce.hrivnak@valpo.edu

Francisco Jiménez-Esteban
Laboratorio de Astrofísica Espacial, Madrid, SPAIN
fje@laeff.esa.es

Erik Josselin
University of Montpellier, GRAAL, FRANCE
josselin@graal.univ-montp2.fr

Sun Kwok
Department of Physics and Astronomy, University of Calgary, CANADA
kwok@iras.ucalgary.ca

Agnès Lèbre
University of Montpellier, GRAAL, FRANCE
lebre@graal.univ-montp2.fr

Brian M. Lewis
Arecibo Observatory, Arecibo, USA
blewis@naic.edu

Ingemar Lundström
Lund Observatory, Lund, SWEDEN
ingemar@astro.lu.se

Thomas Maas
Instituut voor Sterrenkunde, K.U. Leuven, Heverlee, BELGIUM
thomas@ster.kuleuven.ac.be
Arturo Manchado
Instituto de Astrofísica de Canarias, La Laguna, Tenerife, SPAIN
amt@ll.iac.es
Mikako Matsuura
The Institute of Space and Astronautical Science, Sagamihara, JAPAN
mikako@ir.isas.ac.jp
Margaret Meixner
University of Illinois at Urbana–Champaign, Urbana, USA
meixner@astro.uiuc.edu
Alexander B. Men'shchikov
Stockholm Observatory, Saltsjöbaden, SWEDEN
sasha@astro.su.se
Joanna Mikołajewska
N. Copernicus Astronomical Center, Warszawa, POLAND
mikolaj@camk.edu.pl
Maciej Mikołajewski
Centre for Astronomy, UMK, Toruń, POLAND
mamiko@astri.uni.torun.pl
Takashi Miyata
SUBARU Telescope, NAO, Hilo, USA
takashi.miyata@nao.ac.jp
Ralf Napiwotzki
Dr. Remeis-Sternwarte, Bamberg, GERMANY
napiwotzki@sternwarte.uni-erlangen.de
Alain Omont
Institut d'Astrophysique de Paris, Paris, FRANCE
omont@iap.fr
M. Parthasarathy
Indian Institute of Astrophysics, Bangalore, INDIA
partha@iiap.ernet.in
Daniel Proga
Goddart Space Flight Center, Greenbelt, USA
proga@sobolev.gsfc.nasa.gov
Maarten Reyniers
Instituut voor Sterrenkunde, K.U. Leuven, Heverlee, BELGIUM
maarten@ster.kuleuven.ac.be

Angels Riera
Universitat Politècnica de Catalunya, Vilanova i la Geltrú, SPAIN
angels.riera@upc.es

Raghvendra Sahai
Jet Propulsion Laboratory, CIT, Pasadena, USA
sahai@grandpa.jpl.nasa.gov

Detlef Schönberner
Astrophysikalisches Institut Potsdam, Potsdam, GERMANY
DeSchoenberner@aip.de

Mirosław Schmidt
N. Copernicus Astronomical Center, Toruń, POLAND
schmidt@ncac.torun.pl

Yvonne Simis
Sterrewacht Leiden, Leiden, THE NETHERLANDS
simis@strw.leidenuniv.nl

Natasza Siódmiak
N. Copernicus Astronomical Center, Toruń, POLAND
alexan@ncac.torun.pl

Thirupathi Sivarani
Indian Institute of Astrophysics, Bangalore, INDIA
sivarani@iiap.ernet.in

Noam Soker
University of Haifa at Oranim, Tivon, ISRAEL
soker@physics.technion.ac.il

Angela K. Speck
University of Illinois at Urbana–Champaign, Urbana, USA
meixner@astro.uiuc.edu

Grazyna Stasińska
DAEC, Observatoire de Paris-Meudon, Meudon, FRANCE
grazyna.stasinska@obspm.fr

Matthias Steffen
Astrophysikalisches Institut Potsdam, Potsdam, GERMANY
MSteffen@aip.de

Olga Suárez
Universidade da Coruña, SPAIN
olga@mail2.udc.es

Ryszard Szczerba
N. Copernicus Astronomical Center, Toruń, POLAND
szczerba@ncac.torun.pl

Marian Szymczak
Centre for Astronomy, UMK, Toruń, POLAND
msz@astro.uni.torun.pl

Toma Tomov
Centre for Astronomy, UMK, Toruń, POLAND
tomtom@astri.uni.torun.pl

Romuald Tylenda
N. Copernicus Astronomical Center, Toruń, POLAND
tylenda@ncac.torun.pl

Toshiya Ueta
University of Illinois at Urbana–Champaign, Urbana, USA
ueta@astro.uiuc.edu

Griet Van de Steene
Royal Observatory of Belgium, Brussels, BELGIUM
gsteene@oma.be

Hans Van Winckel
Instituut voor Sterrenkunde, K.U. Leuven, Heverlee, BELGIUM
Hans.VanWinckel@ster.kuleuven.ac.be

Eva Villaver
Instituto de Astrofísica de Canarias, La Laguna, Tenerife, SPAIN
villaver@ll.iac.es

Kevin Volk
Department of Physics and Astronomy, University of Calgary, CANADA
volk@iras.ucalgary.ca

David A. Weintraub
Vanderbilt University, Nashville, USA
david.weintraub@vanderbilt.edu

Jan Martin Winters
Max-Planck-Institut für Radioastronomie, Bonn, GERMANY
jwinters@mpifr-bonn.mpg.de

Peter R. Wood
RSAA, Australian National University, Canberra, AUSTRALIA
wood@mso.anu.edu.au

Laimons Začs
Ventspils Int. Radio Astronomy Center, Riga, LATVIA
zacs@latnet.lv

Albert Zijlstra
Department of Physics, UMIST, Manchester, UNITED KINGDOM
aaz@iapetus.phy.umist.ac.uk

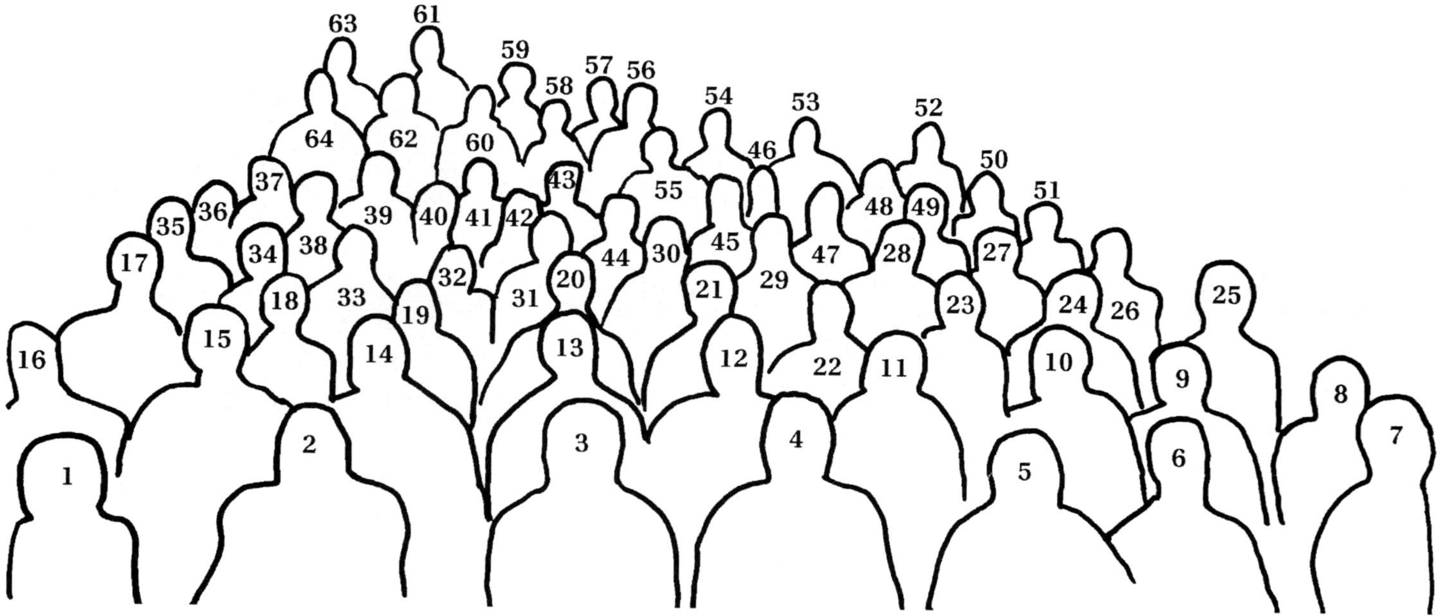

1) D.R. Gonçalves, 2) R. Tylenda, 3) I. Lundström, 4) F. Herwig, 5) N. Soker, 6) M. Parthasarathy, 7) H.E. Bond, 8) G. Van de Steene, 9) R. Sahai, 10) D. Schönberner, 11) S. Kwok, 12) T. Blöcker, 13) A.B. Men'shchikov, 14) A. Manchado, 15) B.J. Hrivnak, 16) K. Gesicki, 17) D. Proga, 18) O. Suárez, 19) J. Mikołajewska, 20) D. Graczyk, 21) R.L.M. Corradi, 22) A. Riera, 23) D. Fong, 24) T. Ueta, 25) K. Volk, 26) A. Frankowski, 27) M. Reyniers, 28) T. Maas, 29) D.A. Weintraub, 30) M. Meixner, 31) A.A. Zijlstra, 32) T. Sivarani, 33) G. Stasińska, 34) S. Hołowacz, 35) R. Szczerba, 36) A.J. Gawryszczak, 37) H. Van Winckel, 38) P. Cox, 39) T. Fujii, 40) Y. Simis, 41) F. Herpin, 42) M. Matsuura, 43) E. Josselin, 44) J.M. Winters, 45) L. Začs, 46) M. Szymczak, 47) R. Napiwotzki, 48) M. Mikołajewski, 49) D. Engels, 50) A.K. Speck, 51) F. Jiménez-Esteban, 52) E. Villaver, 53) P.R. Wood, 54) S.K. Górny, 55) M. Steffen, 56) B. Gertner, 57) S. Hony, 58) D.A. García-Hernández, 59) T. Miyata, 60) N. Siódmiak, 61) A. Omont, 62) A. Lèbre, 63) T. Tomov, 64) B.M. Lewis,

INTRODUCTORY REVIEW

PROTO-PLANETARY NEBULAE AS A PHASE OF STELLAR EVOLUTION

Sun Kwok
Department of Physics and Astronomy
University of Calgary, Calgary, Canada T2N 1N4
kwok@iras.ucalgary.ca

Abstract Proto-planetary nebulae (PPN) represent the last missing link in stellar evolution. As the result of the IRAS sky survey, many PPN candidates have now been identified. In this review, we summarize the observed properties of PPN and the role PPN play in our understanding of chemical synthesis and the shaping of planetary nebulae.

1. INTRODUCTION

Proto-planetary nebulae (or pre-planetary nebulae, PPN) are objects in transition between the asymptotic giant branch (AGB) and planetary nebulae phases of stellar evolution. Stars on the AGB lose mass at a high rate ($10^{-7} - 10^{-4}$ $M_{\odot}$ yr^{-1}) in the form of a stellar wind. Such mass loss eventually depletes the hydrogen envelope of the star and exposes the electron-degenerate carbon-oxygen core. This leads to an increase of the effective temperature of the star and the termination of mass loss. As the small amount of remaining hydrogen above the core continues to be consumed by hydrogen shell burning, the effective temperature of the star will gradually increase. When the stellar temperature reaches approximately 30,000 K, the amount of stellar photons with energies above the Lyman limit will be large enough to photoionize the surrounding material. The emergence of strong recombination lines of hydrogen and helium and collisionally excited lines of metals signifies the beginning of the planetary nebula phase. The evolutionary phase between the end of AGB mass loss and the beginning of photoionization is defined as the PPN phase (Kwok 1993).

In the literature, the term "proto-planetary nebulae" has been used to refer to a variety of objects, including AGB stars with large mass loss rates (e.g. AFGL 3068), PN with bipolar morphology (e.g. M 2–9), symbiotic stars and slow novae (e.g. V1016 Cyg), nebulae around massive stars (e.g. IRC+10420), and even young stellar objects. In some cases, an exact boundary is difficult

R. Szczerba and S.K. Górny (eds.), Post-AGB Objects as a Phase of Stellar Evolution, 3–10.

to draw (e.g., has the mass loss stopped in AFGL 3068?), but obvious cases of confusion should be avoided.

An AGB star with low core mass ($< 0.55\ M_\odot$) at the time of H envelope depletion may evolve too slowly to become PN. A more general term "post-AGB stars" can be used to refer to these objects.

2. THE SEARCH FOR PROTO-PLANETARY NEBULAE

Do PPN exist? In the past it was believed that PN are formed instaneously from red giants by a sudden ejection of the hydrogen (H) envelope and no thought was given to the possibility of the existence of transition objects. The stellar evolution models of Paczyński (1971) and Schönberner (1979, 1981) suggest that the central stars of PN will take a finite amount of time to burn out the H envelope before it reaches a high enough temperature to photoionize the nebula. This transition time is highly dependent on the mass of the central star, and has values of a few thousands years for a central star of $\sim$0.6 $M_\odot$ (Schönberner 1987).

The interacting stellar wind (ISW) model (Kwok et al. 1978) suggests that PN are not created by a sudden ejection but represent the result of a re-arrangement of material ejected during the AGB. The ISW model predicts that remnants of the AGB circumstellar envelope (CSE) should still be observable during the transition phase, therefore making it possible for PPN to be identified. Emissions from dust in the AGB CSE make them bright infrared sources and this infrared signature allows us to distinguish PPN from ordinary stars. Although two PPN were known as the result of the AFGL sky survey, a systematic search for PPN was only possible after the all sky infrared survey by the IRAS satellite.

3. OBSERVATIONAL PROPERTIES OF PROTO-PLANETARY NEBULAE

From our knowledge of the observed properties of evolved AGB stars and young PN, we come to the following expected properties of PPN:

- High luminosity ($> 3,000\ L_\odot$), low surface gravity (luminosity class I or Ia)
- Intermediate temperature (3,000-30,000 K) and spectral class (K to B)
- Evidence of remnant AGB envelope (infrared excess, molecular emission with a broad velocity profile)
- Detached dust envelope and doubled peak spectral energy distribution
- Nebulosity due to scattered light

Since young stellar objects (YSOs) also have infrared excesses, they can be confused with PPN. However, YSOs are of low luminosity and their molecular emission originate from the parent molecular clouds which often have narrow line profiles. The circumstellar dust around YSOs have a broader range of temperatures and are often cooler, whereas the color temperature of the dust component around PPN are higher (100-150 K) and often can be fitted by a single blackbody.

As we will discuss later, even these set of criteria is not a fool-proof way to identify PPN. Chemical abundance and the signature of AGB nucleosynthesis therefore becomes a powerful tool in separating PPN from other objects.

4. CHEMICAL ABUNDANCES AND AGB NUCLEOSYNTHESIS

PPN can be classified as oxygen- or carbon-rich depending on whether the O/C abundance ratio is larger or less than unity. Since the less abundant element is tied up almost exclusively in CO, it is not available to form other molecules. Oxygen-rich PPN can be identified by the presence of the 9.7 μm circumstellar silicate feature whereas carbon-rich PPN show molecular absorption bands of C_2, C_3, and CN.

Stars on the AGB undergo hydrogen and helium shell burning as well as neutron capture processes. Some PPN show large overabundance of s-process elements (Sr, Y, Zr, Ba, La, Ce, Pr, Nd, Sm, and Eu), confirming that these objects have indeed evolve through the AGB phase (van Winckel & Reyniers 2000). Several examples of PPN are given in Table 1.

Table 1 Examples of proto-planetary nebulae

Object	Sp. type	Chemistry	[s/Fe][a]	L_*/D^2 ($L_\odot$/kpc^2)
IRAS 04296+3429	G0 Ia	C	+1.5	5.7(2)
IRAS 07134+1005	F5 I	C	+1.5	1.4(3)
IRAS 18095+2704	F3 Ib	O	−0.1	2.1(3)
IRAS 19500−1709	A2-3 I	C	+1.1	2.3(3)
AFGL 2688	F5 Iae	C		1.5(4)
IRAS 22272+5435	G5 Ia	C	+2.5	3.9(3)
IRAS 23304+6147	G2 Ia	C	+1.6	6.3(3)

[a] van Winckel & Reyniers (2000)

5. MOLECULAR SYNTHESIS IN THE CIRCUMSTELLAR ENVIRONMENT

Infrared spectroscopy of PPN has revealed the presence of organic compounds with aromatic and aliphatic structures. A series of infrared emission features at 3.3, 6.2, 7.7, and 11.3 μm due to aromatic C-H and C-C stretching and bending modes are commonly seen in carbon-rich PPN. In addition, features at 3.4 and 6.9 μm due to stretching and bending modes of aliphatic subgroups are also detected (see Fig. 1). These features, as well as the underlying broad emission plateaus around 8 and 12 μm, can be found in the laboratory infrared spectra of polycyclic aromatic hydrocarbons, quenched carbonaceous composites, hydrogenated amorphous carbons, or coal.

In addition, strong emission features at 21 and 26 μm are also seen in PPN. In particular, the 21 μm feature, recently identified as due to the TiC cluster, is unique to PPN and has not been seen elsewhere in the interstellar medium.

The detection of these carbonaceous compounds suggests that complex organic molecules can be formed over a short ($\sim 10^3$ year) timescale (Table 2). This process is believed to first begin during the late AGB phase, where simple organic molecules such as acetylene (C_2H_2) are formed in the AGB wind. Chemical reactions in the envelope can lead to the formation of benzene (C_6H_6) and possibly other aromatic hydrocarbons. PPN are therefore useful laboratories for the study of astrochemistry.

Table 2 Summary of the changes in relative strengths of unidentified emission features as stars evolve from AGB to PN

IR features (μm)	Origin	Extreme C Stars	PPN	PN
primarily features:		no	yes	strong
3.3, 6.2, 7.7, 11.3	aromatic stretch and bending modes			
secondary features		no	yes	weak
3.4, 6.9	C-H aliphatic stretch and bend			
12.1, 12.4, 13.3	C-H out-of-plane bend with 2, 3, and 4 adjacent H atoms			
broad 8, 12	-	no	yes	weak
broad 21	TiC	weak	strong	no
broad 30	-	yes	yes	yes

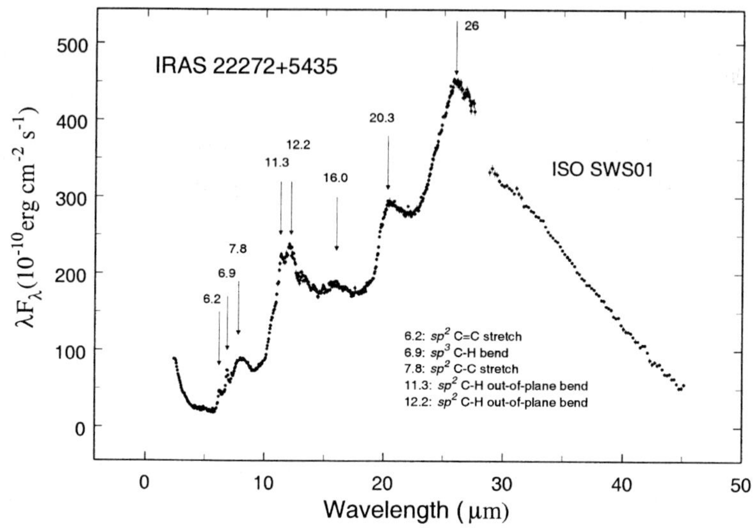

Figure 1 ISO SWS spectrum of the PPN IRAS 22272+5435 showing various stretching and bending modes of aromatic and aliphatic bonds.

6. THE SHAPING OF PLANETARY NEBULAE

Many PN have bipolar morphologies and how the spherically symmetric circumstellar envelopes of AGB stars are transformed into such bipolar shapes is an area of active research (c.f. Kastner et al. 2000). If the AGB envelope has a higher density in the equatorial regions, then when the fast outflow begins it will preferentially push through the polar direction and carve out cavity of bipolar shape (Balick 1987). High resolution optical imaging has detected bipolar reflection nebulae around PPN, suggesting that the shaping process occurs early (Kwok et al. 1996). The detection of concentric arcs in bipolar PPN (Fig. 2) and PN and the fact that the inner-most arcs have typical dynamical ages of 10^3 yr further constraints the shaping timescale (Kwok et al., this volume). The existence of the arcs also implies that most of the AGB envelope is not disturbed by the fast outflow and the fast outflow is confined to a limited solid angle. The detection of a visible disk in IRAS 17106−3046 (Fig. 3) suggests that the outflow may be disk collimated.

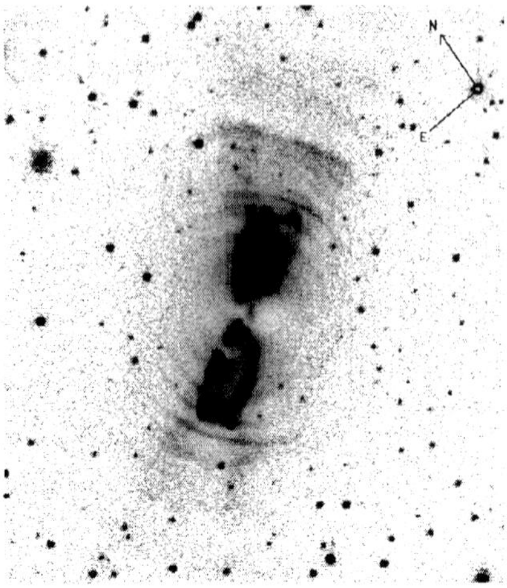

Figure 2 HST F606W image of IRAS 17150−3224 showing a pair of bipolar lobes and 8 concentric rings.

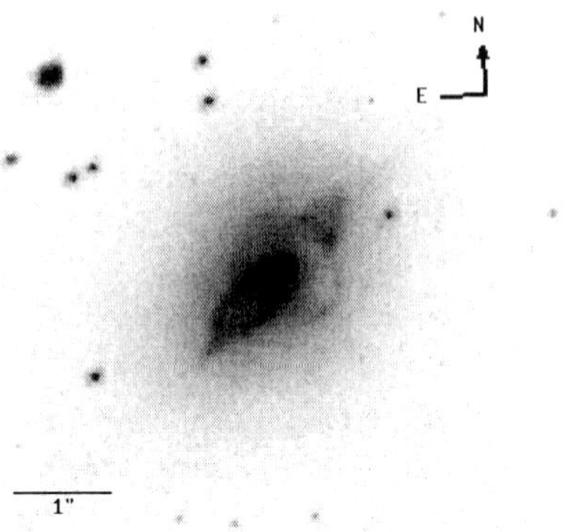

Figure 3 HST F606W image of IRAS 17106−3046 showing a bipolar outflow emerging from a circumstellar disk.

7. RELATED OBJECTS

Besides PPN candidates identified from cool IRAS sources, there are other classes of stars that have been suggested to be objects in the post-AGB phase of evolution.

- High-latitude luminous stars: A number of high-latitude yellow supergiants (e.g. HD 161796: F3 Ib, b=31°; 89 Her: F2Ia, b=23°; Bidelman 1951) are found to have large infrared excesses (Parthasarathy & Pottasch 1986) and have spectral energy distributions similar to those of PPN.

- RV Tauri and UU Her stars: Pulsating variables such as RV Tauri stars have long been known to have infrared excesses and have been suggested to be post-AGB stars.

- Rapidly evolving stars: PPN with massive central stars are expected to evolve rapidly with time and the change in spectral type can be detectable over decades. For example, SAO 85766 (IRAS 18062+2410) changed from A5 (1940) to B1 I (T_*=22,000 K) in 1993 (Arkhipova et al. 1999, Parthasarathy et al. 2000).

- Reflection nebulae around cool stars: The detection of reflection nebulae around objects found in the early infrared surveys (IRC and AFGL) led to the discovery of the first PPN (AFGL 618 and AFGL 2688). Other reflection nebulae with cool central stars and strong infrared excesses

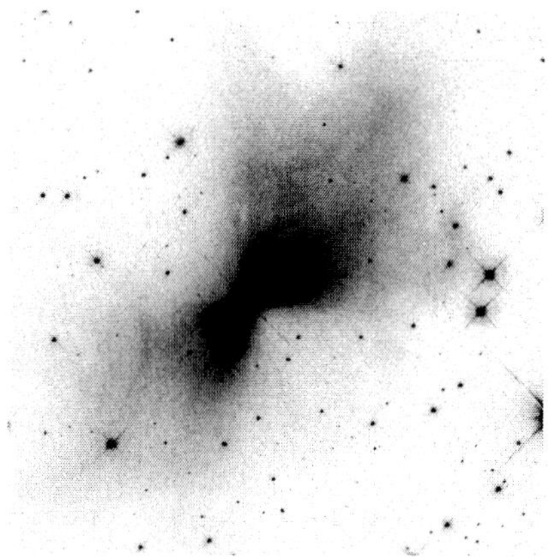

Figure 4 HST WFPC2 F606W image of IRAS 12419−5414, the Boomerang Nebula.

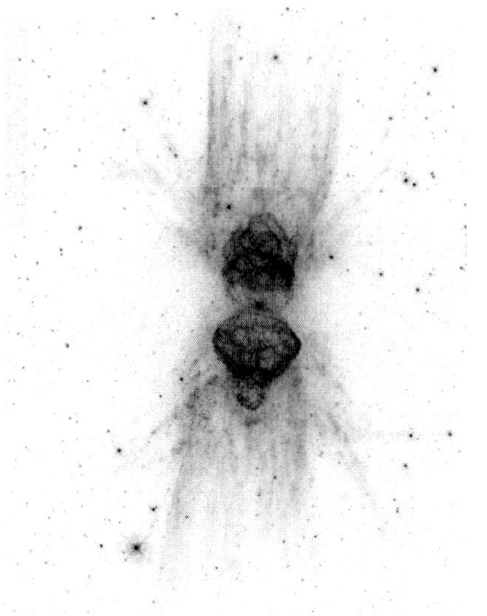

Figure 5 HST WFPC2 [NII] image of Mz 3.

(see, e.g., Calvet & Cohen 1978), include Roberts 22 (T_*=10,000K), Mz 3 (32,000K), M 1−92 (20,000K), etc. The reflection nebulae often show bipolar morphologies (Fig. 3 & 4) similar to PN (Bujarrabal et al. 1998, Sahai et al. 1999). One of the best-studied infrared object with nebulosity is HD 44179 (the Red Rectangle), which has a reflection nebula surrounding a B8-A0 central star. It also has very strong aromatic hydrocarbon features and extended red emission. Another example is IRAS 09371+1212 (The Frosty Leo Nebula, Sahai et al., this volume).

Spectroscopic monitoring of some of the above objects has revealed several members to be binaries (van Winckel 1999, e.g., AC Her: 1994 days, U Mon: 2597 days, HD 44179: 318 days, 89 Her: 2884 days). The infrared excesses in these objects are likely to be due to circumstellar disks surrounding the binary systems rather than from expanding dust envelopes. It is therefore doubtful whether these objects can be considered to be PPN.

8. SUMMARY

PPN play a key role in our understanding of the shaping of PN and the chemical synthesis of large molecules in the interstellar medium. They also serve as probes of nuclear synthesis in the AGB. However, we still face uncertainties on the exact nature of some of the PPN candidates, as well as the role of binaries play in post-AGB evolution.

Acknowledgments

This work is supported by the Natural Sciences and Engineering Research Council of Canada and a Killam Fellowship from the Canada Council for the Arts.

References

Arkhipova, V.P., Ikonnikova, N.P., Noskova, R.I., Klochkova, V.G. 1999, *Astron. Letters*, **25**, 25

Balick, B. 1987, AJ, 94, 671

Bidelman, W. 1951, *ApJ*, **113**, 304

Bujarrabal, V., Alcolea, J., & Neri, R. 1998, *ApJ*, **504**, 915

Calvet, N., & Cohen, M. 1978, *MNRAS*, **182**, 687

Hrivnak, B.J., Volk, K., & Kwok, S. 2000, *ApJ*, **535**, 275

Kastner, J., Soker, N., & Rappaport, S.A. 2000, in *Asymmetrical Planetary Nebulae II: From Origins to Microstructures*, ASP Conf. Ser. 199

Kwok, S. 1993, *Ann. Rev. Astr. Ap.*, **31**, 63

Kwok, S., Purton, C.R., & FitzGerald, M. P. 1978, *ApJ*, **219**, L125

Kwok, S., Hrivnak, B. J., Zhang, C. Y., & Langill, P. L. 1996, *ApJ*, **472**, 287

Paczyński, B. 1971, *Acta Astr.*, **21**, 417

Parthasarathy, M., & Pottasch, S. R. 1986, *A&A*, **154**, L16

Parthasarathy, M., García-Lario, P., Sivarani, T., Manchado, A., & Sanz Fernández de Córdoba, L. 2000, *A&A*, **357**, 241

Sahai, R., Zijlstra, A., Bujarrabal, V., Te Lintel Hekkert, P. 1999, *AJ*, **117**, 1408

Schönberner, D. 1979, *A&A*, **79**, 108

Schönberner, D. 1981, *A&A*, **103**, 119

Schönberner, D. 1987, in *The Late Stages of Stellar Evolution*, S. Kwok & S.R. Pottasch (eds.), p. 337

Van Winckel, H. 1999, in *IAU Symp. 191: Asymptotic Giant Branch Stars*, Le Bertre, T., Lébre, A., Waelkens, C. (eds), ASP, p. 465

Van Winckel, H., & Reyniers, M. 2000, *A&A*, **354**, 135

I

SURVEYS AND OBSERVATIONAL DATA

OBSERVATIONAL DATABASE AND STATISTICAL PROPERTIES OF PROTO–PLANETARY NEBULAE

R. Szczerba, S.K. Górny, M. Zalfresso–Jundziłło

Nicolaus Copernicus Astronomical Center, PAS, Toruń, Poland

Abstract We analyze statistical properties of proto–planetary nebulae in sample of about 220 objects selected from the literature. First we discuss problem of the candidates selection and then present their distribution over the spectral type and on the IRAS color–color diagram. We analyze also galactic distribution of proto–planetary nebulae candidates from the created database.

1. INTRODUCTION

Before turning into planetary nebulae (PNe) stars with low to intermediate initial masses ($\sim$0.8$\div$8 M$_\odot$) evolve along the Asymptotic Giant Branch (AGB). In this phase, mass loss dominates over nuclear burning and eventually terminates the AGB evolution by reducing mass of the hydrogen rich envelope ($M^{\rm H}_{\rm env}$) below a certain limit (of the order of 10^{-3} M$_\odot$ for a core mass of 0.60 M$_\odot$). The large scale mass loss from so depleted atmosphere is no longer possible and the star enters the post–AGB phase of evolution. During the proto–planetary nebula (PPN) stage, due to further mass reduction of the remaining hydrogen rich envelope the effective temperature ($T_{\rm eff}$) increases continuously. Finally, when $T_{\rm eff}$ is high enough ($\sim$25000 K) the ejected AGB envelope starts to be ionized and a planetary nebula develops.

After discovery of the first proto–planetary nebulae, AFGL 618 and the Egg Nebula (AFGL 2688), during the Air Force Sky Survey (Westbrook et al. 1975, Ney et al. 1975) it has become clear that infrared (IR) observations are very important for identification of PPN candidates. Due to a large amount of dust formed at the very end of the AGB evolution, these stars are primarily seen at the IR wavelengths. Later on, when optical depth decreases and post-AGB stars become optically visible it is difficult to distinguish them from ordinary stars without search for the IR excess. Therefore, a big progress in identification of new PPN candidates has been made when data from the InfraRed Astronomical Satellite (IRAS) all sky survey has become available. Here, we report on some

R. Szczerba and S.K. Górny (eds.), Post-AGB Objects as a Phase of Stellar Evolution, 13–20.

statistical properties of PPN candidates selected by us from the literature. A detailed description of our database and its fuller analysis will appear elsewhere.

2. OBSERVATIONAL DATABASE

The most important observational criteria for selection of proto–planetary nebulae candidates have been summarized by Sun Kwok in this volume and will not be repeated here. We would like, however, to mention some additional classes of post-AGB objects which have been included into our database. They are B type sources with very often no or very small infrared excess (e.g. McCausland et al. 1992) and non–variable OH/IR stars which are probably at the beginning of PPNe phase of evolution (e.g. Bedijn 1987). In this way we try to cover the full range of the effective temperatures suspected during this stage of stellar evolution. Altogether our database contains about 220 candidates. Since the details of transition from the AGB to the post–AGB are not well established, we discuss shortly problem of the coolest PPN candidates selection.

2.1. TRANSITION FROM THE AGB TO THE POST-AGB

The critical value of $M^{\mathrm{H}}_{\mathrm{env}}$ below which mass loss at superwind rates is not longer possible is not known from first principles and up to now there are neither theoretical nor observational constraints to establish this value. From the $M^{\mathrm{H}}_{\mathrm{env}}$–$T_{\mathrm{eff}}$ relation (see e.g. Schönberner 1989) which is valid for H–burning models we can conclude only that the mass of hydrogen rich envelope at such a moment *cannot be too high.* Otherwise, lifetime of proto–planetary nebulae would be too long and nebulae will disperse before ionization.

Fig. 1 presents the $M^{\mathrm{H}}_{\mathrm{env}}$–$T_{\mathrm{eff}}$ relation for some evolutionary models with solar metallicity from Vassiliadis & Wood (1993). Models of Schönberner (1979, 1983) and Blöcker (1995) show similar behavior. During the AGB evolution models attain minimal value of T_{eff} below 3000 K (neglecting variations related to thermal pulses) when $M^{\mathrm{H}}_{\mathrm{env}}$ is reduced to about 0.1 (0.5) $\mathrm{M}_\odot$ for the final core mass of 0.568 (0.891) $\mathrm{M}_\odot$. From this moment effective temperature continuously increases with ongoing reduction of the envelope mass. The steep slope of this relation changes to more flat at T_{eff} of about 4000 K up to 7000 K for core masses of 0.568 up to 0.891 $\mathrm{M}_\odot$. It is very likely that at this point mass loss is already (more or less gradually) reduced to much smaller rates which are typical for the post–AGB phase of evolution. If mass loss rates for the flat part of the $M^{\mathrm{H}}_{\mathrm{env}}$–$T_{\mathrm{eff}}$ relation would be still very high then the evolution of the central stars of proto–planetary nebulae would be too fast (small changes in $M^{\mathrm{H}}_{\mathrm{env}}$ would result in large increase of T_{eff}) and PPNe of late types would not be observed. The description of the process of superwind mass loss reduction near the end of the AGB evolution based on the stars' pulsational properties has been proposed by Blöcker (1995) and is investigated in more detail by Tylenda

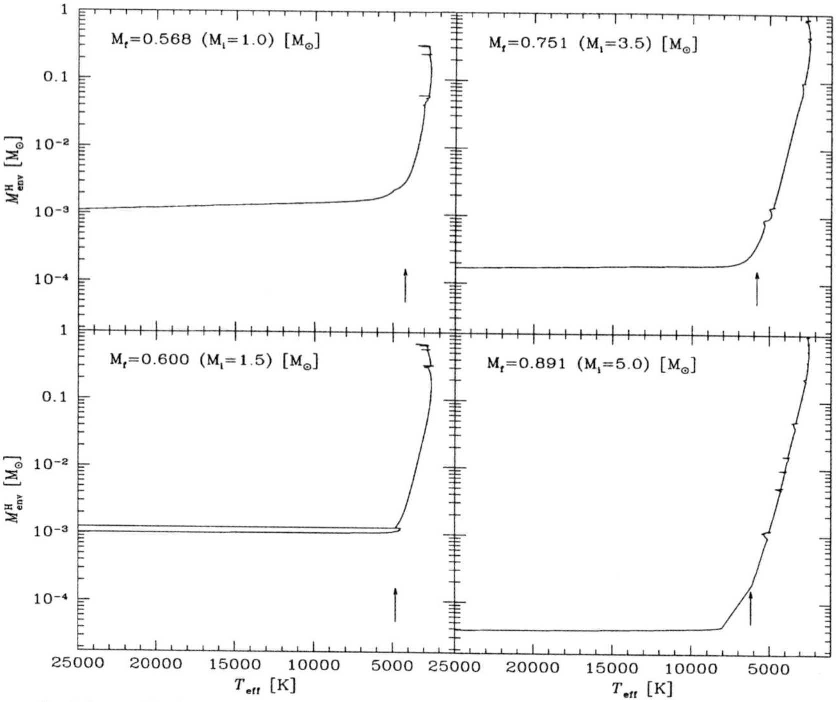

Figure 1 Mass of hydrogen rich envelope versus effective temperature for models of Vassiliadis & Wood (1993). M_f and M_i mean the final and initial mass, respectively. Arrows indicate T_{eff} at which pulsational period, computed according to the formula of Ostile & Cox (1986), drops to 100 days.

et al. (this volume). For reference, arrows which indicate pulsational period (P) of 100 days are marked in each panel of Fig. 1.

Due to the large optical depth of the envelope during PPNe infancy it is not always possible to recognize objects which have ceased process of mass loss at superwind rates. Some clues come, however, from spectral type (if known) and the considerations above. Namely, the coolest PPN candidates should have spectral types not later than early K and certainly not M ! as it is sometimes assumed. The questionable objects, for example AFGL 3068 and the Frosty Leo, are not considered here as PPNe.

3. DISTRIBUTION OVER SPECTRAL TYPE

Theoretical timescales of the post-AGB evolution can be used to predict the distribution of PPNe over spectral type which can be compared with the observed histograms. In this way, two unknown factors (the end of superwind and the post–AGB mass loss rate) which have the most important effect on the evolutionary timescales can be constrained (e.g. van Hoof et al. 1997, Tylenda et al. this volume). Observed distributions over spectral type for different samples

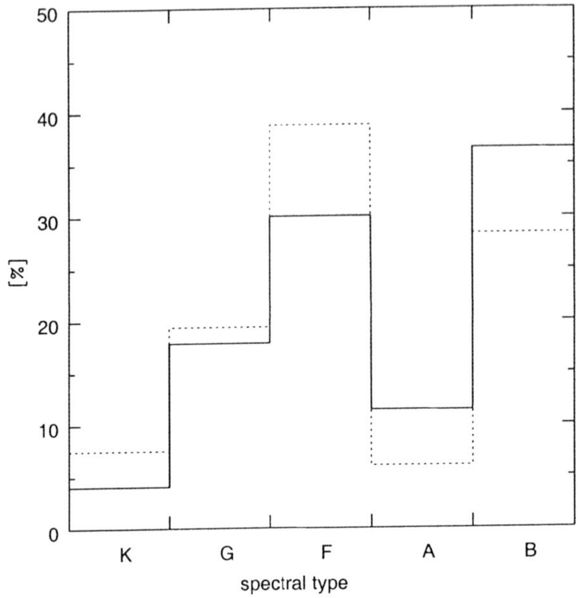

Figure 2 Distribution over spectral type for 123 proto–planetary nebulae candidates from our database (solid line) and a subsample of 68 PPNe with effective temperature determined from the model atmosphere analysis.

of PPNe candidates have been constructed by many authors (e.g. Oudmaijer et al.1993, Manchado et al. this volume). However, these datasets suffer from the selection effects imposed on them just by the sample definition. Our database is attempted to contain *all* known proto–planetary nebulae and should be free of the limitations generated by such procedures.

In Fig. 1 we present distribution over spectral type for all objects from our database (123 sources - 56% of our sample) for which spectral type is known from the observed spectra or $T_{\rm eff}$ has been determined from the model atmosphere analysis. The conversion from effective temperature to spectral type listed by Schmidt–Kaler (1982) has been used. By dashed line in Fig. 1 we present distribution of spectral types for a subsample of 68 PPN candidates discussed by Schmidt (this volume) which have effective temperatures determined from the model atmosphere analysis. Both distributions are similar and it can be seen that our database contains significant part of B–type objects which were *missing* in the Oudmaijer's sample and discrepancy with theoretical predictions discussed by van Hoof et al. (1997) seems to be (at least quantitatively) resolved. For more detailed discussion of theoretical distributions over spectral type for post–AGB objects the reader is referred to van Hoof et al. (1997) and Tylenda et al. (this volume).

4. THE IRAS COLOR–COLOR DIAGRAM

The IRAS color–color diagram has long been proved to be an efficient and simple tool for discussion of the properties and evolutionary status of objects during late stages of stellar evolution. The $\log(F_{60}/F_{25})$ versus $\log(F_{25}/F_{12})$ diagram for all PPN candidates with the IRAS photometry of a good quality (Q=3 or 2) is presented in Fig 3. Altogether positions of about 160 PPNe candidates are marked. For 30 PPNe there is no IRAS data and for similar number of sources quality of the IRAS photometry is bad (Q=1). As it can be seen from the location of proto–planetary nebulae candidates relative to the regions of van der Veen & Habing (1988) there is a lot of objects which have IRAS colors different from those of planetary–nebulae (region V and the redder $\log(F_{25}/F_{12})$ color) or evolved AGB stars (region IIIb and IV).

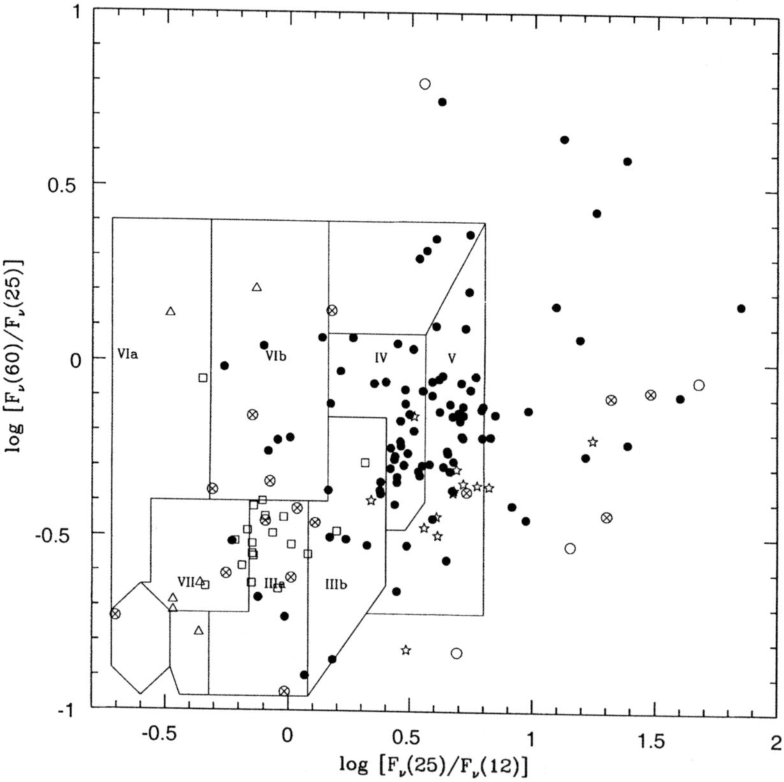

Figure 3 IRAS color–color diagram for proto–planetary nebulae candidates from our database. For reference, the regions defined by van der Veen & Habing (1988) are shown. Symbols used are as follow: squares – RV Tauri stars; triangles – R CrB objects; stars – 21 μm emitters; circles – high galactic latitude hot post–AGB sources; crossed circles – other high galactic latitude supergiants; filled circles – other PPNe candidates.

Szczerba & Steffen (1999) have performed hydrodynamical modeling of the post–AGB phase of evolution for two sets of evolutionary tracks (Vassiliadis & Wood 1993 and Blöcker 1995). They showed that big differences in timescales of post–AGB evolution generated by various assumptions concerning mass loss during transition from the AGB to PNe results in a large spread of the computed sequences over the IRAS color–color diagram. Only models which evolved sufficiently long (of order of 5 10^3 years) before shell ionization were able to explain positions of PPNe candidates with IRAS colors much different from those of planetary nebulae. However, these computations were simplified and performed only for carbon dust. In reality, the influence of dust opacity law and solid state features on the behavior of tracks in the IRAS color-color diagram (see van Hoof et al. 1997) cannot be neglected. Therefore, the firm conclusions cannot be drawn yet as far as the evolution of post–AGB objects on the IRAS color–color diagram is concerned, but certainly PPNe are located not only in the limited region(s) of this diagram.

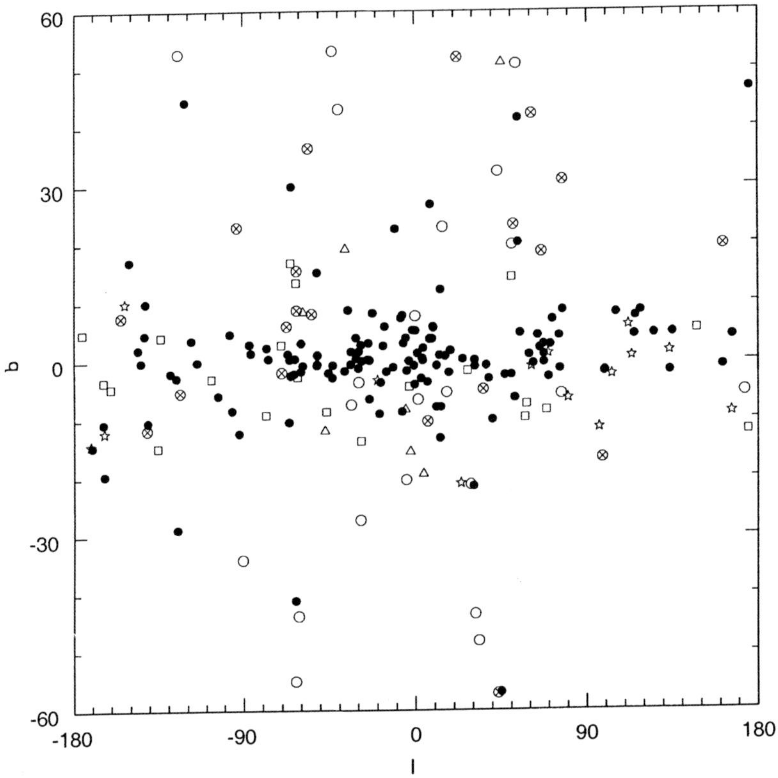

Figure 4 The galactic distribution of PPN candidates. Symbols used are the same as in Fig.3.

5. GALACTIC DISTRIBUTION OF PPN CANDIDATES

The galactic distribution of all proto–planetary nebula candidates in our database is presented in Fig. 4. There is a concentration toward the plane, without any distinct enhancement toward the galactic center. In this sense, the distribution of PPN candidates is different from that of planetary nebulae (Acker et al. 1992). More clearly it is seen in Fig. 5 where comparison between the galactic longitude distributions for post–AGB objects from our sample and planetary nebulae is presented. The simplest explanation is that proto–planetary nebula candidates are not detected at so large distances as planetary nebulae are. Certainly, the infrared fluxes of PPN candidates from the galactic bulge are below IRAS detectability limit of about 0.5 Jy. Another difference between galactic distributions of PPNe and PNe is related to the larger proportion of post–AGB objects at high galactic latitudes. However, these high latitude

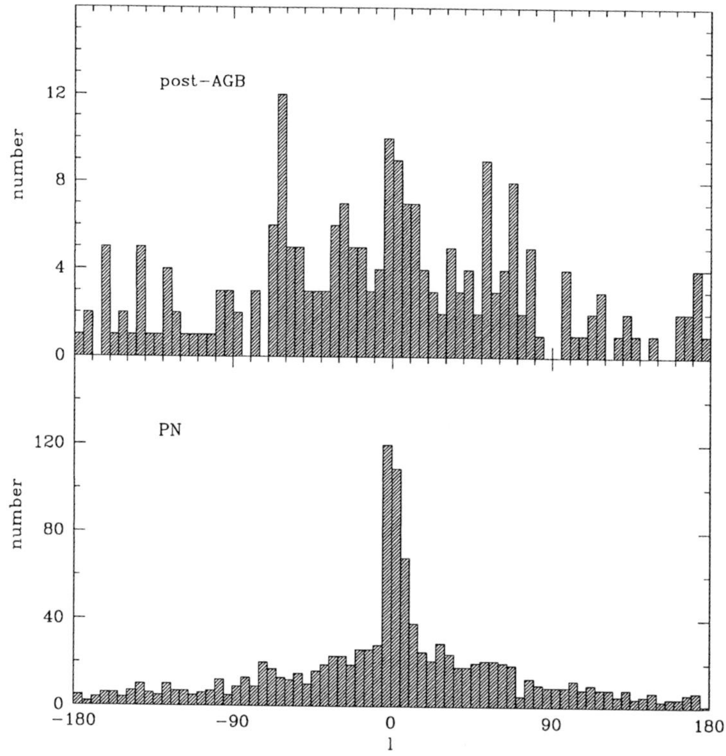

Figure 5 The galactic longitude distribution of post–AGB candidates and PNe.

sources will probably never become PNe as it could be inferred from their low or luck of the far–infrared excess (the envelopes disperse before ionization). From this point of view the name of proto–planetary nebulae is rather spurious for them. They are post–AGB objects but not proto–planetary nebulae. Finally, note that the galactic distribution of PPN candidates shows similar pattern as carbon stars (e.g. Thronson et al. 1987) but do not ressembles distribution for oxygen–rich sources which show a sudden decline at longitudes of about +75 and −90 degrees as it is seen, for example, in the galactic distribution of OH/IR sources (e.g. Chen et al.2001).

6. FURTHER PROSPECTS

Investigation of statistical properties of proto–planetary nebulae (or post–AGB objects, in general) should allow for better understanding of this interesting stage of stellar evolution. For this purpose as wide as possible sample of the candidates should be assembled and analyzed. Study of smaller samples, selected for example on the basis of their IRAS colors alone, cannot answer all questions related to the formation and evolution of these fascinating objects.

Acknowledgments

The authors acknowledge support from grant No. 2.P03D.020.17 of the Polish State Committee for Scientific Research.

References

Acker A., Ochsenbein F., Stenholm B, et al., 1992, Starsbourg-ESO Catalogue of Galactic Planetary Nebulae, ESO, Garching

Bedijn P.J., 1987, A&A 186, 136

Blöcker Th., 1995b, A&A 299, 755

Chen P.S., Szczerba R., Kwok S., Volk K., 2001, A&A 368, 1006

McCausland R.J.H., Conlon E.S., Dufton P.L. et al., 1992, ApJ 394, 298

Ney E.P., Merrill K.M., Becklin E.E., et al., 1975, ApJ 198, L129

Ostile D., Cox A.N., 1986, ApJ 311, 864

Oudmaijer R.D., Waters L.B.F.M., Pottasch S.R., 1993, in Schwarz H.E. (ed.), Second ESO/CTIO workshop, Mass loss on the AGB and beyond, ESO Conference and Workshop Proceedings No.46, Garching, p.122

Schmidt–Kaler Th., 1982, Physical parameters of the stars, In: Schaifers K., Voigt M.H. (eds.), Landolt Börnstein A&A Vol. VI, 2b, Springer, Berlin, p.455

Schönberner D., 1979, A&A 79, 108

Schönberner D., 1983, ApJ 272, 708

Schönberner D., 1989, in Planetary Nebulae, IAU Symp. 131, ed. S.Torres–peimbert, Kluwer, p.463

Szczerba R., Steffen M., 1999, Ap&SS 262, 187

Thronson H.A., Latter W.B., Black J.H. et al., 1987, ApJ 322, 770

van der Veen W.E.C.J., Habing H.J., 1988, A&A 194, 125

van Hoof P.A.M., Oudmaijer R.D., Waters L.B.F.M., 1997, MNRAS 289, 371

Vassiliadis E., Wood P.R., 1993, ApJ 413, 641

Westbrook W.E., Becklin E.E., Merrill K.M., et al., 1975, ApJ 202, 407

OPTICAL SURVEY OF POST-AGB CANDIDATES

Arturo Manchado

Instituto de Astrofísica de Canarias, C/Vía Láctea, E-38200 La Laguna, Tenerife, Spain

amt@ll.iac.es

Olga Suárez

Departamento de Ciencias de la Navegación y de la Tierra E. S. Marina Civil. Universidade da Coruña Paseo de Ronda 51, E-15011 A Coruña, Spain

Pedro García-Lario

ISO Data Centre. Astrophysics Division, ESA. Estación de Seguimiento de Satélites de Villafranca del Castillo. Apartado de Correos 50727, E-28080 Madrid, Spain

Minia Manteiga

Departamento de Ciencias de la Navegación y de la Tierra E. S. Marina Civil. Universidade da Coruña Paseo de Ronda 51, E-15011 A Coruña, Spain

Stuart R. Pottasch

Kapteyn Astronomical Institute, Postbus 800, NL-9700 AV Groningen, The Netherlands

Abstract Over the past 12 years we have been conducting a survey of IRAS (PSC) sources with dust temperatures between 100 and 250 K, which is the temperature of AGB shells after mass loss has ceased. About 1000 PSC sources fulfill this criterion, about half of them identified previously in the literature: most are PNe (49 %) and post-AGB stars (27 %), but there is also a small fraction of young stellar objects and active galaxies. In order to classify the unidentified sources, we took low resolution spectra of these objects, in the range from 3500 to 9000 Å. The region between 3600 and 5050 Å has been used to carry out the spectral classification of most of the unidentified sources. Partial results are presented here: of the 187 objects studied, 34 were PNe, 10 proto-PNe and 103 post-AGB candidates with or without Hα emission. The rest have been identified as young stellar objects (34), LBVs (2), and active galaxies (3).

R. Szczerba and S.K. Górny (eds.), Post-AGB Objects as a Phase of Stellar Evolution, 21–27.

1. INTRODUCTION

Over the last 15 years Post-Asymptotic Giant Branch (Post-AGB) stars have been identified mainly by their characteristic infrared excess (IRE) at mid- and far-infrared wavelengths. At 10 μm a similar IRE was first discovered in planetary nebula (PNe) by Gillet, Low, & Stein (1967) around NGC 7027. Cohen & Barlow (1974) found also a strong IRE at 10 and 18 μm in a sample of 113 PNe. Parthasarathy & Pottasch (1986) identified three high galactic latitude F-supergiants as post-AGB stars based on the mid- and far-IRE detected by the IRAS satellite, very similar to those detected in well known PNe also detected by IRAS. With the advent of the IRAS satellite more than 200 post-AGB stars were identified this way. Van der Veen, Habing, & Geballe (1989) selected 42 Point Source Catalog (PSC) sources based on their IR colors and their LRS spectra as post-AGB candidates, while Hrivnak, Kwok & Volk (1989) identified 8 F-G supergiants as post-AGB stars based only on their LRS spectra. Volk, & Kwok (1989) added 20 possible post-AGB candidates to the list based on their IRAS colors while Trams et al. (1991) found 25 supergiants located at high galactic latitudes and photometrically variable which were suggested to be in the post-AGB stars. More recently, Oudmaijer et al. (1992) selected 414 SAO stars with IRE from the IRAS PSC. However, many of them turned out to be young stellar objects. More recently, Oudmaijer (1996) selected 23 SAO stars with luminosity class I or II and spectral type mainly B as post-AGB candidates. Like this, most of the former studies above mentioned suffer from strong selection effects. Most of the suggested post-AGB candidates were selected to be high galactic latitude stars, therefore their spectral types may be affected by this bias. In order to carry out an unbiased study of the post-AGB phase we have chosen a complete sample of IRAS sources.

2. SELECTION CRITERIA

Our selection criteria is based on the expected dust temperature for a post-AGB. If thermal equilibrium is assumed, the dust temperature of the equilibrium radius, according to Scoville & Kwan (1976) is;

$$T_d = 1.658 f^{-1/5} r^{-2/5} L_*^{1/5}$$

where **f** is the dust emissivity, **r** the equilibrium radius and L_* the star luminosity. Adopting a dust emissivity at 60 μm of 0.004 (Draine, & Lee 1984) for a typical post-AGB radius of 0.01 and 0.1 pc, dust temperatures of 197 and 80 K respectively are derived. Therefore, our selection criteria were aimed to search all the PSC sources within this temperature range. If we define;

$$[12] - [25] = -2.5 log F_{12\mu\text{m}}/F_{25\mu\text{m}}$$

$$[25] - [60] = -2.5 log F_{25\mu\text{m}}/F_{60\mu\text{m}}$$

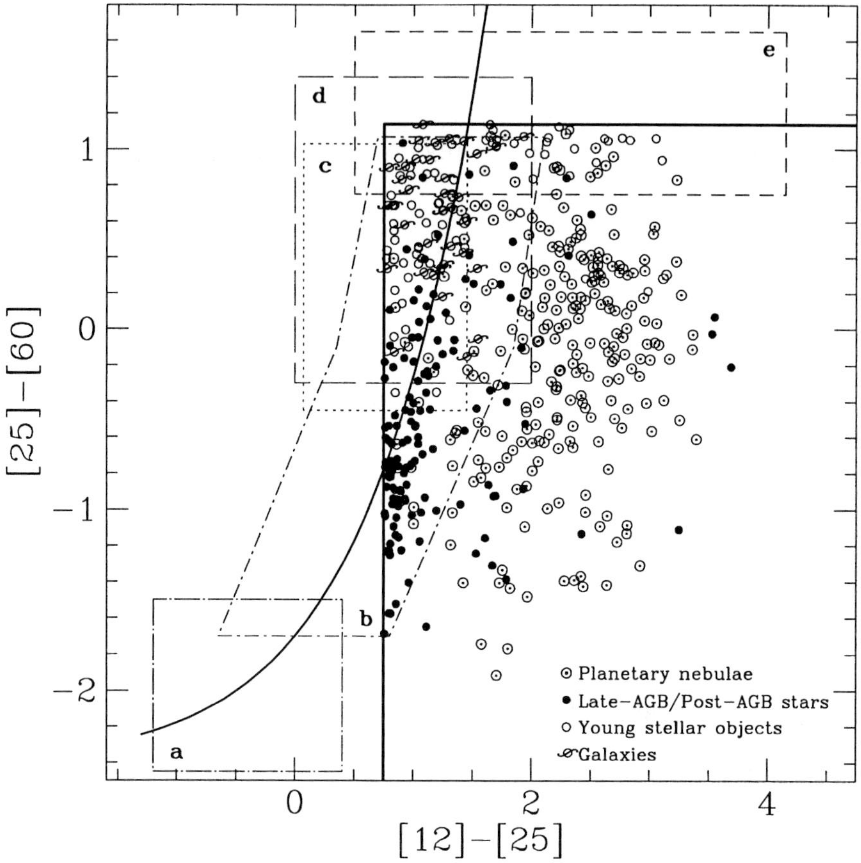

Figure 1 IRAS color positions for different sources; a) optical Mira Variables, b) Variable OH/IR stars, c) T-Tauri and Herbig Ae/Be stars; d) Active galactic nuclei; e) Compact HII regions. Thick solid lines indicate the limits of our sample and the exponential curve represents the evolutionary track expected for O-rich stars in the late AGB.

we want to search all the PSC sources detected in the 12, 25 and 60 μm bands that fulfill the conditions:

$$[12] - [25] > 0.75$$

and

$$[25] - [60] < 1.14$$

There were 1084 sources matching this criteria. 509 of them show a clear identification in the literature: 248 (49 %) were PNe, 96 (19%) OH/IR stars, 88

(17 %) young stellar objects, 43 (8 %) post-AGB stars and 34 (7 %) galaxies. Therefore 76 % of the identified sources were evolved stars. Figure 1 shows the IRAS color diagram with the position of sources under each of these categories.

In order to classify the rest of the sample (575 sources) we carried out multi-wavelength observations over the last 12 years. In a series of papers (Manchado et al. 1989, García-Lario et al. 1990 and García-Lario et al. 1997) near-IR (J,H and K bands) photometry in both hemispheres were conducted, identifying 110 new post-AGB stars. In order to complete the sample and obtain a spectral classification low resolution spectroscopy was carried out over the last 10 years. Here we report the results of the southern hemisphere sample.

3. OBSERVATIONS

Observations were carried out from 1990 to 1995 using the ESO 1.5m telescope (La Silla) with the Bollers & Chivens spectrograph with a spectral resolution between 2 and 4 Å. Due to the PSC error box (16 arcsec in RA and 8 arcsecond in declination), the first step was the association of the PSC source with the optical counterpart, when possible. This was done, first, if the near-IR counterpart was known, by taken the specrtrum of the optical counterpart closer to that position. When no near-IR counterpart was known, the star closest to the nominal PSC position with a visible optical counterpart in the red POSS plate was observed. Misidentifications are expected in less than 3 % of the sources so studied. 217 different IRAS positions were analised this way, finding no optical counterpart for 30 of them.

4. SPECTRAL CLASSIFICATION

With the aim of obtaining an optimum classification of the spectra observed, we chose Silva's catalogue (Silva, 1992) to select reference spectra. The catalogue comprises mean spectra at wavelengths between 3510 and 8930 Å, with a spectral resolution of 11 Å (FWHM). All of them are scaled to flux 100 at 5450 Å.

The spectra of post-AGB stars are very similar to those of supergiant stars (luminosity class I) due to their low surface gravity. For this reason, we used the spectra of this luminosity class as templates for comparison with our target spectra. The resolution of Silva's catalogue stars is two subtypes for each spectral type. Thus, this is the maximum resolution we obtain when making our classification.

The first step in our classification process is to normalize the continuum spectra of both the templates as well as of the post-AGB spectra to classify. This way, we focus on the information present in the spectral lines without taking into account the energy distribution of the spectral continuum, which can be strongly affected by interstellar reddening in some cases.

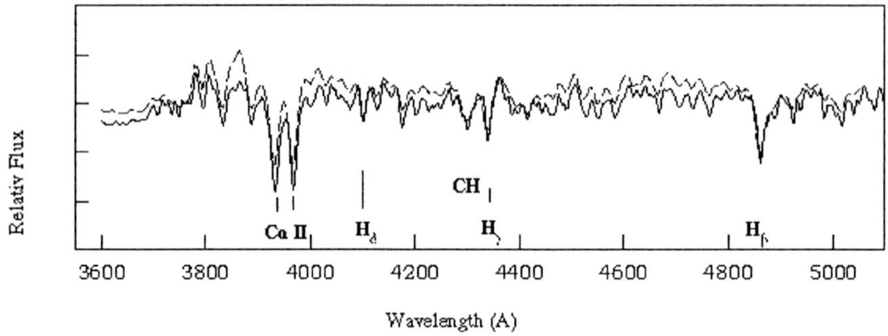

Figure 2 Normalized spectrum of IRAS 13110−5425 (solid line), and the spectrum of a standard F7-9 I spectrum (dashed line).

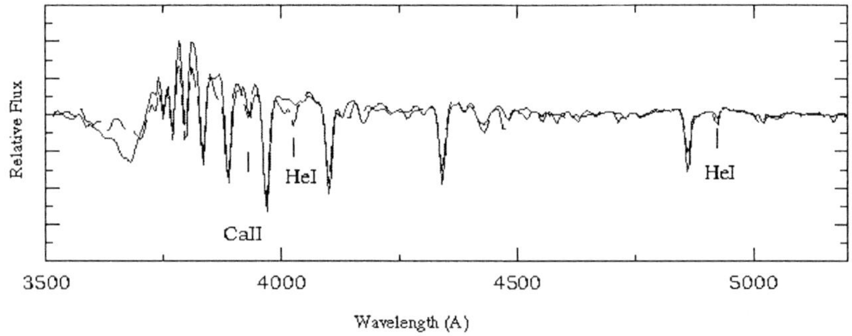

Figure 3 Normalized spectrum of IRAS 17476−4446 (solid line), and the spectrum of a standard B3-5 I spectrum (dashed line).

Once normalized, the next step in classifying our spectra is the direct comparison (by eye) with the spectral templates. Initially a classification with the precision of one spectral type can be made this way considering the whole spectral range from 3500 to 9000 Å. For a more refined classification, we chose the blue region of the spectra, between 3600 and 5050 A, as this spectral region is more sensitive to small spectral type changes. The comparison was made by overimposing the spectra with the different templates belonging to the preliminary spectral type found. The spectral subtype finally assigned was the one which showed a better agreement with the spectra to classify. We always took into account the special caracteristics of post-AGB spectra, which can show emission lines instead of the typical absorption lines present in the spectra of supergiant stars.

In the case of some spectra initially classified as late-type spectra (G, K and M) which showed a strong reddening, we also used a red region (between 5050 and 8900 Å) of the spectrum to make the comparison, because the blue region was too faint to allow the identification of any spectral line.

In Figure 2, we show the example of the spectrum of IRAS 13110−5425, comparing it with that of the catalogue corresponding to a type F7-9 I star. We can see in that Figure that there is a very good agreement between both of them, the Balmer lines, the H and K lines from Calcium and the CH molecular band being the most prominent features in this spectral range.

In Figure 3, we can see the spectrum of a post-AGB star which shows Hα and Hβ emission. In this case, when comparing it with the template, we must take into account that the intensity of the Balmer lines can be lower than it should be, since they will be affected by the emission.

5. RESULTS

As a result of this optical identification and spectral classification exercise, we have identified 3 new galaxies (2 % of the sources in our subsample), 2 (1 %) new LBV's, 34 (18 %) young stellar objetcs, 34 (18%) PNe identified as such for the first time, and 114 (61 %) post-AGB stars. Out from the 114 post-AGB stars, 25 show Hα emission.

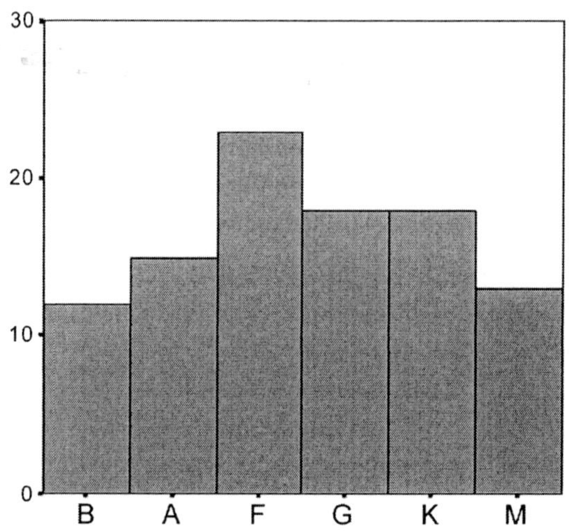

Figure 4 Histogram of the different spectral types assigned to the stars identified as post-AGB stars in our sample.

The distribution of spectral types found among the post-AGB stars in the sample is shown in Figure 4 and cover a wide range from M to B. It is important to note that the sample is large enough to have statistical meaning for all the spectral types. From Figure 4 it can be seen that the peak of the distribution corresponds to spectral types F and G.

References

Cohen, M., Barlow 1974, ApJ, 193, 401

Draine, B.T., Lee, H.M. 1984, ApJ 285, 89

García-Lario, P. Manchado, A.; Suso, S.R., Pottasch, S.R.; Olling, R. 1990, A&AS 82, 497

García-Lario, P., Manchado, A., Pych, W., Pottasch, S.R. 1997, A&AS 126, 497

Gillet, F.C., Low, F.J., Stein, W.A. 1967, ApJ, 149, L97

Hrivnak, B.J., Kwok, S., Volk, K.M. 1989, ApJ 346, 265

Oudmaijer, R. D., van der Veen, W. E. C. J., Waters, L. B. F. M., Trams, N. R., Waelkens, C., Engelsman, E. 1992, A&AS 96, 625

Oudmaijer, R.D. 1996, A&A 306, 823

Manchado, A., Garcia-Lario, P., Esteban, C., Mampaso, A., Pottasch, S. R. 1989, A&A 214, 139

Meixner, M., Ueta, T., Dayal, A., Hora, J. L., Fazio, G., Hrivnak, B.J., Skinner, C. J., Hoffmann, W.F.; Deutsch, L.K. 1999, ApJS 122, 221

Parthasarathy, M., Pottasch, S.R. 1986, A&A 154, L16

Sahai, R. 1999, ApJ 514, L115

Silva, D.R., Cornell, M.E. 1992, ApJS, 81, 865

Scoville, N.Z., Kwan, J. 1976, ApJ 206, 718

Trams, N. R., Waters, L.B.F.M., Lamers, H.J.G.L.M., Waelkens, C., Geballe, T.R., The, P.S. 1991, A&A 87, 361

Ueta, T. Meixner, M., Bobrowsky, M. 2000, ApJ 528, 861

van der Veen, W.E.C.J., Habing, H.J., Geballe, T.R., A&A 1989, 226, 108

Volk, K.M., Kwok, S. 1989, ApJ, 342, 345

HOT POST-AGB STARS

M.Parthasarathy[1], G.Gauba[1], T.Fujii[2], Y.Nakada[2,3]

1. Indian Institute of Astrophysics, Bangalore, India

2. Institute of Astronomy, University of Tokyo, Japan

3. Kiso Observatory, University of Tokyo, Japan

partha@iiap.ernet.in

Abstract From the study of IRAS sources with far-IR colors similar to planetary nebulae (PNe), several proto-planetary nebulae with hot (OB) post-AGB central stars have been detected. These stars form an evolutionary link between the cooler G,F,A supergiant stars that have evolved off the Asymptotic Giant Branch (AGB) and the hot (OB) central stars of PNe. The optical spectra of these objects show strong Balmer emission lines and in some cases low excitation nebular emission lines such as [NII] and [SII] superposed on the OB stellar continuum. The absence of of [OIII] 5007Å line and the presence of low excitation nebular emission lines indicate that photoionisation has just started. The UV(IUE) spectra of some of these objects revealed violet shifted stellar wind P-Cygni profiles of CIV, SiIV and NV, indicating hot and fast stellar wind and post-AGB mass loss. These objects appear to be rapildy evolving into the early stages of PNe similar to that observed in the case of Hen1357 (Stingray Nebula) and SAO85766.

1. INTRODUCTION

From an analysis of the Infrared Astronomical Satellite Point Source Catalog (IRAS PSC) cooler post-AGB stars having G,F,A supergiant like character were first identified (Parthasarathy and Pottasch 1986, Lamers et al. 1986, Pottasch and Parthasarathy 1988, Hrivnak et al. 1989). These stars were found to have circumstellar dust shells with far-IR colors and flux distribution similar to the dust shells of planetary nebulae (PNe). Parthasarathy and Pottasch (1986) interpreted that the dust shells around these stars were the result of mass-loss during the AGB phase of evolution. Some of these stars were found to be at high galactic latitudes, having high space velocities and were metal-poor, indicating that they have evolved from low-mass stars and are now in the post-AGB stage of evolution.

Later, from an analysis of IRAS data, Parthasarathy and Pottasch (1989) found a few peculiar emission line (forbidden) B-type stars. Their supergiant

R. Szczerba and S.K. Górny (eds.), Post-AGB Objects as a Phase of Stellar Evolution, 29–40.

like character, the presence of cold detached dust shells and far-IR colors similar to PNe suggested that they are in a post-AGB phase of evolution. Thus, there seems to be an evolutionary sequence ranging from the cooler G,F,A supergiant-like stars to hotter O-B types, evolving from the tip of the AGB towards young PN stage.

2. HOT POST-AGB CANDIDATES

In Table 1 we provide a list of known and candidate hot post-AGB stars. The IRAS sources listed in the table were selected on the basis of their far-IR colors (F(12μ)/F(25μ)<0.35 and F(25μ)/F(60μ)>0.3). Based on low resolution spectroscopy, they were later identified to be hot post-AGB stars(Parthasarathy 1993b; Parthasarathy, Vijapurkar and Drilling 2000a). A few LSS and MWC sources were identified as hot post-AGB candidates based on recently obtained low resolution spectra with the 2.34m Vainu Bappu telescope at Kavalur (Gauba, Parthasarathy, A&A, in preparation). Most objects in the list have OBe spectral types and show H-alpha in emission. Some also show [SII] and [NII] emission lines. Several objects in the list have low resolution IUE spectra. Study of the UV continuum and spectral features (Gauba et al., A&A, in preparation) also suggest that these are early (OB) type post-AGB stars and are rapidly evolving into early stages of PN similar to Hen1357 (Stingray Nebula) and SAO85766. The E(B$-$V) values listed in the table are derived from the spectral types and observed (B$-$V) magnitudes of the stars.

2.1. HEN1357 = STINGRAY NEBULA

Hen 1357 = Stingray Nebula = IRAS 17119-5926 was the first object that was shown to have evolved from a B1 type post-AGB supergiant into a PN in the extremely short timescale of 20 years (Parthasarathy et al. 1993, 1995). The UV (IUE) spectra of Hen 1357 (Fig. 1), shows the rapid evolution of the central star from 1988 to 1996. The P-Cygni profiles of NV(1240Å) and CIV(1550Å) in the spectra taken in 1988, had indicated a terminal wind velocity of -3500 km/s. The wind appears to have completely ceased by 1994 (Fiebelman 1995, Parthasarathy et al. 1995). We find that the object has faded by a factor of 3 in the UV, from 1988 to 1996. Then, if the luminosity were to remain the same, the temperature of the central star should have increased from 37000K to 50000K in the same period. The fading of the UV continuum and the termination of the stellar wind suggest that the nuclear fuel is almost extinct as a result of the post-AGB mass loss. Alternatively, the fading could be due an episode of high mass-loss and it is possible that the shell hydrogen burning has not stopped or stopped temporarily. The fading suggests that the central star of Hen1357 is rapidly evolving into a DA white dwarf.

Table 1 Hot post-AGB candidates

IRAS	Name	b	Sp.Type	V	B-V	E(B−V)	IRAS Fluxes 12μ	25μ	60μ	100μ
06556+1623	HD51585	+8.89	BQ[ ]	11.5			1.40	7.14	4.35	1.79
07171+1823							0.41L	1.38	0.70	1.00L
	LSS1021	-1.18	B1Vpe	9.15	+0.05	0.31				
	LSS1339	-7.34	B	10.7	+0.2					
09470-4617	LSS1351	+5.59	B1Iae	12.43	+0.57	0.76	2.87	4.03	1.36	1.55L
	LSS1392	-5.96	B	10.43	+0.14					
10082-5647	Hen373	-0.78	Be	12.5			8.70	22.67	56.80	143.40L
10178-5958	Hen401	-2.72	PPN	12.5			4.12	38.33	76.14	41.49
11065-6026	Hen591	-0.31	Be	11.79			12.69	205.30	129.30	60.52L
12584-4837	Hen847	+13.95	Be	10.57	+0.08		36.07	48.75	13.04	3.31
13185-6922	Hen892	-6.92	Be	9.6			0.88	2.07	2.39	2.25L
13266-5551	-55 5588	+6.36	B1Ibe	10.7			0.76	35.90	35.43	11.66
14072-5446	CPD-54 5573	+6.12	A3I	10.9			0.25L	4.78	4.48	12.39L
14331-6435	Hen1013	-4.20	B3Ie	11.5	+0.58	0.72	4.04	108.70	70.71	20.61
14488-5405	CPD-53 5736	+4.49	Be	10.5			6.59	63.74	30.54	13.56
	BD+33 2642	+50.79	B2IVp	10.84	-0.17	0.17				
15154-5258	PNG324.0 +03.5	+3.53	WC	13			2.25	12.93	14.20	8.58L
15513-6600	HD141969	-9.66	BQ	11.93	+0.334		2.16	48.18	43.81	19.58
16206-5956	SAO 243756	-7.49	A3Iabe	9.89	+0.24	0.18	0.36L	11.04	12.30	4.83
	HD137569	+51.93	B5III	7.95	-0.07	0.10				
17041-2709	Hen1336	+7.70	Be	12.17	+1.130		0.25L	0.65	1.02	3.72L
17074-1845	Hen1347	+12.26	B3IIIe	12.28			0.50	12.20	5.66	3.47
	HD326971	-1.27	Be	11.0						
17119-5926	Hen1357	-12.16	B3e	10.95	-0.043		0.65	15.50	8.20	3.52
17203-1534		+11.49	B1IIIpe	12.5			0.32	10.70	6.88	3.37
17311-4924	Hen1428	-9.04	B1IIe	10.677	+0.403	0.663	18.34	150.70	58.74	17.78
17381-1616	LSS4331	+7.50	B1Ibe	12.2			0.25L	5.76	1.25	13.71L
17418-4604	HD161044	-8.83	BQe	10.5			2.02	11.88	13.38	4.72
17423-1755	Hen1475	+5.78	PPN	11.0			7.05	28.31	63.68	33.43
17427-3010	HD316248	-0.72	BQ[]	11.5			14.40	177.90	60.33	256.40L
17460-3114	SAO209306	-1.88	O8III	7.92	+0.23	0.54	6.26	20.82	12.20	220.40L
18023-3409	LSS4634	-6.31	B2IIIe	11.6			0.26L	2.94	1.82	25.64L
18062+2410	SAO85766	+19.79	B1IIIpe	11.6			3.98	19.62	2.90	1.00L
18131-3008	HD167402	-6.39	BOII	8.95	-0.01	0.28	1.31	8.02	2.11	3.91L
18237-0715	MWC930	+2.23	Be	11.0			1.84	4.01	38.24	35.13
18313-1738	MWC939	-4.27	B5e	12.4			9.41	7.28	1.00:	68.70L
18371-3159	LSE63	-11.82	B1Iabe				0.25L	6.31	5.16	1.95
18379-1707	LSS5112	-5.42	B1IIIpe	11.8			1.67	23.76	7.12	3.66L
18442-1144		-4.35	Be				0.81	14.9	13.42	31.80L
19157-0247	LSIV -2 29	-7.22	B1III	11.5			8.88	7.16	2.45	11.05L
19336-0400	PN G034.5-11.7	-11.75	B1Iape	12.5			0.38L	8.18	5.56	2.32
19399+2312	LSII +2317	+0.08	B1III	10.4			1.13	2.27:	23.31L	80.20L
19590-1249	LSIV -12 111	-21.26	B1Ibe	11.4	+0.10	0.30	0.29	10.26	6.45	1.77
20462+3416	LS II +34 26	-5.75	B	11.06	+0.18	0.38	13.68	12.12	5.93	
22023+5249	LSIII+5224	-1.96	B				1.02	24.69	14.52	3.93L
22495+5134	LSIII+5142	-6.77	B	11.99	+0.11		0.54	12.37	7.18	3.12

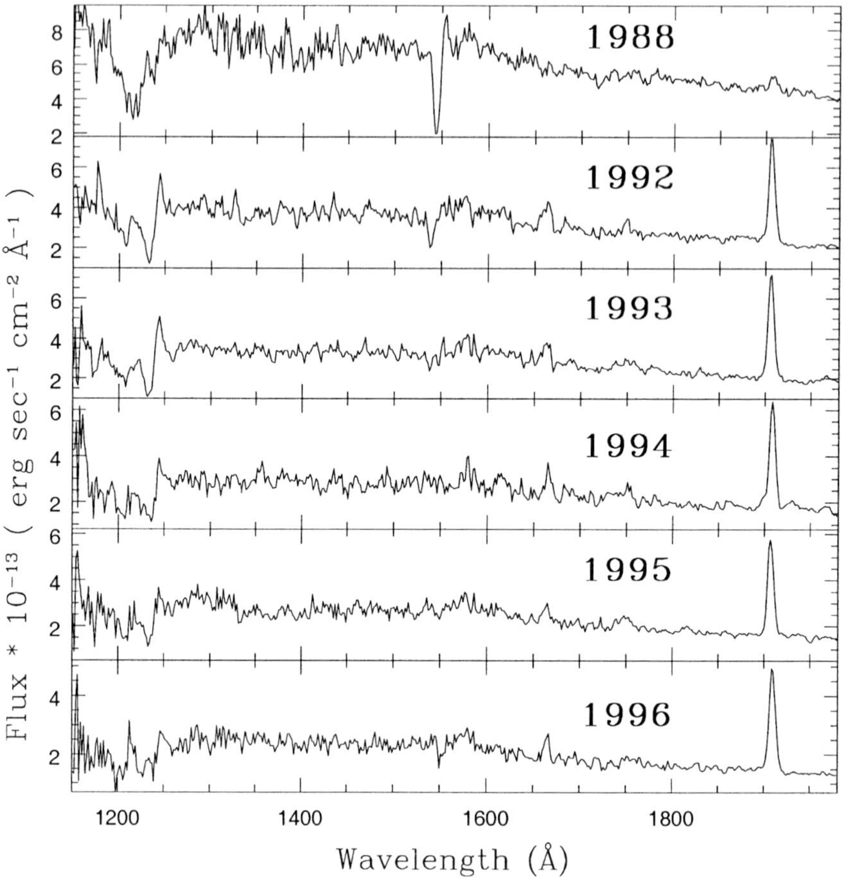

Figure 1 UV(IUE) spectra of Hen1357. Notice the variation in the P-Cygni profiles of NV(1240Å) and CIV(1550Å). Also notice the variation in the strength of [CIII] 1909Å from 1988 to 1993 and the decrease in the continuum flux from 1988 to 1996.

The rapid evolution of Hen1357 indicates that the fading is occuring much more rapidly than predicted, considering that it has taken 2700 years to go from the tip of the AGB to T_{eff}=21000K (Teff in 1971), and about 20 years to turn into a PN. Secondly, if the nuclear shell burning has stopped, it has stopped at a much lower surface temperature than any of the post-AGB evolutionary models suggest. Using a distance of 5.6kpc, Parthasarathy et al. (1993) estimated the luminosity of the central of the star to be 3000$L_\odot$. The core mass luminosity relation then suggests that the core mass should be 0.55$M_\odot$. A 0.6$M_\odot$ core mass post-AGB star takes about 10,000 years to evolve from the tip of the AGB to the white-dwarf cooling track. The luminosity, the core mass, observed rapid evolution and fading of of Hen1357 are not in agreement with the time scales of evolution of low or high mass post-AGB models. Recent HST images of

Hen1357 (Bobrowsky et al., 1998) showed the presence of a 1.67″ X 0.92″ nebula around the central star. The HST images also revealed the presence of collimated outflows and a binary companion to the central star of Hen1357. The object needs to be monitored in UV, optical and IR during the next few decades in order to understand the birth and early evolution of PNe and their central stars.

Recently, Gauba et al. (2000a), studied the UV(IUE) spectra of the central stars of PNe Hb 7 and Sp 3. From the P-Cygni profiles of NV (1240Å) and CIV(1550Å), they estimated the stellar wind velocities in these stars. The strong P-Cygni profiles of NV and CIV indicated the presence of ongoing post-AGB mass-loss in these objects. To understand when and how a strong stellar wind arises during the hot post-AGB phase and its effect on the mass-loss mechanisms in these stars and also to understand if such winds could be responsible for the early evolution of objects like Hen1357, it is important to study the wind profiles and mass-loss rates of central stars of several high galactic latitude planetary nebulae.

2.2. SAO85766

SAO85766 (IRAS 18062+2410) appears to have rapidly evolved in the last 20-30 years (Arkhipova et al. 1999, Parthasarathy et al. 2000b) similar to the case of Hen 1357 (see section 2.1). It was classified as a high galactic latitude, post-AGB star having far infrared colors similar to PNe (Volk and Kwok, 1989, Parthasarathy, 1993b). According to the HDE Catalog, its spectral type in 1940 was A5. The UV colors of this star listed in the Catalogue of Stellar Ultraviolet Fluxes (Thompson et al., 1978) and based on observations with the Ultraviolet Sky Survey Telescope (S2/68) onboard the ESRO satellite TD-1 indicated that in 1973 its spectral type was still A5I. However, the spectral energy distribution of this star obtained in 1985-87 indicated that it is an early B-type star (B1I) (Downes and Keyes, 1988). An analysis of its high resolution spectra, revealed the underabundance of carbon and metals, high radial velocity and the presence of low excitation nebular emission lines (Parthasarathy, 2000b). These characteristics coupled with its high galactic latitude, the presence of a PN type cold detached shell and the variations observed in its spectrum suggest that SAO85766 has evolved rapidly during the last 20-30 years and is now in the early stages of the PN phase. HST WFPC2 imaging of this object may reveal the nebulosity around the central star.

2.3. LS II +34 26 AND LS IV -12 111

Based on its far infrared colors, global energy distribution, low gravity and high galactic latitude, Parthasarathy (1993a) first suggested that LS II +34 26

(IRAS 20462+3416) was in post-AGB stage of evolution. Until then, it was considered to be a young massive Population I star located at 17.8 kpc from the Sun. Later, nebular emission lines in the optical spectrum of LS II +34 26 indicated that the star may be considered to be a very low excitation PN (Parthasarathy 1993d). Differences in the spectrograms obtained by Turner and Drilling in 1977 and 1981 and the recent spectra suggest that the star may be rapidly evolving into the PN stage. A multiwavelength study of the object by García-Lario et al. (1997) confirmed that the central star is surrounded by a very low excitation and compact nebula. Like Hen1357 and SAO85766, LS II +34 26 may be another case of a rapidly evolving hot post-AGB star.

LS IV -12 111 (IRAS 19590−1249) was classified as a high latitude hot post-AGB star based on its IRAS data and high galactic latitude (Parthasarathy 1993a). Analysis of its high resolution spectrum by McCausland et al. (1992), Conlon et al. (1993) confirmed that it is a hot post-AGB star rapidly evolving into a low excitation planetary nebula. McCausland et al. (1992) derived chemical composition of several high latitude hot post-AGB stars. In contrast to the cooler post-AGB stars, LS IV -12 111 showed a severe carbon depletion. Similar carbon depletions have been reported in other hot post-AGB stars at high galactic latitudes. These include LS II+34 26 (García Lario et al., 1997), PG 1323−086, PG 1704+222 (Moehler and Heber, 1998) and SAO85766 (Parthasarathy et al, 2000b).

3. UV(IUE) OBSERVATIONS OF HOT POST-AGB STARS

In Fig. 2 we show the UV (IUE) spectra of a few hot post-AGB candidates. The UV spectra of these stars show a stellar continuum in the wavelength interval 1150 to 3200 Å which is consistent with their OB spectral types. The spectra have been corrected for reddening using the 2200Å feature in the UV spectra. The anomalous extinction observed in the UV (IUE) spectra of some hot post-AGB stars like LS IV -12 111 and LS II +34 26 (Conlon et al., 1993, Parthasarathy 1993c, García- Lario et al., 1997) suggests that part of the reddening is of circumstellar origin and that the interstellar reddening law may not be applicable. The hot post-AGB candidates IRAS 14331−6435 (Hen1013), IRAS 17203−1534 and IRAS 17311−4924 (Hen1428) show SiIV(1394Å , 1404Å) in absorption which is consistent with their classification as possible hot post-AGB stars or proto-planetary nebulae.

Hen1475 and Hen401 (listed in Table 1), were classified as proto-planetary nebulae (PPN) by Parthasarathy and Pottasch (1989) based on their far infrared colors. Parthasarathy and Pottasch (1989) found the ratio of far infrared luminosity to visual luminosity to be very high in these objects. Hen1475 was observed with the HST, the VLA, the NTT, and the MPIA/ESO 2.2 m tele-

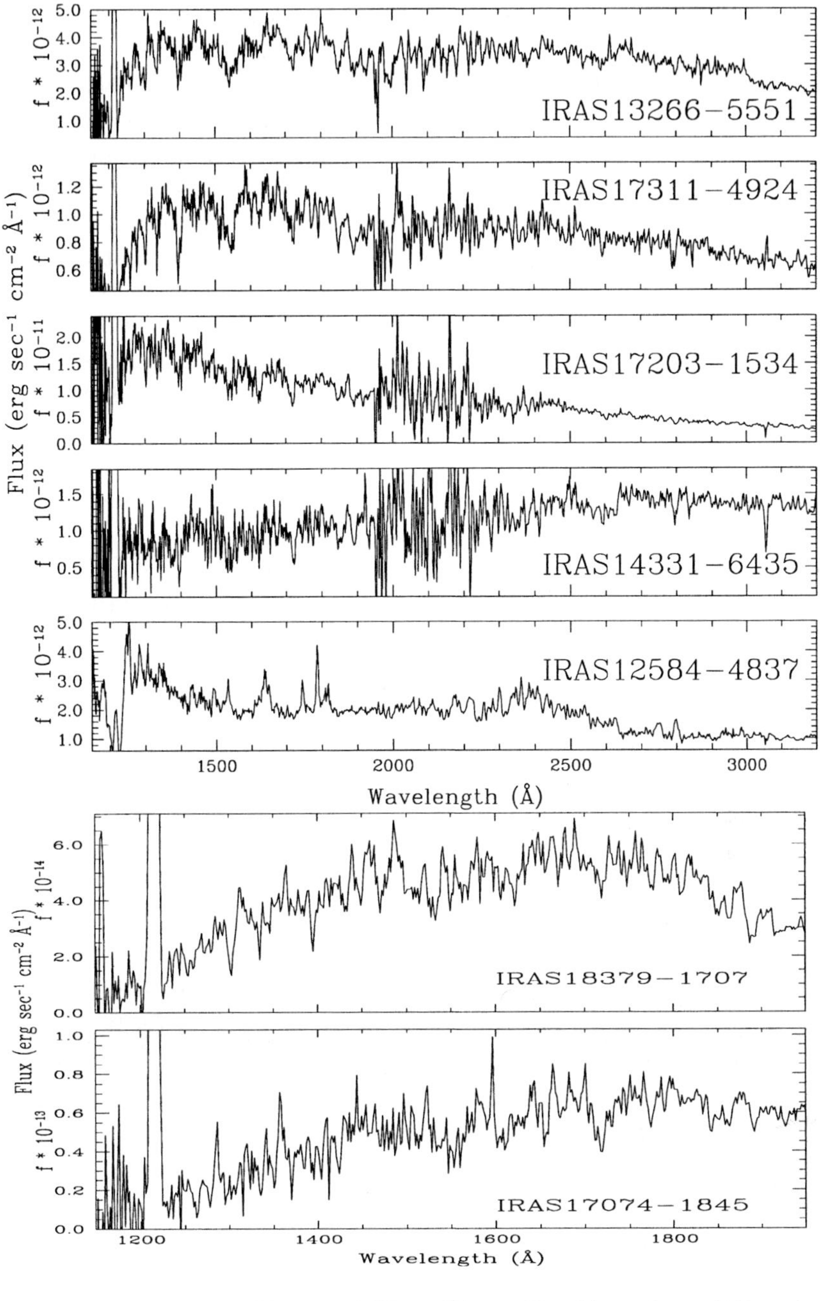

Figure 2 UV(IUE) spectra of hot post-AGB candidates. The objects show a UV continuum which is consistent with their OB spectral types. Some of the objects in the above figure show emission features. A few of the objects also show SiIV and CIV.

scope (Bobrowsky et al. 1995). Based on these observations, the central star of Hen1475 was presumed to be at the origin of a 15″long structure containing symetrically opposing bright knots. The knots were found to have radial velocities of ± 500 km s^{-1}. The bright central component seen in the NTT images was resolved by the HST into a highly asymmetric structure a few arcseconds across. A multiwavelength study of Hen1475 by Riera et al. (1995) established the object as a massive post-AGB star in the transition phase between the end of the AGB and the formation of a new PN. HST WFPC2 images of Hen1475 by Borkowski et al. (1997) confirmed its classification as a point-symmetric PPN. Hen401 was also imaged by the HST WFPC2 (Sahai et al. 1999). The HST images revealed a bipolar reflection nebula around the central star. Two very long (14.5″), cylindrical-shaped bipolar lobes were seen in reflected light, each with a length/width ratio of about 7, probably the largest aspect ratio seen in a PPN so far.

Interestingly, the UV(IUE) spectra of Hen401 (Parthasarathy et al. 2000c) and Hen 1475 show no evidence of the hot (Be) central star capable of exciting the nebula around these stars. The UV continuum observed in the IUE spectra of Hen401 corresponds to a B8 or A0 dwarf star which is too cool to explain the nebular emission lines observed in its optical spectra by García-Lario et al. (1999). This may be explained if the hot central star of Hen1475 and Hen401 respectively, is heavily obscured by a dusty circumstellar disk. HST WFPC2 images of Hen401 have revealed the presence of an equatorial torus (Figure 1 in Sahai et al. 1999). CO observations have revealed the presence of an expanding torus of molecular material around the central star (Knapp et al. 1995) of Hen1475. This torus is believed to intercept most of the stellar radiation, reradiating it through the far-infrared thermal dust emission (Parthasarathy and Pottasch 1989). Based on an analysis of the UV spectra and the range of E(B−V) values derived from the UV spectra, optical spectra and BVRI photometry, Parthasarathy et al.(2000c) concluded that the central star of Hen401 is a binary. The characteristics of Hen401 were found to be very similar to those of the well studied bipolar PPN M 2−9. [OIII] 5007Å was not observed in the optical spectra of Hen401 and Hen1475 indicating that the nebula around these stars is not completely photoionised.

An analysis of the HST WFPC2 image of Hen1475 by Borkowski et al. (1997) revealed what appears to be the collimation of large-scale outflows from the central star into bipolar jets. Collimated outflows have also been observed in Hen1357 and Hen401. The appearance of collimated outflows in Hen1357 suggests that collimation of the stellar wind from the hot central star occurs quite early in the transition phase from the tip of the AGB to the PN. The fast outflows observed in Hen1475 (Riera et al. 1995, Bobrowsky et al., 1995), may accelerate the evolution of this star towards the young PN stage, similar to the case of Hen1357. To identify and study more objects in this stage of evolution, it is important to look at the UV spectra of the objects listed in Table 1 and also to image them with the HST WFPC2. Observations with the HST may reveal the nebulosity around these stars.

4. UV-BRIGHT STARS IN GLOBULAR CLUSTERS

The term "UV-bright" stars is used for globular cluster members which appear more luminous in the ultraviolet at a given temperature than normal stars on the horizontal branch or on the AGB (Zinn et al., 1972). They are found to be in rapid post-horizontal branch evolution. It is still not clear whether these stars are produced through normal stellar evolution or through interactions within a binary system. Radial velocity measurements of these objects are important in order to confirm their cluster membership and also to resolve the question of their origin.

Observationally, few UV-bright stars in globular clusters have been observed (Zinn et al. 1972, Harris et al. 1983 and references therein). This is attributed to the short time scale of post-horizontal branch evolution and to observational selection effects due to crowding of star images in the central regions of the globular clusters. Some interesting and easily identifiable post-AGB stars in globular clusters have been studied spectroscopically. One such object is the hot-post AGB star Barnard 29 in the globular cluster M13 (see section 5)

The post-AGB nature of stars in globular clusters can be firmly established from their position in the cluster's color-magnitude diagram. Studies on UV-bright stars in different globular clusters are being carried out to establish the chemical composition of post-AGB stars as a function of mass and initial metallicity. Since globular cluster distances are well established from Cepheid variables and since extinction at high galactic latitudes is negligibly small, we can unambiguously determine the evolutionary status of post-AGB stars in globular clusters.

5. CHEMICAL COMPOSITION OF HOT POST-AGB STARS

At the dispersions normally used for spectral classification, it may not be possible to distinguish young massive OB stars from the evolved Population II stars. Previously, based on their low resolution spectra, all high latitude B-type supergiants were thought to be massive young Population I stars. However, during the late stages of stellar evolution, stars develop a complex abundance pattern. This coupled with the fact that iron does not undergo processing, provides a means of distinguishing old Population II and young Population I stars based on a quantitative abundance analysis. Recent high resolution studies of these objects (Mc Causland et al. 1992, Parthasarathy et al. 2000b) suggest that some of these are hot-post AGB stars.

Hot post-AGB stars at high galactic latitudes are found to show severe carbon deficiency. Model stellar atmosphere and abundance analyis of the post-AGB star Barnard 29 in the globular cluster M13 (Conlon et al., 1994) using UV (IUE) and optical data revealed an effective temperature $\simeq$ 20000K, log $\simeq$ 3. Barnard 29 was found to be metal deficient with near solar helium abundance. In common with most high latitude field post-AGB stars, carbon was found to be severely depleted.

The observed carbon deficiency is often explained on the basis of one of the following three possibilites. The first being that these stars have not gone through the third dredge up during their AGB phase of evolution. If on the other hand, the elemental abundances of elements heavier than oxygen, indicate the occurence of a third dredge up then the carbon which was produced during the triple-alpha process during a thermal pulsation, must have been removed by another process. Yet another scenario to explain the abundance distribution in hot post-AGB stars was put forth by Napowitzki et al. (1994 and these proceedings). They put forth a gas-dust separation model to explain the abundance pattern of BD+33°2642. However, the limited observational evidence has not been able to firmly establish any of these three scenarios.

Fujii et al. (2000 a,b,c) carried out multi-color photometric monitoring and detailed spectral energy distribution analysis of several cooler post-AGB stars. Similar study of hot post-AGB candidates is in progress. Small amplitude photometric variability seems to be the characteristic feature of all post-AGB candidates (Fujii et al., 2000c). Fujii et al. (2000d) discovered Brackett gamma and 2.12μ molecular hydrogen emission in several hot post-AGB candidates. Near infrared spectroscopy to detect molecular hydrogen and Br_γ in post-AGB stars has also been carried out by García-Hernández et al. (these proceedings). The H_2 emission line may be excited either by fluorescence, through the absorption of UV photons from the central star in the photodissociation region or by shocks. An attempt to understand the origin of 2.12μ molecular hydrogen and a study of H_2 to Br_γ flux ratios in hot post-AGB stars would enable us to understand the origin of fast stellar winds in the hot post-AGB phase and their role in the shaping and formation of planetary nebulae.

6. CONCLUSIONS

From an analysis of IRAS data several hot post-AGB candidates were identified. Their low gravities and extended atmospheres appear to imitate Population I supergiants. Some of these hot post-AGB stars were found to be at high galactic latitudes. High resolution studies of diverse populations of these objects are required to understand for instance, the severe carbon depletion in high latitude hot post-AGB stars; to establish the abundance patterns of these objects based on mass and inital metallicity and to understand the mass-loss phenomena and rapid evolution of objects such as Hen1357 and SAO85766. Continued spectroscopic and photometric studies of hot post-AGB candidates are also important to identify more objects like Hen1357 and SAO85766 which have turned into young PNe in the last few decades. All the objects listed in Table 1 show infrared emission from circumstellar dust shells. Their OB spectral types suggest that the central stars emit strongly in the UV. It is important to study these objects from the UV to the infrared in order to understand the birth and early evolution of PNe.

References

Arkhipova V.P., Ikonnikova N.P. et al. 1999, Astronomy Letters 25, 25

Bobrowsky M., Sahu K.C., Parthasarathy M., García-Lario P. 1998, Nature 392, 469

Bobrowsky M., Zijlstra A.A., Grebel E.K. et al. 1995, ApJ 446, L89

Borkowski K.J., Blondin J.M., Harrington J.P. 1997, ApJ 482, L97

Conlon E.S., Dufton P.L., McCausland R.J.H., Keenan F.P. 1993, ApJ 408, 593

Conlon E.S., Dufton P.L., Keenan F.P. 1994, A&A 290, 897

Downes R.A., Keyes C.H. 1988, AJ 96, 777

Fiebelman W.A. 1995, ApJ 443, 245

Fujii T., Nakada Y., Parthasarathy M. 2000a, A&A (in press)

Fujii T., Nakada Y., Parthasarathy M. 2000b, (in this proceedings)

Fujii T., Nakada Y., Parthasarathy M. 2000c (in this proceedings)

Fujii T., Nakada Y., Parthasarathy M. 2000d (in preparation)

García-Lario P., Parthasarathy M., de Martino D., et al. 1997, A&A 326, 1103

García-Lario P., Manchado A., Pych W., Pottasch S.R. 1997, A&AS 126, 47

García-Hernández D.A., Manchado A., García-Lario P. et al., (these proceedings)

Gauba G., Parthasarathy M., Nakada Y., Fujii T. 2000a, (submitted to A&A)

Gauba G., Parthasarathy M., A&A (in preparation)

Gauba G. et al., A&A (in preparation)

Hrivnak B.J., Kwok S., Volk K.M. 1989, ApJ 346, 265

Harris H.C., Nemec J.M. and Hesser J.E. 1983, PASP, 95, 256

Knapp G.R., Bowers P.F., Young K., Philips T.G. 1995, ApJ 445, 293

Lamers H.J.G.L.M., Water L.B.F.M. et al. 1986, A&A 154, L20

McCausland R.J.H., Conlon E.S., Dufton P.L., Keenan F.P. 1992, ApJ 394, 298

Moehler S. and Heber U. 1998, A&A 335, 985

Napiwotzki R., Heber, Köppen 1994, A&A 292,239

Napiwotzki R., Herrmann M., Hebber U., (these proceedings)

Parthasarathy M., Pottasch S.R. 1986, A&A 154, L16

Parthasarathy M., Pottasch S.R. 1989, A&A 225, 521
Parthasarathy M., García-Lario P. et al. 1993, A&A, 267, L19
Parthasarathy M. 1993a, ApJ 414, L109
Parthasarathy M. 1993b, In : Sasselov D.D(ed.) *Luminous High Latitude Stars*, ASP Conf. Ser., 45, p.173
Parthasarathy M. Pottasch S.R., Clavel J. 1993c, IAU Symposium 155, 398
Parthasarathy M. 1993d, In : J. Corbally et al.(ed.) *The MK Process at 50 years*, ASP Conf. Ser., 60, p.625
Parthasarathy M., García-Lario P., de Martino D. et al 1995, A&A 300, L25
Parthasarathy M., Vijapurkar J., Drilling J.S. 2000a, A&AS 145, 269
Parthasarathy M., García-Lario P., Sivarani T. et al. 2000b, A&A 357, 241
Parthasarathy M., García-Lario P., Gauba G., de Martino D., Nakada Y., Fujii T., et al. 2000c, (submitted to A&A)
Pottasch S.R., Parthasarathy M. 1988, A&A 192, 182
Riera A., García-Lario P., Manchado A., Pottasch S.R., Raga A.C. 1995, A&A 302, 137
Sahai R., Bujarrabal V., Zijlstra A. 1999, ApJ 518, L115
Thompson G.E., Nandy K., Jamar C., et al. 1978, Catalogue of stellar ultraviolet fluxes, SERC
Volk K.M., Kwok S. 1989, ApJ 342, 345
Zinn R.J., Newell E.B. and Gibbson J.B. 1972, A&A 18, 390

PROPERTIES OF POST-AGB STARS WITH IRAS COLORS TYPICAL OF PLANETARY NEBULAE

Griet C. Van de Steene
Royal Observatory of Belgium, Ringlaan 3, 1180 Brussels, Belgium
gsteene@oma.be

Peter A.M. van Hoof
CITA, University of Toronto, 60 St. George Street, Toronto, ON M5S 3H8, Canada
vanhoof@cita.utoronto.ca

Peter R. Wood
RSAA, Australian National University, Canberra ACT 0200, Australia
wood@mso.anu.edu.au

Abstract We have investigated the infrared properties of a sample of objects with *IRAS* colors typical of planetary nebulae. The selected objects have not yet evolved to the planetary nebula stage since they have no detectable radio continuum emission and they are therefore likely post-AGB stars.

1. INTRODUCTION

A most intriguing challenge is to understand how Asymptotic Giant Branch (AGB) stars transform their surrounding mass-loss shells in a couple of thousand years into the variety of shapes and sizes observed in Planetary Nebulae (PNe). There are a number of theories currently being investigated. No consensus about the dominant physical process responsible for the shaping of PNe has emerged so far, but there is agreement that the shaping occurs during the early AGB-to-PN transition stage.

The details of the rate of evolution and the strength of the stellar wind during the AGB-to-PN transition phase are essential ingredients in understanding the shaping of the PN. However, both are very poorly known, either theoretically or observationally. To remedy this we have started to study a sample of *IRAS*-selected post-AGB candidates.

R. Szczerba and S.K. Górny (eds.), Post-AGB Objects as a Phase of Stellar Evolution, 41–44.

2. SAMPLE SELECTION

Objects were selected from the *IRAS* catalogue with far infrared colors typical of PNe. Apart from PNe, only post-AGB stars are typically found in this part of the color-color diagram (van Hoof et al. 1997). Furthermore, we selected objects that were not detected in the radio continuum above a detection limit of 3 mJy (Van de Steene & Pottasch 1993). Hence we may assume that they have not yet evolved to the PN stage.

3. OBSERVATIONS

We obtained JHKL images of the candidates with CASPIR on the 2.3-m telescope at Siding Spring Observatory in Australia. To search for hydrogen emission, we obtained Brγ spectra with IRSPEC on the NTT at ESO. The objects showing Brγ in emission were re-observed in the radio continuum with the Australia Telescope Compact Array.

4. RESULTS

The *IRAS* counterparts were identified on the basis of their position and near infrared colors. Accurate positions were obtained by using the USNO catalog. The weather was photometric and accurate JHKL magnitudes were determined for the *IRAS* counterparts. By combining the JHKL and *IRAS* photometry, various color-color diagrams were constructed. In these diagrams our objects show evidence for a very large range of extinction. The observed trends in the infrared colors are consistent with the expected evolution of the circumstellar shell (van Hoof et al. 1997).

Of 16 positively identified objects, 6 show Brγ in emission, 7 in absorption, and in 3 no clear Brγ absorption or emission was visible. The absorption lines are very narrow in six objects, indicating a low surface gravity. The objects showing Brγ in emission have a strong underlying continuum, unlike normal PNe.

The fact that our objects were mostly selected from the region in the *IRAS* color-color diagram where PNe are typically found may explain our higher detection rate of emission line objects compared to previous studies.

None of the objects showing Brγ in emission were detected above a detection limit of 0.6 mJy/beam at 6 cm and 0.7 mJy/beam at 3 cm, while they should have been easily seen, if the radio emission was optically thin and Case B recombination was applicable. We argue that the Brγ emission is likely due to ionization in the post-AGB wind, present before the star is hot enough to ionize the AGB shell (Van de Steene et al. 2000).

In the near- and far-infrared color-color diagrams, no distinction can be made between objects showing Brγ in emission, Brγ in absorption, or a flat

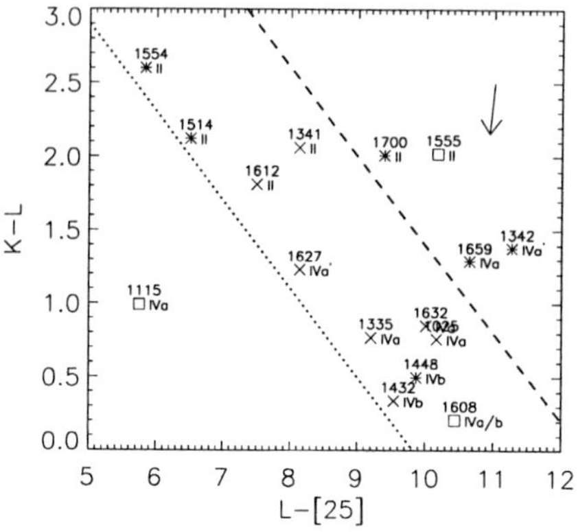

Figure 1 The K-L, L-[25] diagram. The crosses represent objects having Brγ in absorption, the asterisks objects having Brγ in emission, and the squares flat spectrum sources. The objects are labeled by the first four numbers of their *IRAS* name. The SED CLASS of each object is indicated next to its symbol. The arrow shows the effect of reducing A_V by 5 mag.

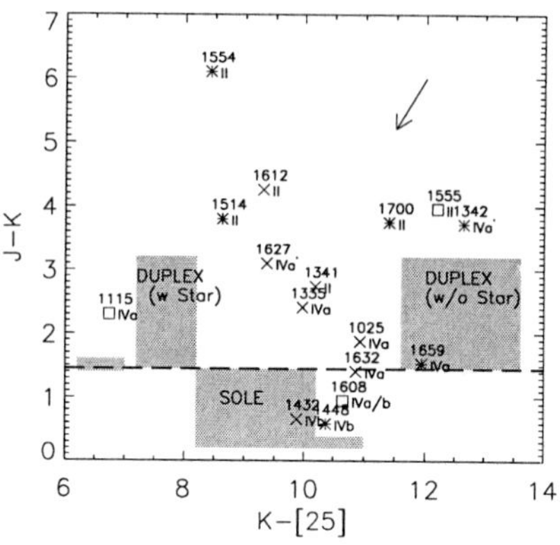

Figure 2 The J-K, K-[25] diagram, as proposed by Ueta et al. (2000). The grey regions show the location of the sample of Ueta et al. (2000). These regions are labeled according to their definitions as STAR-OBVIOUS LOW LEVEL ELONGATED (SOLE) and DUST PROMINENT LONGITUDINALLY EXTENDED (DUPLEX) with and without star. The dashed line separates these two kinds of objects.

spectrum. Whether the positions of the objects in the color-color diagrams can be directly related to the temperature and core mass of the central star needs further investigation, but doesn't seem likely based on our data.

5. COLOR-COLOR DIAGRAMS

In Figs. 1 and 2 we show two of the color diagrams we have constructed. The K−L and J−K colors follow the evolution of the circumstellar extinction. The L−[25] and K−[25] colors relate the stellar component with the peak of the dust emission, a larger color indicating a more prominent dust shell. We note that the L−[25] color is relatively insensitive to extinction.

In Fig. 1, stars obscured by their dust shells lie in the upper half of the diagram (larger K−L) while stars visible through their dust shells lie in the lower half. The objects located to the right of the dashed line are extended and some show bipolar morphology in the K-band. Probably they have thicker, cooler circumstellar dust shells than objects located below the dashed line (more 25 μm-band flux) and/or their central star temperatures are higher (less L-band flux).

For our limited sample, all but one of the objects are found to the right of the dotted line. This may reflect the evolution: K−L decreases when mass loss ceases and the inner edge of the circumstellar shell expands, and L becomes fainter as the central star becomes hotter at constant luminosity, causing L−[25] to increase.

In Fig. 2 there is virtually no overlap between the sample of Ueta et al. 2000 and our sample. This may be because of different selection procedures for the two samples. Ueta et al. 2000 selected known post-AGB candidates from the literature and imaged their nebulosities with WFPC2 in the optical. We selected objects from the PNe region in the *IRAS* color-color diagram: very few of them have bright optical counterparts.

All our objects are located within 5 degrees of the galactic plane, compared with only 11 of the 27 objects of Ueta et al. 2000 (3 *SOLE*, 6 *DUPLEX*, and 2 stellar). The arrow in the diagram indicates an extinction correction of A_V = 5 mag. Clearly, larger interstellar extinction alone cannot explain why our samples appear different. If they are at similar distances, it is plausible that we have more massive central stars in our sample, and that they are evolving faster across the HR diagram. Further investigations will be needed to understand the differences between both samples.

References

Ueta T., Meixner M., Bobrowsky M., 2000, ApJ 528, 861

van Hoof P.A.M., Oudmaijer R.D., Waters L.B.F.M., 1997, MNRAS, 289, 371

Van de Steene G.C., Pottasch S.R., 1993, A&A, 274, 895

Van de Steene G.C., Wood P.R., van Hoof P.A.M., 2000, ASP Conf. Ser. Vol. 199, 191

$BVRIJHK$ PHOTOMETRY OF POST-AGB CANDIDATES: EVOLUTION OF POST-AGB CENTRAL STARS

Takahiro Fujii and Yoshikazu Nakada
Institute of Astronomy, University of Tokyo, Mitaka, Tokyo 181-0015, Japan
fujii@ioa.s.u-tokyo.ac.jp, nakada@kiso.ioa.s.u-tokyo.ac.jp

M. Parthasarathy
Indian Institute of Astrophysics, Bangalore - 560034, India
partha@iiap.ernet.in

Abstract Post-AGB stars are generally believed to be on their way to the planetary nebula stage. However, there has been a question about the evolutionary path for the low-mass ones. We present $BVRIJHK$ photometry of post-AGB candidates selected on their IRAS colors. Their spectral energy distributions showed that 16 out of 27 objects were very likely to be the post-AGB stars. Two objects were remarkable in that their central stars were of rather late spectral type while their dynamical ages of their detached dust shell were considerably older than the others. We believe they are low mass post-AGB stars dispersing their dust shells which will not become planetary nebula but will directly evolve to white dwarfs.

1. INTRODUCTION

The rapid evolution of post-AGB stars crossing the HR diagram horizontally has recently attracted much interest of many astronomers. The various spectral type of the central stars has been considered to represent their temporal change in their stellar temperature. However, a simple arrangement of post-AGB stars in the order of their spectral type was interpreted by the fact that each post-AGB star has its own evolutionary speed probably determined by its stellar mass. In order to overcome this complexity, we have classified a sample of post-AGB stars with the help of the stellar temperature and the dynamical age of the detached shell. Since they span a wide range in the spectral type of the central star as well as in the optical thickness of the dust shell, the analysis of their spectral energy distribution (SED) needed photometric data from the visible to

R. Szczerba and S.K. Górny (eds.), Post-AGB Objects as a Phase of Stellar Evolution, 45–48.

the far-infrared wavelengths. Here, we present the BVRIJHK observations of some post-AGB candidates selected on the IRAS colors.

2. OBJECTS AND OBSERVATIONS

Total of 27 post-AGB candidates were selected from the IRAS point source catalog. They are all within the range of $F_{12\mu m}/F_{25\mu m} < 0.3$ and $F_{25\mu m}/F_{60\mu m} > 0.3$.

$BVRI$ photometry was performed at Kiso Observatory using the 105cm Schmidt telescope attached with the optical CCD camera. The JHK observations were carried out using the same telescope with the near-infrared camera. After standard data processing, the natural magnitudes were transformed to either the Johnson-Cousins system or the CTIO system.

The investigation of the spectral type of the central star, the IRAS colors, and the shape of the SED left 16 objects as likely post-AGB stars. Others were rejected because they were most likely pre-main sequence stars.

Table 1 Derived parameters of post-AGB stars

IRAS Name	T_{star} [a] [K]	T_{dust} [b] [K]	t_{dyn} [c] [yr]	$M_{dust+gas}$ [d] [$M_\odot$]	A_V [e] [mag]	Distance [kpc]
02143+5852	6900	205	338	0.740	1.66	16.9
04296+3429	5550	193	354	0.386	4.49	6.80
05040+4820	8750	97.3	2460	0.254	1.31	4.01
05089+0459	3650	189	303	1.08	1.18	13.9
05113+1347	4590	185	358	0.240	2.34	8.53
05238−0626	7380	183	464	0.0396	0.555	9.31
05341+0852	7060	210	323	0.0900	4.12	7.42
05381+1012	4850	183	377	0.0724	0.255	9.91
06338+5333	6250	267	167	0.00150	0.00	5.81
06530−0213	7700	182	480	0.129	5.40	4.88
07131−0147	3330	171	371	0.554	1.74	13.9
07171+1823	13600	203	488	0.111	0.00	22.8
07253−2001	6900	206	335	1.27	0.00	22.3
07430+1115	4850	201	300	0.242	2.51	7.46
08187−1905	6630	135	937	0.168	0.378	5.07
23304+6147	5200	173	453	0.366	4.33	4.88

[a] effective stellar temperatures obtained from the spectral type cited from the literature. [b] The dust temperature determined at the inner boundary of the dust shell. [c] transformed from R_{in} by being divided by the assumed expansion velocity of the dust shell (15 km s^{-1}). [d] correspond to past AGB mass-loss rate $\dot{M}$ ($= M_{dust+gas} \times 10^{-4} M_\odot yr^{-1}$). [e] interstellar extinction derived from the model fitting.

3. ANALYSIS AND DISCUSSION

A post-AGB star is generally associated with a detached dust shell which was ejected during its TP-AGB phase. Except for the case of the very early stage, the optical thickness of the shell should be small because of the attenuation effect by the expansion of the shell. Therefore, we adopted a simple dust shell model to analyze the SEDs of the observed objects; the temperature distribution inside the shell was assumed as the one for the isolated dust particle around a naked star. Other assumptions are as usual; the dust shell is spherically symmetric; the central star emits the blackbody radiation; the luminosity was fixed as $8\,000 L_\odot$. The extinction law of the dust shell was assumed to follow the power-law function of wavelength. An extinction law of the model carbon material was adopted for some carbon-rich objects, but the obtained spectra did not show much difference with the original models.

The four quantities, inner radius of the shell, total dust mass, distance to the object, and the interstellar extinction, were treated as free parameters in the fitting procedure of the observed SED to the model SED. The derived set of parameters are shown in the Table 1, where the dust mass has been transformed to the total mass of the shell by assuming the gas to dust ratio of 200. The SEDs with model fitting curve are shown in Figure 1.

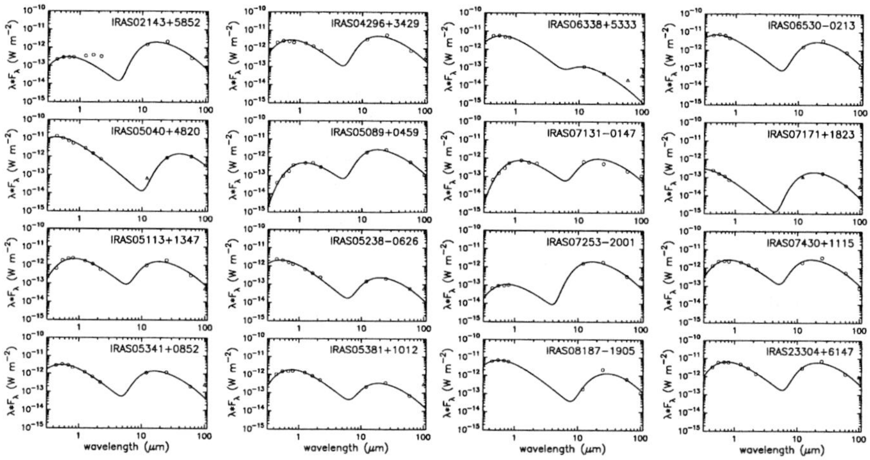

Figure 1 SEDs of post-AGB stars with calculated SED curve using parameters in Table 1 (full lines). For IRAS 02143+5852, near-infrared data points were not used for calculation because of its near-infrared excess.

Figure 2 shows the increase of the stellar temperature as a function of dynamical time of the dust shell. Although the previous model of the post-AGB stars started when the temperature of the central star reaches as high as 5 000 – 6 000 K, six of our objects were cooler than this. Two of them were cited as late as M-type while the others were of G-type spectra. However, within the uncertainty of the spectral classification, our observation can be in accord with the theoretical evolutionary tracks of the post-AGB stars. The same kind of figures are shown by van der Veen et al. (1989) and Schönberner & Blöcker (1993). Figure 2 is consistent with their figures. It is interesting the two peculiar stars,

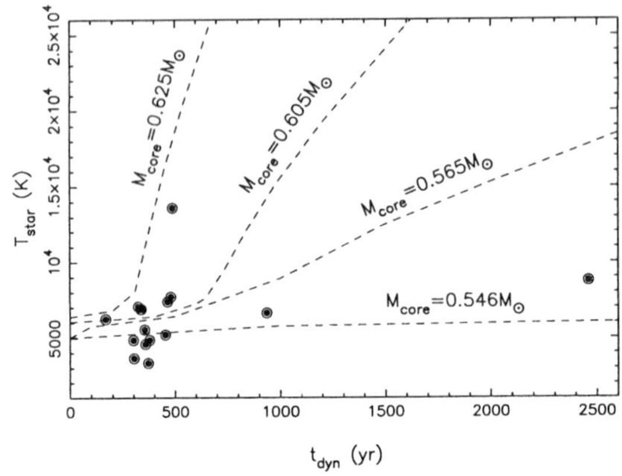

Figure 2 The increase of the stellar temperature as a function of dynamical time of the dust shell, assuming constant expansion velocity (15 km s^{-1}). The Dashed lines are theoretical evolution of hydrogen-burning post-AGB models with core mass 0.546–$0.625 M_\odot$ from Schönberner (1983) and Blöcker (1995).

IRAS 05040+4820 and IRAS 08187−1905, show a rather late spectral type of the central star in spite of their large dynamical age of the detached shell. These dust shells most likely will disperse into the interstellar space before the stellar temperature rises to ionize the surrounding gases. We note they are on the evolutionary track of $M_{core} = 0.55 M_\odot$ indicating the low mass for their parent stars, probably one solar mass or less. Renzini (1981) predicted the fate of a low mass star to become a white dwarf bypassing the PN stage. Scarcity of PNe in the globular clusters supports his hypothesis, but no direct evidence has not been found as far as we know. The above two IRAS sources are the first sample of Renzini's *lazy* AGB remnants.

References

Blöcker, T., 1995 A&A 299, 755

Schönberner, D., 1983, ApJ 272, 708

Schönberner, D., Blöcker, T., 1993, in *Luminous High-Latitude Stars*, ed. D. Sasselov, ASP Conf. Ser. 45, p.357

Renzini, A., 1981, in *Physical Processes in Red Giants*, ed. Iben, I. & Renzini, A., Dordrecht:Reidel, p.431

van der Veen, W.E.C.J., Habing, H.J., Geballe, T.R., 1989, A&A 226, 108

NEAR IR OBSERVATIONS OF THE ARECIBO SAMPLE OF OH/IR STARS

F. Jiménez-Esteban

Laboratorio de Astrofísica Espacial y Física Fundamental, Ap. 50727, E-28080 Madrid, Spain.

Universidad Alfonso X El Sabio, 28691-Villanueva de la Cañada, Madrid, Spain.

fje@laeff.esa.es

D. Engels

Hamburger Sternwarte, Gojenbergsweg 112, D-21029 Hamburg, Germany.

P. García-Lario

ISO Data Centre. Astrophysics Division, ESA, Ap. Correos 50727, E-28080 Madrid, Spain.

Abstract We have obtained near IR J, H and K images of the full Arecibo sample of OH/IR stars ($\approx$ 350 sources) at the Calar Alto 1.23m telescope equipped with MAGIC. These data will be used to identify first the near IR counterparts and then the sources optically on the POSS plates, whenever possible, and to improve their coordinates. About 70% of the observations were made under photometric conditions, which permits us to combine the near IR with the IRAS fluxes and to study the spectral energy distribution for a great part of the sample. One focus is the identification of PPNe in a region of the IRAS color-color diagram which is populated mostly by **variable OH/IR stars**.

At the present time approximately 75% of the observations have been reduced. Due to their unusual colors in the near IR we propose three new PPNe candidates.

1. THE SAMPLE

To study the properties of OH/IR stars statistically the Arecibo sample of OH/IR stars is the best sample available, because it is the most sensitive with regard to the maser flux and the only complete one. The sample was selected from the IRAS Point Source Catalog requiring IR fluxes S(25)>2 Jy and appropriate colors in the IRAS color-color diagram (Lewis 1994). The sky surveyed was restricted by the limited viewing zone of the telescope ($0 < \delta < 37°$). About one third of all IRAS sources fulfilling these criteria turned out to be

R. Szczerba and S.K. Górny (eds.), Post-AGB Objects as a Phase of Stellar Evolution, 49–52.

OH/IR stars with maser peak flux densities ≥50 mJy. The final sample consists of 385 OH/IR stars, from which 80% have longitudes $30° < l < 90°$ (i.e. $18^h <$ R.A. $< 23^h$). Note, that the term "OH/IR star" is used here to designate IRAS selected AGB stars with OH maser emission, which comprises not only the optically invisible objects, but also objects which would have been classified as Mira variables, if discovered optically. A minority of the Arecibo sample might actually be M-Supergiants or PPNe. Before our observations, images were available only for a small part of the sample near IR from various imaging programs of obscured IRAS sources, and less than 5% are well studied variables with already known periods.

2. OBSERVATIONS AND REDUCTIONS

In June 1999 we observed during 10 nights at the Calar Alto Observatory (Almería, Spain) using the 1.23m telescope equipped with the IR camera MAGIC (Herbst et al. 1993). All sources in the Arecibo samples accessible at this epoch of the year were observed in the J (1.25μm), H (1.65μm) and K (2.20μm) near IR bands. The MAGIC camera uses a 256x256 pixel NICMOS3 array which provides an approximate field of 5'x5'. The typical integration time was 1 min for J and H, and 30 seconds for the K band. Standard stars were observed during the night at different airmasses to calculate the atmospheric extinction.

Data reduction have been made using the IRAF software package. The main difficulty is the fast temporal and spatial variation of the background. In order to avoid this problem we took for each object and every filter 5 images with the target centered at different positions in the frame (the so-called "moving sky" technique). Taking the median average of the 5 images leads to a temporally and spatially well evaluated sky level.

3. PHOTOMETRY

70% of the observations were made under good photometric conditions. The flux of a source in a given filter is derived from the average of the five independent flux measurements made with the source located at five different positions in the frame. This way the typical error is less than 0.2 magnitudes in the final data. Limiting magnitudes were 14.5 mag in J and H, and 13.5 in K. At the moment we have obtained photometric magnitudes for 75% of the sources in the Arecibo sample.

4. RESULTS

J, H, and K photometry was obtained for 285 OH/IR stars from the Arecibo sample. For a number of sources the images were saturated in some filters and for several the identification has not been finished yet. As an example

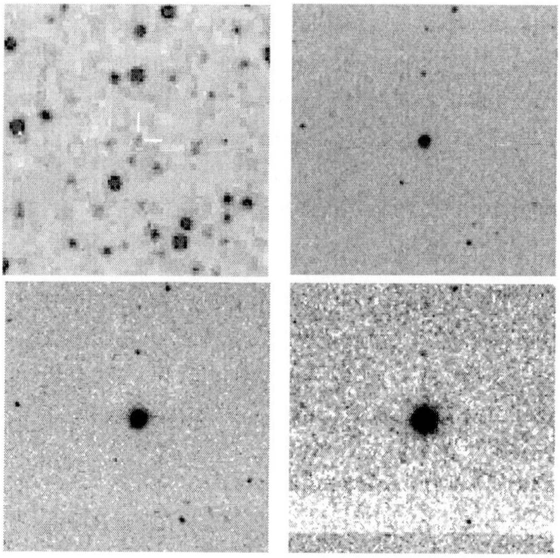

Figure 1 Sample images from the OH/IR star IRAS 17308+0822. The visual image (upper left) is taken from the digitized POSS plates, the J (upper right), H, and K (lower panel) images from the observations in June 1999.

for the IR images obtained we show in Figure 1 those corresponding to IRAS 17308+0822. It clearly shows the necessity to study this kind of objects in the near IR because some of them might be (almost) completely invisible in the optical.

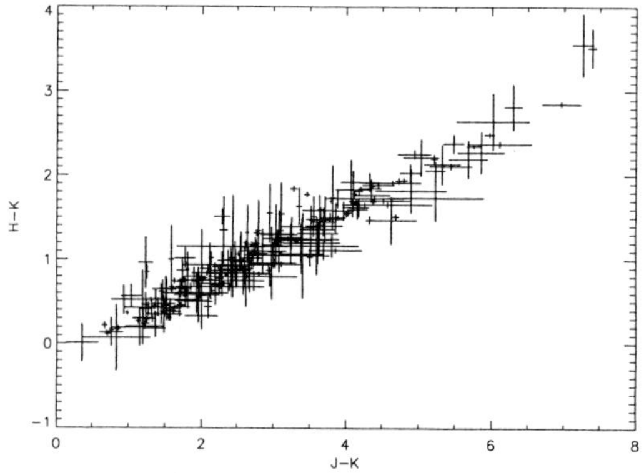

Figure 2 Near IR color diagram for the complete set of available data. The IR colors have been plotted together with their errors.

Figure 2 shows the (J–K, H–K) color-color diagram. The near IR colours show a clear linear relation with the scatter mainly produced by the error in the photometry. In Figure 3 we combined the near-IR data with the IRAS mid-IR data. Note that the amplitudes of the variations expected in the K brightness are of the order of 1-2 magnitudes which can explain part of the large scatter observed. The scatter expected in the IRAS [12]–[25] color due to variability and photometric uncertainties is expected to be very small in comparison. Multi-epoch observations are required to obtain mean K magnitudes and to obtain a better definition of the color-color sequence in this diagram.

5. NEW PPNE CANDIDATES

Usually PPNe candidates are identified by their cold circumstellar shells ([12]–[25] $>$ 2.5) in the IRAS color-color diagram. There might be others, however, which have just very recently left the AGB and still show relatively warm envelopes hidden in the region populated by variable OH/IR stars. PPNe candidates in this region might be identified by their non- or irregular variability or by the detection of a bright central star in the near-IR (or even in the optical). With single-epoch observations only available we cannot follow the first criterion but using the NIR-IRAS color diagram (Figure 3) we were able to single out objects, which are relatively bright in the near-IR compared to others with similar IRAS colors. The three objects marked in Figure 3 are IRAS 18548+0823, 19035+0801, and 20160+2734. IRAS 19035+0801was already suspected to be a possible new planetary nebula by Preite-Martinez (1988), while IRAS 20160+2734 is the SR variable AU Vul with a period of 145 days.

For all three candidates only few observations are available. They were neither detected in the OH main lines (1665/1667 MHz) (Lewis 1997) nor in the H_2O maser line at 22 GHz (Engels & Lewis 1996). Additional observations are planned in order to confirm their evolutionary stage.

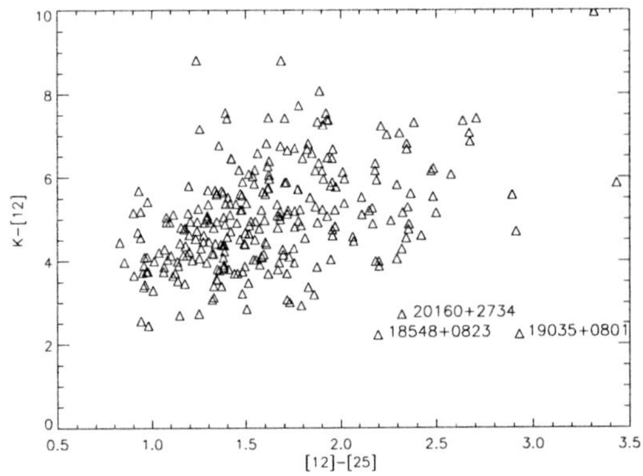

Figure 3 Near- and Mid-IR color diagram for the complete set of available NIR data.

References

Engels D., Lewis B.M., 1996, A&AS 116, 117.
Herbst T.M., Beckwith S.V., Birk C., 1993, Proc. SPIE Vol.1946, p.605.
Lewis B.M., 1994, ApJS 93, 549.
Lewis B.M., 1997, ApJS 109, 489.
Preite-Martinez A., 1988, A&AS 76, 317.

HST IMAGING OF PROTO-PLANETARY NEBULAE AND VERY YOUNG PLANETARY NEBULAE–TOWARDS A NEW UNDERSTANDING OF THEIR FORMATION

Raghvendra Sahai

Jet Propulsion Laboratory, California Institute of Technology, MS183-900, 4800 Oak Grove Drive, Pasadena, CA 91109

sahai@grandpa.jpl.nasa.gov

Abstract With the advent of the Wide Field Planetary Camera 2 (WFPC2) onboard the Hubble Space Telescope (HST) in Dec 1993, a growing number of young planetary and proto-planetary nebulae (PNe and PPNe) have been imaged with unprecedented high angular resolution and dynamic range. These objects represent the last phases of the evolution of intermediate-mass ($\sim$1-8 $M_{\odot}$) stars.

Our WFPC2 Hα SNAPshot imaging survey of young planetary nebulae (selected only on the basis of their low excitation) show that these objects are highly aspherical, with complex *multipolar* morphologies. The central star is often systematically offset from the geometrical symmetry centers of various nebular components. The bright aspherical nebulae are often found to be surrounded by faint, roughly round halos – signatures of the progenitor AGB envelopes produced by isotropic mass-loss. A number of these halos include numerous concentric arcs, evidence for quasi-periodic modulation of the mass-loss on time-scales of a few hundred years. Our detailed imaging studies of PPNe (objects in transition between the AGB and PN evolutionary phases) shows the emergence of complex morphological structures during the PPN phase.

The complexity, organization and symmetry of the morphological structures we find is forcing radical changes in, and inspiring fresh theoretical efforts to advance, our understanding of the mass-loss processes during late stellar evolution. In this paper, we review the HST data, and show some of the highlights of our imaging studies. Although the origins of many of the morphological features remains puzzling, we find that the data support a model for PN formation in which the primary agent for shaping PNe are high-speed collimated outflows or jets which operate during the late AGB and/or early post-AGB evolutionary phase. The multipolar shapes indicate that these outflows are bipolar and undergo episodic changes in their orientation and/or multiple collimated outflows occur with different orientations. Our discovery of a very highly-collimated, knotty bipolar jet in a planetary nebula, and its amazing morphological similarity

R. Szczerba and S.K. Górny (eds.), Post-AGB Objects as a Phase of Stellar Evolution, 53–63.

to a low-mass YSO provides strong empirical evidence for a common physical mechanism for generating collimated outflows in protostars and evolved stars.

1. INTRODUCTION

Planetary Nebulae evolve from intermediate-mass ($\sim 1 - 8M_{\odot}$) stars, following intense mass loss (10^{-5}–10^{-4} $M_{\odot}$ yr^{-1}) during the Asymptotic Giant Branch (AGB) phase of stellar evolution. When most of the stellar envelope of the AGB star has been expelled, the hot ($T_{eff} \gtrsim$ 25,000K) carbon-oxygen (white-dwarf) core is exposed, and UV radiation ionizes the surrounding circumstellar envelope to produce a PN.

The extended circumstellar envelopes (CSEs) of most AGB stars appear largely spherically symmetric, and "filled" (i.e. not shell-like) (e.g. Bowers et al. 1983, Sahai & Bieging 1993, Neri et al. 1998). However, drastic changes in the mass-loss geometry are observed during post-AGB evolution – planetary nebulae show bright rims and well defined shell-like structures, and a dazzling variety of morphologies, ranging from simple round and elliptical shapes, to much more complicated bipolar ones (e.g. Balick 1987, Schwarz, Corradi & Melnick 1992). Our challenge is to determine the physical mechanisms responsible for the dramatic transformation in the symmetry of the ejected matter as the central star evolves from the AGB to a PN nucleus (i.e. a white dwarf).

According to the *current paradigm* – the Generalised Interacting Stellar Winds (GISW) model – the bright rims seen in PNe are shells of swept-up matter resulting from a fast ($\gtrsim$ 1000 $km\ s^{-1}$) stellar wind interacting with the slowly expanding CSE of the progenitor AGB star (Kwok 1982) and the asphericity of PNe is due to the presence of an equatorial density enhancement in the AGB CSE, which restricts the expansion of the swept-up shell at low latitudes (Balick 1987). Although detailed hydrodynamic calculations (e.g. Soker and Livio 1989; Icke, Balick and Frank 1992; Frank et al. 1993) show that a wide variety of *axisymmetric shapes* can be produced with the GISW model, the latter has been unable to satisfactorily address a number of important issues. These are (i) the nature and origin of the asymmetry in the AGB envelopes, (ii) the presence of point-symmetric structures, "quadrupolar" structures (e.g. Corradi & Schwarz 1995, Manchado, Stanghellini, & Guerrero 1996) and ansae.

The detection of fast ($\sim$ 100 $km\ s^{-1}$) bipolar outflows in a few PPNe and late-AGB stars implies that the physical phenomena which shape PNe occur during their earliest formation stages, and are probably transitory in nature. Hence, studies of *young PNe & PPNe*, which are most likely to retain strong signatures of the physical phenomena crucial to their development, are of paramount importance. But, young objects are *compact*, and important structural details are usually blurred beyond recognition in ground-based optical images. Although radio interferometers enable imaging with high resolution ($\sim$0".1), images of PNe obtained, e.g. with the Very Large Array (e.g. Aaquist & Kwok 1990) have insufficient dynamic range and resolution.

2. HST OBSERVATIONS

HST's stable PSF and high angular resolution allow the imaging of reflection or emission-line nebulosity with very high dynamic range, and makes it the facility of choice. We summarise below our HST imaging surveys of a large number of young PNe (Sahai & Trauger 1998, Sahai 2000a) and detailed studies of individual PPNe. Although these studies are in progress, the data obtained so far provide important clues to the processes which shape PNe.

All PNe were imaged through the Hα (F656N) filter using the 800x800 pixel^2 Planetary Camera (PC) of WFPC2, which has a plate scale of $0''.0456$/pixel (some of the brightest objects are were also imaged in [NII]λ6586). All PPNe were imaged through one or more wide-band continuum filters F555W (V), F606W (0.6 μm), F814W (I), mostly with the PC, and in two cases with the Wide Field Cameras (800x800 pixel^2 with a plate scale of $0''.0996$/pixel). In selected PPNe, (i) polarimetric imaging with WFPC2 and/or NICMOS (Near-Infrared Camera and MultiObject Spectrometer on HST), and (ii) narrow-band emission-line imaging [H_2 (F212N/F215N) with NICMOS, and Hα and [SII]$\lambda\lambda$6717,6731 (F673N) with WFPC2] was also carried out. Both short and long exposures were obtained in pairs, allowing us to correct for saturation effects and remove cosmic-ray events. For a majority of objects, a two-point dither was used for improving image quality.

2.1. DETAILED STUDIES OF PROTOPLANETARY NEBULAE

We have carried out detailed studies of several PPNe by imaging their reflection nebulosities – in these objects, the central star (with spectral types K,G,F,A) is not yet hot enough to substantially ionise the circumstellar matter. All PPNe imaged by us [e.g. (1) CRL 2688 or Egg Nebula (Fig.1): Sahai et al. 1998a,b; (2) IRAS 16342−3814 (Fig.1), Hen 3-401 (Fig.1), & Roberts 22: Sahai et al. 1999a,b,c; (3) IRAS 04296+3429 (Sahai 1999) and (4) IRAS 09371+1212 or Frosty Leo Nebula (Fig.1): Sahai et al. 2000] or other groups (e.g. IRAS 17150−3224: Kwok et al. 1998; IRAS 16594−4656 & 17245−3951: Hrivnak et al. 1999; IRAS 22272+5435, 17423−1755, & 06530−0213: Ueta, Meixner & Bobrowsky 2000) show bipolar or multipolar morphologies. In particular, the best-resolved of these, CRL 2688 and Frosty Leo (Fig.1), show the presence of features attributable to the presence of multiple low-latitude jet-like outflows. CRL 2688 is also a prime example of a small list of PPNe and PNe which show the presence of roughly concentric arcs in their extended halos – evidence for a quasi-periodic modulation of the AGB mass-loss rate on time-scales of a few hundred years.

2.2. IMAGING SURVEYS OF YOUNG PLANETARY NEBULAE

Our sample of young PNe was inspired by the class of very-low-excitation objects discovered in an objective prism emission-line survey by Sanduleak & Stephenson (1972, 1973). The only strong emission lines in these objects were the lower members of the Balmer series and [O II] λ3727, which was comparable or stronger than [O III] λ5007, 4959. We cross-correlated these low-excitation objects with the recent compilation of galactic PNe by Acker et al. (1992), and found that almost all of them were identified as PNe. We therefore compiled our sample from the Acker et al. (1992) catalog, selecting for (a) a relatively large Hα-to-[O III] flux ratio (>1.5), (b) an average Hα intensity $\gtrsim 6 \times 10^{-13}\, erg\ s^{-1}\ cm^{-2}\ arcsec^{-2}$, (c) an optical or radio diameter (whichever was larger) $\geq$1 arcsec. *Since our sample has been uniformly selected on the basis of excitation criteria, it represents an unbiased sample for determining the morphological characteristics of young PNe.* Three of the brightest PNe which meet our selection criteria – BD+30°3639 (Harrington et al. 1997), MyCn 18 (Sahai et al. 1999d), and M 1$-$92 (Bujarrabal et al. 1998) – have been studied in detail using HST.

The Hα/[O III] flux ratio is expected to be large in young PNe since (i) the relatively low stellar T$_{eff}$ ($\sim$25000–40000 K) results in a low nebular excitation state (ii) the nebulae are compact with larger dust optical depths towards their central regions, resulting in a larger selective extinction of the shorter wavelength, more centrally concentrated [O III] line compared to Hα.

The most notable conclusions from our HST observations are:

- There are no round PNe or PPNe; most objects are *multipolar*
- The variety of morphologies seen is much larger than known from ground-based surveys
- Point-symmetry is widely manifest
- Many PNe have internal geometric components (e.g. rings, inner hourglass)
- The central star is often offset from the centers of these components
- Faint round halos are found around the central aspherical nebula in many PNe and PPNe, (evidence for isotropic mass-loss on the AGB)
- Multiple concentric rings are found in the halo in several objects (evidence for episodic changes in the AGB mass-loss rate)

Thus, although our HST images show that PPNe/PNe shapes are characterised by *complex symmetries* they are not *chaotic*. In addition, subtle but crucial features of the geometrical shapes are revealed for the first time, specially the presence of “corners" and “parallelogram" shapes. Thus, many objects, whose bright shells have been classified as “elliptical" from the ground are

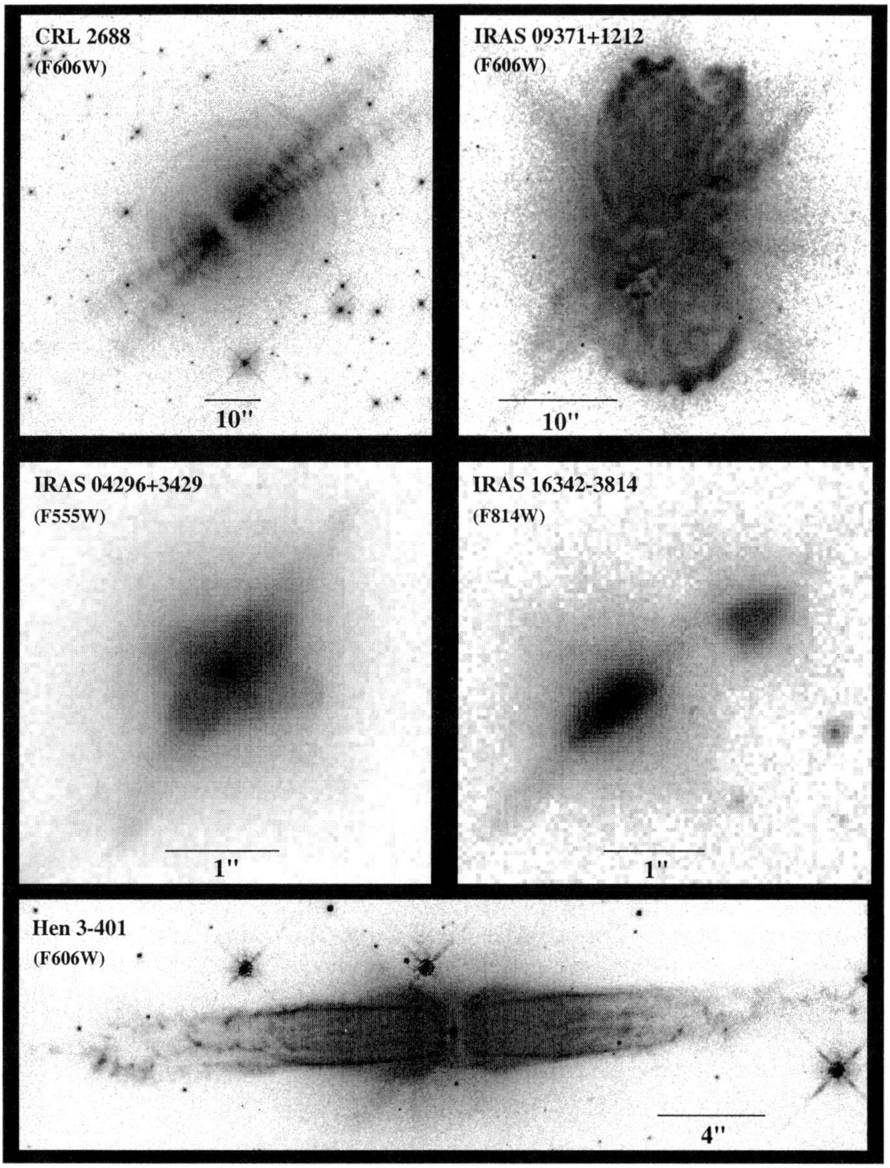

Figure 1 HST/WFPC2 images (in reverse grey-scale) of selected proto-planetary nebulae. Sharp features have been modestly emphasized in IRAS 09371+1212 and Hen 3−401 as described in Sahai & Trauger (1998). Note the wide variety of morphologies; the concentric arcs and "searchlight beams" in CRL 2688 (details in Sahai et al. 1998a) and the collimated jet-like lobes near the nebular waist in IRAS 09371+1212 (details in Sahai et al. 2000).

actually (point-symmetric) parallelograms [e.g. IC 418 (Fig.4) & NGC 7027 (Latter et al. 2000)]. Finally, our HST survey has led to the serendipitous discovery of a very highly-collimated, pulsed, bipolar jet in the PN Hen 2−90 (Fig. 4) (Sahai & Nyman 2000). The amazing morphological similarity of Hen 2−90 to a low-mass YSO provides strong empirical evidence for a common physical mechanism for generating collimated outflows in protostars and evolved stars.

3. A NEW MODEL FOR PRODUCING ASPHERICAL PLANETARY NEBULAE

Our HST images of young PNe strongly suggest that *the key to understanding the shaping of aspherical PNe is a physical mechanism which can generate multipolar and point-symmetric structures.* Long-slit spectroscopic imaging of several bright PNe shows that point-symmetric features in the images are also kinematically point-symmetric (e.g. Guerrero, Vazquez, & Lopez 1999). Bipolar rotating episodic jets (BRETs) have been invoked for creating point-symmetric structures in individual PNe (Lopez, Meaburn & Palmer 1993). We (Sahai & Trauger 1998) propose a two-step mechanism in which, first, a high-speed (few 100 $km\ s^{-1}$) collimated or jet-like outflow (or outflows) operating during the late-AGB or early post-AGB evolutionary phase carve an imprint within an intrinsically spherical AGB mass-loss envelope. This imprint *sets the stage* for the development of an aspherical planetary nebula. Subsequent expansion of a hot, tenuous, even faster stellar wind inside this imprinted AGB envelope, then produces the observed PN, whose shape and structure depends in detail on how the characteristics (direction, strength, opening angle) of the jet-like outflows change with time. Extended dense equatorial structures, if present, also provide an impediment to the expansion of the fast wind in the equatorial plane. However, all of our PNe show complex aspherical shapes, regardless of whether or not such structures are visible in the images, which suggests that the equatorial structures are not the primary agent for producing aspherical PNe. Departures from point-symmetric structure can result in two ways: (1) departure from bipolar symmetry in the jets (which in the extreme would result in a one-sided jet), and (2) departures from symmetry in the progenitor AGB envelope. Such departures could lead to irregularly shaped nebular structures, which are also observed.

3.1. MECHANISMS FOR MAKING WOBBLING JET-LIKE OUTFLOWS

Mechanisms for producing jet-like outflows, directly or indirectly, require the presence of a close binary companion. The most promising ones are those in which the outflows are (i) accretion disk-driven (e.g. Morris 1987; Soker 1996; Soker & Livio 1994), or (ii) magneto-hydrodynamically collimated by a stellar

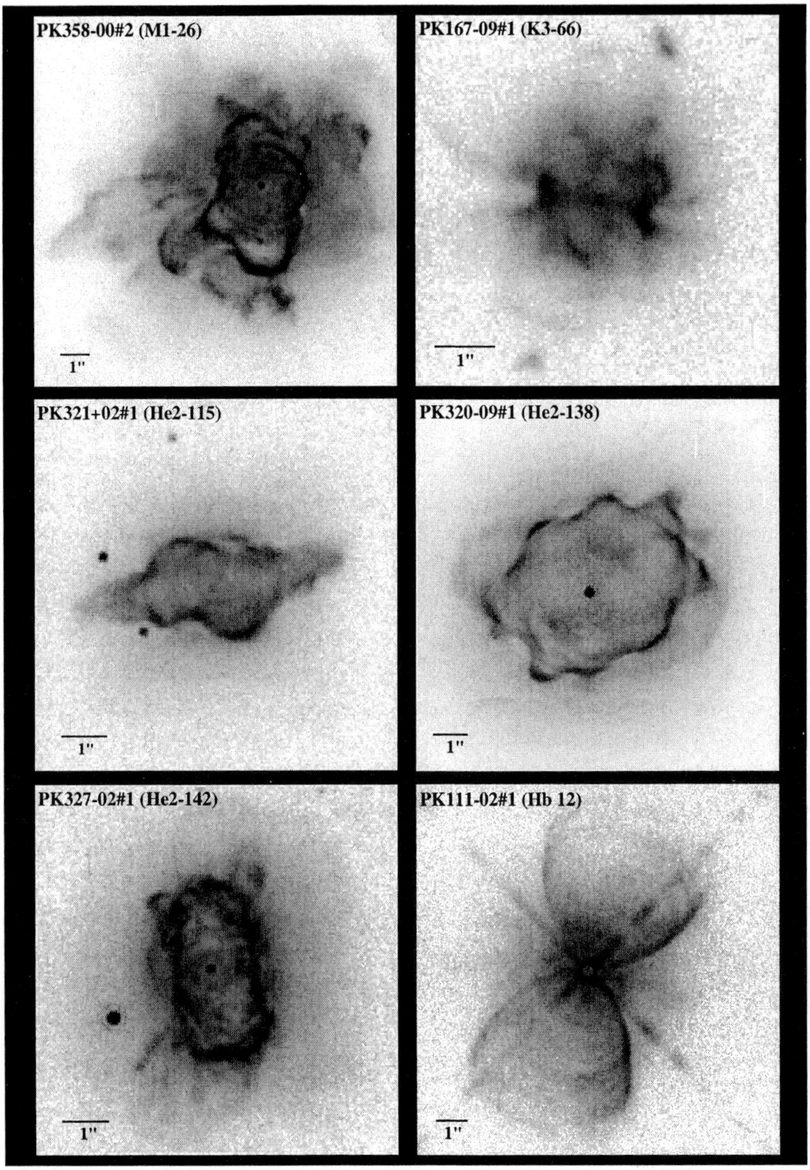

Figure 2 HST/WFPC2 Hα images (in reverse grey-scale) of young planetary nebulae from our survey. Note the wide variety of morphologies. Sharp features have been emphasized as described in Sahai & Trauger (1998).

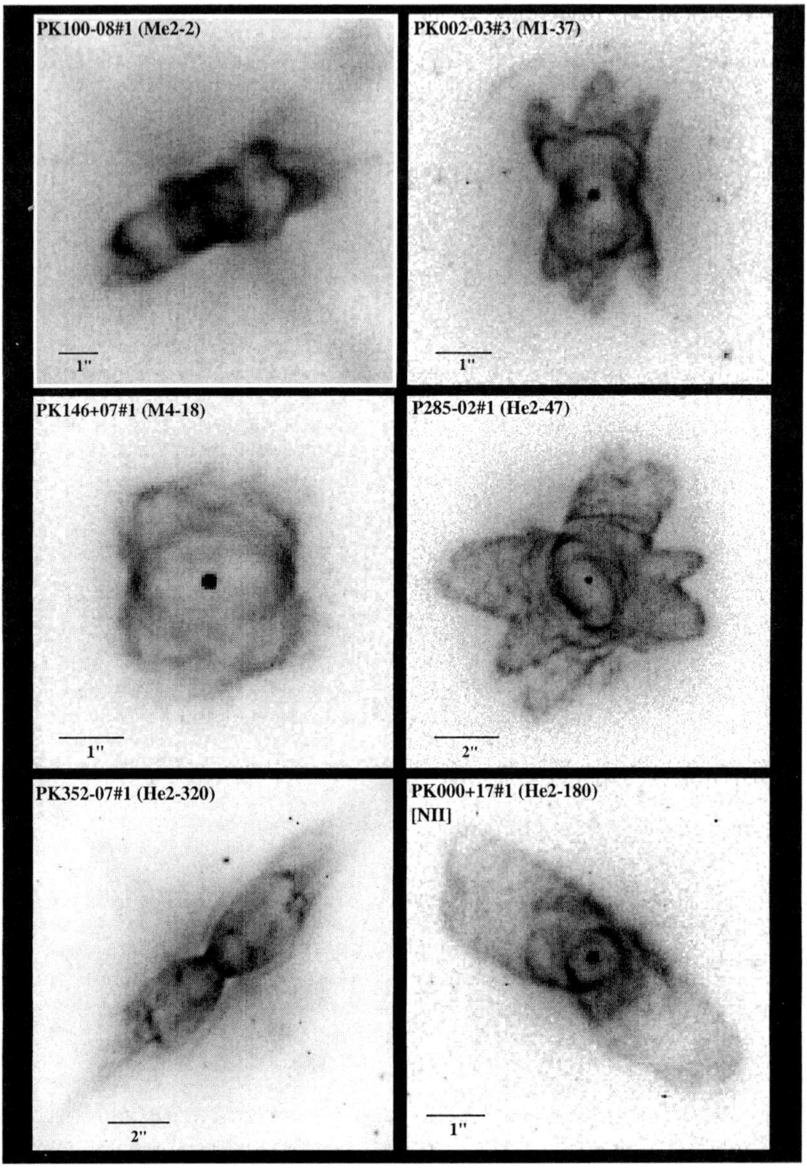

Figure 3 Same as in Fig. 2, except for PK000+17#1, which is imaged in [NII]λ6586. Note the plethora of lobes in the "Starfish Twins" M 1−37 & He 2−47 (details in Sahai 2000b).

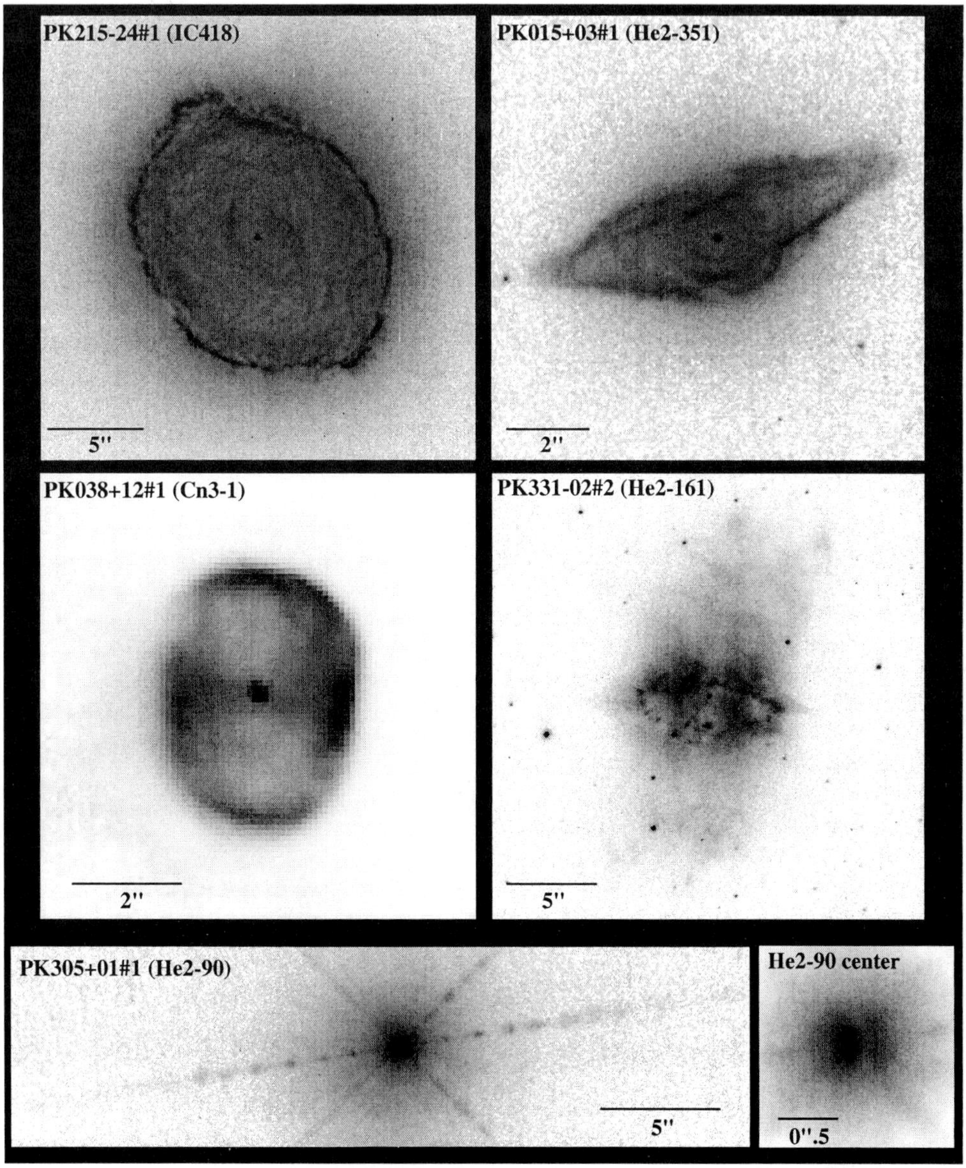

Figure 4 Same as in Fig. 2. Note the amazing morphological similarity of the highly-collimated, pulsed, bipolar jet and central bipolar nebula with obscuring edge-on disk seen in Hen 2−90 to the features seen in low-mass YSOs like HH 30 (details in Sahai & Nyman 2000).

toroidal magnetic field (e.g. Garcia-Segura 1997, Rozyczka & Franco 1996). In case (i), the jet-axis may change due to a radiation-induced instability which can warp the accretion disk, causing it to precess or wobble (Livio & Pringle 1997), whereas in case (ii), precession of the rotation axis of the star due to a binary companion can change the direction of the outflow. The variety and complexity of the structures seen in our sample of young PNe suggest that the changes in the properties of these outflows are episodic, and sometimes frequent. This may be difficult to achieve in a simple binary system. We suggest that the changes in the properties of these jets, and possibly their production, may be related to the presence of multiple sub-stellar companions (such as brown dwarfs and giant Jupiters). The recent detections of Jupiter-mass companions around Sun-like stars (e.g. Marcy 1998), specially those in close orbits (orbital radius <0.2 AU) (e.g. Butler et al. 1997) should spur new theoretical efforts to investigate the affects of such bodies on the formation and shaping of PNe.

Acknowledgments

Financial support for this work has been provided by NASA through a LTSA grant (no. 399-30-61-00-00) and grants GO-06353.01-95A and GO-08345.01-97A from the Space Telescope Science Institute.

References

Aaquist, O.B. & Kwok, S. 1990, A&ASS, 84, 229

Acker, A., Ochsenbein, F., Stenholm, B., Tylenda, R., Marcout, J. & Schohn, S. 1992, Strasbourg-ESO Catalog of Galactic Planetary Nebulae, European Southern Observatory, Garching

Balick, B. 1987, AJ, 94, 671

Bowers, P.F., Johnston, K.J., & Spencer, J.H. 1983, ApJ, 274, 733

Bujarrabal, V., Alcolea, J., Sahai, R., Zamorano, J., & Zijlstra, A.A. 1998, A&A, 331, 361

Butler, R.P., Marcy, G.W., Williams, E., Hauser, H., & Shirts, P. 1997, ApJ, 474, L115

Corradi, R.L.M. & Schwarz, H.E. 1995, A & A, 293, 871

Frank, A., Balick, B., Mellema, G., & Icke, V. 1993, ApJ, 404, L25

Garcia-Segura, G. 1997, ApJ, 489, L189

Guerrero, M.A., Vasquez, R., & Lopez, J.A. 1999, AJ, 117, 967

Harrington, J.P., Lame, N.J., White, S.M., & Borkowski, K.J. 1997, AJ, 113, 2147

Hrivnak, B.J., Kwok, S. & Su, K.Y.L. 1999, ApJ, 524, 849

Icke, V., Balick, B., & Frank, A. 1992, A&A, 253, 524

Kwok, S. 1982, ApJ, 258, 280

Kwok, S., Su, K.Y.L. & Hrivnak, B.J. 1998, ApJ, 501, L117

Latter, W.B., Dayal, A., Bieging, J.H., Meakin, C., Hora, J.L., Kelly, D.M. & Tielens, A.G.G.M. 2000, ApJ, 539, 783

Livio, M. & Pringle, J. 1997, ApJ, 486, 835

Lopez, J.A., Meaburn, J., & Palmer, J.W. 1993, ApJ, 415, L135

Manchado, A., Stanghellini, L., Guerrero, M.A. 1996, ApJ, 466, L95

Marcy, G.W. 1998, Nature, 391, 127

Morris, M. 1987, PASP, 99, 1115

Neri, R., Kahane, C., Lucas, R., Bujarrabal, V., & Loup, C. 1998, A & AS,130, 1

Rozyczka, M. & Franco, J. 1996, ApJ, 469, L127

Sahai, R. 1999, ApJ, 524, L125

Sahai, R. 2000a, in *Asymmetrical Planetary Nebulae II: From Origins to Microstructures*, Eds. J.H. Kastner, N. Soker & S.A. Rappaport, ASP, 199, 209

Sahai, R. 2000b, ApJ, 537, L43

Sahai, R. & Bieging, J.H. 1993, AJ, 105, 595

Sahai, R. & Nyman, L-Å 2000, ApJ, 538, L145

Sahai, R. & Trauger, J.T. 1998, AJ, 116, 1357

Sahai, R. et al. 1998a, ApJ, 492, L163

Sahai, R. et al. 1998b, ApJ, 493, 301

Sahai, R., Te Lintel Hekkert, P., Morris, M., Zijlstra, A. & Likkel, L. 1999a, ApJ, 514, L115

Sahai, R., Bujarrabal, V. & Zijlstra A. 1999b, ApJ, 518, L115

Sahai, R., Zijlstra, A., Bujarrabal, V., Te Lintel Hekkert, P. 1999c, AJ, 117, 1408

Sahai, R. et al. 1999d, AJ, 118, 468

Sahai, R. Bujarrabal, V., Castro-Carrizo, A. & Zijlstra, A. 2000, A&A, 360, L9

Sanduleak, N. & Stephenson, C.B. 1973, ApJ, 185, 899

Sanduleak, N. & Stephenson, C.B. 1972, ApJ, 178, 183

Schwarz, H.E., Corradi, R.J.M. & Melnick, J. 1992, A & AS, 96, 23

Soker, N. 1996, ApJ, 468, 774

Soker, N. & Livio, M. 1994, ApJ, 421, 219

Ueta, T., Meixner, M. & Bobrowsky, M. 2000, ApJ, 528, 861

MORPHOLOGICAL PROPERTIES OF PPNS: MID-IR AND HST IMAGING SURVEYS

Toshiya Ueta and Margaret Meixner
University of Illinois at Urbana-Champaign
ueta@astro.uiuc.edu, meixner@astro.uiuc.edu

Abstract We will review our mid-infrared and HST imaging surveys of the circumstellar dust shells of proto-planetary nebulae. While optical imaging indirectly probes the dust distribution via dust-scattered starlight, mid-IR imaging directly maps the distribution of warm dust grains. Both imaging surveys revealed preferencially axisymmetric nature of PPN dust shells, suggesting that axisymmetry in planetary nebulae sets in by the end of the asymptotic giant branch phase, most likely by axisymmetric superwind mass loss. Moreover, both surveys yielded two morphological classes which have one-to-one correspondence between the two surveys, indicating that the optical depth of circumstellar dust shells plays an equally important role as the inclination angle in determining the morphology of the PPN shells.

1. INTRODUCTION

While circumstellar dust shells (CDSs) of asymptotic giant branch (AGB) stars display high spherical symmetry (e.g. Habing & Blommaert 1993), planetary nebulae (PNs) are renowned for their spectacular axisymmetry (e.g. Zuckerman & Aller 1986). To understand *when* and *how* such a drastic morphological conversion takes place, the proto-planetary nebula (PPN) phase, a brief transitionary phase between the AGB and PN phases (e.g. Kwok 1993) has been extensively investigated. In fact, PPNs are the most ideal space laboratories for such studies: AGB mass loss histories are imprinted on the PPN shells in the most pristine form because the PPN central stars are not hot enough to generate disturbing fast winds and/or ionizing UV photons as PN central stars.

Therefore, by conducting imaging surveys on a large number of PPN CDSs, we can pinpoint the epoch of morphological transformation and obtain insights about yet unknown mechanisms for the AGB mass loss. To achieve our goals, we employed mid-infrared (IR) and optical wavelength ranges, both of which have strengths and weaknesses that would complement one another. Diffraction-limited mid-IR images give marginal resolution ($1''$) but allow one

R. Szczerba and S.K. Górny (eds.), Post-AGB Objects as a Phase of Stellar Evolution, 65–70.

to directly probe dust mass distribution through thermal dust emission. On the other hand, optical images are of high resolution ($0''.1$), while they only allow an indirect mass probing through dust-scattered starlight escaping from the dust shells.

2. MID-IR IMAGING SURVEY

The aim of the mid-IR survey is to map out distribution of warm dust grains at the inner edge of the PPN CDSs. So far we have observed about 70 PPN candidates with 8 to 25 micron filters. We have been using IRTF and UKIRT ($\sim 1''$ resolution with $\sim$ 3m apertures) and MMT (sub-arcsecond resolution with 6.5m aperture). Out of 72 objects, we have only resolved 22 object due to marginal resolution. However, 16 of these 22 showed clear elongation, and moreover, elongations were morphologically characterized into two groups: toroidal and core/elliptical types (Meixner et al. 1999).

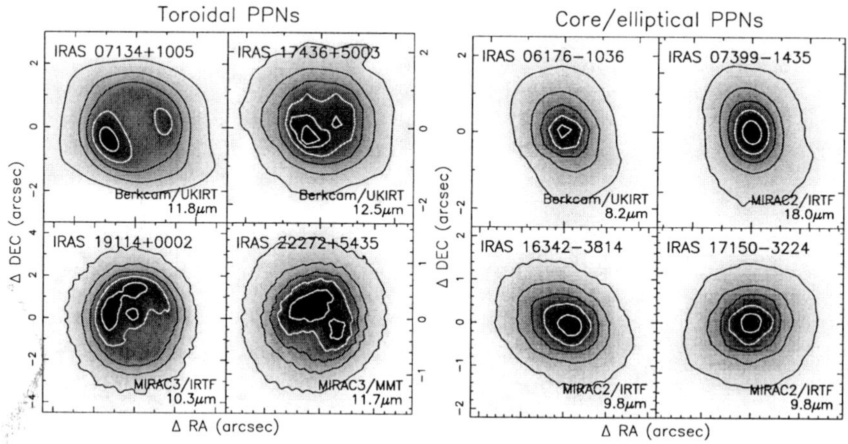

Figure 1 Two types of mid-IR morphologies: toroidal (left) and core/ellitical (right) PPNs. Indicated are object names, instruments, wavelengths, and sizes. The size of the emission core with respect to the entire nebula is the main difference.

The toroidal PPN morphology is characterized by a large emission core which shows some evidence of a dust torus. The resolved peaks in the emission core are interpreted as limb-brightened edges of an optically thin dust torus. i.e., the inner structure is revealed through optically thin dust shells. The morphology of the core/elliptical PPNs is characterized by a relatively compact emission core which is surrounded by a larger emission plateau. This type of PPNs seem to keep their inner structure hidden beneath optically thick dust shells.

3. HST IMAGING SURVEY

The goal of the optical imaging survey was to detect faint circumstellar nebulosities caused by dust-scattered starlight. We observed 27 PPN candidates using HST/WFPC2 during cycle 6 with typical resolution of $0''.1$. 21 out of 27 objects showed nebulosities, and all of which showed two kinds of axisymmetry (Ueta et al. 2000): the first group being star-obvious low-level elongated nebulae (SOLE) and the second group being dust-prominent longitudinally extended nebulae (DUPLEX).

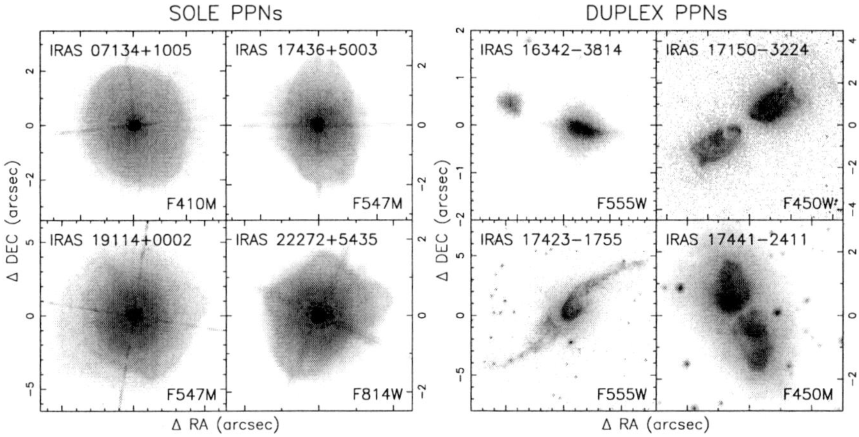

Figure 2 Two types of optical morphologies: SOLE (left) and DUPLEX (right) PPNs. Indicated are object names, WFPC2 filters, and sizes. The main difference is the visibility of the central star due to dust obscuration, which manifests itself in the nebula shapes.

SOLE PPNs have very faint and smooth, ellitpically elongated nebulosities with the central stars prominently visible at the very center of the nebulae. The morphology of DUPLEX PPNs is characterized by their bipolar nebulosities. Depending on the inclination angle of the object, you may or may not see the central star but you will always recognize some evidence for a dust lane between lobes.

4. COMBINED RESULTS

Both surveys showed that PPN dust shells have a striking preference towards axisymmetry. This is consistent with previous studies on the axisymmetry in PPNs (e.g., Trammel et al. 1994; Hrivnak et al. 1999). It is, therefore, most likely that the epoch of morphological transformation should start before the end of the AGB phase. If we restrain ourselves from invoking any exotic mechanisms, a standard AGB evolution scenario (e.g., Iben 1995 for a review) suggests one event which might trigger the tranformation, the superwind. The bulk of the PPN dust shell is considered to be ejected during this brief but violent phase of

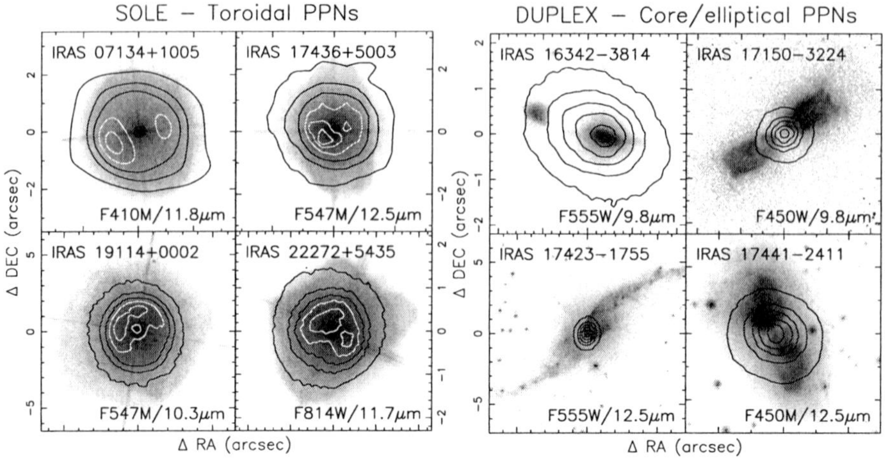

Figure 3 Two PPN types?: grayscale HST images overlaid with Mid-IR contours showing a one-to-one correspondence.

mass loss at the end of the AGB phase, and superwind may be violent enough to cause such a drastic morphological change.

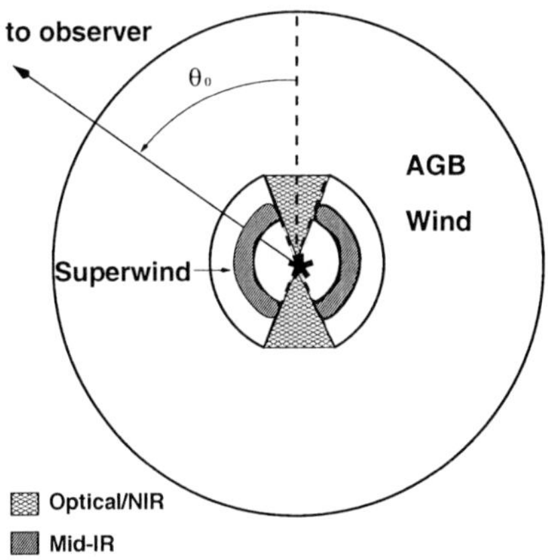

Thus, we propose a zeroth-order model for the structure of the PPN dust shells, shown on the left. In this model, an axisymmetric superwind shell is embedded in an otherwise spherical AGB wind shell. Mid-IR thermal emission comes from the inner edge of the superwind shell, which manifests itself as a dust torus we have observed (Fig. 1), while optical bipolar lobes are evidently dust-scattered starlight escaping the dust torus through these bicone openings.

Now, how does this model explain the observed dual morphologies in these surveys? Interestingly, there is a one-to-one correspondence between two dual morphologies found in both surveys (Fig. 3): toroidal mid-IR shells are always associated with SOLE type optical nebula, and core/elliptical mid-IR shells are always associated with DUPLEX type optical nebulosities.

While the extent of both mid-IR and optical nebulae is comparable in SOLE-Toroidal PPNs, in DUPLEX-Core/elliptical PPNs, optical nebulae can reach far beyond the mid-IR emitting regions, which roughly cover the core of the dust torus. The optical morphologies have been explained by the inclination angle effect, in which an intrinsically bipolar nebula can be viewed as a bipolar nebula near edge-on or an ellipitically elongated nebula near pole-on depending on the the inclination angle of the object. This explanation fails to explain the mid-IR morphology of SOLE-Toroidal PPNs. Within the framework of the inclination angle interpretation, SOLE-Toroidal PPNs must be oriented nearly pole-on. However, their mid-IR morphology clearly indicates the presence of rather edge-on dust toroids and can not be satisfactorily explained by the inclination angle effect alone.

The two-layered PPN shell model we proposed can explain both mid-IR and optical morphologies if we consider the PPN dust shells that can have a range of optical depth (Fig. 4). In this optical depth effect interpretation, the optical depth of the SOLE-Toroidal PPN dust shells are low enough that starlight can scatter into all directions making an elliptical looking nebulosities while mid-IR images showing two emission peaks. On the other hand, in DUPLEX-Core/relliptical PPNs, the optical depth is very high and starlight can escape into only the bicone openings of the dust shell making bipolar nebulosities while mid-IR emission showing only one, compact emission core.

Although the overall morphology of PPN dust shells is still dependent on the inclination angle of the shells, the results from our surveys evidently demonstrated that the optical depth effect plays an equally important role in defeining the morphology of the PPN dust shells.

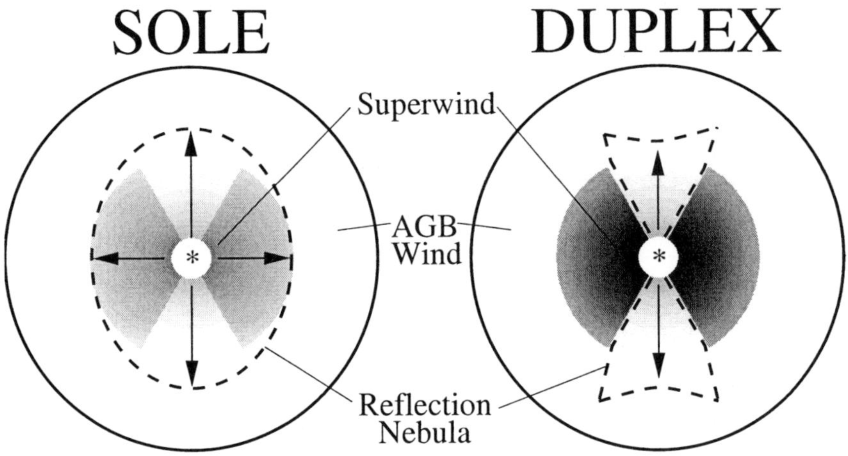

Figure 4 Optical depth interpretation of the mid-IR/optical dual morphologies of the two PPN types. SOLE-Toroidal (left) is a manifestation of an optically thin PPN dust shell while DUPLEX-Core/elliptical (right) is a result of an optically thick PPN dust shell.

5. FUTURE PROSPECTS

As we have seen, well-resolved mid-IR images are very effective means to directly probe the structure of dust distribution. With large aperture telescopes coming online, we are now able to obtain sub-arcsecond resolution mid-IR images, and detailed spatial information at the innermost regions of PPN CDSs will permit refined understanding of the AGB mass loss. Also, spatially detailed data will help constraining input geometrical parameters for radiative transfer calculations done in a fully two-dimensional grid. These model calculations are important because they allow us to quantify the axisymmetric nature of PPN CDSs by, for example, the pole-to-equator density ratio. We will also be able to constrain the inclination angle from such model calculations, clarifying ambiguities between effects of the inclination angle and optical depth of the PPN dust shells.

Acknowledgments

These surveys have been done in collaboration with W. F. Hoffmann, P. Hinz (Steward Obs./Univ. of Arizona), A. Dayal (IPAC/JPL), J. L. Hora, G. Fazio (Harvard/CfA), L. K. Deutsch (Boston Univ.), B. J. Hrivnak (Valparaiso Univ.), C. J. Skinner (deceased), and M. Bobrowsky (Challenger Center for Space Science Education).

References

Habing, H. & Blommaert, J.A.D.L. 1993, in IAU symp. 155: Planetary Nebulae, eds. R. Weinberger & A. Acker (Dordrecht: Kluwer), 243

Hrivnak et al. 1999, ApJ, 513, 421

Iben, I. Jr. 1995, Phys. Rev., 250, 2

Kwok, S. 1993, ARA&A, 31, 63

Meixner et al. 1999, ApJS, 122, 221

Trammel, S.R., Dinerstein, H.L., & Goodrich, R.W. 1994, AJ, 108, 984

Ueta, T., Meixner, M., & Bobrowsky, M. 2000, 528, 861

Zuckerman, B. & Aller, L.H. 1986, ApJ, 301, 772

MSX SOURCES IN THE LMC:
DUST-ENSHROUDED AGB STARS AND POST-AGB STARS

P.R. Wood
Research School of Astronomy and Astrophysics, Australian National University
wood@mso.anu.edu.au

M. Cohen
Radio Astronomy Laboratory, University of California, Berkeley
mcohen@astro.berkeley.edu

Abstract A region of size $3^\circ \times 3.5^\circ$ covering most of the LMC bar has been searched for infrared sources using data from the MSX satellite. In the most sensitive band at 8.28 μm, 504 point sources were detected. Subsequent near-infrared $JHKL$ imaging has shown that 125 of the MSX sources are red supergiants or foreground stars, 230 are dust-enshrouded AGB stars and 25 appear to be post-AGB stars (including three RCB stars). There are no K detections for 124 objects to a limit of K=16. MACHO light curves exist for about half the AGB and post-AGB stars, and spectra have been taken for most of the post-AGB stars and a selection of the AGB stars and red supergiants. The evolution and properties of the post-AGB stars are discussed.

1. INTRODUCTION

The Magellanic Clouds are particularly important sites for the study of evolved stars and mass loss because they contain large samples of objects at the same well-determined distance and they are close enough that individual stars can be easily resolved and studied with modest telescopes. In particular, the 33.7 cm diameter telescope of the Midcourse Space Experiment (MSX) (Price 1995; http://www.ipac.caltech.edu) was able to detect large numbers of point sources in the LMC at 8.28 μm. The positional accuracy of the LMC MSX point sources is $\sim 2''$ so that identification of these sources with near-infrared objects identified at K is secure in almost all cases. Compared to the IRAS point source survey of the LMC, MSX found about 10 times more objects with about 10 times the positional accuracy.

R. Szczerba and S.K. Górny (eds.), Post-AGB Objects as a Phase of Stellar Evolution, 71–76.

At the present time, most phases of stellar evolution in the Magellanic Clouds have been well-investigated and broadly understood using optical studies. The stars which have not yet been studied in detail are the dust-enshrouded AGB stars, the post-AGB stars in the transition from the AGB to the planetary nebula phase, and the dust-enshrouded supergiants. A relatively small number of candidate stars of these types have been selected from IRAS data by Reid et al. (1990) and Schwering (1998), and a number of follow-up studies of these and other IRAS point sources have been made. For example, near-infrared photometry, spectra, variability studies and OH maser detections can be found in Wood et al. (1992), Wood (1998), van Loon et al. (1998) and references therein, and Trams et al. (1999). However, it should be stressed that these surveys based on IRAS point source detections are very incomplete and they contain only the brightest of the dust-enshrouded, evolved stars. Wood (1998) estimates that there should be $\sim$640 LMC AGB stars in the final, dust-enshrouded superwind phase, compared to roughly $\sim$50 known from IRAS studies.

Here, we present the results from a near-infrared photometric and optical spectroscopic study of 504 MSX point sources in a region of size $3^\circ \times 3.5^\circ$ which is centred on the LMC bar and which contains a large fraction of the total population of old and intermediate-age stars in the LMC. The observations were made with the 2.3m telescope of the Australian National University at Siding Spring Observatory. The CASPIR infrared array camera (McGregor et al. 1994) was used for the near-infrared observations and the Dual Beam Spectrograph (Rodgers et al. 1988) was used for the spectra. In addition, the MACHO data base was searched for light curves of the MSX sources.

2. LMC MSX SOURCES IN THE (K,J-K) DIAGRAM

Fig. 1 shows the (K,J-K) diagram for point sources discovered at 8.28 μm in the LMC by MSX. In addition, the figure shows a sample of field stars, these being obtained from the CASPIR images centred on the MSX sources. Note that the area from which the MSX sources were selected is $\sim$60 times larger than the area from which the field stars were selected, so that the MSX sources are highly over-represented in Fig. 1.

In Fig. 1, the common field stars on the first giant branch (FGB) lie in the well populated sequence below $K \approx 12.4$. From the tip of the FGB, a sequence of AGB stars curves across the figure to J-$K \approx 6$. The FGB and AGB sequences are most likely dominated by intermediate-age stars of age $\sim 3 \times 10^9$ yr (masses $\sim 1.5 M_\odot$) since there seems to have been a burst of star formation in the LMC $\sim 3 \times 10^9$ years ago, with some continuation since (e.g. Bertelli et al. 1992). The high-mass red supergiants ($M > 10 M_\odot$) which lie at the top of the diagram are clearly the result of continuing star formation in the LMC.

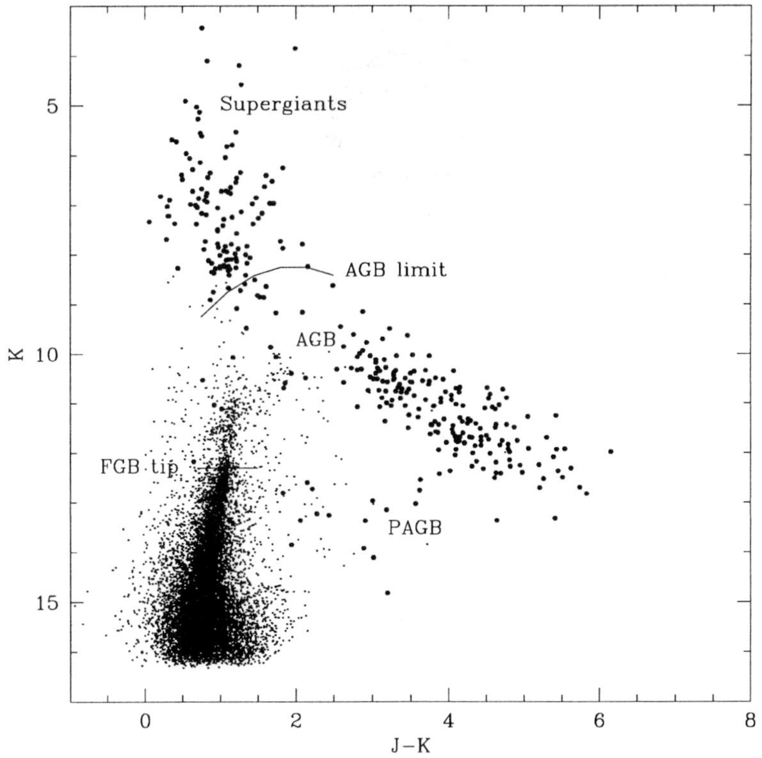

Figure 1 The (K,J-K) diagram for field stars (small dots) and MSX sources (large dots) in the LMC. The area from which the field stars were selected is ~60 times smaller than the area from which the MSX sources were selected. The maximum luminosities for the FGB and AGB are shown: these lines separate FGB stars from AGB stars and AGB stars from supergiant stars, respectively. Objects thought to be post-AGB stars fall in the area labelled PAGB.

The reddest stars on the AGB sequence are those with the highest current mass loss rates and consequently the thickest surrounding dust shells. It is not clear that the red end of the AGB sequence in Fig. 1 really represents the end-point of AGB evolution. Since the J limit of the CASPIR photometry is ~19 mag., we are unable to make a J measurement for the reddest sources with $K > 13$. In fact, there are ~40 MSX sources with a K detection but no J detection: these stars lie beyond the red end of the AGB sequence in Fig. 1 (this can be compared to the ~213 MSX sources on the AGB with both K and J detections).

The most interesting feature of Fig. 1 from the point of view of these proceedings is the sequence of 25 stars with $K \sim$ 13-14 extending back across the (K,J-K) diagram from the end of the AGB sequence towards the blue part of the diagram: this is what we believe to be the first-ever detection of a sequence of post-AGB stars transiting from the very end of the superwind stage on the AGB to the planetary nebula stage.

It may be thought that these objects are not genuine post-AGB stars but a simple spatial coincidence of a very red AGB star with one of the common FGB stars: this would give a combined image with an intermediate colour. However, for the following reasons, we do not believe this to be the case and we believe there is good supporting evidence that these stars are genuine post-AGB stars with cool dust shells which are currently expanding to reveal the stellar photosphere:

- The typical K magnitude of the post-AGB star candidates is similar to or fainter than the K magnitude at the end of the AGB sequence. If the post-AGB star candidates are simply the spatial superposition of a dust-enshrouded AGB star and a normal red giant or main sequence star, then the K magnitude of the dust-enshrouded AGB star would have to be at the very low end of the range of K magnitudes for stars on the AGB sequence. This seems very unlikely - why would the more numerous and brighter dust-enshrouded AGB stars with $K \sim 12$ (say) not also be coincident with blue stars? There clearly are no post-AGB candidates with $K \sim 12$.

- Spectra of the post-AGB star candidates reveal that 10 of the 25 show Hα emission. Since Hα emission is generally rare in stars but is a common feature in post-AGB stars (Van de Steene et al. 2000), the Hα emission suggests that these stars are indeed post-AGB stars. We certainly would not expect to see Hα emission from the image of a very red AGB star and a random blue star spatially aligned on the sky.

- MACHO light curves of the post-AGB star candidates show that three, perhaps four, of them are R Coronae Borealis (RCB) stars (Fig. 2). In addition, two of the candidates show monotonic secular changes in magnitude at rates much faster than normal evolutionary timescales. Such an unusual collection of variable stars in a sample of 25 candidates argues strongly that these stars are not mere chance alignments of blue and red stars on the sky. Most interestingly, the existence of RCB stars in this group of stars suggests that the RCB stars are post-AGB stars rather than stars that result from binary mass transfer (see Lambert and Rao 1994 for a discussion of possible origins of RCB stars). We note that the spectrum and light curve of one of the RCB stars shown here (MSX053349.0-701325) were discussed by Alcock et al. (1996).

- The spectral energy distributions of many of the post-AGB candidates are often bimodal, showing the presence of both a warm photosphere and a cool dust shell, as expected for a post-AGB star.

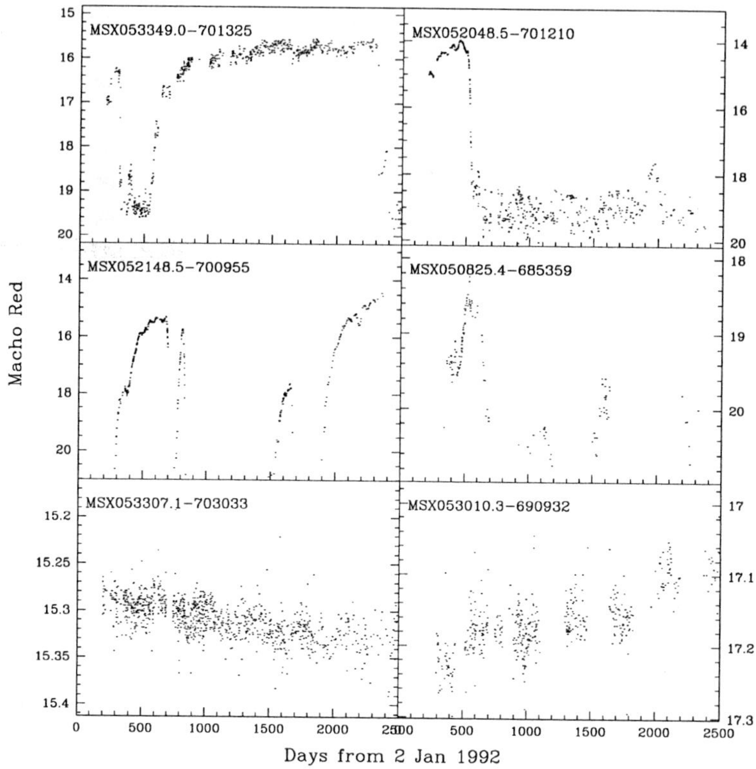

Figure 2 MACHO light curves for six of the MSX sources that are post-AGB star candidates. The first three stars, and possibly the fourth, are clearly RCB stars while the last two stars show monotonic secular changes in magnitude with time (the scatter of the points over short time intervals is noise, and the annual observation cycle can be seen in the distribution of points for the last object).

Taken together, the above results provide strong support for the suggestion that the MSX sources which lie spread across the (K,J-K) diagram at $K \sim 13$ are indeed post-AGB stars evolving from the end of the AGB to the planetary nebula stage.

This set of post-AGB stars will be very useful for studying the evolution of post-AGB stars. In particular, the timescale of evolution can be derived. By comparing the luminosity function of FGB stars with a theoretical luminosity function, we derive a death rate of 1 star every 40 years in the $3° \times 3.5°$ region studied. Given that there are 25 post-AGB stars in this area, the lifetime derived for the post-AGB stars is thus 1000 years.

Another feature of the post-AGB stars apparent in Fig. 1 is that their K magnitude does not appear to vary greatly over the $\sim$1000 years of the post-AGB lifetime. This result, combined with the magnitudes in other bandpasses, provides a direct observational characterization of the change in the spectral energy distribution with time through the post-AGB phase: such information will provide strong constraints on models for the evolution of the dust shells of post-AGB stars (e.g. van Hoof et al. 1997).

3. SUMMARY

Near-infrared photometry, MACHO light curves and spectral information have been obtained for 504 MSX sources in the LMC. The MSX sources are clearly divided into red supergiants, dust-enshrouded AGB stars and post-AGB star candidates. There are also 124 sources undetected in the near-infrared. Light curves and spectra provide strong evidence that the post-AGB star candidates are indeed genuine post-AGB stars. Full details of the observations described here will be presented elsewhere.

Acknowledgments

The authors are grateful for access to the MACHO data base from which the light curves shown here were derived. The MSX LMC point source catalog is available from the IRSA (IPAC, Pasadena, California).

References

Alcock C. et al. (the MACHO collaboration), 1996, ApJ 470, 583

Bertelli G., Mateo M., Chiosi C., Bressan A., 1992, ApJ 388, 400

Frogel, J. et al. 1990, ApJ, 352, 96

Lambert D.L., Rao N.K., 1994, J. Ap.&Astr. 15, 47

McGregor P., Hart J., Downing M., Hoadley D., Bloxham G., 1994, Experimental Astronomy 3, 139

Price, S.D. 1995, Space Sci. Rev., 74, 81

Rodgers A.W., Conroy P., Bloxham G., 1988, PASP 100, 626

Trams N.R. et al., 1999, A&A 346, 843

Van de Steene G.C., Wood P.R., van Hoof P.A.M., 2000, in it Asymmetrical Planetary Nebulae II: from Origins to Microstructures, eds. J.H. Kastner, N. Soker & S.A. Rappaport, ASP Conf. Ser. 199, 191

van Hoof P.A.M., Oudmaijer R.D., Waters L.B.F.M., 1997, MNRAS 289, 371

van Loon J. et al., 1998, A&A 329, 169

Wood P.R. et al., 1992, ApJ 397, 552

Wood P.R., 1998, A&A 338, 592

POST-AGB STARS IN GLOBULAR CLUSTERS AND GALACTIC HALOS

Howard E. Bond
Space Telescope Science Institute
bond@stsci.edu

David R. Alves
Space Telescope Science Institute

Abstract We discuss three aspects of post-AGB (PAGB) stars in old populations. (1) *HST* photometry of the nucleus of the planetary nebula (PN) K 648 in the globular cluster (GC) M15 implies a mass of $0.60\,M_{\odot}$, in contrast to the mean masses of white dwarfs in GCs of $\sim 0.5\,M_{\odot}$. This suggests that K 648 is descended from a merged binary, and we infer that single Pop II stars do not produce visible PNe. (2) Yellow PAGB stars are the visually brightest stars in old populations ($M_V \simeq -3.3$) and are easily recognizable because of their large Balmer jumps; thus they show great promise as a Pop II standard candle. Two yellow PAGB stars in the GC NGC 5986 have the same V magnitudes to within ± 0.05 mag, supporting an expected narrow luminosity function. (3) Using CCD photometry and a u filter lying below the Balmer jump, we have detected yellow PAGB stars in the halo of M31 and in its dwarf elliptical companion NGC 205. With the Milky Way zero point, we reproduce the Cepheid distance to M31, and find that NGC 205 is $\sim$100 kpc further away than M31. The star counts imply a yellow PAGB lifetime of about 25,000 yr, and their luminosities imply masses near $0.53\,M_{\odot}$.

1. INTRODUCTION

In this paper we will discuss several aspects of post-AGB (PAGB) stars in old stellar systems, in particular those in globular clusters (GCs) and in the halos of the Milky Way and other galaxies.

R. Szczerba and S.K. Górny (eds.), Post-AGB Objects as a Phase of Stellar Evolution, 77–82.

Specifically, we will discuss observations with the *Hubble Space Telescope* (*HST*) of the planetary nebula in the GC M15, ground-based observations of two PAGB stars in the GC NGC 5986, and ground-based searches for PAGB stars in the halos of nearby galaxies and their possible application as standard candles for measuring extragalactic distances.

2. K 648 IN M15

Since this work has now been published in detail (Alves, Bond, & Livio 2000), we give only a fairly brief summary here.

Küstner 648 in M15 is one of just four planetary nebulae (PNe) known in Galactic GCs. It was the first to be found (Pease 1928), and it was more than six decades before the other three were discovered: in M22 (Gillett et al. 1989) and Pal 6 and NGC 6441 (Jacoby et al. 1997). As Jacoby et al. point out, this number is lower than expected if every star now evolving in a GC makes a PN.

There are several lines of argument that PAGB evolution in GCs is slow compared to the timescale for dissipation of a PN into space following its ejection at the tip of the AGB. First, recent *HST* detections of white dwarfs in nearby GCs (Renzini et al. 1996; Cool, Piotto, & King 1996; Richer et al. 1997) indicate that the remnant masses are very low ($0.50 \pm 0.02\,M_\odot$; see Alves et al. 2000 for details). PAGB stars of masses $\lesssim 0.55 M_\odot$—the lowest for which calculations have been made—already have theoretical evolutionary timescales that are much longer than the dissipation timescale for PNe (Schönberner 1983; Vassiliadis & Wood 1994), and the timescales should be even longer at masses of $\sim$0.5 $M_\odot$. In support of this expectation, star counts of PAGB stars in the M31 halo (see below) indicate quite long lifetimes for the portion of the post-AGB evolution from effective temperatures of 5,000 to 10,000 K. It is thus unlikely that typical halo stars will ever be able to ionize visible PNe.

The question then shifts from "Why are there so few PNe in GCs?" to "Why are there *any* PNe at all?". Jacoby et al. (1997) proposed that the observed GC PNe were formed through close-binary interactions, although they did not propose any specific mechanisms.

In order to test this hypothesis, we obtained observations of K 648 and its central star with *HST* and its WFPC2 over an interval of 7 days, in order to search for variations in the brightness of the nucleus due to heating effects in a close binary system. As discussed in our paper, we did not find any variations, and thus have no direct evidence that the nucleus is at present a close binary.

However, the argument can be pressed further by considering the luminosity of the central star. By combining our accurate measurement of the star's V magnitude with its known effective temperature ($40,000 \pm 3,000$ K, based on spectroscopic observations cited in our paper), we find $\log(L/L_\odot) = 3.74 \pm$

0.08. We can now compare its position in the HR diagram with theoretical PAGB tracks, as shown in Fig. 1.

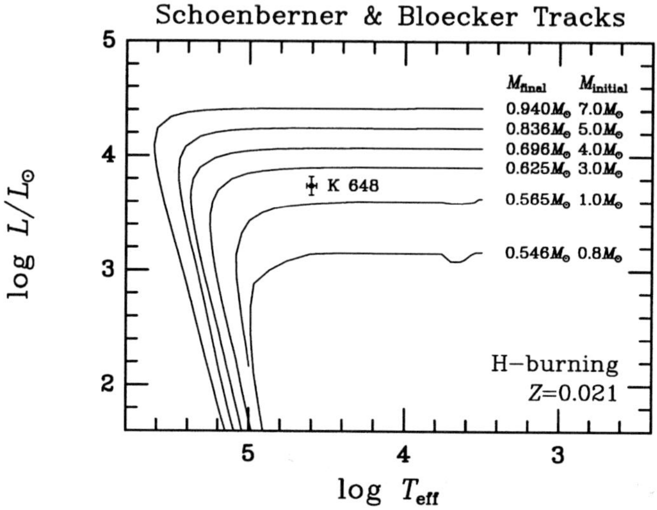

Figure 1 Position of central star of K 648 in the HR diagram, compared with theoretical post-AGB tracks of Schönberner (1979, 1983) and Blöcker (1995). The initial and final masses for each track are indicated at the right. The implied mass of the nucleus is $0.60 \pm 0.02\,M_\odot$.

Although the PAGB tracks plotted in Fig. 1 are for solar metallicity, they still illustrate that the mass of the central star of K 648 is significantly above that of the stellar remnants currently evolving in GCs, which, as noted, are about $0.5\,M_\odot$. Vassiliadis & Wood (1994) find a PAGB luminosity/core-mass relation of $L/L_\odot = 56694(M_c/M_\odot - 0.5)$, nearly independent of metallicity, and applicable to both hydrogen- and helium-burning tracks. Using this formula, we find that the mass of the K 648 central star is $0.60 \pm 0.02\,M_\odot$ (the error estimate includes only observational errors, not any errors in the theory).

Since the K 648 central star is more massive than the normal stellar remnants in GCs, we conclude that the progenitor experienced mass augmentation at some point in its prior evolutionary history, most likely from a close-binary interaction. A plausible scenario is one in which a close binary on the main sequence underwent a W UMa-type contact stage, followed by coalescence (at which time it would probably have appeared as a blue straggler in the color-magnitude diagram). The star subsequently evolved as a more massive object than the bulk of the single stars (which have masses $\sim 0.8\,M_\odot$), allowing it to produce a $0.6\,M_\odot$ remnant which evolved rapidly enough to ionize its ejected AGB envelope before the envelope had time to dissipate.

Our detailed study of K 648 supports the hypothesis that all halo PNe, including those in GCs and in the halo field, are descendants of close binaries, and that single stars in old populations do not make PNe.

3. THE POST-AGB A-F STARS IN NGC 5986

This work has now also been published (Alves, Bond, & Onken 2001), so again we will give only a short summary here.

Population II stars evolving off the AGB and passing through spectral types F and A are the visually brightest members of old populations (due to the behavior of the bolometric correction with effective temperature). These stars should have a narrow luminosity function (LF), because essentially a single main-sequence turnoff mass is feeding the PAGB region of the HR diagram. Moreover these "yellow" PAGB stars are easily recognized because of their enormous Balmer jumps. For these reasons (summarized in more detail by Bond 1997), we believe that yellow PAGB stars in old populations show great promise as a new standard candle for measuring extragalactic distances, especially to early-type galaxies that contain no Cepheid variables. A zero-point calibration for the luminosities of Pop II yellow PAGB stars may be set using those in Galactic GCs.

The little-studied southern GC NGC 5986 is remarkable because it contains two candidate A-F type PAGB stars, discovered some years ago during a photographic grism survey (Bond 1977). Cluster membership was confirmed by their radial velocities and by Strömgren photometry showing very low surface gravities. These two stars are potential primary zero-point calibrators for PAGB absolute magnitudes.

We have obtained CCD photometry of NGC 5986, from which we have derived a reddening and distance modulus of $E(B-V) = 0.29 \pm 0.02$ and $(m-M)_0 = 15.15 \pm 0.10$, respectively (see Alves et al. 2001 for details). The two yellow PAGB stars have the same V magnitudes to within ± 0.05 mag, confirming the expectation of a very narrow LF and supporting the idea that they may constitute a useful new standard candle. By including also the absolute magnitude of the bright A-type star ROA 24 in the GC ω Cen (González & Wallerstein 1992), we find a mean $M_V(\text{PAGB}) = -3.28 \pm 0.07$. Fig. 2 shows the locations of the two PAGB stars in the color-magnitude diagram (CMD) of NGC 5986.

We now have underway a CCD survey of about 100 Galactic GCs, using our new *uBVI* photometric system (see below), in order to find more examples of yellow PAGB stars and strengthen the zero-point determination.

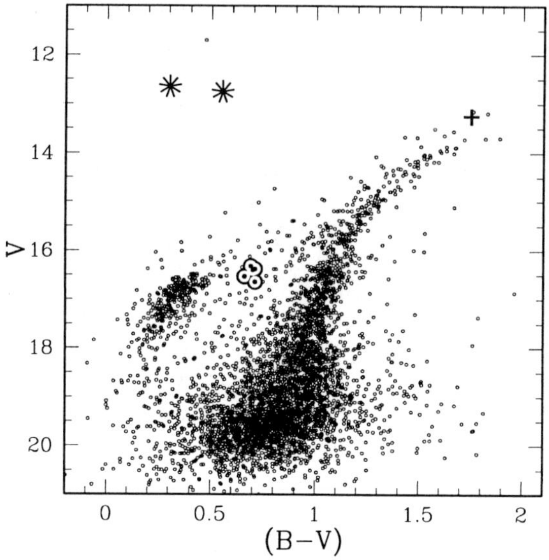

Figure 2 CMD of NGC 5986, from Alves, Bond, & Onken (2001). The cluster's two PAGB stars are shown as asterisks; they are the brightest stars in the cluster, with absolute magnitudes near $M_V = -3.3$. Also marked with circled dots are the cluster's RR Lyrae variables, some 4 mag fainter than the PAGB stars.

4. YELLOW PAGB STARS IN HALOS OF LOCAL GROUP GALAXIES

As described above, yellow PAGB stars show promise as a standard candle, and are easily detectable using a photometric system that is sensitive to their large Balmer jumps. We have been developing a new CCD photometric system that combines the Gunn u filter (lying almost entirely below the Balmer jump) with the standard Johnson-Kron-Cousins *BVI* bandpasses.

In a preliminary analysis of frames taken in the halo of M31, Bond and Laura Fullton have discovered a sequence of stars with $0 < (B - V) < 0.5$ and large $u - B$ color indices, which are strong candidates for PAGB stars belonging to M31 . Their mean V is 20.88 ± 0.06. If we use the preliminary zero-point of $M_V = -3.3$ based on NGC 5986 and ω Cen, as outlined above, we obtain an M31 distance modulus of $(m - M)_V = 24.2$, in reasonably good agreement with that based on Cepheids and other indicators, but requiring only about an hour of 4-m telescope time!

Bond & Fullton (1997) have also obtained *uBVI* frames of NGC 205, a dwarf elliptical companion of M31. Here we find a sequence of yellow PAGB stars some 0.35 mag fainter than those belonging to the superposed halo of M31, suggesting that NGC 205 is $\sim$100 kpc further away than M31 itself (in fair agreement with distances from PNe and RR Lyrae variables, although these observations have somewhat greater uncertainties due to the small number of PNe in NGC 205 and the faintness of the RR Lyr stars).

The star counts in the M31 halo fields imply a PAGB lifetime of $\sim$25,000 yr (for the portion of the evolution from $B - V = 0.5$ to 0.0 only), confirming as noted in our comments on K 648 that PAGB evolution is extremely leisurely in old populations. The luminosities of the M31 PAGB candidates imply a mean mass of $\sim 0.53 M_{\odot}$, in reasonable agreement with the masses inferred for white dwarfs in old populations.

Acknowledgments

We gratefully acknowledge support from NASA (grant NAG 5-6821) for the work on PAGB stars in galactic halos, and from STScI (grant GO-6751) for the work on K 648We thank Mario Livio, Chris Onken, and Laura Fullton for their contributions to these projects.

References

Alves, D.R., Bond, H.E., & Livio, M. 2000, AJ, 120, 2044

Alves, D.R., Bond, H.E., & Onken, C. 2001, AJ, 121, 318

Blöcker, T. 1995, A&A, 299, 755

Bond, H.E. 1977, BAAS, 9, 601

Bond, H.E. 1997, in The Extragalactic Distance Scale, eds. M. Livio, M. Donahue, & N. Panagia (Cambridge: Cambridge University Press), 224

Bond, H.E., & Fullton, L.K. 1997, BAAS, 29, 843

Cool, A.M., Piotto, G., & King, I.R. 1996, ApJ, 468

Gillett, F.C. et al. 1989, ApJ, 338, 862

González, G., & Wallerstein, G. 1992, MNRAS, 254, 343

Jacoby, G.H. et al. 1997, AJ, 114, 2611

Pease, F.G. 1928, PASP, 40, 342

Renzini, A. et al. 1996, ApJ, 465, L23

Richer, H.B., et al. 1997, ApJ, 484, 741

Schönberner, D. 1979, A&A, 79, 108

Schönberner, D. 1983, ApJ, 272, 708

Vassiliadis, E., & Wood, P. 1994, ApJS, 92, 125

II

STELLAR EVOLUTION AND WINDS

THE EVOLUTION OF AGB STARS TOWARDS PLANETARY NEBULAE

Detlef Schönberner and Matthias Steffen
Astrophysikalisches Institut Potsdam
An der Sternwarte 16
D-14482 Potsdam
DeSchoenberner@aip.de, MSteffen@aip.de

Abstract We combined hydrodynamical simulations with stellar evolutionary calculations based on prescribed mass-loss rates in order to get a better understanding of how structure and kinematics of circumstellar matter develop with time during the late AGB and the following post-AGB evolution. We show (i) that the final density profile of the circumstellar shells produced by AGB mass loss falls off steeper than $\rho \sim r^{-2}$, (ii) that stages of high mass-loss rates are periodically interrupted by shorter phases of much lower outflow rates and speeds due to the luminosity dips associated with thermal pulses, and (iii) that during the post-AGB evolution ionization destroys the initial density and velocity distribution set up at the end of the AGB.

1. INTRODUCTION

Density and velocity distributions of circumstellar matter around objects that are about to leave the asymptotic giant-branch (AGB) are ruled by the preceding mass-loss history. The flow of matter from the stellar surface is driven by radiation pressure on dust grains which transfer the acquired momentum to the gas component by collisions. Theoretical studies of such ‘dust-driven’ winds suggest that the mass-loss rate depends strongly on stellar luminosity and effective temperature (Bowen 1988; Arndt, Fleischer and Sedlmayr 1997), whereas the wind velocity does not vary much (Habing, Tignon and Tielens 1994; Ivezić and Elitzur 1995; Steffen, Szczerba and Schönberner 1998). Typical rates are between about 10^{-7} and 10^{-4} $M_{\odot}$/yr, with outflow velocities from 5 to 25 km/s, depending on the parameters of the AGB star.

Since the flight time of a gas parcel through such dusty circumstellar wind envelopes is comparable to the period of luminosity variations between thermal pulses, the associated mass-loss and velocity variations determine the envelope’s structure. However, density and velocity are intertwined such that only

R. Szczerba and S.K. Górny (eds.), Post-AGB Objects as a Phase of Stellar Evolution, 85–92.

dynamical simulations of AGB wind envelopes will result in meaningful initial structures for the following post-AGB evolution.

The evolutionary speed of an AGB remnant across the HR-diagram is controlled by its internal constitution (i.e. whether hydrogen or helium burning supplies most of the surface luminosity), by its mass (= core mass), and by mass loss. A thorough discussion of this issue is given in Schönberner and Blöcker (1993) and cannot be repeated here. An illustration of how post-AGB evolutionary time scales depend on the remnant mass is presented in Fig. 1.

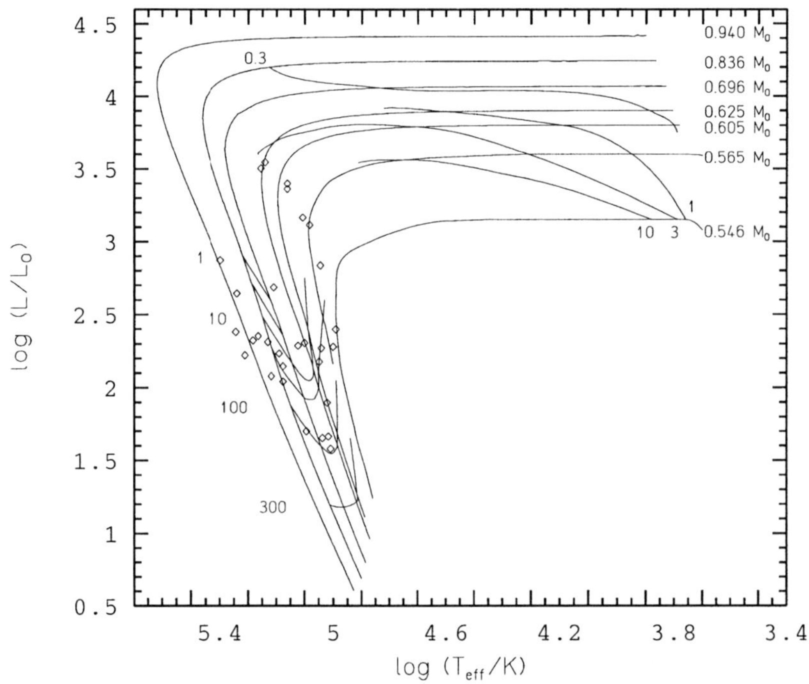

Figure 1 Evolutionary tracks of hydrogen-burning post-AGB models with isochrones (in 10^3 yrs), together with a sample of central stars of 'optically thick' PNe (Blöcker 1995b). The post-AGB ages are counted from positions close to the AGB coinciding with the beginning of the plotted tracks. Note that these remnants stem from different progenitors in the mass range 1 to 7 $M_\odot$ whose evolutionary histories have been consistently followed from the main sequence through all the later phases including mass loss and thermal pulses (see Blöcker 1995a for the details).

Based on the zero points given in Fig. 1, the proto-planetary nebula phase ranges from $\simeq$ 100 000 yrs for 0.55 $M_\odot$ to $\simeq$ 12 yrs for 0.94 $M_\odot$. The very short transition time for the higher-mass remnants is a consequence of the zero points assumed by Blöcker (1995b). They can be considerably larger if a different mass-loss description in the vicinity and the tip of the AGB is used (e.g. Vassiliadis and Wood 1994)[1].

The evolution of the AGB remnant in temperature and luminosity drives in turn the evolution of the circumstellar wind envelope by *two* processes, viz. by the concomitant changes of the stellar radiation field and the wind power. The relative importance of both processes varies with the remnant's age (or effective temperature). The number of hydrogen-ionizing photons emitted per second follows directly from the evolutionary path: at first it increases with effective temperature until later the luminosity decrease dominates. The maximum photon flux occurs between 60 000 and 70 000 K.

The properties of the mass-loss during the post-AGB evolution are more difficult to evaluate. In the vicinity of the AGB the rate is certainly orders of magnitudes below that on the tip of AGB (see above), and is usually assumed to be that of Reimers' (1977). In the central-star domain the theory of 'line-driven' winds predicts rates of $\approx 10^{-8}\,\mathrm{M}_\odot$/yr and $V \simeq (2\ldots4)\,V_{\rm esc} \simeq 1\,000\ldots10\,000$ km/s. The wind power, $P = \dot{M}V^2/2$, reaches its maximum close to the turn-around point at maximum effective temperature and declines then rapidly.

It should, however, be noted that the stellar wind does not interact directly with the nebular/AGB material. Instead, the wind's kinetic energy thermalizes through a shock and is added to the internal energy content of hot, shocked wind material expelled at earlier times. The thermal pressure of this 'bubble' of hot but very tenuous gas drives the inner edge of the planetary. Though it is actually the time integral over the wind power that determines the energy content of the bubble, the maximum bubble pressure coincides roughly with the maximum wind power of the central star.

2. THE EARLY POST-AGB EVOLUTION

During the early evolution off the AGB one is faced with the problem of selecting the proper initial density and velocity structures since these are determined by the AGB mass-loss history. We thus have chosen to model the AGB and post-AGB evolution of the circumstellar wind envelopes by combining stellar evolution with hydrodynamical simulations (Schönberner et al. 1997; Steffen, Szczerba and Schönberner 1998; Schönberner and Steffen 1999; Schönberner et al. 2000).

In these models, the stellar outflow is assumed to be spherically symmetric, and the equations of hydrodynamics are solved for the gas and the dust component, coupled by momentum exchange due to dust-gas collisions. We used a modified version of the code developed by Yorke and Krügel (1977), making

[1] A rapid decrease of the mass-loss rate by orders of magnitudes is usually assumed to mark the beginning of the post-AGB evolution.

use of the following simplifications: (i) Radiation transfer is considered only for the dust component, i.e. exchange of photons between dust grains and the gas is neglected, (ii) the dust temperature is computed from radiative equilibrium, and the gas (neutral hydrogen) is assumed to have the same (local) temperature, (iii) the dust consists of single-sized grains, either based on oxygen or carbon chemistry, adopting a fixed dust-to-gas ratio at the dust condensation point. We introduced time-dependent values of stellar mass, luminosity, effective temperature *and* variable mass loss (as shown in Fig. 2) with a constant flow velocity equaling the local sound speed, $\approx$ 3 km/s, at the inner boundary. A more detailed description of this fully implicit radiation hydrodynamics code can be found in Steffen et al. (1997) and Steffen, Szczerba, and Schönberner (1998). It should finally be mentioned that all the simulations shown here are based on one evolutionary sequence, viz. that with 3 $M_\odot$ on the main sequence, ending up as a white dwarf of 0.605 $M_\odot$ (also shown in Fig. 1).

An illustration of our hydrodynamical simulations for oxygen-based grains ("astronomical silicates") is given in Fig. 2 which covers the last 50 000 years on the AGB together with the post-AGB evolution. Mass-loss rate and effective temperature (or radius) of the star are coupled according to the prescription of Blöcker (1995a,b), and the most prominent feature is a rapid decrease of the mass-loss rate by orders of magnitude within about 100 years around effective temperatures of 6 000 K, setting the zero point of the post-AGB evolution (cf. Fig 1). The consequence is a rapid detachment and thinning of the dust shell since the density of any newly formed dust is strongly reduced and gives no detectable signature. This is illustrated by the sequence of spectral energy distributions in the lower panel of Fig. 2 which covers a time interval of only about 500 years. The gradual disappearance of the strong silicate absorption feature with increasing shell detachment is clearly seen. At the same time, the previously totally obscured AGB remnant becomes visible. Our modelled spectral energy distributions resemble very much those of known proto-planetary nebulae (Hrivnak, Kwok and Volk 1989), indicating that the mass-loss decrease at the end of the AGB evolution as chosen by Blöcker (1995b) is close to reality!

The mass-loss variation (upper panel of Fig. 2) leads to a density structure clearly different from the usual assumption of a $\rho \propto r^{-2}$ law (middle panel): The density dip near $r = 10^{18}$ cm is caused by the last thermal pulse about 30 000 years ago (cf. upper panel), while the rapid density increase towards the inner parts of the shell ($\rho \propto r^{-3}$) is due to the recent increase of the mass-loss rate. The outflow velocity remains rather constant, $\approx$ 9 km/s, except for a slight decrease to about 6 km/s in the low-density region near $r = 10^{18}$ cm. Consequently, the faster high-density wind compresses the slower low-density matter into a thin shell. More details about the further evolution of such wind compressed gas shells are given by Steffen and Schönberner (this volume).

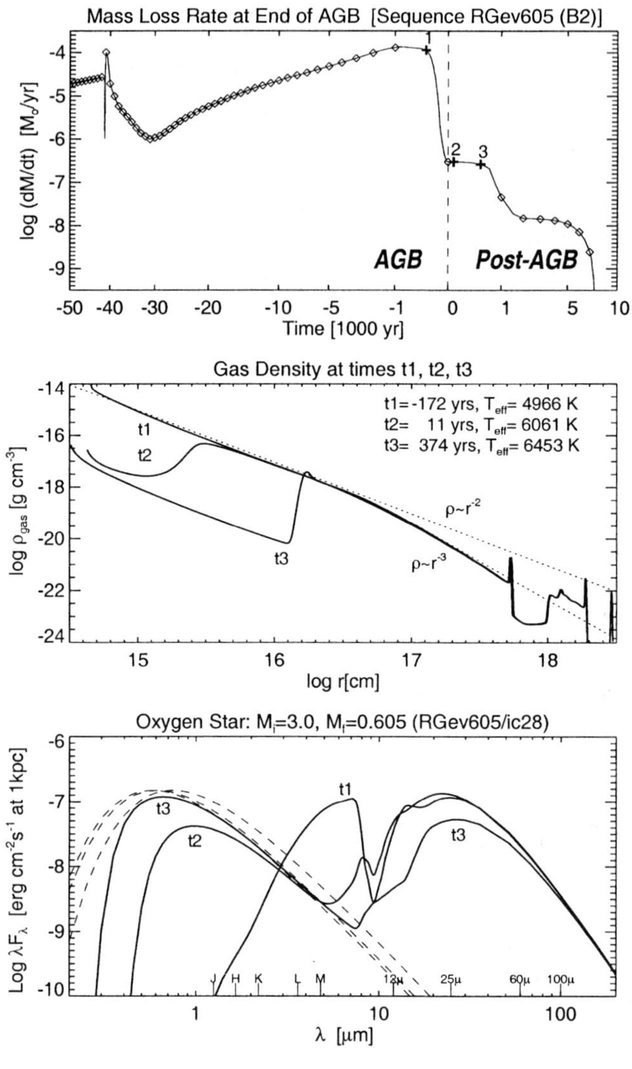

Figure 2 **Top:** Adopted mass-loss rate for the late AGB and early post-AGB evolution ending up with a remnant of 0.605 $M_\odot$ (Blöcker 1995b). **Middle:** Radial gas density distributions for a model with silicate grains at the three selected times. Note the density 'spikes' at $r = 5\ 10^{17}$ and $r = 2\ 10^{18}$ cm resulting from two-wind interaction on the AGB (cf. Steffen and Schönberner, this volume). **Bottom:** Spectral energy distributions at the same times. The dashed lines are the corresponding intrinsic spectra of the central star.

Our hydrodynamics models are in remarkable agreement with recent empirical results on the density structures of cool wind envelopes: Sahai et al. (1998) estimated $\rho \propto r^{-2.7}$ for the envelope of CRL 2688 , whereas Volk, Xiong and Kwok (2000) found $\rho \propto r^{-2.2...-3}$ for the extreme carbon stars investigated.

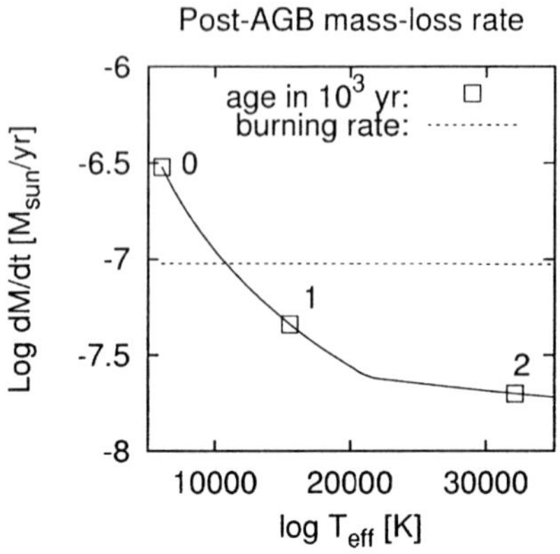

Figure 3 Mass-loss and burning rate during the first 2 000 years of post-AGB evolution of the 0.605 $M_\odot$ remnant plotted as a function of the effective temperature. Time marks indicate post-AGB ages. Note that mass loss dominates over nuclear burning only during the very early post-AGB evolution ($T_{\rm eff} \leq$ 10 000 K)

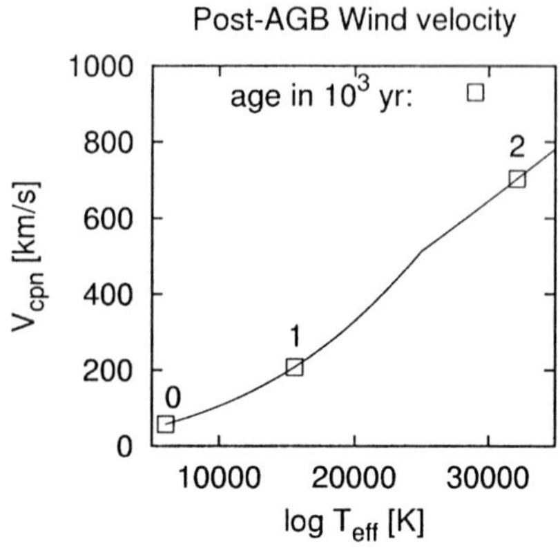

Figure 4 Central-star wind speed during the early post-AGB evolution as adopted for the 0.605 $M_\odot$ remnant (Marten and Schönberner 1991).

In order to investigate the transformation of a cool AGB wind envelope into a planetary nebula, we used the model structure shown in Fig. 2 at time $t2$ as input for another radiation hydrodynamics code. This one-component explicit code is based on a second-order Godunov-type advection scheme and considers time-dependent ionization, recombination, heating and cooling of six elements (H, He, C, N, O, Ne) with all of their ionization stages. At the inner boundary, $r_{\rm i} = 5\,10^{14}$ cm, the time-dependent values of wind speed and density according to the prescription of Marten & Schönberner (1991) are introduced (cf. Figs. 3 and 4). The star is assumed to radiate as a black body. More details can be found in Perinotto et al. (1998).

The few selected snapshots presented in Fig. 5 illustrate how a hot, ionized HII region develops inside the AGB wind envelope, with a quite specific density and velocity structure. The mechanical power of the central star wind has only a minor effect on the envelope's shape. Only when the central star becomes hot enough, ionization is able to reshape the envelope's density and velocity field, destroying thereby any memory of the former shape set up during the late AGB evolution. Specifically, the thermal pressure of the ionized matter accelerates and compresses neutral material into a dense outwards moving shell and brakes – and even inverts for some time – the flow further in. The increasing pressure

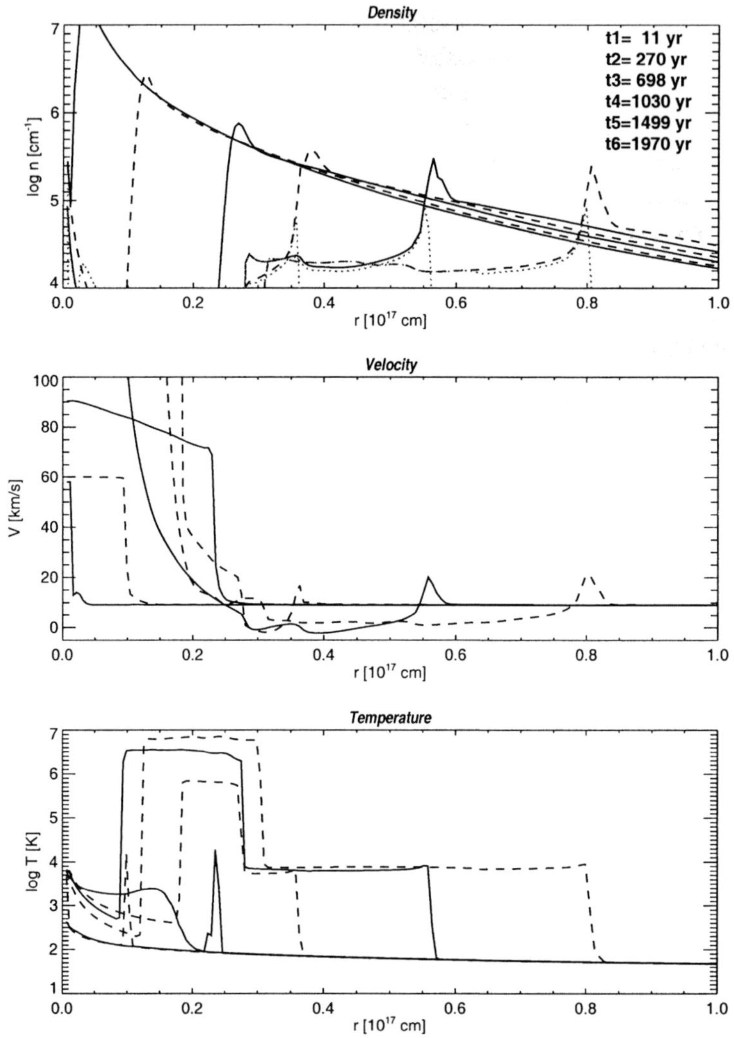

Figure 5 **Top:** Number density of heavy particles and electrons (dotted) at six selected post-AGB times. **Middle:** Velocities at the same times. **Bottom:** Temperature distributions at the same times. Note how the well cooled thin interface between the central-star wind and the AGB material develops into an extended very hot, tenuous region, the so-called 'hot bubble', once the wind velocity exeeds $\approx$ 150 km/s.

exerted by the shocked central-wind material finally prevents the ionized matter to flow inwards. As a consequence, at an age of about 1 000 years ($t4$ in Fig. 5) a region of rather constant density and low flow velocity develops rapidly. A young planetary nebula is born, bounded by still neutral AGB matter.

2.1. FINAL REMARKS

Our simulations demonstrate that the whole physical system, consisting of the star *and* the wind envelope, has to be treated as consistently as possible, with due consideration of the final AGB mass-loss history. Given the models as presented here, the post-AGB structures are (i) set by the mass-loss properties of the final AGB evolution, and (ii) are hardly modified by the rather modest stellar wind. Once ionization becomes important, the original density and velocity structure will be destroyed completely within the ionized region.

References

Arndt, T.U., Fleischer, A.J., and Sedlmayr, E., 1997, A&A 327, 614

Blöcker, T., 1995a, A&A 297, 727

Blöcker, T., 1995b, A&A 299, 755

Bowen, G.H., 1988, ApJ 329, 299

Habing, H.J, Tignon, J., and Tielens, A.G.G.M., 1994, A&A 286, 523

Hrivnak, B.J., Kwok, S., and Volk, K., 1989, ApJ 346, 265

Ivezić, Z., and Elitzur, M., 1995, ApJ 445, 415

Marten, H., and Schönberner, D., 1991, A&A 248, 590

Perinotto, M., Kifonidis, K., Schönberner, D., and Marten, H., 1998, A&A 332, 1044

Reimers, D., 1977, *Problems in stellar atmospheres and envelopes*, Berlin: Springer, 229

Sahai, R., et al., 1998, ApJ, 493 301

Schönberner, D., and Blöcker, T., 1993, *Luminous high-latitude stars*, A.S.P. Conf. S., Vol. 45, 337

Schönberner, D., and Steffen, M., 1999, *Optical and infrared spectroscopy of circumstellar matter*, A.S.P. Conf. S., Vol. 188, 281

Schönberner, D., Steffen, M., Stahlberg, J., Kifonidis, K., and Blöcker, T., 1997, *Advances in stellar evolution*, Cambridge: University Press, 146

Schönberner, D., Steffen, Stahlberg, J., Kifonidis, K., and Blöcker, T., 2000, *The carbon star phenomenon*, IAU Symp. No. 177, Kluwer Acad. Publ., 469

Steffen, M., Szczerba, R., Menshchikov, A., and Schönberner, D., 1997, A&AS 126, 39.

Steffen, M., Szczerba, R., and Schönberner, D., 1998, A&A 337, 149

Vassiliadis, E., and Wood, P.R., 1994, ApJS 92, 125

Volk, K., Xiong, G.-Z., and Kwok, S., 2000, ApJ 530, 408

Yorke, H.W., and Krügel, E., 1977, A&A 54, 183

MASS LOSS CHARACTERISTICS OF DUST FORMING LONG-PERIOD VARIABLE STARS

Jan Martin Winters[1,2,3] & Thibaut Le Bertre[1]

[1] *UMR 8540, Observatoire de Paris, 61 av. de l'Observatoire, F-75014 Paris, France*

[2] *MPI für Radioastronomie, Auf dem Hügel 69, D-53121 Bonn, Germany*

[3] *TU Berlin, Institut für Astronomie und Astrophysik, PN 8-1, Hardenbergstr. 36, D-10623 Berlin, Germany*

winters@mesioc.obspm.fr

Abstract We present results obtained from a grid of model calculations for dust forming circumstellar shells around carbon-rich long-period variable stars. These models consistently treat the time-dependent hydrodynamics, chemistry, dust formation and the radiative transfer problem. We show that variations of the mass loss rates observed on different time scales are well reproduced by our models.

1. INTRODUCTION

In order to investigate the mass loss mechanism of dust forming long-period variable stars (LPVs), we built a grid of $\sim$ 150 carbon-rich hydrodynamic models covering a wide range of physical situations. This grid allows to characterize the effects of the model parameters, i.e. the stellar mass, luminosity, and temperature, the period and amplitude of the pulsation, and the C/O element abundance ratio, on the behavior of AGB winds and on the mass loss rates. A detailed analysis of these calculations is given in Winters et al. (2000a). Here we focus on the general results concerning the outflow characteristics in terms of the wind momentum and on the temporal variation of the mass loss rate on different time scales. To represent the global results of the time-dependent model calculations in a compact way, we introduce a time averaged, dimensionless radiative acceleration $\langle\alpha\rangle_t$, determined in the sub-sonic region of the outflow where dust is condensing. α is the ratio between the radiative acceleration $a^{\rm rad}$ and the gravitational deceleration $a^{\rm grav}$:

$$\alpha \equiv \frac{a^{\rm rad}}{a^{\rm grav}} = \frac{\frac{4\pi}{c}\chi_H H}{\rho G \frac{M_r}{r^2}} = \frac{4\pi\chi_H r^2 H}{c\rho G M_r} = \frac{1}{4\pi c G}\frac{L_*}{M_r}\frac{\chi_H}{\rho}$$

R. Szczerba and S.K. Górny (eds.), Post-AGB Objects as a Phase of Stellar Evolution, 93–100.

(c: speed of light, G: gravitation constant, L_*: stellar luminosity, M_r: mass contained in a sphere of radius r, χ_H: flux-mean extinction coefficient, ρ: mass density, H: Eddington-flux).

2. DEPENDENCE OF THE MODELS ON THE RADIATIVE ACCELERATION

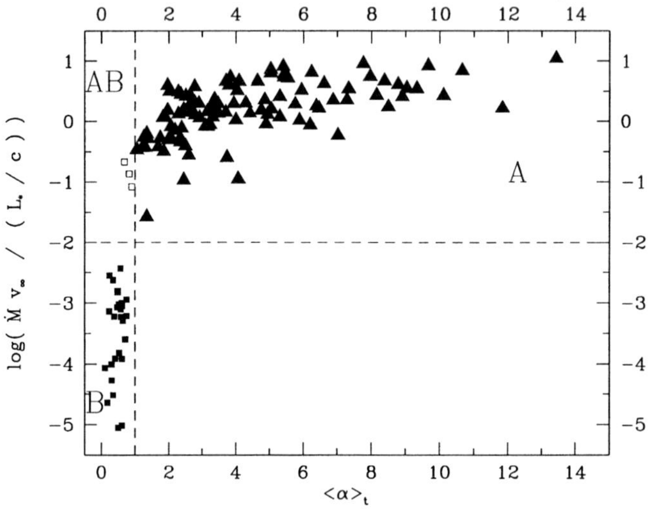

Figure 1 Momentum ratio $\beta = \dot{M}v_\infty/(L_*/c)$ vs. time averaged radiative acceleration $\langle\alpha\rangle_t$

Depending on the value of $\langle\alpha\rangle_t$ we find two regimes for the stellar outflows (see Fig. 1). The first one (regime A, $\langle\alpha\rangle_t > 1$) is characterized by stable winds with a layered structure of the circumstellar dust shell, outflow velocities in the range of $5 - 40\,\mathrm{km\,s^{-1}}$, and a large rate of mass loss, $\dot{M} \geq 3\,10^{-7} M_\odot \mathrm{yr}^{-1}$. For these outflows we find good correlations between near-infrared colors and mass loss rates (see Winters et al. 2000a). In the second regime (B, $\langle\alpha\rangle_t < 1$), the winds are slow, they carry a low mass loss rate ($\dot{M} < 3\,10^{-7} M_\odot \mathrm{yr}^{-1}$), and they do not present a layered structure. The outflows displaying this second behavior come, e.g., from models with low stellar luminosity, high temperature, or short period. For them there is no correlation between colors and mass loss rates. We have explored the influence of the different parameters (stellar mass, luminosity, temperature, C/O-ratio, pulsation period and -amplitude) on the behavior of the winds. We find that, all other parameters being kept identical, there is a narrow range of values for each parameter within which the models abruptly change from regime B to A, and that once a model is stabilized in the A mode the changes in the values of each parameter have only a smooth effect on the wind characteristics.

In Fig. 1 we represent the global model results in a diagram displaying the ratio β of the momentum carried by the wind ($\dot{M}v_\infty$) to the momentum contained in the radiation field (L_*/c) versus the radiative acceleration $\langle\alpha\rangle_t$. From analytical considerations (e.g., Gail & Sedlmayr 1987) one finds, that there exists a minimum mass loss rate for dust driven winds of the order of $10^{-7}M_\odot\mathrm{yr}^{-1}$, a value which is well reproduced by consistent numerical calculations for stationary dust-driven winds (Dominik et al. 1990) and also by the time-dependent models presented here. In Winters et al. (2000a) the behavior of the mass loss rate $\dot{M}$ and of the outflow velocity v_∞ are discussed separately in terms of the radiative acceleration $\langle\alpha\rangle_t$. It turns out that those models for which the wind is not dominated by radiation pressure (i.e. $\langle\alpha\rangle_t < 1$) always produce low outflow velocities ($v_\infty < 5\,\mathrm{km\,s^{-1}}$). By plotting β on the ordinate in Fig. 1, the separation of the wind models into zone A (dust dominated) and zone B (pulsation dominated) is strongly emphasized by the combined effect of the low mass loss rate and the simultaneously low outflow velocity for the B-models. Once $\langle\alpha\rangle_t$ drops below unity, there is a sudden and very pronounced drop in the momentum ratio β.

Additionally to the models in regions A and B, we find 3 models in the present grid with $\langle\alpha\rangle_t < 1$ and $\beta \approx 10^{-1}$ (region AB). These models show rather regular long term variations of the shell structure and therefore also of $\dot{M}$ and of $\langle\alpha\rangle_t$. Since $\langle\alpha\rangle_t$ is determined in the sub-sonic wind region around $5R_*$ whereas the mass loss rate is calculated simultaneously outside of $30R_*$, there is a phase shift between the values of $\langle\alpha\rangle_t$ and $\dot{M}$. The models found in region AB have just been caught in a state of high $\dot{M}$ and low $\langle\alpha\rangle_t$.

3. TEMPORAL VARIATION OF THE MASS LOSS RATE

The different types of models found in our grid show temporal variations of the mass loss rate (and, accordingly, of the shell structure) on very different time scales. Therefore, the terms 'mass loss rate' $\dot{M}$ and 'outflow velocity' v_∞ used here refer to temporal averages (usually over 20 pulsation periods) of the respective quantities, defined here as radial mean values in a shell region where dust formation and wind acceleration are complete (in practice, the radial mean values are taken over the shell volume between $30\,\mathrm{R}_*$ and $60\,\mathrm{R}_*$). The temporal variation of the radially averaged mass loss rate, $\langle\dot{M}\rangle_r$, is shown in Fig. 2 for the three different types of models. The A-type models typically show variations of the mass loss rate by about 20% on a time scale of a few pulsation periods (an effect of the *multiperiodicity* discussed by Fleischer et al. 1995). On the contrary, the B-models develop outflow structures which resemble very much those of stationary winds and, therefore, do not display relevant temporal variations of, e.g., the mass loss rate. Finally, the AB models produce rather

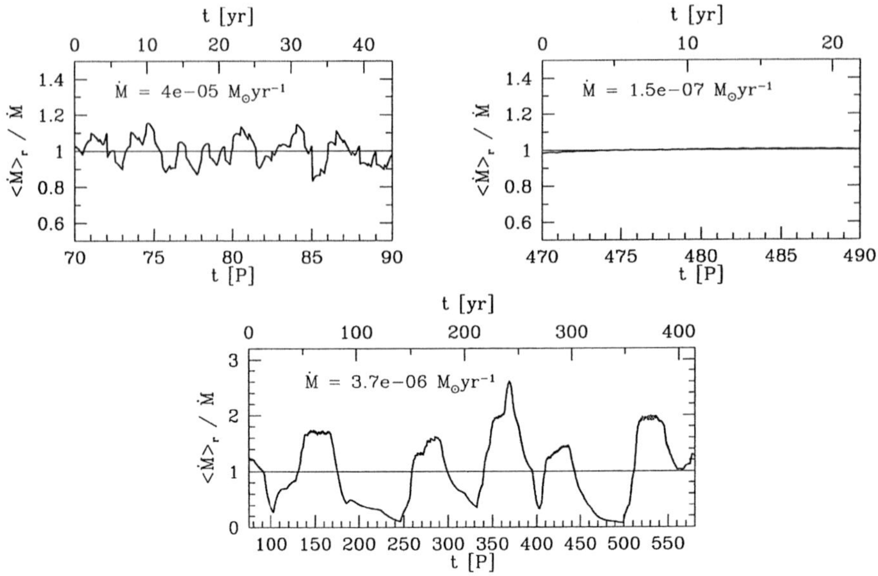

Figure 2 Temporal variation of the mass loss rate of an A-model (upper left), a B-model (upper right) and an AB-model (lower diagram)

large variations of the mass loss rate on a typical time scale of the order of 100 yr (note the different scaling in the lower diagram of Fig. 2!). In the three models presently identified as being of the AB type, the outflow velocities are relatively low, $v_\infty \approx 5\,\mathrm{km\,s^{-1}}$ and virtually constant in time.

Apart from the mass loss variations described here, recent evolutionary calculations (Schröder et al. 1999) revealed, for a very limited range of initial stellar masses, a high mass loss rate episode which is connected to a thermal pulse. When the hydrogen-burning shell resumes its energy production after the He shell flash, the star is driven into the A-regime for a time interval of about 1000 yr.

4. EVIDENCE FOR MASS LOSS VARIATIONS FROM OBSERVATIONS

Recent observations indicate the occurrence of variations of the mass loss rates on rather different time scales. For the prominent tip-AGB carbon star IRC+10216, e.g., one finds present-day temporal variations of the shell structure on a time scale of about 10 yr (e.g., Sada 1993, Osterbart et al. 2000, Tuthill et al. 2000, Winters et al. 2000b), resembling the typical *multiperiodicity* time scale found in our A-type models. Imaging of the dust shell around IRC+10216 in scattered galactic light revealed an additional variation of the mass loss rate on a $\approx$ 100 yr time scale leading to the formation of the so-called multiple layers (Mauron & Huggins 1999, 2000). Also for IRC+10216 one finds from different observational methods (CO rotational line profiles and mapping, CO ro-vibrational line profiles, near-infrared continuum fluxes, 1.3 mm continuum mapping), that the (average) mass loss rate of this object has decreased by a factor of about 5 within the last $\sim$ 5000 yr (e.g., Sahai 1987, Truong-Bach et al. 1991, Keady & Ridgway 1993, Groenewegen et al. 1998). Another type of mass loss variation, which occurs on a $\approx$ 1000 yr time scale and which most probably is connected to the luminosity variation after a thermal pulse, is evidenced by the presence of thin detached shells around some AGB stars (Olofsson et al. 1996). This kind of mass loss variations explains that, in fact, different observational methods to derive mass loss rates may give very different results. For instance, using CO rotational line data Loup et al. (1993) find $\dot{M} \sim 10^{-5}\,\mathrm{M_\odot\,yr^{-1}}$ for R Scl, whereas Le Bertre (1997), using IR continuum data, finds $\dot{M} \sim 10^{-7}\,\mathrm{M_\odot\,yr^{-1}}$, a factor of $\sim$ 100 less. The explanation for this discrepancy lies in a drastic reduction of the R Scl mass loss rate about 100 years ago. The situation is even more complex if one considers the multiple molecular winds recently brought to evidence from composite CO line profiles (Knapp et al. 1998). In these cases, the mass loss rate variations can be uncovered because the properties of the winds, in particular the outflow velocities, are very different. One may suspect that, in other situations where the changes are less radical, such effects would not be detected, but still would affect the determination of $\dot{M}$. This is particularly true for the CO rotational line method which probes a shell of typically 10^{17} cm in size, formed typically within 3000 years. The IR method is less sensitive to this problem because it is the most internal parts of the circumstellar dust shells ($\sim 10^{15}$ cm in size) which dominate the near-IR continuum emission.

A feature common to all methods for deriving mass loss rates from observations is that an integral of the shell structure along the line of sight is involved (e.g., in terms of column densities, optical depths, or integrated emissivities) and therefore, always some radial averaging is implicitly performed. The radial averaging procedure adopted in the present work for determining the mass loss rate thus very much resembles this aspect of the observational methods.

5. COMPARISON WITH OBSERVATIONS

The variety of different behaviors of the models found in our grid makes it tempting to draw some comparisons to actually observed properties of AGB stars:

A-models: This type of models has been extensively discussed in earlier papers (see Winters et al. (2000a) and references therein) and it has been demonstrated that they reproduce quite well different observed properties of dust forming LPVs. Here we only stress that the effect of multiperiodicity found in these models offers a possible explanation for some of the irregular light curves found, e.g., from the MACHO survey (see Wood et al. (1999) and Wood & Cohen, this volume).

B-models: From the ISOGAL survey, Omont et al. (1999) discovered a population of dust forming sources in the Galactic Bulge with very low mass loss rates, in the range $\dot{M} \approx 10^{-9} M_{\odot}\mathrm{yr}^{-1} \ldots 10^{-7} M_{\odot}\mathrm{yr}^{-1}$. These objects are believed to be in an evolutionary state on the early AGB or near the tip of the Red Giant Branch. The typical luminosity of these sources is relatively low ($L_* \sim 1400 - 3200 L_{\odot}$). The low mass loss rates that we find for the B-type models, which emerge, e.g. for low luminosities, could provide a paradigm for the mass losing low luminosity giants in the Galactic Bulge.

Young (1995) performed a volume limited (d $\leq$ 500 pc) survey of Miras. For most sources which have been detected, the CO lines can be fitted with a one–component wind. Several sources have expansion velocities $\leq$ 5 km s^{-1}, and one, T Ari, has a $v_{\rm exp}$ = 2.7 km s^{-1}, typical of our B models. Furthermore, Kerschbaum et al. (1996) measured $v_{\rm exp}$ = 1.9 km s^{-1} for the Semiregular variable L^2 Pup, a value again typical of the B-models. In a survey of 45 AGB stars observed in CO with good spectral resolution and good signal-to-noise ratio, Knapp et al. (1998) found in 9 cases evidence for the presence of 2 winds with different expansion velocities. In general, the slow wind has an expansion velocity smaller than 5 km s^{-1} and a mass loss rate smaller than 5 10^{-7} M$_{\odot}$ yr^{-1}, and the opposite for the fast wind. It is tempting to assign the slow wind to the B-type and the fast wind to the A-type models. The presence of winds with very different properties, simultaneously around the same star, can be understood in the framework of our model where a slight change in the stellar parameters may induce a switch from the A to the B mode, or vice versa.

AB-models: One might speculate that the long-term variations of the mass loss rate resulting from the AB-type models could lead to a layered density structure in the extended circumstellar shell which resembles the characteristics of the multiple layers recently found in some carbon-rich proto-planetary nebulae (e.g., Sahai et al. (1998) and Sahai, this volume) but also in the shell of IRC+10216 (Mauron & Huggins 1999, 2000). Although the outflow velocities of our present AB-models are almost constant in time as would be required for such an interpretation, in the present 3 models it is only about 5 km s^{-1}, thus a factor of 3 lower than the outflow velocity measured, e.g., for IRC+10216.

Transition from region A to B: As mentioned in the previous sections, a slight change in the stellar parameters (e.g., the stellar luminosity, or the temperature, or a combination of both) may induce a switch of the model from the A-type to the B-type and vice versa. It is exactly this effect, which leads to the abrupt onset and switch-off of the short ($\approx$ 1000 yr) high mass loss rate episode in our evolution calculations for low initial-mass stars (Schröder et al. 1999).

In this case the switch is mainly induced by the luminosity variation following a thermal pulse. This behavior might provide a mechanism to explain the formation of detached shells observed, e.g., around R Scl (see previous Sect.) or around TT Cyg (Olofsson et al. 2000). It is worth stressing that in the context of an excursion in the HR diagram, our models imply that the star is again in the B-mode, after the shell has been expelled. Therefore, we would expect a low present-day mass loss rate ($\dot{M} < 3\,10^{-7} M_{\odot}\mathrm{yr}^{-1}$) of the central star, carried by a low outflow velocity wind ($v_{\infty} < 5\,\mathrm{km\,s}^{-1}$). In Table 1 the values observed for 3 of these objects are summarized; they are in perfect agreement with this model prediction.

Table 1 Properties of 3 detached-shell objects

Object	$\dot{M}_{\mathrm{present}}$ $[M_{\odot}\mathrm{yr}^{-1}]$	Ref.	$v_{\infty,\mathrm{present}}$ $[\mathrm{km\,s}^{-1}]$	Ref.
R Scl	$\sim 10^{-7}$	(1)	?	-
TT Cyg	$3\,10^{-8}$	(2)	3.8	(2)
S Sct	?	-	4.8	(3)

References: (1) Le Bertre (1997), (2) Olofsson et al. (2000), (3) Olofsson et al. (1996)

Acknowledgments

The calculations were performed on the Cray parallel and vector computers of the Konrad--Zuse-Zentrum für Informationstechnik Berlin. Subsequent data analysis has been performed on the WAP–cluster of the physics department of the TU–Berlin. This work has been supported by the BMBF (grant 05 3BT13A 6) and by the PROCOPE program under grants D/9822849 (DAAD) and F/99001 (MAE).

References

Dominik C., Gail H.-P., Sedlmayr E., Winters J.M., 1990, *A&A* 240, 365

Fleischer A.J., Gauger A., Sedlmayr E., 1995, *A&A* 297, 543

Gail H.-P., Sedlmayr E., 1987, *A&A* 177, 186

Groenewegen M.A.T., van der Veen W.E.C.J., Matthews H.E., 1998, *A&A* 338, 491

Keady J.J., Ridgway S.T., 1993, *ApJ* 406, 199

Kerschbaum F., Olofsson H., Hron J., 1996, *A&A* 311, 273

Knapp G., Young K., Lee E., Jorissen A., 1998, *ApJS* 117, 209

Le Bertre T., 1997, *A&A* 324, 1059

Loup C., Forveille T., Omont A., Paul J.F., 1993, *A&AS* 99, 291

Mauron N., Huggins P.J., 1999, *A&A* 349, 203
Mauron N., Huggins P.J., 2000, *A&A* 359, 707
Olofsson H., Bergman P., Eriksson K., Gustafsson B., 1996, *A&A* 311, 587
Olofsson H., Bergman P., Lucas R., et al., 2000, *A&A* 353, 583
Omont A., Ganesh S., Alard C., et al., 1999, *A&A* 348, 755
Osterbart R., Balega Y.Y., Blöcker T., Men'shchikov A.B., Weigelt G., 2000, *A&A* 357, 169
Sada P.A.V., 1993, *Temporal variations in the circumstellar shell IRC+10216*, PhD thesis, New Mexico State University, Las Cruces, USA
Sahai R., 1987, *ApJ* 318, 809
Sahai R., Trauger J.T., Watson K.R., et al., 1998, *ApJ* 493, 301
Schröder K.-P., Winters J.M., Sedlmayr E., 1999, *A&A* 349, 898
Truong-Bach, Morris D., Nguyen-Q-Rieu, 1991, *A&A* 249, 435
Tuthill P., Monnier J., Danchi W., Lopez B., 2000, *ApJ*, in press
Winters J.M., Le Bertre T., Jeong K.S., Helling Ch., Sedlmayr E., 2000a, *A&A* 361, 641
Winters J.M., Keady J.J., Gauger A., Sada P.V., 2000b, *A&A* 359, 651
Wood P.R., Alcock C., Allsman R.A., et al., 1999, In Le Bertre T., Lèbre A., Waelkens C., editors, *IAU Symp. 191: AGB stars*, p. 151
Young K., 1995, *ApJ* 445, 872

PULSATIONS IN PROTO-PLANETARY NEBULAE AS SEEN FROM LIGHT CURVE STUDIES

Bruce J. Hrivnak et al.
Dept. of Physics & Astronomy, Valparaiso University, U.S.A.
bruce.hrivnak@valpo.edu

Abstract We have carried out a monitoring program of ~40 proto-planetary nebula (PPN) candidates over the past seven years to investigate light variability. Almost all are found to vary in light, with typical amplitudes of 0.15−0.40 mag. Periods are found for several of the PPNs, in the range of 50−150 d. However, most are non-periodic or perhaps have multiple periods. Even those with periods are found to vary in amplitude. An inverse correlation is found between the time scale of variability and the temperature. These variations are attributed to pulsations.

1. INTRODUCTION

We have been carrying out a program for the past ten years to study variability in proto-planetary nebulae (PPNs). This was initially begun by monitoring radial velocities of PPNs and has been carried on more recently by monitoring light variations in a larger sample of PPNs.

Why study variability? Such a program can provide an additional source of information about the nature of these objects. Variability can arise from either binarity or pulsations. Suggestions of binarity in the central stars of planetary nebulae (PNs) and PPNs have arisen because many of these have a bipolar nebulosity, and a binary companion has been shown to be an effective mechanism to produce such a morphology (Livio & Soker 1988). Some central stars of PNs are known to be binaries with short periods (P<3 d; Bond 2000); by studying PPNs with F−G spectral type stars and thus many sharp spectral lines, one can investigate longer-period binaries. Thus a study of binarity can be a useful observational constraint on suggested mechanisms to shape the nebulae. It is known that some post-AGB stars vary in light due to pulsations, such as RV Tau stars (F−K, P=50−150 d) and UU Her stars (F, P=40−100 d). From the pulsations we can hope to learn about the physical properties of the stars. While variability has previously been noted in a few individual PPNs, there has been no previous study of variability in a sample of PPNs.

R. Szczerba and S.K. Górny (eds.), Post-AGB Objects as a Phase of Stellar Evolution, 101–104.

2. LIGHT CURVE STUDY

A light curve monitoring program of ~40 PPN candidates is being carried out. This was begun in 1994 at the Valparaiso University Observatory (0.4-m telescope and CCD camera) and is continuing through the present time, primarily with the assistance of undergraduate students. Observations are made through the V filter to monitor light variations and occasionally through the R filter to investigate color variability. The precision of the differential measurements is $\leq$0.01 mag, except in the faintest cases. The PPNs range in brightness from 7 to 14 mag and in spectral type from K to O.

The following preliminary results have been found from this light curve study. (1) Almost all of the PPN candidates vary in light, with a typical range of 0.15 to 0.40 mag. (2) Periodic variability was found in seven PPNs, with periods ranging from 45 to 150 d (see Figs. 1, 2).

These have spectral types of F and G. (3) Even in those with periods, the amplitudes vary. (4) Short time-scale variability (<10 d) was found in many of the objects, especially those of early spectral types, O–B. (5) Several have unusual light curves, such as the appearance of a periodic variation in one season and not the next or an unusual shape to the light curve (see Fig. 3).

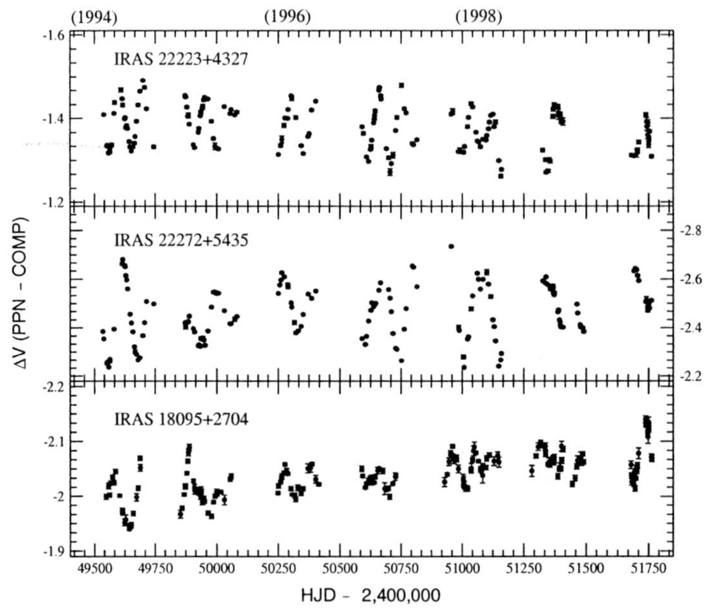

Figure 1 Differential V light curves for periodic variables. The periods are as follows: IRAS 22223+4327, 89 d; IRAS 22272+5435, 133 d; IRAS 18095+2704, 97 d.

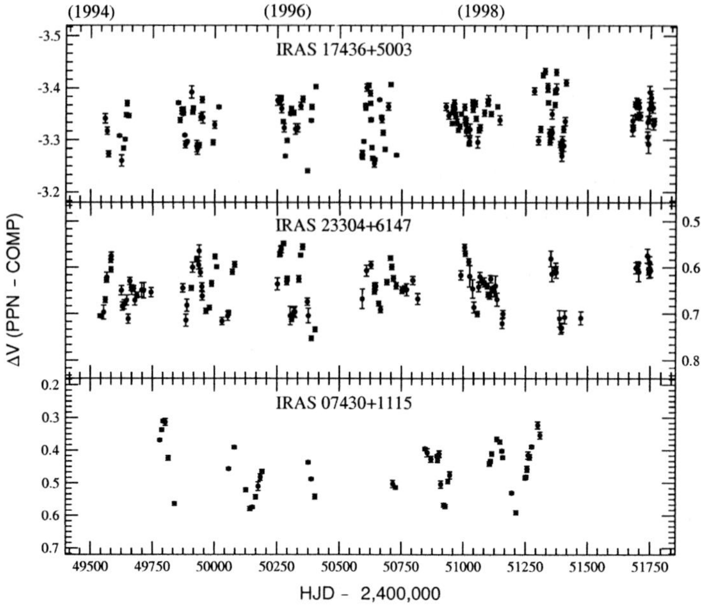

Figure 2 Differential V light curves for periodic variables. The periods are as follows: IRAS 17436+5003, 52 & 45 d; IRAS 23304+6147, 84 d; IRAS 07430+1115, ~145 d.

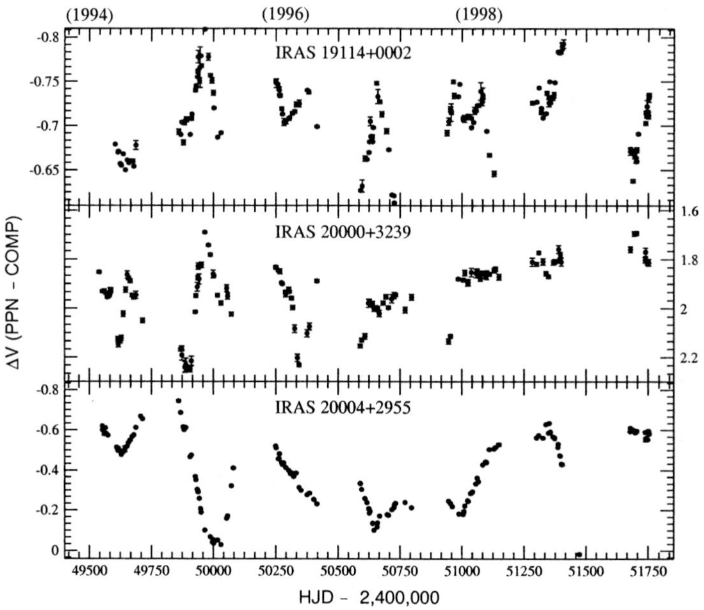

Figure 3 Differential V light curves for three PPNs with non-periodic variability: IRAS 19114+0002, IRAS 20000+3239, IRAS 20004+2955.

3. RADIAL VELOCITY STUDY

A radial velocity monitoring program was carried out from 1991 to 1994 at the Dominion Astrophysical Observatory, in collaboration with S. Morris and A. Woodsworth. The 1.2-m Coudé telescope was used with the radial velocity spectrometer. Nine of the brightest PPNs (V=7−10 mag), with spectral types F and G, were observed approximately 35 times each. The typical precision of the observations was 0.65 km s^{-1}.

The preliminary results for this radial velocity study are as follows. (1) All nine vary in velocity, with a peak-to-peak variation of ~10 km s^{-1}. (2) Three are periodic: IRAS 18095+2704, P=109 d; IRAS 22223+4327, 89 d; IRAS 22272+5435, 127 d. (3) The periodic ones have a semi-amplitude of ~2.5 km s^{-1}, but the amplitudes vary. More details of the radial velocity study are given by Hrivnak & Lu (2000).

4. CONCLUSIONS

- (Almost) all PPNs vary in light and velocity.
- The nature of the variability indicates that it is due to pulsation.
- No binaries were discovered. (However, there may be selection effects involved here.)
- The variability does not have a simple periodic form; there appear to be varying amplitudes, varying periods, or multiple periods.
- For the three cases with both light and velocity periodicity, the light and velocity periods are approximately the same.
- There appears to be a general trend of longer-term variability associated with later spectral types and shorter-term variability with earlier spectral types.

Acknowledgments

Support for this research has been provided by the NSF (AST-9315107 AST-9900346) and by NASA through the JOVE program (NAG8-232).

References

Bond, H. 2000, in Asymmetrical Planetary Nebulae II: From Origins to Microstructures, eds. J.H. Kastner, N. Soker, & S. Rappaport (San Francisco: ASP), 115

Hrivnak, B.J., & Lu, W. 2000, in IAU Symp 177, The Carbon Star Phenomenon, ed. R.F. Wing, (Dordrecht: Kluwer), 293

Livio, M., & Soker, N. 1988, ApJ, 329, 764

ATMOSPHERIC DYNAMICS OF POST-AGB STARS: NEW RESULTS ON THE NATURE OF THE PULSATION OF HD 56126

A. Lèbre[1], A. Fokin[2], D. Barthès[3], D. Gillet[4], N. Mauron[1]
[1]*University of Montpellier, GRAAL, France*
[2]*Institute of Astronomy, Academy of Sciences, Moscow, Russia and visiting scientist at University of Montpellier, GRAAL, France*
[3]*University of Barcelona, Spain*
[4]*Haute-Provence Observatory, LAM, France*
contact: lebre@graal.univ-montp2.fr

Abstract We present a complete analysis of spectroscopic and photometric observations of HD 56126, a variable carbon-rich post-AGB star. In order to improve the knowledge of its pulsation and of its fundamental parameters, we have computed a set of linear and non-linear pulsating models relevant to the post-AGB stars. From two different observational data sets, we find a main pulsating period $P = 36.8 \pm 0.2$ days. The spectroscopic and photometric variations we analysed appear to be consistent with a radial pulsation (the dominant pulsation mode is the first overtone) yielding shock waves which propagate throughout the atmosphere and generate a complex, asynchronous motion of the outer layers.

1. INTRODUCTION

HD 56126 (= SAO 96709 = IRAS 07134+1005) is a F5Iab carbon-rich post-AGB star. A lot of observations have already been accumulated on this Pop. II object, reporting a metal deficiency, an excess of s-elements in its chemical composition, a strong infrared excess due to a dust shell (with a torus like shape of the envelope) and the detection of the famous 21 μm and 30 μm emission features (see Lèbre et al. 1996 and references therein).

There is no convincing evidence that this star is a binary like several other post-AGB stars, but binarity cannot be definitely ruled out. Some of the fundamental parameters of HD 56126 are still not well determined and suffer from a relatively large uncertainty (as $\log g$, $T_{\rm eff}$), or they are intrinsically variable (or both). In particular, the luminosity (or the distance), and the mass are not well determined, and studying the pulsation properties may help to narrow the possible range of these parameters.

R. Szczerba and S.K. Górny (eds.), Post-AGB Objects as a Phase of Stellar Evolution, 105–110.

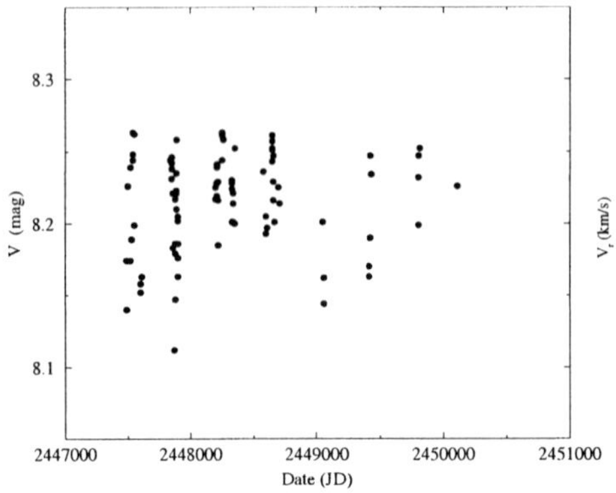

Figure 1 Photometric measurements.

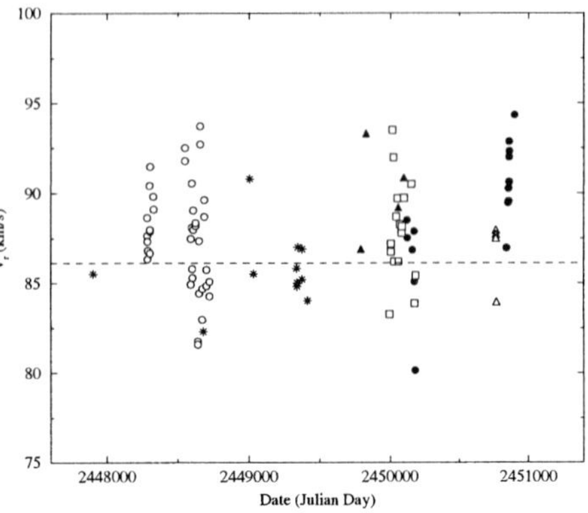

Figure 2 Photospheric radial velocities.

In a previous work (Lèbre et al. 1996) we presented the results of a first monitoring of the photospheric radial velocity (hereafter RV). Spectroscopic variations of the Hα and NaD line profiles were established and also investigated. For metallic lines, RV variations were reported (ΔVr $\sim$ 10–15 km s^{-1}). These data first suggested that strong atmospheric motions were occurring, possibly with shock wave propagation (amplitude $\sim$ 40–47 kms^{-1}). By means of Fourier and CLEAN analysis of the RV time series, we also derived several possible periods of pulsation, the main period being around 27 days. At the same time, Bogaert (1994) established the photometric variability of HD 56126, from the small and irregular variations ($\Delta V = 0.06 - 0.15$) observed in the photometric light curve[1] displayed in Fig 1 (the typical uncertainty on V magnitude is about 0.005 mag.). A period ($\sim$ 40d) was also proposed by Bogaert (1994) from the analysis of these 87 V-band Geneva photometric measurements made over 7 years.

Concerning the RV data, it was obviously necessary to increase both the number of points and the total duration of the series. Thus, we collected new high resolution spectra[2] (regularly taken during 6 months). These new spectroscopic observations confirmed important profile variations of the Hα line and revealed that this line was always exhibiting one or two emission wings. Profile variations of the Hβ line were also presented for the first time (Barthès et al. 2000). These variations could be significant over time intervals as short as $\sim$7 days. Moreover, complementary data (spectra and RV points) were kindly

[1]Based on photometric data obtained with the 70cm Swiss Telescope at ESO/La Silla.

[2]Based on spectroscopic observations performed at the Observatoire de Haute-Provence (France) and at E.S.O. (La Silla, Chile).

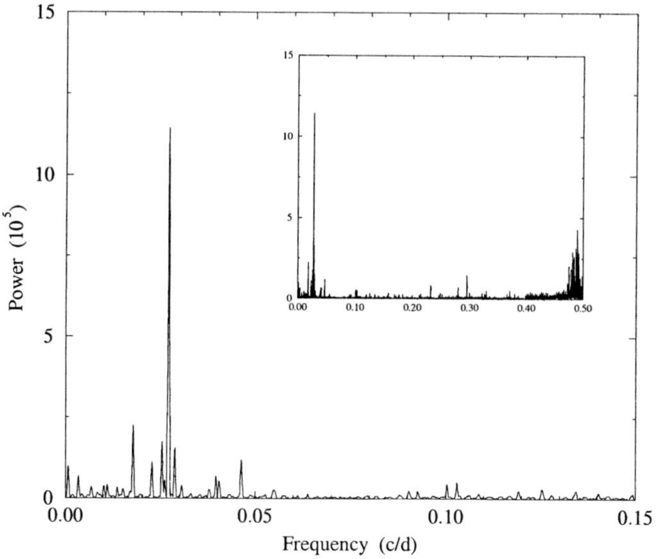

Figure 3 CLEAN spectrum of V mag.

provided by Dr. E. Bakker. The resulting set of heliocentric RV data, spanning ~ 8 years, is shown in Fig 2. The horizontal dashed line indicates the central radial velocity of molecular circumstellar profiles ($V_{helio} = 86.1 \pm 1.3$ km s^{-1}). Different symbols stand for different contributions to the data set (see Barthès et al. 2000) and the typical accuracy on RV data is ≤ 1 km s^{-1}.

2. PERIOD SEARCHING AND LINEAR NONADIABATIC ANALYSIS

In order to determine the pulsation period(s) of the star, we have analysed these photometric and spectroscopic data using spectral and CLEAN-based methods. The CLEAN spectra for photometric and spectroscopic data are displayed in Fig 3 and Fig 4, respectively. In spite of irregular sampling and seasonal gaps in both data sets, a very clear main period of about 36.8 ± 0.2 days is found (see the remarkable coincidence of the main component location). Weaker secondary oscillations, probably related to the irregularity of the pulsation, are also detected. There is no prominent Fourier component at $\sim 2P$ and the pulsation of HD 56126 is certainly quite different from that of RV Tauri variables.

In order to understand the nature of the pulsation of HD 56126 linear nonadiabatic analysis has been performed (Barthès et al. 2000). The only radial linear mode that fits the 36.8 period is the first overtone. This LNA analysis puts constraints on the stellar fundamental parameters of HD 56126, and taking into account the AGB core mass–luminosity relation, we find : $\log g = 0.38$–0.49,

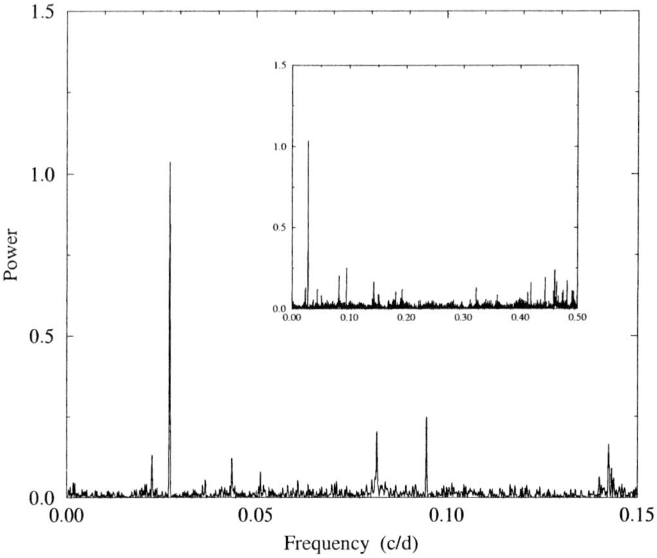

Figure 4 CLEAN spectrum of RV data.

$M = 0.57–0.63\,M_{\odot}$, $T_{\rm eff} = 5700–6100$ K and $L = 5250–8160\,L_{\odot}$. Owing to its significantly higher temperature, HD 56126 is probably more evolved than post-AGB RV Tauri's.

3. NON-LINEAR PULSATING MODELS FOR THE POST-AGB STARS

Our spectroscopic analysis of HD 56126 provided us with precious informations on its atmospheric dynamics and shock waves propagation. Thus, adopting the photospheric parameters we inferred for this object, we performed new non-linear modelling relevant to the post-AGB stars, improving the work already undertaken by Jeannin et al. (1996). Our goal was to reconcile the presence of shock waves with the very small variations detected in the light curve. Indeed the construction of non-linear pulsational model must be able to reproduce the photometric variability (with small and irregular variations of Δ V $\sim$ 0.06–0.15) and the dominant dynamical period ($\sim$ 37 d).
In Figs 5 and 6, we present a new and extended grid of non-linear models for X = 0.7, Z = 0.004 and M = 0.8 $M_{\odot}$. Each model is represented for a given couple (L, $T_{\rm eff}$). In Fig 5, we display the computed bolometric light curves, while in Fig 6 we display their corresponding power spectra. In Fig 6, period value of the main component is indicated in the most interesting models (regarding the pulsational behavior of HD 56126).

In Figs 5 and 6, one can notice the systematic changes occurring, with decreasing L and increasing $T_{\rm eff}$, on the morphology and amplitude of the light

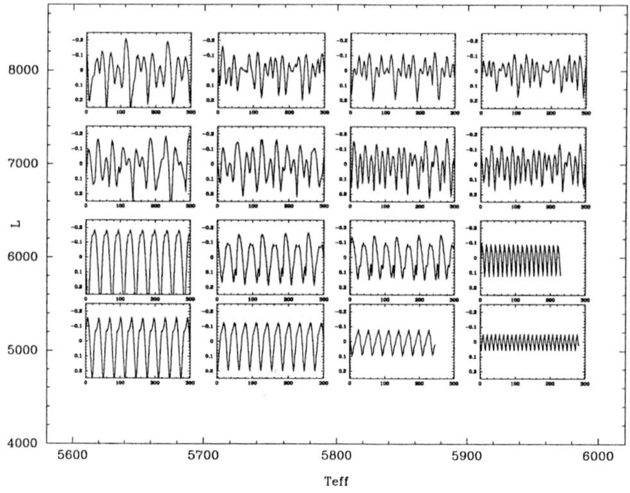

Figure 5 Grid of computed bolometric light curves (X = 0.7, Z = 0.004, M = 0.8 $M_\odot$).

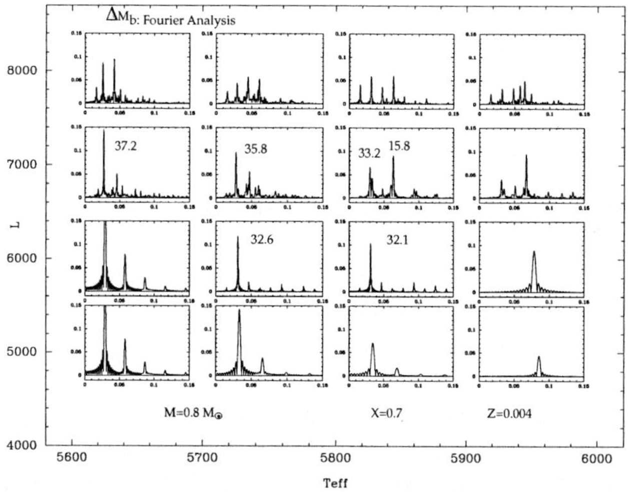

Figure 6 Power spectra of computed light curves (same parameters as in Fig 5).

curve as well as on its associated power spectrum (relative strenght of components and period values). Along those sequences, models evolve from dynamically unstable models to periodic or quasi-periodic models.

We have also computed grids of model series for different chemical composition (X = 0.7 and Z = 0.003) and different mass (M = 0.7 $M_\odot$ and M = 0.6 $M_\odot$). Among all those grids (Fokin et al. 2001, in preparation for A&A) and regarding the pulsational behavior of HD 56126, we find our best models for X = 0.7, Z = 0.004, M = 0.8 $M_\odot$, L = 6000–7000 $L_\odot$ and $T_{eff} \leq 5850$ K. In Fig 7, we

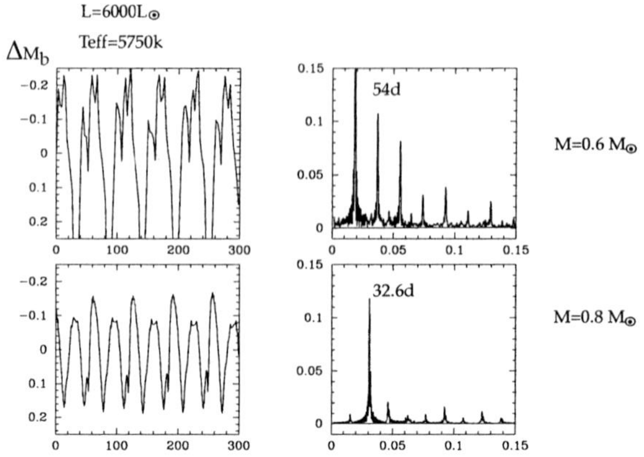

Figure 7 Behavior of the stellar mass parameter (M = $0.6 M_\odot$ and M = $0.8 M_\odot$).

present light curves and associated power spectra for two models (M = $0.6 M_\odot$ and M = $0.8 M_\odot$) relevant to HD 56126 (L = 6000 $L_\odot$ and T_{eff} = 5750 K). The amplitude is drastically increasing with the decrease of the mass, and thus the models computed at M = $0.6 M_\odot$ clearly fail to reproduce the pulsational nature we inferred for HD 56126.

4. SUMMARY

Spectral analysis of HD 56126 has been performed on 89 RV data and 87 V measurements. One main period P = 36.8 d $\pm$ 0.8 d has been found, and weak secondary periods also exist but need to be confirmed. This reflects an irregular pulsation for HD 56126 and the occurrence of complex atmospheric dynamics with shock waves propagation. Non linear modelling has been performed and the best models fitting the observational characteristics of HD 56126 are found for X = 0.7, Z = 0.004, M = $0.8 M_\odot$, L = 6000–7000 $L_\odot$ and $T_{eff} \leq 5850$ K. Models computed with smaller mass clearly fail to reproduce the pulsation nature of HD 56126 which appears to be different from the RV Tauri's one.

Acknowledgments

The authors wish to thank Dr. E. Bakker and Drs. G. Burki and C. Waelkens for providing them with RV data and Geneva photometry of HD 56126.

References

Bogaert E., 1994 Ph.D. Thesis, Leuven University, Belgium
Barthès D., Lèbre A., Gillet D., Mauron N., 2000 A&A 359, 168
Jeannin L., Fokin A., Gillet D., Baraffe I., 1996 A&A 314, L1
Lèbre A., Mauron N., Gillet D., Barthès D., 1996 A&A 310, 923

LIGHT VARIABILITY OF COLOR SELECTED IRAS SOURCES:

VARIABLES IN LATE AGB AND EARLY POST-AGB PHASE

Takahiro Fujii and Yoshikazu Nakada
Institute of Astronomy, University of Tokyo, Mitaka, Tokyo 181-0015, Japan
fujii@ioa.s.u-tokyo.ac.jp, nakada@kiso.ioa.s.u-tokyo.ac.jp

M. Parthasarathy
Indian Institute of Astrophysics, Bangalore - 560034, India
partha@iiap.ernet.in

Abstract The transition from AGB to post-AGB stars was investigated using the H-band monitoring of these objects for four years. The Mira-type pulsations were found only in the restricted area in the IRAS two color diagram. The non-variable sources occupy the region close to the AGB variables. A new class of post-AGB variables was discovered beyond the non-variables. They are probably crossing the instability strip on their rapid evolution to the planetary nebula stage.

1. INTRODUCTION

It is generally believed that the mass-loss from an AGB star is associated with a strong pulsation of its envelope and that a post-AGB star no longer shows light variation caused by a large amplitude pulsation. Non-variable OH sources are thought to be a good sample of post-AGB stars which terminated the pulsation very recently. However, few systematic studies of the last stage of the AGB pulsation has been done for an enough number of IRAS sources. Here we report our four years monitoring of color-selected IRAS sources to determine the termination line of the Mira-type light variation on the IRAS two color diagram.

The mass-loss during the post-AGB phase is another interesting phenomenon for the evolution of a low to intermediate mass star. It is not clear if the post-AGB stars are pulsating or their mass-loss is even connected to the pulsation of their thin envelope.

R. Szczerba and S.K. Górny (eds.), Post-AGB Objects as a Phase of Stellar Evolution, 111–114.

We have extended our light monitoring to the post-AGB stars and found that some stars were actually pulsating with a small amplitude. It looks that the stellar pulsation ceases for a while after the star leaves the AGB but at some point it starts again with a smaller amplitude than their precursors, AGB stars.

2. OBSERVATIONS AND ANALYSIS

A total of 72 targets were selected from the IRAS point source catalog. The selection criteria were 1) $40° < l < 320°$ and $|b| > 2.5°$, 2) $Dec. > -30°$, 3) $F_{12\mu m} > 1.5$ Jy, 4) IRAS quality index is 333Any and 5) IRAS colors inside a thick box in Figure 1.

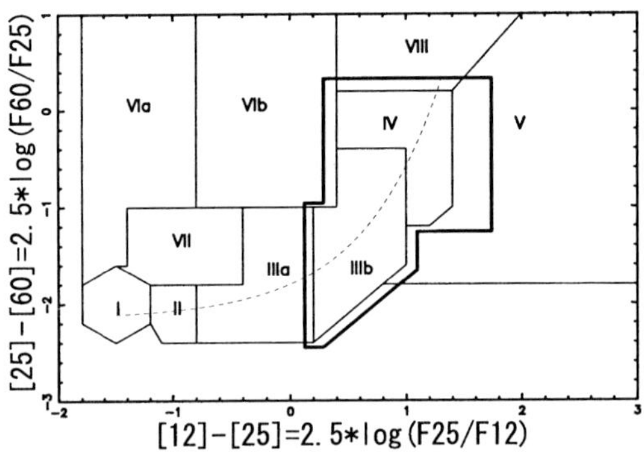

Figure 1 The color range of our samples (thick box).

The photometric observations were carried out in the J, H and K-bands with a near-infrared camera attached to the 105 cm Schmidt telescope at Kiso Observatory, the University of Tokyo. The light variations were monitored by the H-band observations. The H-band images were taken once or twice a month during the observable seasons for two to four years. The H magnitudes were determined relative to the surrounding stars in the same frame with a typical error of 0.03 mag. Out of 72 candidates, the variability was found in 39 objects.

These variables were classified into two groups, the AGB variables and the post-AGB variables except a few objects like a T Tau or an eclipsing binary system. Each group has its own SED and variability: the AGB variables have a single peaked SED, a period longer than 500 days and an amplitude of ~1.0 magnitude, while the post-AGB variables, IRAS 04296+3429, 05113+1347, 07134+1005, 07430+1115 and 22272+5435, have a double peaked SED, a period shorter than 150 days and an amplitude smaller than 0.4 magnitude in the H-band. The variability of some of these stars was reported recently by Barthés et al. (2000) and Hrivnak & Lu (2000).

3. PROPERTY OF VARIABLES

Figure 2 shows the position of the observed objects in the IRAS two-color diagram. The post-AGB variables are clearly separated from the AGB variables. The spectral type of the post-AGB variables in this study are all G – F type. This probably indicates that they are on the instability strip.

Non-variables are located midway between the AGB variables and the post-AGB variables. A simple interpretation of this configuration would be that the pulsation stops at the edge of the variable box leading to the non-variable post-AGB stage but a different type of pulsation starts again when the star enters into some unstable zones. However, there are some non-variables deep inside the area of the AGB variables. It is very difficult to explain the existence of such stars by a simple dust shell model.

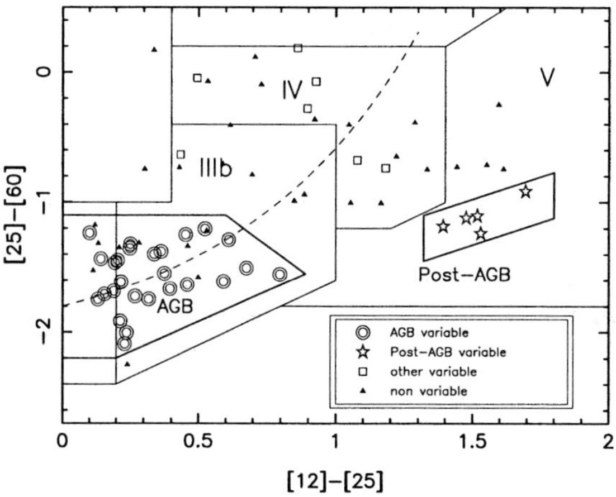

Figure 2 The positions of the variables and non-variables in the IRAS color-color diagram.

It is interesting to compare our post-AGB variables with RV Tau stars which are believed to be in the post-AGB phase. The IRAS colors of the RV Tau stars are confined to a rectangle bounded by [12]-[25]= 1.00 - 0.55, [25]-[60]= -1.8 - -1.0, much bluer than the domain of post-AGB variables in Figure 2.

Figure 3 shows another two color diagram which combined near- and far-infrared magnitudes. The RV Tau stars were also plotted in Figure 3. The near-infrared magnitudes of RV Tau stars were taken from Lloyd Evans (1985) and Goldsmith et al. (1987).

Apparently, the post-AGB variables and the RV Tau stars are separate in the diagram. This separation could be explained from the fact that they have similar spectral types which cause the closeness in $J - H$ while they have different dust shell temperatures which produce the different $H-[25]$.

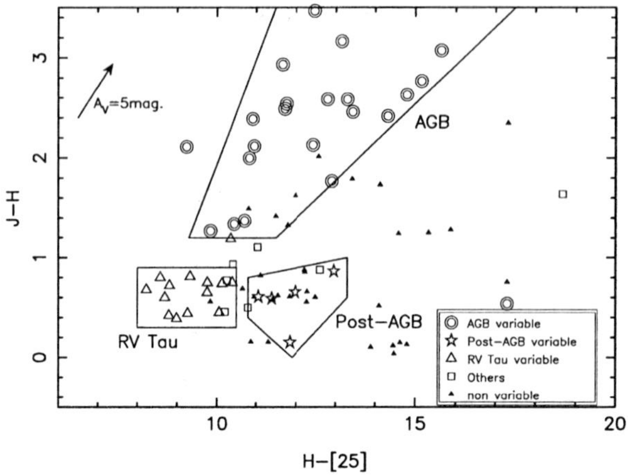

Figure 3 The positions of the variables in the near- and far-infrared color-color diagram. The horizontal axis means $H+2.5\log(\mathrm{F}_{25\mu m})$. The colors of RV Tau stars are cited from the literatures.

Recent studies of RV Tau variables indicate that they are binary systems (Percy 1993, Van Winckel et al. 1999) and they seem to have the same period-color relation similar to that of Type II Cepheids (Pollard et al. 2000). The bluer IRAS colors, i.e. higher dust temperature, of RV Tau stars are probably due to the presence of a circumstellar dusty disk or torus around the central star. On the other hand, the post-AGB variables in this study are perhaps single stars in the instability strip.

References

Barthés, D., Lébre, A., Gillet, D., Mauron, N., 2000, A&A 359, 168

Goldsmith, M.J., Evans, A., Albinson, J.S., Bode, M.F., MNRAS 1987, 227, 143

Hrivnak, B.J., Lu, W., 2000, in *The Carbon Star Phenomenon*, ed. Wing, R.F.,Proc. IAU Symp. 177, p. 293

Lloyd Evans, T., 1985, MNRAS 1985, 493

Percy, J.R., 1993, in *Luminous High-Latitude Stars*, ed. Sasselov, D., ASP Conf. Ser. 45, p.295

Pollard, K.R. et al., 2000, in *The Impact of Large-scale Surveys on Pulsating Star Research IAU Colloquium 176*, eds. Szabados, L. and Kurtz, D.W., ASP Conf. Ser. 203, p. 89

Van Winckel, Waelkens, C., Fernie, J.D., Waters, L.B.F.M., 1999, A&A 343, 202

CIRCUMSTELLAR ARCS IN AGB AND POST-AGB STARS

Sun Kwok, Kate Y.L. Su, and J.A. Stoesz

Department of Physics & Astronomy, University of Calgary, Calgary, Canada T2N 1N4

kwok@iras.ucalgary.ca

1. INTRODUCTION

One of the most unexpected discoveries of the *Hubble Space Telescope (HST)* observations of planetary nebulae (PN) is the revelation of concentric arcs in the nebulae. As of 2000, circumstellar arcs have been found in three PN (NGC 7027, Bond 2000; NGC 6543, Bond 2000, Terzian & Hajian 2000; Hb 5, Balick & Wilson 2000), and three proto-planetary nebulae (PPN) (AFGL 2688, Sahai et al. 1998; IRAS 17150−3224, Kwok et al. 1998; IRAS 17441−2411, Su et al. 1998). Even though some of these nebulae have bipolar shapes, the arcs are remarkably circular. These arcs are likely to have originated from the remnants of the circumstellar envelopes of the asymptotic giant branch (AGB) progenitors, and therefore are manifestations of the mass loss process on the AGB. This hypothesis is confirmed by the ground-based detection of similar arcs in the circumstellar envelopes of the carbon star IRC+10216 Mauron & Huggins 1999. In this paper, we report the detection of circumstellar arcs in the *HST* image of IRC+10216 and summarize the properties of circumstellar arcs observed in PN and PPN.

2. DATA REDUCTION AND ANALYSIS

The *HST* data were retrieved through Canadian Astronomy Data Centre (CADC) *HST* archive. The data were automatically recalibrated at CADC using the latest recalibration software. This process uses the recommended calibration reference files, which may be different from the calibration files used in the original calibration, and produces better calibrated data than the original archived data.

The data usually contain multiple exposures. If the exposures were taken at the same position, we used the task "crrej" in IRAF/STSDAS package to remove the cosmic rays and to combine the exposures into a final image. If

R. Szczerba and S.K. Górny (eds.), Post-AGB Objects as a Phase of Stellar Evolution, 115–120.

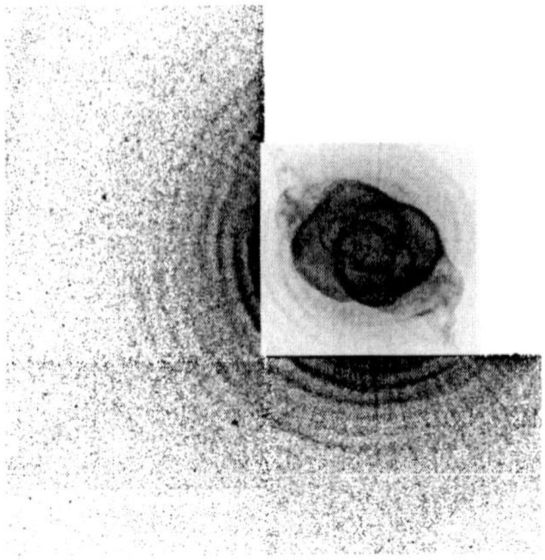

Figure 1 The HST WFPC2 F502N image of NGC 6543. The arcs are circular in shape and are centered at the central star. The main nebula of NGC 6543 is centered on the PC chip, but the arcs can be seen extending over the other three WF chips. Five arcs are detected on the PC image and at least 6 more can be seen in the mosaic image

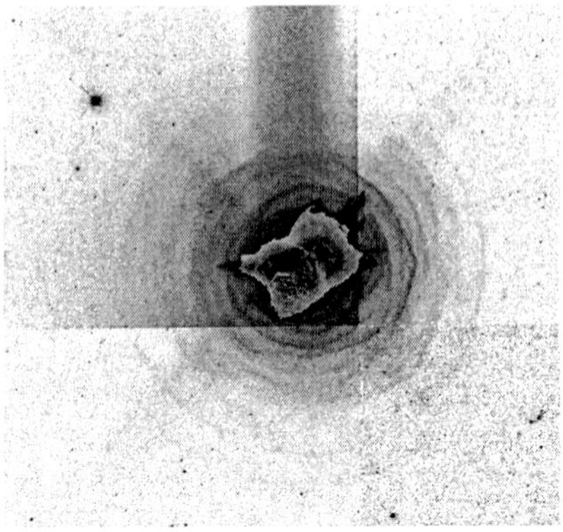

Figure 2 A composite of the HST WFPC2 F814W and F656N images of NGC 7027. Circumstellar arcs can be seen in both the wide band F555W and F814W and narrow-band Hα images of NGC 7027. At least 10 arcs are detected in the Hα image, almost all in the neutral envelope. None of the arcs seem to be centered on the central star.

the exposures were taken centering at different positions, we used DRIZZLE package in IRAF/STSDAS to combine the dithered exposures. If the arcs are extended over more than one chip, we made a mosaic image to view the global structure of the arcs. In order to enhance the detected arc structures, we applied an unsharp mask filter to the image (usually to the one with the longest exposure). The unsharped images are helpful in determining whether the arcs are concentric or intersecting, as well as in making quantitative measurements of the arcs

3. RESULTS

The measured positions of the arcs are summarized in Table 1. Most of the arcs are also resolved in the radial direction, with widths ranging from $0.28''$ in IRAS 17150−3224 to $0.92''$ in NGC 6543. Using the expansion velocities derived from CO observations, the time intervals between the arcs can be estimated. These values, as well as the dynamical age of the outer most arc, are summarized in Table 2.

The processed images of NGC 6543, NGC 7027, Hb 5, IRAS 17150−3224, and AFGL 2688 are given in Figs. 1-5 respectively.

Fig. 6 shows a radial intensity profile of AFGL 2688 integrated over all azimuth angles after the removal of obvious field stars. For a stellar wind

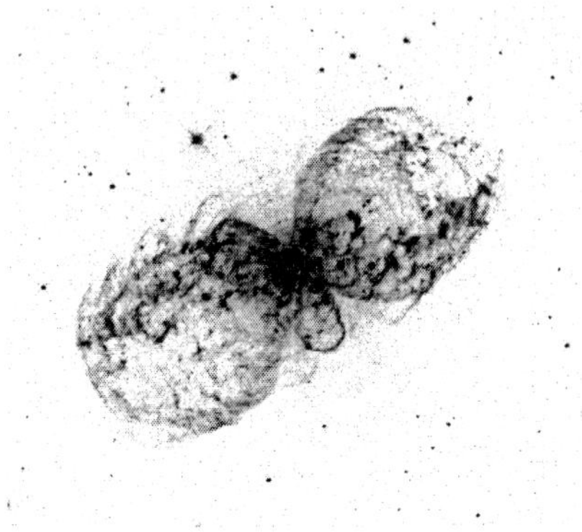

Figure 3 HST WFPC2 F658N image of Hb 5 after having been processed by an unsharp mask. Arcs are seen in F565N (Hα), F658N ([NII]), and F673N ([SII]) images of Hb 5, but not in F502N [OIII] and F631N ([OI]). Arcs are most prominent in the [NII] image, and. at least 6 concentric arcs can be identified.

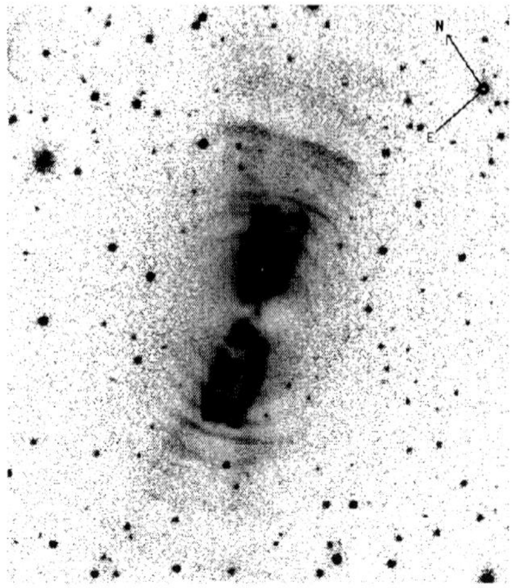

Figure 4 HST 606W image of IRAS 17150-3224 after having processed an unsharp mask. The arcs are brighter in the direction of the bipolar lobes, and can be detected over an angle of $\sim 80°$ on each side. Like AFGL 2688, the arcs lie on almost concentric circles and the spacings between the arcs are very even.

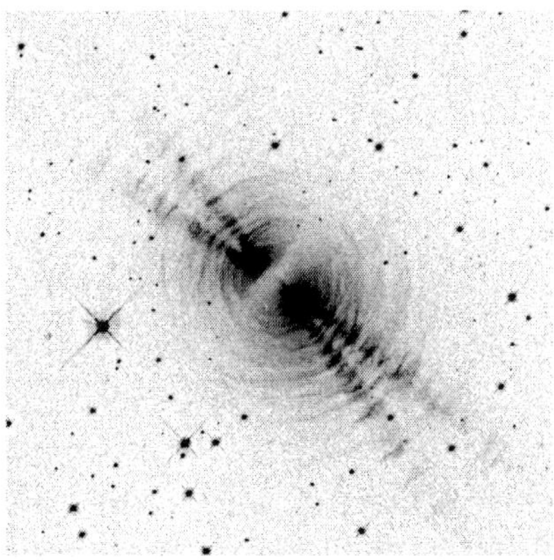

Figure 5 HST F606W image (exposure time 860 s) of AFGL 2688 after having been processed by an unsharp mask. A total of 24 approximately concentric rings centered on the crossing point of the two searchlight beams can be seen.

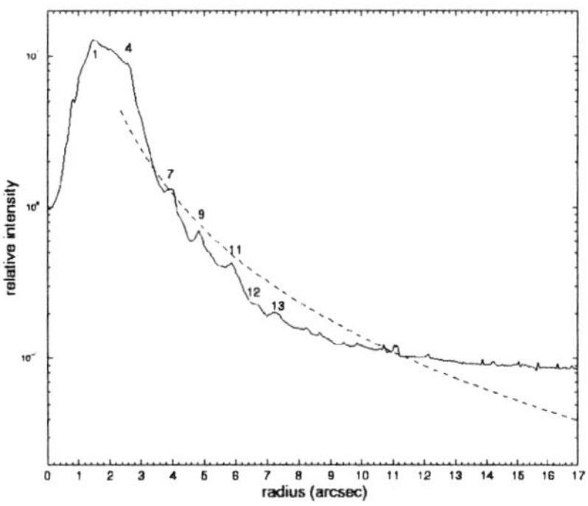

Figure 6 Radial intensity profile (integrated over all azimuth angles) of AFGL 2688. The numbered peaks correspond to positions of the arcs. For comparison, a power law distribution $I \propto p^{-2.4}$ is plotted as a dashed line.

Table 1 Positions (in arcsec) of circumstellar arcs

Arc	AFGL2688	17150−3224	17441−2411	Hb 5	NGC6543	NGC7027
1	1.43	1.31	1.98	2.30	12.51	13.20
2	1.90	1.79	2.58	2.90	14.67	17.50
3	2.25	2.28		3.63	17.43	19.90
4	2.62	2.85		4.92	20.19	22.90
5	3.20	3.34		5.75	22.77	25.20
6	3.55	4.05		6.85	24.90	30.00
7	3.80	4.72			27.70	35.55
8	4.26	5.54			31.50	
9	4.87				35.30	
10	5.28				38.80	
11	5.86				41.60	
12	6.71					
13	7.36					
14	8.26					
15	8.95					
16	9.50					
17	10.14					
18	10.93					
19	11.80					
20	12.90					
21	13.76					
22	14.97					
23	16.90					
24	18.04					
Total	24	8	2	6	11	7

where both the density and the stellar flux are decreasing with radius as r^{-2}, the expected scattered light intensity profile for a spherically symmetric stellar wind is p^{-3} where p is the impact parameter. The observed scattered light intensity profile is consistent with a stellar wind structure.

For 4 of the arcs (7, 9, 11, and 13), the brightness ratio between the arcs and the background can be measured. These ratios are found to be about 1.3. If the brightness is proportional to density, these values suggest that the density enhancements in the arcs are at least 30%. If we include the consideration of projection effects where the observed arc intensity is averaged over intershell

Table 2 Physical properties of the circumstellar arcs

Object	Averaged arc sep (arcsec)	V (km/s)	D (kpc)	Δt (yr)	t_{max} (yr)
IRC+10216	6.85	14	0.2	470	...
AFGL 2688	0.69	20	1	160	4300
IRAS 17150−3224	0.57	10	2	540	5300
IRAS 17441−2411	0.06	15	3	570	2500
Hb 5	0.82	15[a]	1	260	2200
NGC 6543	3.00	15[a]	1[b]	950	13200
NGC 7027	3.63	16	0.7[c]	760	7400

[a] assumed. [b]Reed et al. (2000) [c]Hajian et al. (1993)

regions along the line of sight, the actual density enhancement is likely to be larger than 30%.

The circumstellar dust envelope of IRC+10216 was first detected in scattered light by Crabtree et al. (1987), and recent deep ground-based V and B images by Mauron & Huggins (1999) show the existence of circumstellar arcs in the envelope. We have "drizzled" the 6 dithered F606W WF3 exposures (5×800s+ 1×700s) of IRC+10216 and applied an unsharp mask to the combined image. At least 9 arc segments are clearly delineated resultant image (Fig. 7). Although the arcs appear to be circular in shape, they are clearly not concentric, and none are centered around the position of the central star (Fig. 8). When we tried to fit circles to the arcs, we found that the centers are spread over an area of $9.6'' \times 7.2''$. These arcs are suggestive of a 3-dimensional spiral projected onto the sky.

4. SUMMARY

The best explanation for the circumstellar arcs is that they are projections of spherical shells left from the AGB. The regularity of the arc separations suggests a periodic density enhancement of the circumstellar shell. This could be due to the dust-gas instability in the outflow (Deguchi 1997), or perturbations due to orbital motions of a binary companion Mastrodemos & Morris 1999. The fact that the arcs are highly circular suggests that mass loss on the AGB is largely spherically symmetric, and the transition from AGB to PPN to PN has not completely disrupted these shells. It is therefore plausible that the shaping of PN can be accomplished by collimated outflows over a small solid angle.

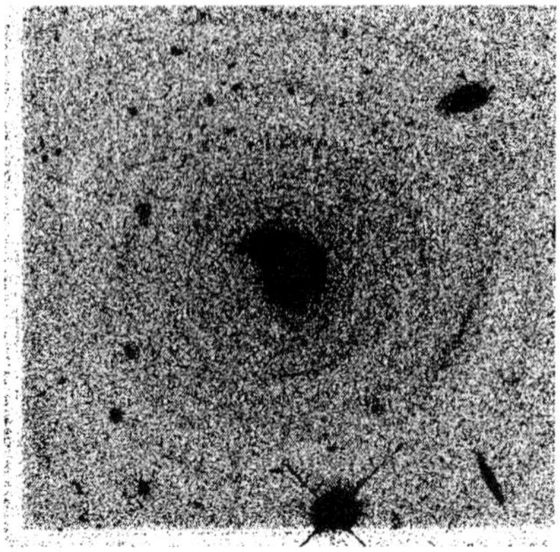

Figure 7 The HST WFPC2 F606W image of IRC+10216 after having been processed by an unsharp mask.

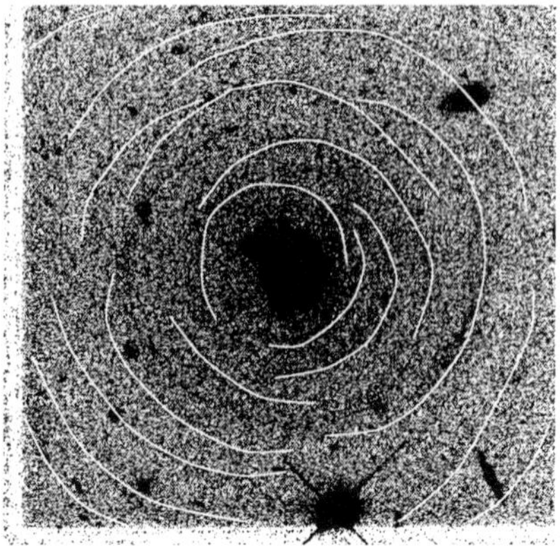

Figure 8 A schematic outline of the arcs in IRC+10216.

References

Balick, B. & Wilson, J. 2000, AJ, in press

Bond, H. 2000, *Asymmetrical Planetary Nebulae II: From Origins to Microstructures*, J.H. Kastner et al. (eds.) ASP Conf. Ser. 199, p. 115

Crabtree, D.R., McLaren, R.A., & Christian, C.A. 1987, *Late Stages of Stellar Evolution*, S. Kwok & S.R. Pottasch (eds.) (Reidel:Dordrecht), p. 145

Deguchi, S. 1997, in *IAU Symp. 180: Planetary Nebulae*, H.J. Habing H.J.G.L.M. Lamers (eds.), Kluwers, p. 151

Hajian, A.R., Terzian, Y., & Bignell, C.R. 1993, AJ, 106, 1965

Kwok, S., Su, K.Y.L., & Hrivnak, B.J. 1998, ApJ, 501, L117

Mastrodemos, N., & Morris, M. 1999, ApJ, 523, 357

Mauron, N., & Huggins, P. J. 1999, A&A, 349, 203

Reed, D.R., Balick, B., Hajian, A.R., Klayton, T.L., Giovanardi, S., Casertano, S., Panagia, N., Terzian, Y. 2000, AJ, in press

Sahai, R. et al. 1998, ApJ, 493, 301

Su, K.Y.L., Volk, K., Kwok, S., & Hrivnak, B.J. 1998, ApJ, 508, 744

Terzian, T. Hajian, A. R. 2000, *Asymmetrical Planetary Nebulae II: From Origins to Microstructures*, J.H. Kastner et al. (eds.) ASP Conf. Ser. 199, p. 33

SIMULATIONS OF QUASI-PERIODIC SHELLS IN DUST FORMING AGB WINDS

Yvonne Simis
Sterrewacht Leiden
Postbus 9513
2300 RA Leiden
The Netherlands
simis@strw.leidenuniv.nl

Vincent Icke
Sterrewacht Leiden
Postbus 9513
2300 RA Leiden
The Netherlands

Carsten Dominik
Sterrenkundig instituut "Anton Pannekoek"
Kruislaan 403
1098 SJ Amsterdam
The Netherlands

Abstract We present a numerical code for time dependent, two-fluid hydrodynamics applied to dust driven winds on the AGB. Our calculations include gas chemistry, grain nucleation and growth, and a new implementation of the viscous momentum transfer between grains and gas, based on microscopic dynamics. This allows us to perform calculations in which no assumptions about the completeness of momentum coupling are made.

Here, we present the first results of the hydro code, which are based on an input model for the object IRC+10216 (Winters et al. 1994). We find that the periodic shells which are observed around IRC+10216 (Mauron & Huggins 1999, 2000) may have a hydrodynamic origin.

R. Szczerba and S.K. Górny (eds.), Post-AGB Objects as a Phase of Stellar Evolution, 121–126.

1. INTRODUCTION

During the last decade, a wealth of small scale structure has been observed in PNe and P-AGB objects. Filaments, FLIERS, knots, blobs and shells are clear indicators of complex physical processes going on. Many theories for the formation of these structures have been developed, quite a number of these require "external" sources, e.g. magnetic fields or (sub)stellar companions. We, however, think there is still much to win by improving hydrodynamical models, in which "smooth" outflow may often be a consequence of assuming e.g. stationary flow or position coupling of gas and grains. Our hydrodynamical calculations apply to AGB winds. We try to find out if, using physically realistic models and in particular two fluid flow, deviations from smooth outflow can arise at the AGB. This turns out to be the case.
It is clear that stationary calculations, in which gas and dust move outward as a single fluid, are no longer sufficient to explain what is observed. The very complex, and non-linear interaction of various physical processes, e.g. chemistry, grain formation and growth, radiation transport and dynamics, should be taken into account when modeling the AGB wind. Various groups have developed codes in which various combinations of the relevant physical processes have been combined. We, for the first time, investigated the interaction of two-fluid flow in time dependent hydrodynamics, combined with chemistry and grain nucleation and growth.

2. MODELING METHOD

From the moment the circumstances in the AGB atmosphere are suitable for the formation of solids, their presence will have large influence on the dynamics. Accelerated grains, due to radiation pressure, will frequently collide with the gas, thereby transferring momentum and driving the typical stellar wind. Hence, a self-consistent numerical simulation of these winds should comprise calculation of the dynamics, (grain)chemistry, radiative processes and gas-grain interactions. With our hydro code, we solve the time-dependent continuity equations for mass and momentum and calculate the composition of the gas and dust using respectively dissociation equilibrium (Dominik 1992) and the moment method for grain nucleation and growth (Gail et al. 1984, Gail & Sedlmayr 1988). Gas and grains are described by their own sets of hydro equations, hence we perform two-fluid hydrodynamics. Matter exchange between both phases is taken into account by the self-consistent implementation of the grain nucleation and growth physics, similar to the moment method. In order to calculate the momentum transfer from grains to gas, due to collisions, gas-grain collisions are studied on a microscopic level. This provides an analytical expression for the behavior of the drift velocity (the relative velocity of grains w.r.t. the gas) as a function of time. Using this expression, one is no longer forced to make assumptions concerning the amount of drift, e.g. assuming equilibrium drift. We perform our calculations on an Eulerian grid, assuming spherical symmetry. Our code is second order accurate and is of the FCT/LCD type (Boris 1976, Icke 1991), using a predictor-corrector mechanism. For a detailed description of the hydro code and the physics involved the reader is referred to Simis et al. (2001).

3. RESULTS

To start the calculation, the stationary profiles for IRC+10216 (Winters et al. 1994) were used. The corresponding stellar parameters will serve as boundary conditions during the time dependent calculation: $M_* = 0.7 M_\odot$, $L_* = 2.4 \cdot 10^4 L_\odot$, $T_* = 2010$K. The carbon to oxygen ratio is $\epsilon_C/\epsilon_O = 1.40$. In order to study the influence of grain drift on the wind, we have carried out three different calculations. First, a single-fluid calculation, in which we assume position coupling (Gilman 1972), hence gas and grain are always assumed to have the same velocity. Second, a calculation in which we imposed equilibrium drift (Dominik 1992), so that both fluids are momentum coupled (Gilman 1972). Finally, we did a full two fluid calculation, using our expression for the momentum transfer rate between grains and gas, which both have their own momentum equation now.

Figure 1 shows the mass loss rate through a sphere at 197 R_* ($R_* = 9.2 \cdot 10^{13} cm$) as a function of time for the three calculations. It is immediately clear that, if one allows (equilibrium) drift, the mass loss rate shows periodic variations which do not appear in the single fluid calculation. The typical time scale for the variability is a few hundred years. This is in agreement with the time scale of the shells that were observed around IRC+10216 by Mauron & Huggins (1999, 2000). Note that we did not include any periodic boundaries (e.g. pistons to mimic stellar pulsation) or other time dependent effects in our calculations. The results are simply obtained by using the stationary initial and boundary conditions (Winters et al. 1994) in the time dependent code, allowing grain drift.

In Figure 2 we give a qualitative comparison between the observed shells (Mauron & Huggins 1999, 2000) around IRC+10216 and our computations. The left frame shows the composite B + V band image, with an average profile subtracted to enhance the contract. This observed intensity is proportional to the dust column density. Therefore, the right frame shows the dust column density of our computation (including drift). One sees that there is a rough resemblance between both images. Note that our research focuses on investigating the physics of AGB wind and that we do not aim to exactly model observations, hence we did not attempt to adjust the stellar parameters to improve the fit. Comparing models with and without drift, we have already concluded that grain

Figure 1 Mass loss rate as a function of time for the position coupled (single fluid) model (upper frame), the momentum coupled (equilibrium drift) model (middle frame) and the free drift model (lower frame).

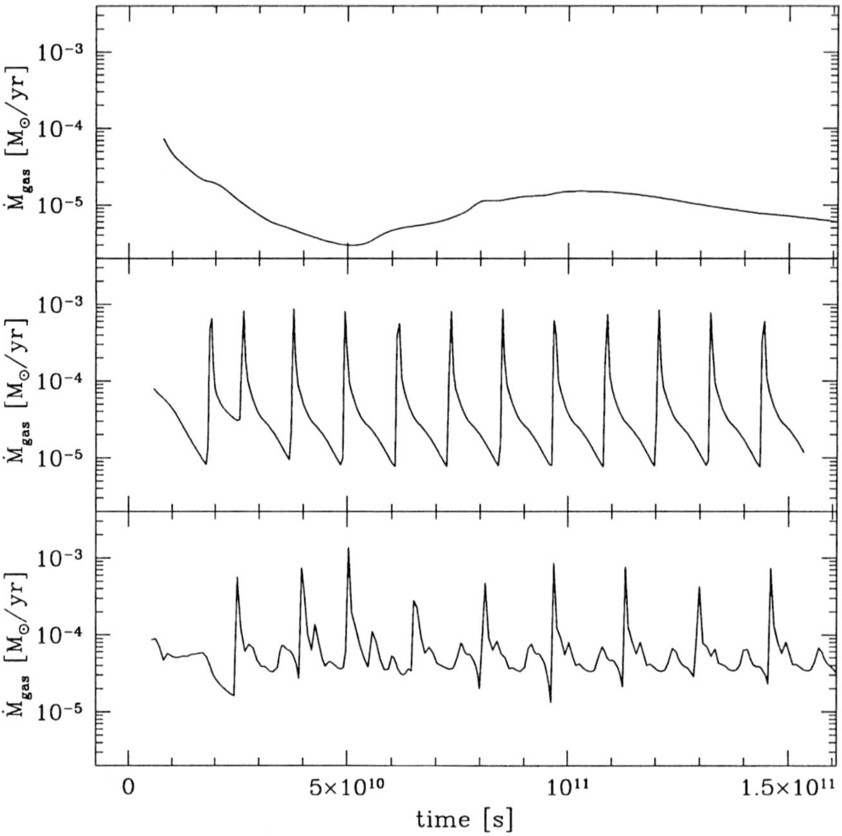

drift is an important ingredient to the formation of the periodic mass loss rate changes. More precisely it is crucial that the momentum input due to radiation pressure on grains is not assumed to directly and completely go to the gas component, as is the case in single fluid, or position coupled, models. The fact that grains can drift means that some momentum can be "stored" in grains. Because the momentum transfer rate (the drag force) from grains to gas is proportional to the square of the drift velocity the transition from inefficient drag to efficient drag (and vice versa) can be quite fast. Also, the sensitivity of the (grain) chemistry to the densities and the hence quick reaction of the grain nucleation rate to changes in the densities, plays an important role. A schematic overview of the interaction of these processes and how they may lead to mass loss variations is

Figure 2 Left: composite B+V image of IRC+10216, an average radial profile is subtracted to enhance the contrast. Right: calculated dust column density, also with an average profile subtracted.

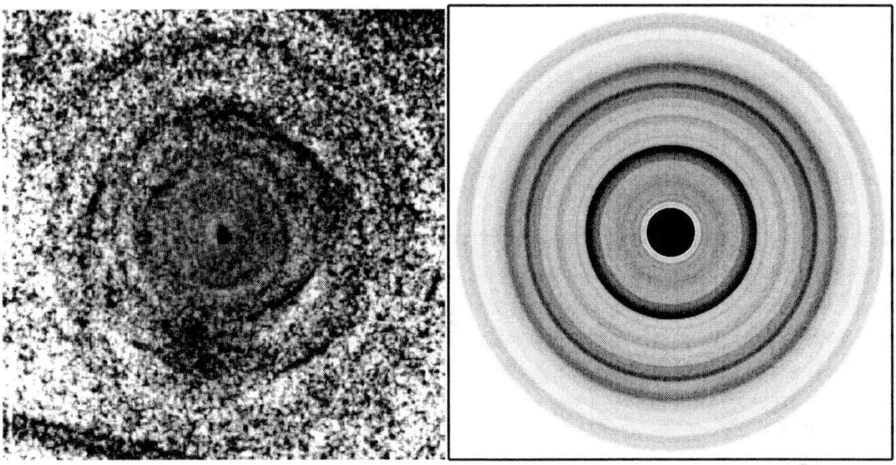

given in Figure 3.

The time scale of the mass loss variations correspond to the time needed by the rarefaction wave to cross the part of the subsonic region in which grain nucleation takes place. This rarefaction wave moves from the sonic point inwards as soon as a shock wave arises due to the sudden increase of the velocity, as a result of the sudden increase in momentum transfer. For a more detailed explanation on the mechanism the reader is referred to Simis et al. (2001).

4. CONCLUSIONS

We found that in two fluid hydrodynamical calculations of AGB winds mass loss variations at a time scale of a few hundred years can appear. These variations lead to the formation of concentric shells which resemble the shells observed around some P-AGB objects and PNe. Although further investigation of the process is needed one can conclude that it seems worthwhile to invest in the improvement of hydrodynamical models, to investigate deviations form smooth flow around late type stars.

Acknowledgments

We thank Jan Martin Winters for providing us with the initial stationary profile for IRC+10216.

Figure 3 Schematic overview of physical processes controlling the periodic changes of the mass loss rate.

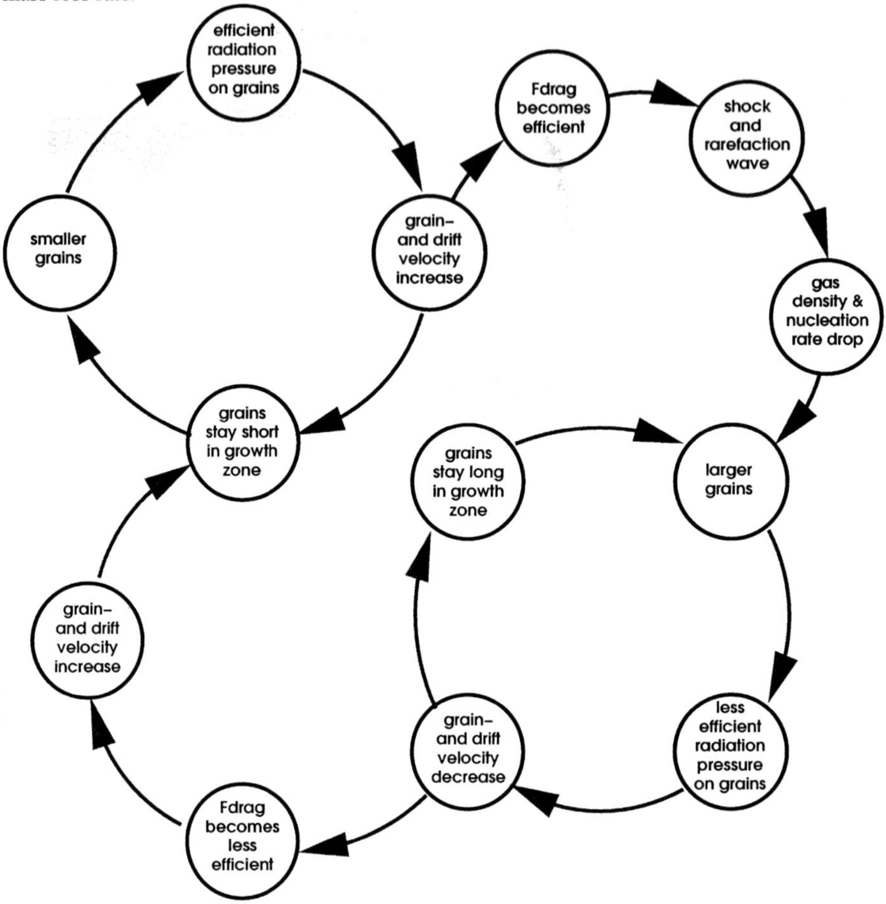

References

Boris J.P., 1976, NRL Mem. Rep., 3237
Dominik C., 1992, Ph.D. thesis, Technischen Universität Berlin
Gail H.-P., Keller R. & Sedlmayr E., 1984, A&A, 133, 320
Gail H.-P., & Sedlmayr E., 1988, A&A, 206, 153
Gilman R.C., 1972, ApJ, 178, 423
Icke V., 1991, A&A, 251, 369
Mauron N. & Huggins P. J., 1999, A&A 349, 203
Mauron N. & Huggins P. J., 2000, A&A 359, 707
Simis Y.J.W., Icke V. & Dominik C., A&A, submitted
Winters J.M., Dominik C. & Sedlmayr E., 1994, A&A, 288, 255

POLARIMETRY AND DIRECT IMAGING OF DETACHED SHELLS AROUND CARBON STARS

David González Delgado and Hans Olofsson
Stockholm Observatory, 133 36 Saltsjöbaden, Sweden
delgado@astro.su.se

Abstract This poster presents polarimetry and direct images in the atomic Na D (λ_1= 5890 Å, λ_2= 5896 Å) and K I (λ= 7699 Å) resonance lines, showing shells around the nearby bright carbon stars R Scl and U Ant. They confirm results obtained in CO radio line observations. A simple analytical model shows that the observed intensities are consistent with resonance line scattering of stellar light provided that about 1% of the K and Na has remained in neutral form.

In the case of R Scl, the direct images, both in K and Na, show an almost perfectly circular disk with essentially uniform intensity out to a sharp outer radius of $\approx$ 21". Crude Wollaston polarimetry also shows that this shell is polarised up to $\approx$ 30%. The K image of U Ant shows a geometrically thin shell of radius $\approx$ 43". However, the Na image suggests the presence of at least two other shells inside the $\approx$ 43" shell. Polarimetry imaging of U Ant still remains to be done.

These narrow shells might be one of the few observable direct consequences of strong and rapid variations in the mass loss rate, induced by thermal pulses, that must have been involved in their origin.

1. INTRODUCTION

Mass loss is the single most important phenomenon that affects the late evolution of low- to intermediate-mass stars on the asymptotic giant branch (AGB), and it is a key process for the chemical evolution of galaxies. The primary observational consequence of mass loss is the formation of a circumstellar envelope (CSE). In a molecular CO mm line study of CSEs around bright carbon stars, Olofsson et al. (1993,1996) found five stars (R Scl, U Cam, U Ant, S Sct, TT Cyg) with clearly detached gas shells whose radii range from $\approx$10" to 70". Interferometric results on U Cam (Lindqvist et al., 1999) and TT Cyg (Olofsson et al., 2000) show that the shells are remarkably narrow, <2", and remarkably spherical. These stars, as well as a few for which reanalyzed IRAS data show dust shells (U Hya, Waters et al., 1994; Y CVn, Izumiura et al., 1996), are the only AGB stars with clear evidence for episodic mass loss. These stars could be the rare examples of the "thermal pulse" induced mass loss eruptions

R. Szczerba and S.K. Górny (eds.), Post-AGB Objects as a Phase of Stellar Evolution, 127–130.

that were proposed by Olofsson et al. 1990, but we cannot exclude effects of interacting winds (i.e., the CO shells being the shock zone). Recent modeling of this phenomenon has been performed by Schröder et al. 1999 and Steffen & Schönberner 2000. However, the molecular radio line emission observations and models have their problems, e.g. uncertainties due to excitation and photodissociation. Hence there is an urgent need for other ways of observing the detached shells.

2. OBSERVATIONS AND DATA REDUCTION

The observations were carried out using the EFOSC1 (0".61 pix^{-1}) and EFOSC2 (0".32 pix^{-1}) focal reducer-type cameras on the ESO 3.6m telescope during two different runs in 1994 and 1999, respectively. We use narrow resonance line filters (Na D and K I), and a coronographic mask to occult the bright stars during the exposures, in order to avoid the photospheric light that, scattered in the earth's atmosphere, would completely swamp the scattered light from the detached envelopes. The observations of a template star provide us with the possibility to remove nearly all the remaining photospheric radiation present in the images despite the use of the mask.

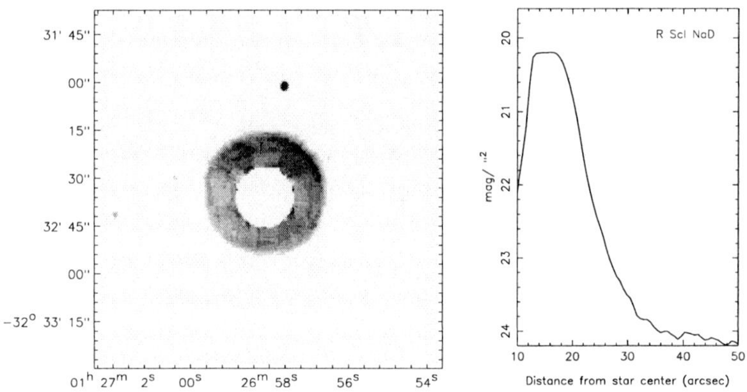

Figure 1 Direct image and azimuthally averaged profile of R Scl in the Na D filter

The reduction of the direct images is done in two steps. First, we perform an ordinary photometric reduction (bias subtraction, flatfield correction, etc.) Second, we subtract the template star to our target star. The stars positions are difficult to locate due to the presence of the mask in the images, but indirect methods are used to centre target and template stars. The template star is then normalised to the target star intensity and smoothed with a median filter. The subtraction of the template star leaves us only with emission stemming from the circumstellar envelopes around the target stars. It should be emphasized that parts of the fine structure in these images are due to bad optics at the telescope and image processing, so we concentrate on the global properties of the emission.

In the case of the polarimetry images, the use of a Wollaston prism makes the reduction somewhat complicate. The field of view is separated in two sets of alternate bands. Each set is observed at two different polarization angles in a same exposure. There is no need for a template star subtraction since the calculation of the Stoke parameters requires the subtraction of images at different polarization angles, which at the same time produces the removal of the unpolarised photospheric light.

The observations shown in this poster were obtained during the first run. This run only aimed at testing whether it was possible to detect resonance line scattered light in large detached envelopes. Standard stars were not observed so an alternative method is used to calibrate the images. Fitting the intensity profile of the template star to an empirical King profile results in an estimate of the mag/square arcsecs in terms of counts/pixel. The observations done during the second run are of much higher S/N-ratio and properly calibrated.

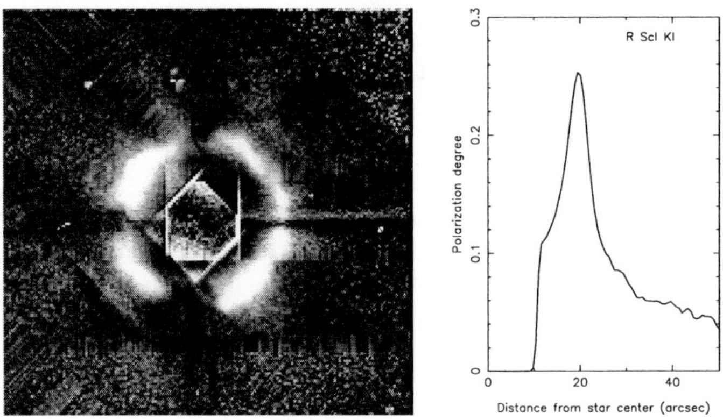

Figure 2 Polarimetry image and azimuthally averaged profile of R Scl in the K I filter

3. RESULTS

The template-subtracted images of R Scl reveal brightness distributions in the form of essentially uniform intensity disks, Fig. 1. The azimuthally-averaged radial profiles are relatively constant within an outer radius of $\approx$ 21" in both resonance filters. The polarimetry images show that this shell is highly polarised ($\approx$30% in the Na D lines and $\approx$25% in the K I line, Fig. 2).

The template-subtracted observations of U Ant in the K I filter show an almost circular ring with a diameter of $\approx$ 85". This is most likely the result of limb brightening in a geometrically thin shell. In the Na D filter observations, we also detected circumstellar emission. However, the morphology is different, in the sense that the brightness stays at nearly constant level much closer to the star, Fig. 3. Higher S/N-ratio images done in this filter suggest a multiple-shell structure.

The comparison between the observed circumstellar-to-stellar ratios and those estimated with an analytical model, appropriate for the analysis of scattering of central stellar light in the optically thin regime in a geometrically thin shell, allow us to draw some conclusions on the optically thickness of the observed shells. In this model, we use that $\approx$1% of the K and Na is in neutral form. Thus, the images of R Scl point towards an optically thick shell in both filters. The observations of U Ant suggest that the Na D lines in this shell are optically thick, while the K I line is at least partly optically thin, and this would explain the different appearances of the shell in these two filters.

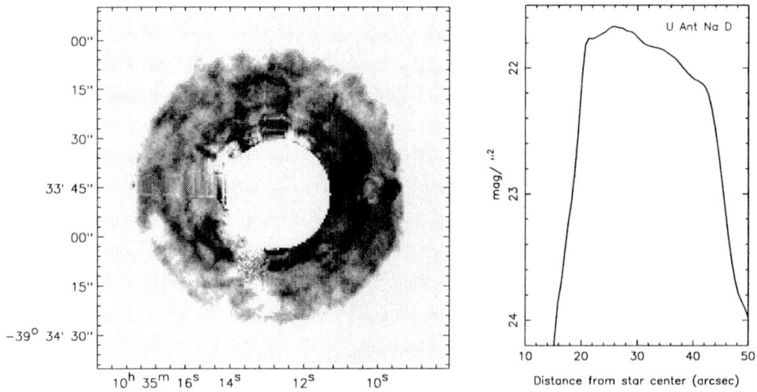

Figure 3 Direct image and azimuthally averaged profile of U Ant in the Na D filter

References

Izumiura H., Hashimoto O., Kawara K., et al., 1996, A&A 315, L221
Lindqvist M., Olofsson H., Lucas R., et al., 1999, A&A 351, L1
Olofsson H., Carlström U., Eriksson K., et al., 1990, A&A 230, L13
Olofsson H., Eriksson K., Gustafsson B., et al., 1993, ApJS 87, 267
Olofsson H., Bergman P., Eriksson K., et al., 1996, A&A 311, 587
Olofsson H., Bergman P., Lucas R., et al., 1996, A&A 353, 583
Schröder K.-P., Winters J.M., Sedlmayr E., 1999, A&A 349, 898
Steffen M., Schönberner D., 2000, A&A 357, 180
Waters L.B.F.M., Loup C., Kester D.J.M., et al., 1994, A&A 281, L1

EVOLUTION OF THIN GAS SHELLS ALONG THE AGB AND BEYOND

Matthias Steffen & Detlef Schönberner
Astrophysikalisches Institut Potsdam
msteffen@aip.de, deschoenberner@aip.de

Abstract Based on numerical simulations with two different computer codes, we have studied the time-dependent hydrodynamics of circumstellar gas/dust shells of AGB stars in their final stages of evolution. We discuss two different mechanisms capable of producing very thin shells of greatly enhanced gas density in the dusty outflows from these stars and illustrate their observable signatures. Our model calculations demonstrate that the thin shells produced on the AGB survive the transition to the post-AGB phase and can still be detected in the haloes of some Planetary Nebulae.

1. INTRODUCTION

It is well known that stellar evolution on the asymptotic giant branch (AGB) is accompanied by substantial mass loss ($\dot{M} \approx 10^{-7} \ldots 10^{-4}$ $M_{\odot}$/yr) through slow winds ($V_{\rm exp} \approx 5 \ldots 25$ km/s) driven by radiation pressure on dust. Observational evidence indicates that the dusty outflows are largely spherically symmetric, but further details, like the time-dependence of the mass loss rate and its dependence on initial stellar mass and metallicity, $\dot{M}(t, M_{\rm ini}, Z)$, are not easily inferred and remain largely unknown.

At the same time, *fully self-consistent theoretical models* of AGB mass loss are not yet available, although considerable progress has already been made for carbon-rich AGB stars (Arndt et al. 1997, AFS97; Winters et al. 2000). Up to now, calculations of AGB stellar evolution are based on *empirical* (e.g. Vassiliadis & Wood 1993, VW93) or *semi-empirical* (e.g. Blöcker 1995, B95) prescriptions of the mass loss rate as a function of stellar parameters. For a given track of a star with $M_{\rm ini} = 3\,M_{\odot}$ due to Blöcker (1995), we have evaluated *a posteriori* different mass loss formulae for comparison in Fig. 1.

While there is a common general trend of increasing mass loss rate with time, the detailed temporal behavior differs considerably: the variation of $\dot{M}$ over a thermal pulse cycle amounts to 1, 2, and 4 orders of magnitude according to the

R. Szczerba and S.K. Górny (eds.), Post-AGB Objects as a Phase of Stellar Evolution, 131–138.

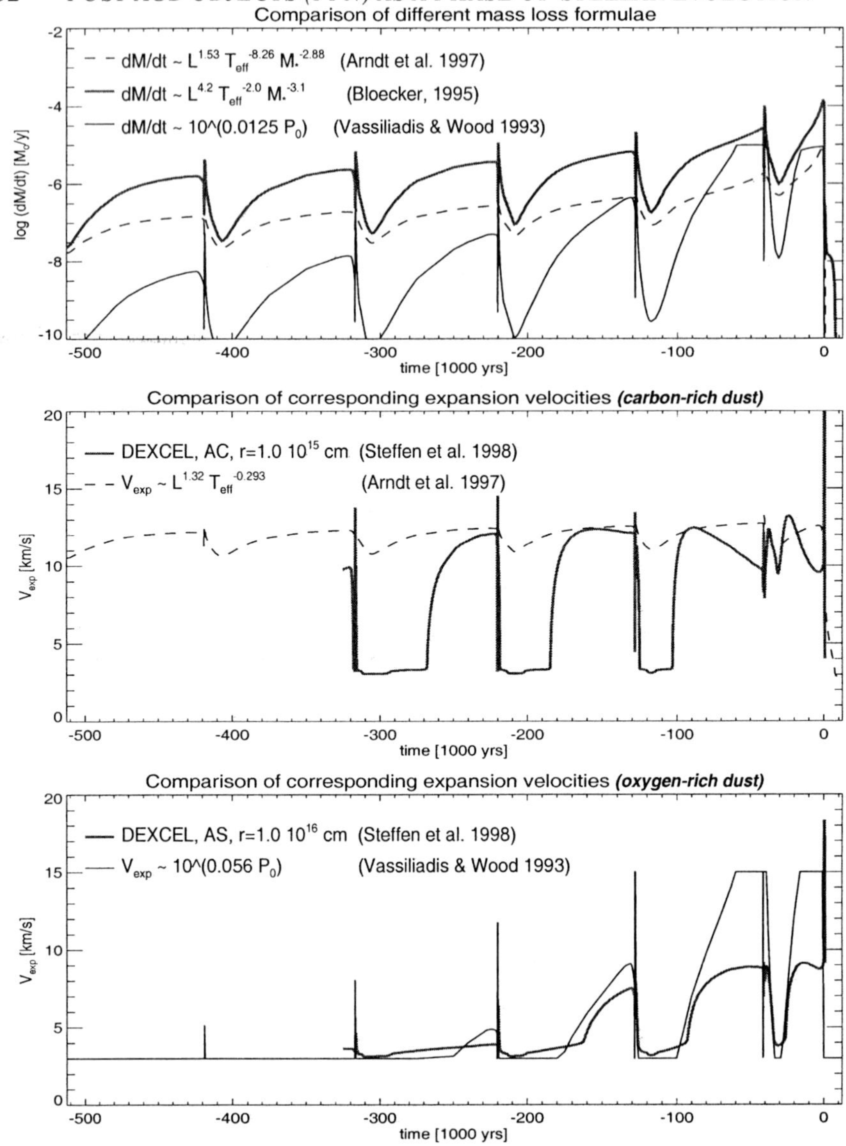

Figure 1 Comparison of different descriptions of AGB mass loss rate ($\dot{M}$, top) and expansion velocity ($V_{\rm exp}$, lower). The semi-empirical mass loss formula by Blöcker (1995) was used to compute the evolutionary track of a 3 $M_\odot$ star. The theoretical relations by Arndt et al. (1997), which are valid only for carbon-rich dust, and the empirical formulae by Vassiliadis & Wood (1993) (with the pulsation period in the fundamental mode $P_0 = 8.5\,10^{-3}\,(R/R_\odot)^{1.94}\,(M/M_\odot)^{-0.9}$) were then applied to this given track to evaluate $\dot{M}$ and $V_{\rm exp}$. Adopting the same Blöcker (1995) track, the expansion velocities by Steffen et al. (1998) have been obtained from two-fluid hydrodynamics, assuming amorphous carbon (middle) and silicate (bottom) dust grains, respectively. The figures cover the final 500 000 years on the AGB.

relations by AFS97, B95, and VW93, respectively. Similarly, the expansion velocity varies by as little as 20% for the AFS97 model and by as much as a factor of 5 for the VW93 description. Of course, the evolutionary track is inconsistent with the AFS97 and VW93 mass loss rates. The following investigation is based on the B95 mass loss formula and corresponding track, a compromise between the two extremes VW93 and AFS97.

2. TIME-DEPENDENT HYDRODYNAMICAL MODELING OF AGB WIND ENVELOPES

A two-component radiation hydrodynamics code, designed to model a stellar wind driven by radiation pressure on dust grains, was employed to simulate the dynamical response of the AGB wind envelope and the emergent dust emission to temporal changes of the stellar luminosity and mass loss rate, using the results of detailed stellar evolution calculations by Blöcker (1995) as a time-dependent inner boundary condition.

A completely independent one-component, second-order Godunov-type hydrodynamics code, which is particularly well suited to resolve shock fronts, was used for verification by recomputing part of the sequences with significantly higher spatial resolution, utilizing the density and velocity obtained with the two-component code at a given radial position $\rho(r_0, t)$ and $V_{\rm exp}(r_0, t)$ as an inner boundary condition. Since the one-component code includes an elaborate treatment of time-dependent ionization/recombination, it can be used to extend the simulations from the AGB into the post-AGB and PN phase of evolution.

Details concerning the numerical modeling are given in Steffen et al. (1998) and Steffen & Schönberner (2000, SS00).

3. FORMATION OF THIN SHELLS ON THE AGB

The numerical simulations outlined above have shown that two-wind interaction can occur in the dusty outflows from AGB stars and is capable of producing very thin shells of greatly enhanced gas density moving outward through the circumstellar wind envelope. We have identified two distinct mechanisms giving rise to two-wind interaction: (i) a mass loss 'eruption' as a result of a helium-shell flash, and (ii) a rapid increase of the outflow velocity after the stellar luminosity and mass loss rate exceed the critical limit for efficient wind acceleration by radiation pressure on dust (for details see SS00).

As an example for the **first scenario**, which seems to work primarily for carbon-rich dust, consider the helium-shell flash at $t \approx$ -317 000 yr (Fig. 1, upper panels). Within a few hundred years, $\dot{M}$ increases by a factor of 30, while at the same time $V_{\rm exp}$ jumps from 3 to 15 km/s. As a consequence of this 'eruption', a thin gas shell of enhanced density is created, moving outward at a faster rate than the matter expelled before. As the shell propagates outward it sweeps up

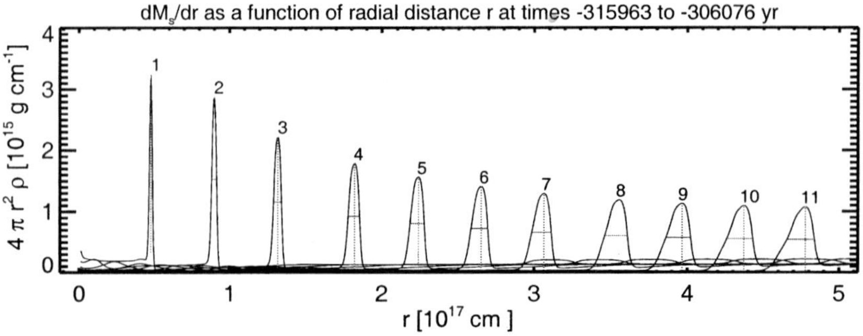

Figure 2 Evolution of a shell of enhanced gas density as a consequence of the helium-shell flash occurring at $t = -317\,000$ yr (see Fig. 1, upper two panels), according to hydrodynamical calculations by Steffen & Schönberner (2000) for *amorphous carbon* dust grains ($a = 0.05\ \mu$m). The quantity shown is $4\pi\, r^2 \rho$, labels '1' to '11' refer to different times, 1 000 to 11 000 years after the flash. The compressed shell moves outward with a velocity of $V_{\rm shell} \approx 13.8$ km/s, the mass inside the shell growing to $M_{\rm shell} \approx 0.01\ {\rm M}_\odot$ during the shown time interval. The *relative* thickness of the shell is approximately constant at $R_{\rm shell}/r \approx 0.04$.

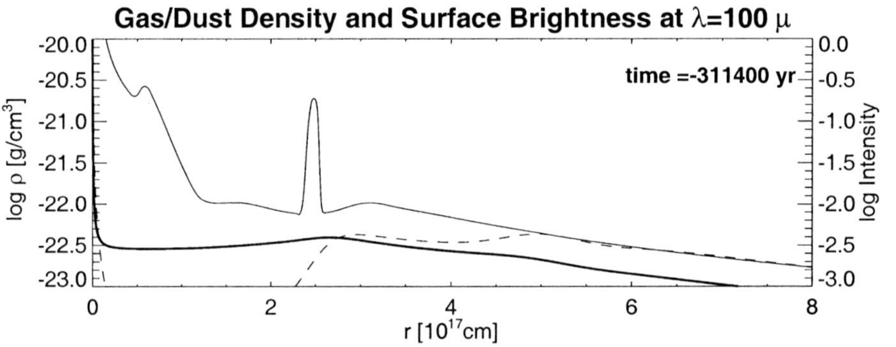

Figure 3 Radial distribution of gas density (thin, left hand scale) and infrared dust emission at $\lambda 100\ \mu$m (thick, right-hand scale), about 5 600 years after the helium-shell flash (between position '5' and '6'). The dust density (multiplied by 1 000, dashed line) clearly indicates the presence of a *detached dust shell*. Note that gas and dust have decoupled in the inner regions due to drift and due to the assumption of incomplete dust condensation when $\dot{M} < 3\,10^{-7}\ {\rm M}_\odot$/yr.

the slower gas and grows in mass ($M_{\rm shell} \approx 0.008\ {\rm M}_\odot$ at $r = 2.5\,10^{17}$cm), while its relative thickness remains almost constant at $R_{\rm shell}/r \approx 0.04$ (see Fig. 2). Since mass loss rate and outflow velocity decrease rapidly soon after the 'eruption', a so-called 'detached dust shell' develops. Fig. 3 shows the situation about 5 600 years after the flash: a very thin gas shell sits at the inner edge of an extended detached dust shell. We propose that the sequence of events described here explains the origin of the geometrically thin CO shells discovered recently around some optically visible carbon stars (Olofsson et al. 2000 and references therein). In particular, the physical parameters derived for the CO shell around the well-studied carbon star TT Cygni agree surprisingly well with the results of the model presented here. Even the computed IRAS colors, at the time corresponding to the observed position of the CO shell ($r = 2.7\,10^{17}$ cm), almost perfectly match the observation!

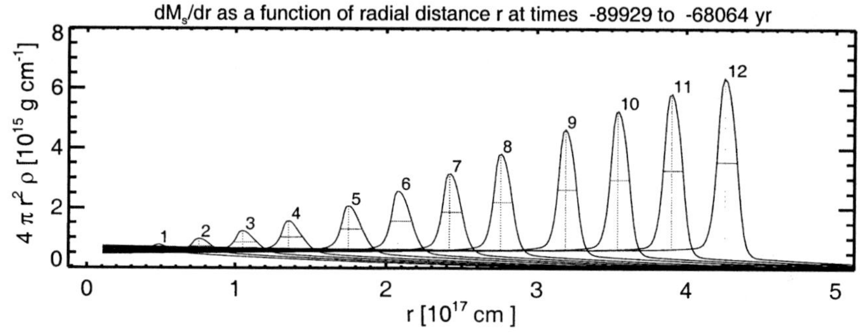

Figure 4 Creation and evolution of a shell of enhanced gas density after the mass loss rate has exceeded the critical limit for the onset of efficient gas/dust coupling, resulting in a faster outflow which interacts with the slower previous wind. In this example the faster wind starts at $t = -90\,000$ yr (see Fig. 1, bottom) according to hydrodynamical calculations by Steffen & Schönberner (2000) for dust grains composed of *astronomical silicates* ($a = 0.05$ μm). Labels '1' to '12' refer to different times, roughly 0 to 22 000 years after the faster wind starts. The compressed shell moves outward with a velocity of $V_{\rm shell} \approx 6.0$ km/s, the mass inside the shell grows to $M_{\rm shell} \approx 0.05$ M$_\odot$ during the shown time interval. The *absolute* thickness of the shell is approximately constant at $R_{\rm shell} \approx 1.5\,10^{16}$ cm.

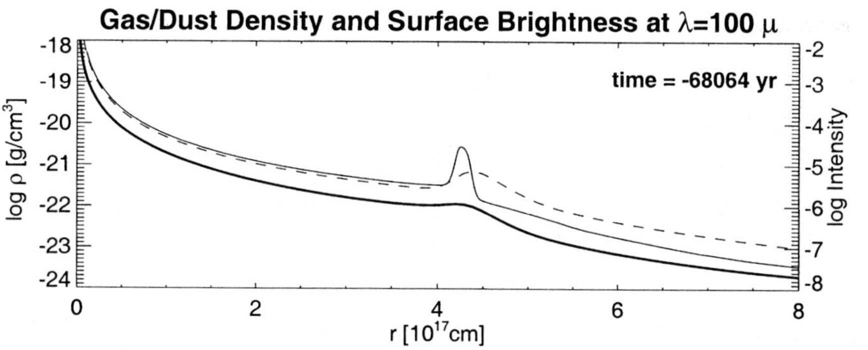

Figure 5 Radial distribution of gas density (thin, left hand scale) and infrared dust emission at λ100 μm (thick, right-hand scale), about 22 000 yr after the onset of the faster, 'dust-driven' outflow. The feature at $r \approx 4\,10^{17}$ cm is the compressed shell seen at time '12' in Fig. 4. The dashed line outlines the dust density (multiplied by 200). Note that gas and dust distributions are shifted somewhat relative to each other because dust moves faster than gas.

According to our simulations, the **second scenario** is realized both for carbon-based and oxygen-based dust. As one out of many possible examples, consider the rapid velocity increase near $t \approx$ -90 000 yr in the oxygen-rich sequence (Fig. 1, lower panel). Even though $\dot{M}$ does not change much, the resulting two-wind interaction creates a compressed gas shell of enhanced density, which propagates outward and sweeps up the slower gas. We find a shell mass of $M_{\rm shell} \approx 0.03$ $M_\odot$ at $r = 2.5\,10^{17}$ cm, considerably larger than in the previous example because of the higher mass loss rates prevailing at this later phase. The shell's absolute geometrical thickness remains almost constant with distance from the star ($R_{\rm shell} \approx 1.5\,10^{16}$ cm, see Fig. 4). Fig. 5 shows the situation 22 000 years after onset of the faster wind: a thin gas shell is located in the middle of an extended attached dusty envelope. The narrow region of enhanced dust density ($r = 4.4\,10^{17}$ cm) corresponds to the compressed gas shell, although it is slightly broader and somewhat shifted due to drift effects.

Our models suggest that the occurrence of thin gas shells of the 'second kind' should be a common phenomenon on the AGB. Yet, they have not been observed so far, although their relative density contrast is comparable to that of the shells believed to correspond to the observed CO shells. Possibly, the much higher densities in and emission from the inner parts of the attached wind envelope makes a detection difficult.

4. POST-AGB AND PN PHASE OF EVOLUTION

At the end of the AGB evolution ($t = 0$), mass loss declines rapidly by orders of magnitude. The transition to the post-AGB phase according to our model sequence with silicate dust is illustrated in Fig. 2 by Schönberner & Steffen (this volume). Note that two narrow peaks stand out in the gas density distribution. The one at $r \approx 2\,10^{18}$ cm can be identified with the compressed gas shell whose origin and properties were discussed in Figs. 4 and 5, the one near $r \approx 5\,10^{17}$ cm is a similar shell related to the sharp velocity increase near $t = -25\,000$ yr. After roughly 1 000 years of post-AGB evolution, the situation is not much different, as shown in Fig. 6: the dusty envelope of the proto planetary nebula (PPN) is still neutral (effective temperature of the central star $T_{\rm eff} = 14\,375$ K), being density bounded by the narrow gas/dust shell near $r = 5.5\,10^{17}$ cm. The infrared emission of the PPN should therefore exhibit a sharp, somewhat limb-brightened outer edge.

After another 4 100 years of evolution, the star's effective temperature has increased to $T_{\rm eff} = 102\,600$ K, and the circumstellar matter is fully ionized out to $r \approx 1.4\,10^{18}$ cm. The dramatic effects of ionization and the impact of the fast stellar wind that accompany the birth and evolution of the Planetary Nebula (PN) are explained in more detail by Schönberner & Steffen (this volume). Here we only point out that, although broadened somewhat by ionization, the compressed shell created during the final thermal pulse cycle on the AGB still exists and constitutes the outer edge of the halo surrounding the PN. In H_α

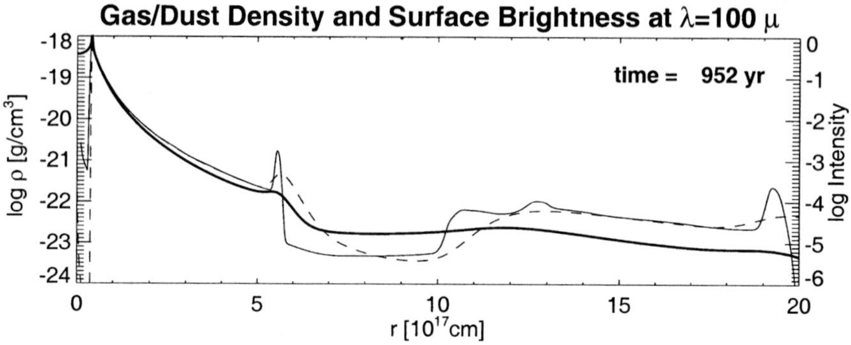

Figure 6 Radial distribution of gas density (thin, left hand scale) and infrared dust emission at $\lambda 100\ \mu$m (thick, right-hand scale), for the same sequence as shown in Fig. 5, about 950 years after the end of the AGB evolution. The feature at $r \approx 2\,10^{18}$ cm is the compressed shell seen in Fig. 4 at $r \approx 4\,10^{17}$ cm. The density peak near $r \approx 5.5\,10^{17}$ cm is the signature of another two-wind interaction which started at $t \approx -25\,000$ yr. The dashed line outlines the dust density distribution (multiplied by 200).

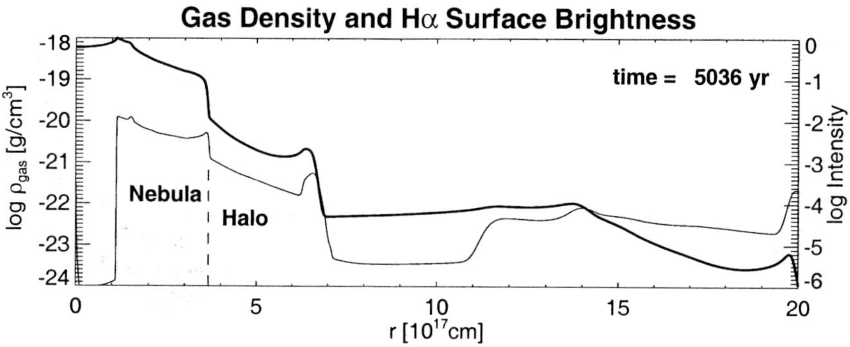

Figure 7 Radial distribution of gas density (thin, left hand scale) and H_α line emission (thick, right-hand scale), for the same sequence as shown before, about 5 000 years after the end of the AGB evolution. The circumstellar matter has been ionized up to $r \approx 1.4\,10^{18}$ cm, forming a *Planetary Nebula* surrounded by a *halo*. The limb brightening of the halo at $r \approx 6.5\,10^{17}$ cm is caused by the compressed shell seen in Fig. 6 at $r \approx 5.5\,10^{17}$ cm. The density peak near $r \approx 2\,10^{18}$ cm, the signature of the previous thermal pulse cycle, is not yet fully ionized.

emission, this density structure gives rise to a clearly limb-brightened halo (see Fig. 7).

Indeed, it is well-known that the haloes of some Planetary Nebulae show exactly this kind of surface brightness distribution. One prominent example is NGC 6826: the H_α emission measured by Plait & Soker (1990) closely agrees with the model results displayed in Fig. 7. A similar example is NGC 6891, investigated by Guerrero et al. (2000).

700 years further into the model sequence ($T_{\rm eff} = 124\,000$ K), the second shell at $r \approx 2\,10^{18}$ cm, originating from the next to last thermal pulse cycle, becomes fully ionized as well. Its H_α surface brightness then reaches about 1/10 the value of the inner shell ($r \approx 5.5\,10^{17}$ cm) and might be detectable on very deep exposures.

5. CONCLUSIONS

We have demonstrated that an ordinary helium-shell flash on the AGB can induce a mass loss 'eruption' associated with a short period of increased outflow velocity, giving rise to two-wind interaction and the subsequent formation of a thin compressed gas shell propagating outward through the dusty wind envelope. These thin shells of the 'first kind' are located at the inner edge of an extended detached gas/dust shell. Very likely, this scenario explains the origin of the CO shells observed around some semiregular, optically visible carbon stars, the most prominent example being TT Cygni .

Thin gas shells can also be formed during the phase of steadily recovering stellar luminosity and mass loss rate when exceeding the critical luminosity. These gas shells of the 'second kind' are comparable to those of the 'first kind' in geometrical thickness, density contrast and mass. However, they sit in the middle of an extended attached gas/dust shell, possibly impeding their detection in the outflows of AGB stars.

According to our hydrodynamical simulations, the thin compressed shells formed near the end of the AGB evolution easily survive the transformation of the AGB wind into a Planetary Nebula, and can give rise to a density bounded, limb-brightened halo. Indeed, the halo structure of a number of observed Planetary Nebulae supports this scenario. We argue that limb-brightened haloes are the signature of two-wind interaction triggered during the final thermal pulse cycle on the AGB.

References

Arndt, T.U., Fleischer, A.J., Sedlmayr, E., 1997, A&A, 327, 614 (AFS97)
Blöcker, T., 1995, A&A 297, 727 (B95)
Guerrero, M.A., Miranda, L.F., Manchado, A., Vazquez, R., 2000, MNRAS 313, 1
Olofsson, H., Bergman, P., Lucas, R., et al., 2000, A&A 353, 583
Plait, P., Soker, S., 1990, AJ 99 (6), 1883
Steffen, M., Szczerba, R., Schönberner, D., 1998, A&A 337, 149
Steffen, M., Schönberner, D., 2000, A&A 357, 180 (SS00)
Vassiliadis, E., Wood, P.R., 1993, ApJ 413, 641
Winters, J.M., Le Bertre, T., Jeong, K.S., Helling, Ch., Sedlmayr, E., 2000, A&A 361, 641

MODELLING OF THE OBSERVED POST-AGB EVOLUTION

R. Tylenda, K. Jeske, R. Szczerba, N. Siódmiak
Department for Astrophysics, N.Copernicus Astronomical Centre
Rabiańska 8, 87-100 Toruń, Poland
tylenda@ncac.torun.pl

Abstract We have done calculations modelling the post-AGB evolution. Our modelling takes into account the evolution of the central star and hydrodynamics of the dust-gas nebula formed from the AGB and post-AGB winds. The results allow us to construct and discuss a histogram of post-AGB objects with the spectral type. We show that important selection effects can be present in the histogram due to large optical thickness of the nebula in early post-AGB phases. A preliminary comparison of the modelled histograms with observations suggests that the typical mass of post-AGB objects is close to 0.6 $M_\odot$.

1. INTRODUCTION

The post-AGB objects, or proto-planetary nebulae, represent an evolutionary phase in which low and intermediate mass stars are in a fast transition from the AGB to the planetary nebulae. When the mass of the H-rich envelope of an AGB star becomes sufficiently low the photosphere starts shrinking. As the luminosity remains constant the effective temperature increases and the star evolves horizontaly on the HR diagram towards the region occupied by the planetary nebula nuclei.

The evolution of a star in the post-AGB phase proceeds at a constant luminosity determined primarily by the mass of the star. For a given stellar mass the effective temperature, $T_{\rm eff}$, is a function of the mass of the H-rich envelope, $M_{\rm env}$ (Schönberner 1983). Thus the evolution of $T_{\rm eff}$ is governed by the evolution of $M_{\rm env}$. $M_{\rm env}$ decreases during the post-AGB phase due to nuclear burning and mass loss. The rate of nuclear burning is well known from stellar evolutionary models and it determines the stellar luminosity. Unfortunately, the mass loss rate cannot be determined from first principles and is usually estimated from approximate considerations or semiempirical formulae. Therefore, the theoretical evolution of $T_{\rm eff}$ during the post-AGB phase is subject to uncertainties and depends on the adopted evolution of the mass loss rate.

R. Szczerba and S.K. Górny (eds.), Post-AGB Objects as a Phase of Stellar Evolution, 139–146.

We have done a number of numerical simulations of the post-AGB evolution for different stellar masses and different values of parameters describing the evolution of the mass loss rate. In the present paper we present some of the results and compare them with the observations.

2. BASIC ASSUMPTIONS AND NUMERICAL TECHNIQUES

2.1. CENTRAL STAR

The evolution of a star in the post-AGB phase can be easily obtained if one knows the relation between $M_{\rm env}$ and $T_{\rm eff}$. In the present study we have adopted the relations for H-burning post-AGB stars from Blöcker (1995). In this way for a given stellar mass from the adopted mass loss evolution and the nuclear burning rate we can determine the evolution of $M_{\rm env}$ and thus of $T_{\rm eff}$.

We use the same mass loss rate on the AGB as in Blöcker (1995). We also adopt the same procedure as in Blöcker for modelling the transition between the AGB wind and the post-AGB wind. Thus as the pulsational period (calculated according to Ostlie & Cox 1986) of an AGB star drops to a certain value, P_1, the AGB wind rate is adopted to decrease linearily and reaches the post-AGB rate when the period is P_2. Most of the Blöcker's caluclations have been done for $P_1 = 100$ days and $P_2 = 50$ days. In our calculations the value of P_1 is treated as a model parameter. However, for avoiding too many free parameters, we keep the ratio $P_2/P_1 = 0.5$.

The post-AGB mass loss, i.e. when the pulsation period drops below P_2, has been treated in the same way as in Górny et al. (1994). Thus the mass loss rate has been calculated from the formula:

$$\dot{M} = \alpha \frac{L}{c v_\infty} \tag{21.1}$$

where L is the stellar luminosity, c is the speed of light and v_∞ is the terminal velocity of the wind related to the escape velocity in the same way as in Górny et al. (1994). α is a parameter usually taken to be lower than 1. We note that Eq. (21.1) with $\alpha \simeq 0.1$ gives the mass loss rate close to that from the Reimers formula adopted in Blöcker (1995).

2.2. NEBULA

The circumstellar nebula formed by the stellar wind is modelled in a way very similar to that in Steffen et al. (1998). Briefly, our hydrodynamical code uses spherical symmetry. The dynamics includes radiation pressure on dust, transfer of momentum between dust and gas, gas pressure, and gravity of the central star. The radiation transfer takes into account absorption and scattering on dust. Dust is assumed to be formed when the gas temperature drops below the condensation temperature (800 K for "astronomical silicates", 1300 K for amorphous carbon). In order to have a realistic model of the nebula at the beginning of post-AGB we start the hydrodynamic calculations 50 000 yrs before the end of AGB.

3. RESULTS

We have done the calculations for the central star masses of 0.605, 0.625 and 0.836 $M_{\odot}$. The central star evolution is based on the results of Blöcker (1995) with modifications depending on the adopted mass loss evolution as discribed in Sect. 2.1. The P_1 parameter defining the end of the AGB wind has been varied between 75 and 200 days. The α parameter in the post-AGB mass loss formula (Eq. 21.1) has been allowed to vary between 0.0625 and 0.5. The modelling of the envelope has been done assuming amorphous carbon grains (Rouleau & Martin 1991) or "astronomical silicates" (Loar & Draine 1993).

In total about 100 models have been calculated. This paper presents a preliminary analysis of some results. Because of lack of space we concentrate primarily on effects of varying P_1 for the models with the central star mass of 0.605 and 0.836 $M_{\odot}$. The results for 0.625 $M_{\odot}$ are only briefly discussed. The value of α as well as the dust composition (amorphous carbon or silicates) have rather minor effects for the global results discussed in this paper. Therefore, we consider only the models with a fixed dust composition, i.e. amorphous carbon, and a fixed value of α, i.e. $\alpha = 0.125$ (which is more or less equivalent to the Reimers wind).

The primary goal of our research project is to better understand the post-AGB evolution and to better interpret the observational data for the objects in this stage. Below we discuss one of the principal observational characteristics for the post-AGB objects, i.e. distribution of the spectral types of the central stars.

Fig. 1 shows the expected distributions of post-AGB objects with the spectral type derived from models having the same central star mass, i.e. 0.605 $M_{\odot}$ (Fig. 1a) or 0.836 $M_{\odot}$ (Fig. 1b). Each bin gives the percentage of the time spent by the given model central star within the effective temperatures corresponding to the particular spectral type. Different lines correspond to different values of P_1, i.e. 75, 100, 150 and 200 days.

Note that the histograms in Fig. 1 have been constructed without any specification of the beginning and end of the post-AGB phase imposed on them. Only the effective temperature of the central star had to be within the considered range of spectral types. Thus the bins corresponding to late types (K and G) may contain objects with the pulsation period greater than P_1, i.e. still evolving with the AGB wind.

As can be seen from Fig. 1a for $P_1 = 200$ days about 85% of the post-AGB objects of 0.605 $M_{\odot}$ is expected to be of type G. Decreasing P_1 lowers the

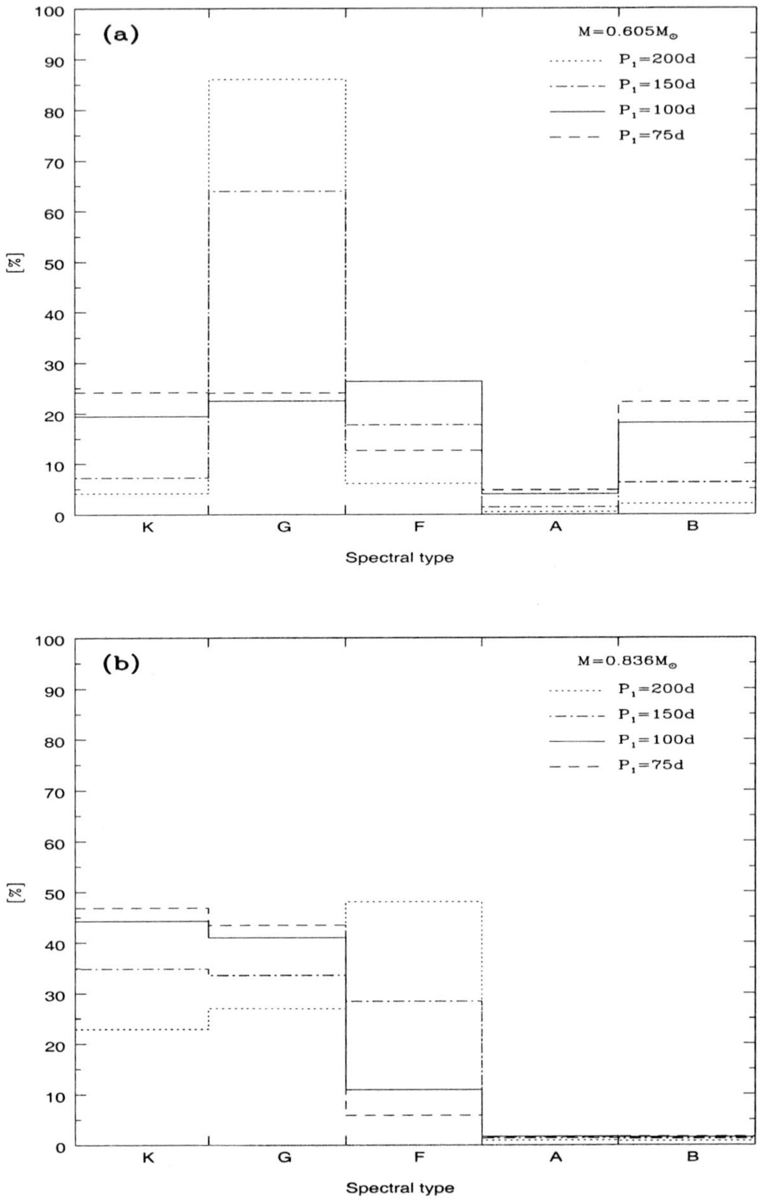

Figure 1 The histograms of post-AGB objects with the central star spectral type. (**a**) - results of models with the 0.605 $M_\odot$ central star. (**b**) - results of models with the 0.836 $M_\odot$ central star. All the calculations have been done with $\alpha = 0.125$ in Eq. (21.1) and amorphous carbon dust. Different histograms correspond to different values of P_1 defining the moment of transition from the AGB to post-AGB wind.

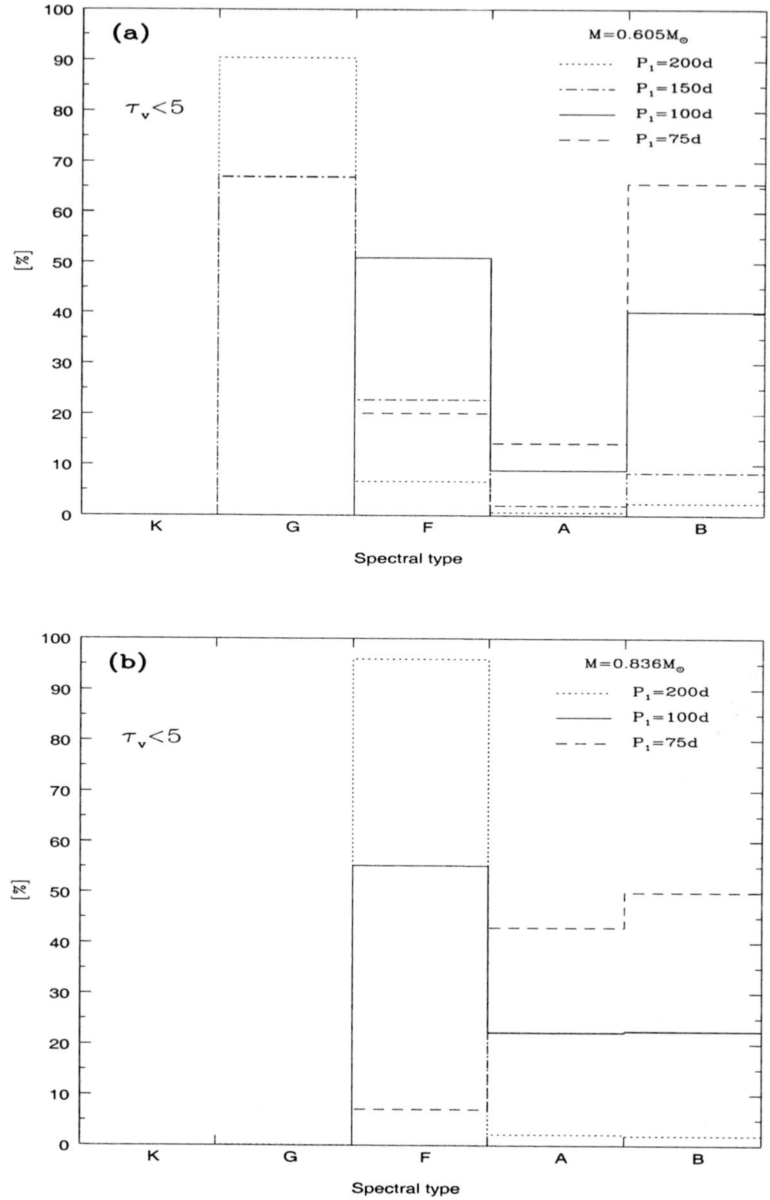

Figure 2 The same as Fig. 1 but with observational selection due to opaque post-AGB nebulae. Only models having nebulae with the optical thickness in V lower than 5 are included in the histograms.

G-type bin and increases the populations in other spectral types. In the case of $P_1 = 200$ days the end of the AGB wind happens at the effective temperature close to the border between types K and G. At this point the mass loss rate starts decreasing which slows down the evolution. As a result the star spends a long time within the G type. A decrease in P_1 extends the AGB wind to higher effective temperatures which results in a faster evolution across the G type. This reduces the G-type bin in Fig. 1a and, correspondingly, increases the relative populations in other spectral types.

Fig. 1b shows the same as Fig. 1a but for the 0.836 $M_\odot$ model. Here the end of the AGB wind takes place when the star is in the F type, even for $P_1 = 200$ days. Thus a decrease of P_1 modifies the relative population in type F in the same way as in the case of type G for the 0.605 $M_\odot$ model (Fig. 1a).

One could attempt to compare the histograms in Fig. 1 with observations. However, these histograms have been obtained just from the evolution of the stars. They do not account for the fact that the post-AGB objects have dusty envelopes which in early phases can hide the stars from direct observations thus unabling spectral classification. This fact can introduce important selection effects in observational diagrams of the kind shown in Fig. 1.

From our models we have found that the central star becomes visible in the optical when the optical thickness of the nebula in the V band drops below $\sim$5. Fig. 2 shows the same as in Fig. 1 but exluding the phases in our models when the nebula has an optical thickness in V larger than 5. Comparing Fig. 2 with Fig. 1 one sees that the optical selection effects can importantly modify the observed populations in late spectral types. In the case of the 0.605 $M_\odot$ model no K type stars can be observed and for short periods ($P_1 \lesssim 100$ days) even the G types disapear (cf. Figs. 1a and 2a). The 0.836 $M_\odot$ model predicts no objects with K and G type stars independently of the value of P_1 (cf. Figs. 1b and 2b). For $P_1 = 75$ days even the F type is affected in this massive model.

4. COMPARISON WITH OBSERVATIONS AND DISCUSSION

One can compare the histograms resulting from our modelling with the observed population of the spectral types for the post-AGB objects, e.g. with Fig.2 in Szczerba et al. (this volume). However, before doing this several remarks have to be made.

First of all, the histograms presented in the present paper have been derived from modelling the evolution for a single stellar mass and the quiescent H-burning. In reality, the observed population certainly contains stars of different masses. Which burning mode, i.e. H or He, and in what proportion of the observed objects is operating is not known although it is believed that the H-burning dominates.

Second, the observed histogram, of the sort presented in Szczerba et al. (this volume), certainly suffers from observational selection. The most important is probably due to the dusty nebulae which hide the central stars in early phases. As shown in this paper the effects of this kind can be modelled. However, observational ways of discovering the post-AGB objects can also introduce selection effects which are certainly much more difficult for appropriate accounting for in the modelling.

Although the dust selection effects can be simulated while modelling the spectral type histogram they are model dependent. Generally speaking, they depend on the history of mass loss in the late AGB and post-AGB. In the present study we have shown that the adopted moment of the transition from the AGB wind to the post-AGB wind (parametrized by the value of P_1) is important in this respect. However, the post-AGB dusty nebula is primarily formed by the intense wind before the AGB end. Therefore the dust effects in the post-AGB must depend on the adopted modelling of the AGB wind. This has not been done in the present study. We note, however, that the selection effects in a histogram similar to those in Fig. 2 but for the 0.625 $M_{\odot}$ star are much weaker than for 0.605 $M_{\odot}$. In principal, the result should be opposite as a more massive star is expected to have a more intense mass loss and to evolve faster. The reason is that Blöcker (1995) have modified his mass loss formula while calculating the 0.605 $M_{\odot}$ model and, as a result, the AGB mass loss rate in this model is significantly higher than in the more massive one.

The selection effects due to the dusty nebulae are most important for late type post-AGB stars. For early types the nebulae are generally well dispersed and they cannot hide the stars from observing them in the optical. Thus if one does not want to discuss the dust selection effects which depend on the uncertain mass loss history he should concentrate on an analysis of the early type post-AGB stars. Our models show that the distribution in the early types first of all depends on the stellar mass. In particular, the ratio of the population in the B type to that in the A type is predicted to be $\sim$4.5 for the 0.605 $M_{\odot}$ model, $\sim$3.5 for the 0.625 $M_{\odot}$ star and drops to $\sim$1.0 if the stellar mass is 0.836 $M_{\odot}$ (these values depend very weakly on α and P_1). The same ratio in the observational sample of Szczerba et al. (this volume) is $\sim$3.5. Thus we can conclude that the typical mass of post-AGB stars is close to 0.6 $M_{\odot}$. This is compatible with results from model atmosphere analyses of the post-AGB stars as discussed in Schmidt (this volume). This is also compatible with mean masses estimated for the planetary nebula nuclei (e.g. Stasińska et al. 1997) and white dwarfs (e.g. Bragaglia et al. 1995).

Following the above conclusion we can compare the observed histogram from Fig. 2 of Szczerba et al. (this volume) with the results of our 0.605 $M_{\odot}$ model as presented in Fig. 2a. As can be seen from Fig. 2a, and as discussed above, the relative population in late spectral types is very sensitive to the adopted value of P_1. Our model results when plotted on the IRAS colour-colour diagram suggest that large values of P_1 are rather excluded by observations. The models with $P_1 = 150$ and 200 days spend most of the time in regions of the IRAS diagram where very little or no post-AGB objects are observed. Returning to Fig. 2a we see that the 0.605 $M_{\odot}$ model with $P_1 \lesssim 100$ days predicts no K and G type objects. This is in obvious contradiction with Fig. 2 in Szczerba et al. Several reasons for this discrepancy can be given. First, as discussed above, the AGB mass loss rate adopted by Blöcker (1995) in the 0.605 $M_{\odot}$ model was exceptionally high. A fainter AGB wind would produce a thinner nebula in the post-AGB and the central star would be observable earlier, i.e. at lower effective temperatures. Indeed, our calculations done for the 0.625 $M_{\odot}$ model (in which the AGB mass loss rate is smaller than in the 0.605 $M_{\odot}$ model) with $P_1 = 100$ days predict a sizeable fraction of objects in type G. Second, the presence of late type (K and G) objects in the observed sample may indicate that the masses of some post-AGB objects are below 0.6 $M_{\odot}$. These low mass stars are expected to have had less intense AGB winds and to evolve slower in the post-AGB phase. Thus they would become observable close to the end of AGB. Third, mass loss might become non-spherical at the end of AGB and we can observe a certain number of late type stars through thinner and thus more transparent nebular regions.

Acknowledgments

The research reported in this paper was partly supported from the grant 2 P03D 020 17 financed by the Polish State Committee for Scientific Research.

References

Blöcker T., 1995, A&A 299, 755
Bragaglia A., Renzini A., Bergeron P., 1995, ApJ 443, 735
Górny S.K., Tylenda R., Szczerba R., 1994, A&A 284, 949
Loar A., Draine B., 1993, ApJ 402, 441
Ostlie D., Cox A.N., 1986, ApJ 311, 864
Rouleau F., Martin P.G., 1991, ApJ 337, 526
Schönberner D., 1983, ApJ 272, 708
Stasińska G., Górny S.K., Tylenda R., 1997, A&A 327, 736
Steffen M., Szczerba R., Schönberner D., 1998 A&A 337, 149

THE EFFECT OF THE TRANSITION TIMES IN THE FORMATION OF PLANETARY NEBULAE

Eva Villaver
Instituto de Astrofísica de Canarias, C/Vía Láctea, 38205 La Laguna, Tenerife, Spain
villaver@ll.iac.es

Arturo Manchado
Instituto de Astrofísica de Canarias, C/Vía Láctea, 38205 La Laguna, Tenerife, Spain

Guillermo García-Segura
IA-UNAM, Apdo Postal 877, Ensenada, 22800 Baja California, México

Abstract We have explored the effect of transition timescales in planetary nebulae (PNe) formation. We have performed hydrodynamical simulations of the evolution of the stellar winds predicted by theoretical stellar evolutionary models adopting two choices for the evolution during the transition time. The evolution of the star from the AGB to the PN stage has been followed for different main-sequence stellar masses (1, 1.5, 2, 2.5, 3.5 and 5 $M_{\odot}$). We find that the sizes and velocities of the PNe shells are very sensitive to the adopted timescale during the transition time.

1. INTRODUCTION

During AGB evolution the matter that leaves the star as a high-density wind forms a circumstellar envelope which expands into the surrounding interstellar medium. At one stage the high mass-loss rates experienced during the AGB terminates. From then onward, until the star reaches high enough effective temperatures to ionize the previously ejected material forming the PN, the star undergoes a short-lived phase known as the transition time (Kwok 1993).

From the point of view of theoretical stellar evolutionary models, the main difference in timescales during this period comes from the fact that the point of departure from the AGB is arbitrarily established; different authors use very different criteria. Vassiliadis & Wood (1994) considered the AGB phase to be

R. Szczerba and S.K. Górny (eds.), Post-AGB Objects as a Phase of Stellar Evolution, 147–152.

completed when the effective temperature of the star becomes hotter than a reference AGB temperature by 0.3 dex, while Blöcker (1995) established the end of the AGB when the envelope mass drops below several per cent of the total stellar mass ($\approx 10^{-2}\ M_{\odot}$). This leads to a differences of more than one order of magnitude in the predictions for the transition timescales between the two authors. The huge discrepancy is caused because in the models of Vassiliadis & Wood (1994), the change from AGB to post-AGB mass-loss rates occurs closer to the AGB, where the envelope has a higher mass to be removed than in those of Blockër (1995), and hence the transition time evolution slows down. Numerical simulations offer a unique opportunity to constrain the theoretical transition times.

2. THE NUMERICAL MODELS

We have followed the evolution of the circumstellar gas from the thermal pulse phase during the AGB stage until the PN formation. Hydrodynamical simulations which use realistic winds from stellar evolution calculations have been performed. In an attempt to obtain a complete picture for the PN scenario, we have investigated the evolution of the of gas ejected for the whole range of stellar masses that, according to stellar evolution, give rise to PN formation. Six models taken from Vassiliadis (1992) with masses 1, 1.5, 2, 2.5, 3.5 and 5 $M_{\odot}$with solar metallicity have been used.

We have adopted two different choices in order to follow the stellar evolution from the AGB to the PN stage. The objective is to explore the consequences of the transition time period in the formation of PNe. In the first instance, the transition times given by Vassiliadis & Wood (1994) were used. The transition time in their calculation represents the time elapsed from the end of the AGB phase, where the post-AGB mass-loss mechanism is activated, until the star reaches a temperature of 10 000 K. A second choice is to consider a sudden transition from the AGB to the PN stage. That is, the star reaches a temperature of 10 000 K just after the AGB stage.

The hydrodynamical computations have been undertaken with the fluid solver ZEUS-3D, used as a hydro-code. The simulations have been carried out in 1D spherical coordinates with the time-dependent wind parameters set in the five innermost zones of the grid for the whole evolution of the star. An outflow boundary condition has been set in the outer radial direction. At time zero the grid is filled with an homogeneous neutral ISM with density 1 cm^{-3}.

2.1. ENERGY INPUT FROM THE STAR DURING THE TRANSITION TIME

From an observational point of view the mass-loss rate and wind velocities during the transition domain are essentially unknown. In order to compute the wind parameters during the transition time, we have adopted the simplest approximation, which is to assume that both the luminosity and the effective temperature of the star changes linearly from the values at the end of the AGB stage to their values at the beginning of the next stage of the evolution, i.e. when the central star of the PN becomes visible. To drive the wind we have assumed that the driving of the outflow occurs via radiation pressure on resonance lines (cf. Pauldrach et al. 1988).

For the second choice (zero transition time), the central star wind mechanism is switched on immediately after the AGB stage. Therefore, wind velocities experience bigger jumps and the star begins to emit a sizeable amount of ionizing photons.

2.2. RESULTS

In Fig. 1 we have plotted the evolution of the radii of the main shell, the brightest one, during the PNe phase. The radii have been computed from the position of the maximum of the Hα brightness profiles. The zero time point represents the time at which the effective temperature of the star reaches 10 000 K. The radii are plotted in the left-hand panel of Fig. 1, in the case where the PN evolution starts after the AGB stage (the transition time is not considered) and in the right-hand panel when transition times are considered.

When transition times are considered, there is a correspondence between the stellar mass and the radius reached by the inner shell: the larger radius corresponds to the higher mass. This is mainly determined by the injection of energy from the star that increases with the stellar mass. The curves show an inflexion point followed by a change in slope. The position of the inflexion point is determined by the point where the shell, driven by the hot bubble, evolves into the brightest shell. Before this point the brightest shell is that caused by the compression associated with the ionization front (IF). We can see how the radii are too large (except for the 1 $M_{\odot}$ model) at the beginning of the PN phase for being real PN radii.

When the evolution during the transition time is not considered in the models, different amounts of time evolution have been skipped in each stellar case, during which either the gas ejected previously during the TP–AGB has not been pushed away from the star or there is no geometrical dilution. Consequently, both the wind and the radiation field from the central star encounter a higher density structure.

From the time the main shell is driven at the expense of the thermal pressure of the hot bubble, the main shell velocity increases rapidly with time. As the central star evolves towards higher effective temperatures the wind velocity increases and so does the thermal energy of the hot bubble, as it depends on the shock velocity. Lower velocities are reported during the PN phase in the case

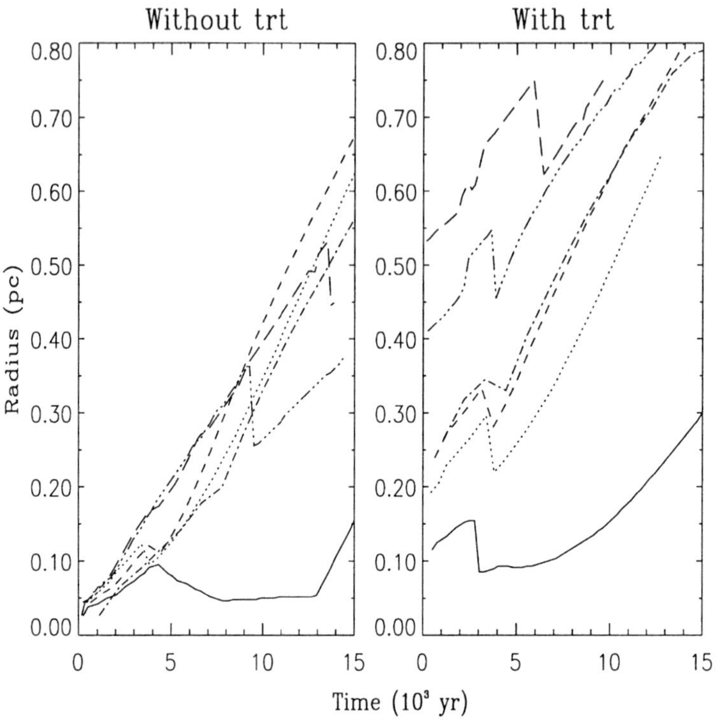

Figure 1 Radius of the main PN shell versus time. The solid, dotted, dashed, dash-dotted, dash-dot-dot-dotted and long-dashed lines represent the 1, 1.5, 2, 2.5, 3.5 and 5 $M_{\odot}$ models, respectively. Right: radii for the models with transition times, left: models without transition times.

where no transition times are considered during the early stages of evolution because the higher densities in this case prevent a faster expansion. The pressure gradients generated by the IF are lower. The models that evolve during the transition time phase (which have lower densities) are pushed more efficiently. For models with transition times the main shell sizes are much larger but the expansion velocity is only slightly higher.

During early stages of PN evolution the nebula remains optically thick to H ionization radiation. The inclusion of the whole range of stellar masses that lead to PN formation reveals that this phase ends at effective temperatures that increase with stellar mass. Moreover, the addition or not of the transition time in the simulations leads to twice as large an effective temperature for the transition since the densities are larger, and therefore the IF moves more slowly. In Table (1) we summarize the effective temperatures of the star for which the nebula becomes density-bounded. Cases I and II refer to when transitions times

Table 1 Effective temperatures and radii

Mass	T_{eff} *Case I*	R_i	T_{eff} *Case II*	R_i
1	15000	0.138	20000	0.091
1.5	25000	0.229	39000	0.098
2	67000	0.291	130000	0.103
2.5	145000	0.330	128000	0.178
3.5	18000	0.509	207000	0.068
5	$\sim$	$\sim$	$\sim$	$\sim$

are taken into account and when they are not, respectively. These values are approximate because they are subject to time errors of the order of $\sim$ 500 yr, which is the average temporal resolution we have between the outputs of the simulations. Note that the 5 $M_\odot$ stellar model is the only one that provides us with an ionization-bounded PN.

The use of models without transition times introduces two important characteristics. Firstly, the nebula remains ionization-bounded for a longer time. Therefore, the central star has evolved to higher effective temperatures because of the slower velocity of the IF, where it encounters higher densities. Secondly, the sizes of the main PNe shells are smaller.

Since the sizes of the main shells are directly related to the evolutionary timescale during the transition time, we find that the adopted timescales are too long at the end of the transition time to account for the observed sizes in PPN (Ueta, Meixner & Bobrowsky 2000) even when distance uncertainties are taken into account.

We conclude that the transition times adopted by Vassiliadis & Wood (1994) are too long. Vassiliadis and Wood pointed out that their definition for the end of the AGB stage was derived for models with $M_i \leq 2.5\ M_\odot$. Thus the transition times for the higher-mass models (3.5 $M_\odot$ and 5 $M_\odot$) are more uncertain than those for lower masses. Also, as we discussed in the introduction, the transition times are mainly determined by the arbitrarily established point of departure from the AGB. The choice made by Vassiliadis & Wood (1994) gives PN radii in our simulations that are unrealistically large for all the masses, with the exception of the 1 $M_\odot$ stellar case. We find that the transition to post-AGB mass-loss rates should occur further away from the AGB than computed by these authors. The variation of the characteristic timescales during the transition time, i.e. the nuclear, thermal and wind timescales, have been explored by Stanghellini & Renzini (2000) by assuming different remnant envelope masses at the end of the AGB phase. The transition times obtained vary between 50 and 8000 yr, depending on the stellar mass and on the assumed envelope mass. Shorter timescales during this period, such as these ones or those computed by Schönberner (1981,1983) and Blöcker & Schönberner (1990) seem to be a better assumption. Our results agree with Tylenda & Stasińska (1994), who found, based on comparison of theoretical evolutionary tracks for post-AGB stars with Galactic PN data, that the transition times given by Vassiliadis & Wood (1994) are too long.

3. CONCLUSIONS

We conclude that the transition times adopted by Vassiliadis & Wood (1994) are too long, and that the transition to post-AGB mass-loss rates should occur further away from the TP–AGB. The transition time is shorter if higher mass-loss rates are considered; higher mass-loss rates should provide a more powerful wind to drive the inner shell over a shorter timescale.

Acknowledgments

We thank M. L. Norman and the laboratory for Computational Astrophysics for the use of ZEUS-3D. The work of EV and AM is supported by grant PB97-1435-C02-01 from Spain's DGES .

References

Blöcker, T. & Schönberner, D. 1990, A&A 240, L11
Blöcker, T. 1995, A&A 299, 755
Kwok, S. 1993, ARA&A, 31, 63
Pauldrach, A., Puls, J., Kudritzki, R. H., Méndez, R. H. & Heap, S. R. 1988, A&A, 207, 123
Schönberner, D. 1981, A&A 103,119
Schönberner, D. 1983, ApJ, 272, 708
Stanghellini, L. & Renzini, A. 2000, ApJ, 542, 308
Tylenda, R. & Stasińska, G. 1994, A&A, 288, 897
Ueta, T., Meixner, M & Bobrowsky, M. 2000, ApJ, 528, 861
Vassiliadis, E. 1992, Thesis, Australian National University
Vassiliadis, E., & Wood, P. 1994, ApJS, 92, 125

ARE MICROSTRUCTURES IN THE OUTER SHELLS OF PNE FOSSIL CONDENSATIONS OF THE AGB WIND?

D.R. Gonçalves[1], R.L.M. Corradi[2], E. Villaver[1], A. Mampaso[1] and M. Perinotto[3]
[1]*Instituto de Astrofísica de Canarias - E-38205 La Laguna, Tenerife, Spain*
[2]*Isaac Newton Group of Telescopes - E-38700 Santa Cruz de La Palma, Spain*
[3]*Università di Firenze - Largo Eurico Fermi 5, 50125 Firenze, Italy*

Abstract Some planetary nebulae (PNe) show microstructures of low ionization located in their outer regions, i.e. external to the bright rim defined by the fast vs. slow wind interaction region. Morphological and kinematic analysis of some of these PNe show low ionization knots which clearly share the expansion of the outer shell of the host nebulae (e.g. IC 2553). Sometimes, these knots possess an accentuated degree of symmetry. Given that in situ instabilities are not thought to form symmetrical features, we explore the idea that these symmetric pairs of knots are fossil condensations of the AGB wind being modified by interaction with the expanding ionization front. This very appealing possibility would imply that the mass loss of the pre-PN stages has episodes of highly collimated and symmetric ejections. In other PNe knots of low ionization in the outer shells are not symmetric (e.g. NGC 5882) and could be formed by either simple in situ instabilities and/or fossil condensations.

1. LOW-IONIZATION STRUCTURES IN PNE

In addition to the main large-scale components of a PN – bright rims, attached shells and haloes – small-scale structures, very prominent in low ionization lines (particularly [N II]) have been observed in a number of PNe (see the catalogue of Corradi et al. 1996). These low ionization microstructures (LIS) appear with a variety of morphologies: knots, tails, filaments, jets and jet-like structures (see Gonçalves et al. 2000, for a detailed study of all these kinds of LIS), attached to or detached from, the main shells of the nebulae. Kinematic studies have revealed features with velocities much higher than those of the environment and the bright rims (such as the pairs of knots studied by Balick and collaborators, the FLIERs; Balick et al. 1998), as well as pairs or isolated knots which share the expansion of the outer shell in which they are embedded.

R. Szczerba and S.K. Górny (eds.), Post-AGB Objects as a Phase of Stellar Evolution, 153–156.

We focus the discussion on those microstructures of low ionization which both: i) are located outside the bright rims (i.e. in the outer shells), and ii) move with velocities not substantially different from those of their environments. We now discuss physical processes for their formation.

2. ON THE ORIGIN OF OUTER SHELL LIS

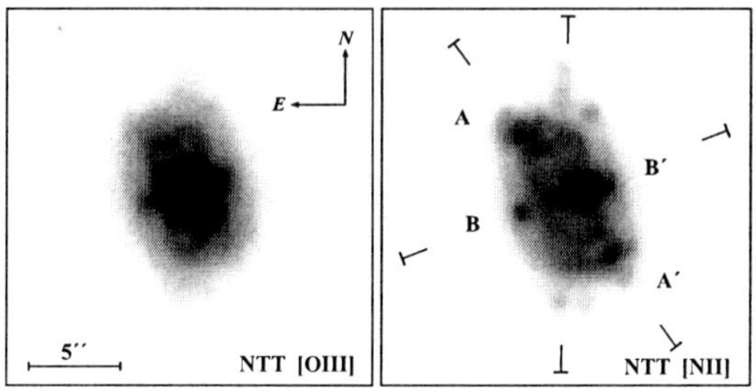

Figure 1 NTT images of IC 2553, on a logarithmic intensity scale. The positions of the slit used for spectroscopy are indicated by short lines on either side of the nebula. Point-symmetric low ionization knots are labelled with capital letters (A-A' and B-B') (Corradi et al. 2000).

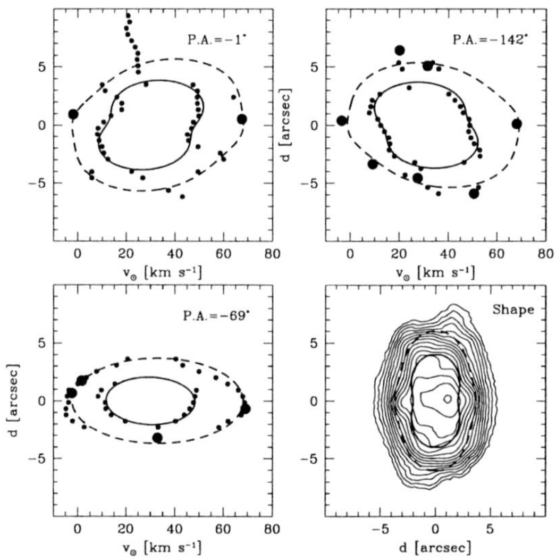

Figure 2 Kinematic modelling of IC 2553 using the description in Solf & Ulrich (1985). The [N II] heliocentric velocities (small dots) at three slit positions through IC 2553. Large dots indicate the velocity of the low ionization knots. The shape of the nebula is shown as as plot contours of the [O III] image. Model fits for the inner and outer shells are the solid and dashed lines, respectively (Corradi et al. 2000).

First, consider the image and spectra of IC 2553 (Figures 1–2), which show clearly that its LIS are associated with the outer shells of these PNe, since they are at the same projected distance from the centre and move with the same velocity as that of the surrounding shell. Second, note that their location and velocities exclude the possibility that they are formed following the interacting stellar wind scenario – which may give rise to the bright inner shells. In addition, the high degree of symmetry of LIS (symmetrical or point-symmetrical pairs

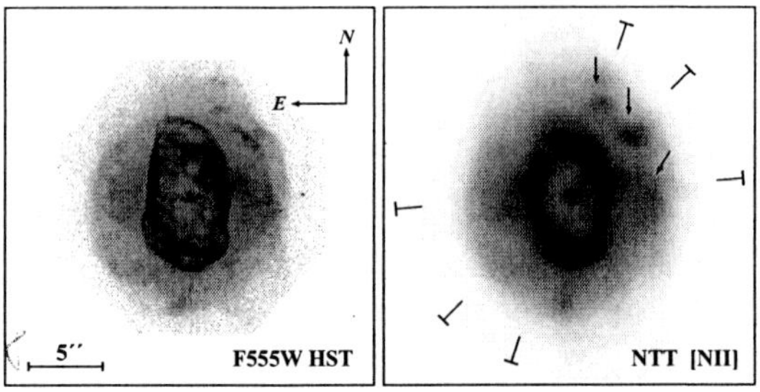

Figure 3 The HST and *NTT* images of NGC 5882on a linear scale. The positions of the slits used for spectroscopy are indicated by short lines, while arrows mark the position of several low ionization knots (Corradi et al. 2000).

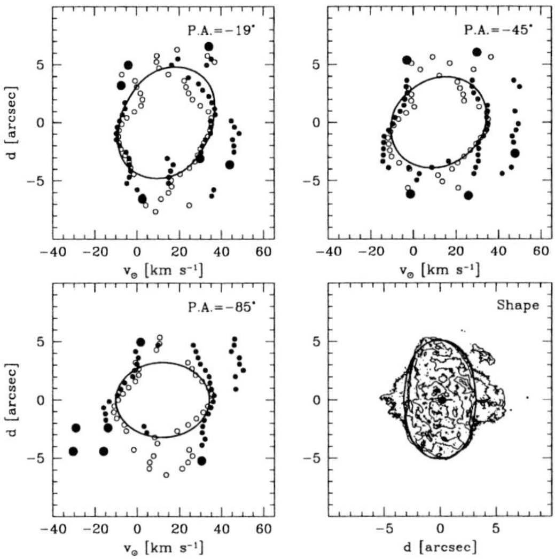

Figure 4 Kinematic modelling for NGC 5882. The representation is as in Figure 2, except that beside the [N II] radial velocities (full circles) we display the Hα ones (empty circles). Large dots indicate the [N II] velocities of the low ionization knots. The shape of the nebula is shown as plot contours of the *HST* image. Model fits for the inner shell are plotted as solid lines (Corradi et al. 2000).

of knots, similarly to He 1−1, NGC 6572 and NGC 7662) also excludes LIS formation by in situ instabilities.

These facts point to their origin as fossil condensations which appear before the onset of the slow–fast wind interaction, i.e. during the AGB (or previous) mass loss episodes. Fossil condensations (usually assumed to be neutral clumps of gas), in the case of IC 2553 with a linear size of 5×10^{16} cm, are expected to survive up to a few 10^3 yr (Mellema et al. 1998; Soker & Reveg 1998). This time is comparable with the age of the outer shell – whose formation is attributed to the expansion of the ionization front through the AGB remnant – of this and other PNe.

This scenario implies highly anisotropic mass ejection episodes prior to PN formation. To date, this anisotropic mass loss, in which mass is ejected in different directions forming pairs of (point-) symmetric features, has been ascribed to non-radial pulsations of the red giant, and/or to interaction in a binary system (Livio & Pringle 1997; Soker & Harpaz 1999).

The case of NGC 5882, and the other PNe showing isolated LIS moving with the outer shells (Hu 2−1, IC 4593 and NGC 7662), are easier to explain, since the constraint of a symmetric knot distribution does not apply. For these non-symmetric knots, the hypothesis of knots originating by in situ instabilities is also viable. Such instabilities might be related to the expansion of the slow wind through the circumstellar matter, as well as to the expansion of the ionization front (García-Segura & Franco 1996).

Acknowledgments

DRG wishes to acknowledge the grant 98/7502-0 of the Brazilian Agency FAPESP, and RLMC and AM acknowledge the Spanish grant DGES PB97-1435-C02-01.

References

Balick, B., Hajian, A. R., Terzian, Y., Perinotto, M., & Patriarchi, P. 1998, AJ, 116, 360

Corradi, R. L. M., Gonçalves, D. R. Villaver, E., Mampaso, A., & Perinotto, M. 2000, ApJ, 542, in press (astro-ph/0003238)

Corradi, R. L. M., Manso, R., Mampaso, A., & Schwarz, H. E. 1996, A&A, 313, 913

García-Segura, G., & Franco, J. 1996, ApJ, 469, 171

Gonçalves, D. R., Corradi, R. L. M., & Mampaso, A. 2000, ApJ, 547, in press (astro-ph/0010282)

Livio, M., & Pringle, J. E. 1997, ApJ, 486, 835

Mellema, G., Raga, A. C., Cantó, J., Lundqvist, P., Balick, B., Steffen, W., & Noriega-Crespo, A. 1998, A&A, 331, 335

Soker, N., & Harpaz, A. 1999, MNRAS, 310, 1158

Soker, N., & Reveg, O. 1998, AJ, 116, 2462

Solf, J., & Ulrich, H. 1985, A&A, 148, 274

THE INFRARED [WC] STARS

Albert A. Zijlstra
UMIST, department of Physics, Manchester M60 1QD, UK
a.zijlstra@umist.ac.uk

Abstract A number of late [WC] stars have unique infrared properties, and together define a class of IR-[WC] stars. They have unusual IRAS colours, and most or all show a double chemistry, with both a neutral (molecular) oxygen-rich and an inner carbon-rich region. Their dense nebulae indicate recent evolution from the AGB, suggesting a fatal-thermal-pulse (FTP) scenario. Although both the colours and the stellar characteristics predict fast evolution, it is shown that this phase must last for 10^4 yr. The stellar temperatures of the IR-[WC] stars appear much higher in low metallicity systems (LMC, Sgr). This may be indicative of an extended 'pseudo' photosphere. It is proposed that re-accretion of ejected gas may slow down the post-AGB evolution and so extend the life time of the IR-[WC] stars.

1. INTRODUCTION

The [WC] stars a group of hydrogen-poor central stars of planetary nebulae, showing strong carbon emission lines. The spectra are similar to those of massive WR stars, and like their massive counterparts they have strong stellar winds. The PNe surrounding the [WC] stars do not differ from other PNe. The [WC] star are thought to be related to a late thermal pulse. This can be have occured during three possible phases: (1) on the white dwarf cooling track (a very late thermal pulse or VLTP); (2) during the hydrogen-burning post-AGB evolution (a late thermal pulse or LTP); (3) at the very end of the AGB, with the pulse terminating the AGB evolution (a fatal thermal pulse or FTP). The VLTP model is preferred from stellar modelling (e.g. Herwig et al. 1999).

Several compact [WC] PNe are extremely bright infrared emitters. I will show that their nebulae have unexpected characteristics, and do not have a counterpart among non-[WC] PN.

2. INFRARED FLUX AND COLOURS

Figure 1 shows the IRAS colour–colour diagram for all [WC] stars with 3-band detections. In Fig. 1a the 40 [WC] stars (filled circles) are shown together

R. Szczerba and S.K. Górny (eds.), Post-AGB Objects as a Phase of Stellar Evolution, 157–161.

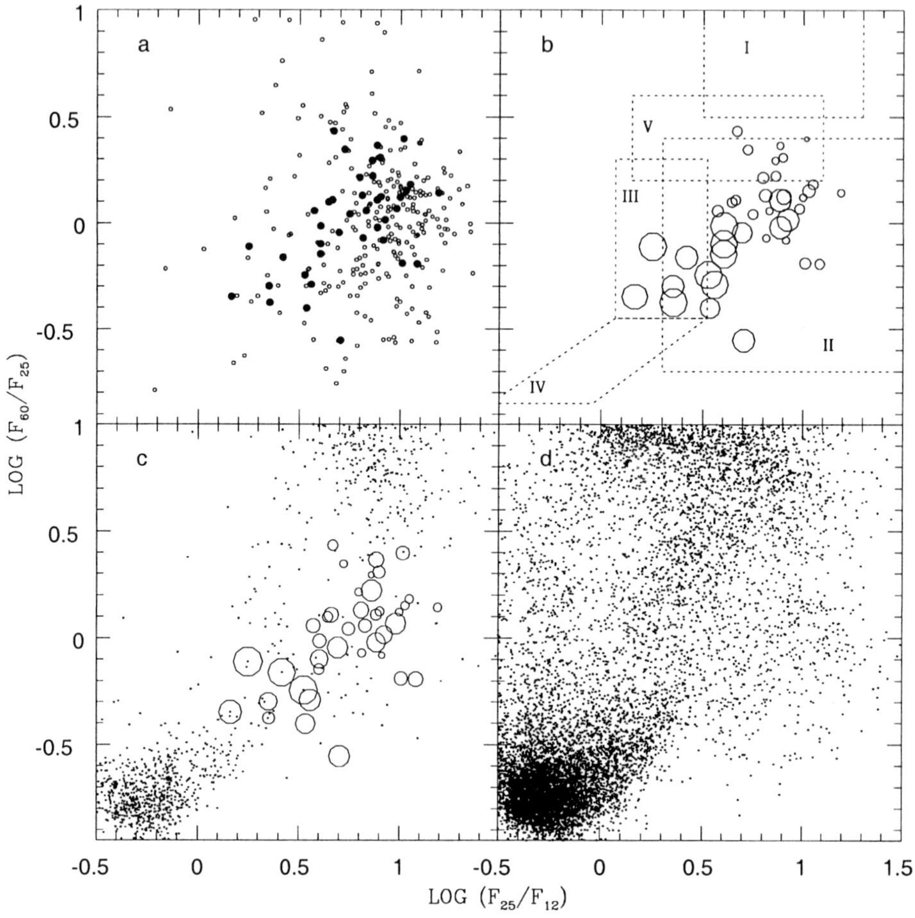

Figure 1 Infrared IRAS colours. The axes are defined in terms of 12μm, 25μm and 60μm IRAS flux densities, without colour correction. Panel a: the open circles show all PNe detected in all three bands; the filled circles are all the [WC] stars. Panel b: the [WC] stars; the size of the circles is proportional to the subclass (class 11 are the largest circles). The boxes show the expected location of HII regions (I), PNe (II), PostAGB stars (III), Miras (IV) and bipolar outflow sources (V) (from Zijlstra et al. 2000). Panel c: the open circles are the [WC] stars; the size of the symbols is now proportional to the integrated IRAS flux. The dots represent all IRAS sources with integrated flux comparable to the IR-brightest [WC] stars. Panel d: All IRAS sources with good detections in all three bands.

with all other PNe (open circles). The two groups cover the same region of the diagram, but the [WC] stars have on average bluer F_{25}/F_{12} colours. Figure 1b shows the [WC] stars, but now the size of the circles is proportional to the [WC] subclass; i.e. the largest circles correspond to the coolest central stars. The coolest stars show the hottest dust, implying their dust is nearest to the star. The difference in distribution in Fig. 1a appears to be caused by these late [WC] stars only.

In Fig. 1c the size of the open circles is proportional to the log of the integrated IRAS flux. The objects with the hottest dust are also very bright in IRAS. Although this trend is affected by the unknown distances (scaling as d^{-2}), the dust emission declines so strongly as the nebula expands that the distance spread is a secondary parameter. The dots in Fig. 1c are all entries in the IRAS point source catalogue with flux quality 3 at 12, 25 and 60μm, and with total in-band flux $F_{ir} > 10^{-11}\,\mathrm{W\,m^{-2}}$ which is similar to the brightest [WC] stars. The bright [WC] stars form a well-defined group among IRAS sources, with little confusion with other types of objects. For comparison, Fig. 1d shows all IRAS sources with flux quality 3 at 12, 25 and 60μm (for all values of total flux). The AGB and its termination point are clearly visible, with the bright [WC] stars located near this point, in a gap in the diagram.

The gap in the distribution of bright IRAS sources in Fig 1c, at $(0.5, -0.3)$, corresponds to the AGB–Post-AGB transition where evolution is extremely fast ($\sim 100\,\mathrm{yr}$). The bright [WC] stars are seen to straddle this gap.

The brightest IRAS PNe are in fact almost all [WC] stars, located near the Post-AGB gap of Fig. 1c. We define this group as the IR-[WC] stars: [WC] stars with IRAS in-band flux $F_{ir} > 0.8 \times 10^{-11}\,\mathrm{W\,m^{-2}}$ and IRAS colours near or in the Post-AGB gap.

The IR-[WC] stars have subclasses ranging from 9 to 11. This very limited range of subclass indicates a likely common origin of the IR-[WC] stars. The IR-[WC] stars have no obvious counterpart among normal PNe.

3. PROPERTIES AND PROBLEMS

Several PNe around [WC] stars show a chemical dichotomy (Waters et al. 1998, Cohen et al. 1998), with evidence for both carbon-rich dust (PAHS) and oxygen-rich dust (crystalline silicates). The first member of the class (IRAS 07027−7934) was found to have a strong OH maser (Zijlstra et al. 1991). Only the IR-[WC] stars sofar show this effect: it is not seen in any other PN or [WC] star. The origin of the double chemistry is not known. It appears likely that the presence in PNe of strong PAH features in combination with OH masers and/or crystalline silicates is uniquely related to the IR-[WC] stars.

The IR colours, the IR fluxes and the chemical dichotomy all point at very short post-AGB time scales of the order of 10^2–10^3 yr. The apparent conclusion is that the IR-[WC] stars were on the AGB before that time. The time when the star became hydrogen-poor may be even more recent. An estimate for the life time of the phase can also be obtained from the observed number of these objects. From this we find a *minimum* life time for an IR-[WC] star of 5×10^3 yr, with a likely life time in excess of 10^4 yr. This is in contradiction with the implied short nebular time scales. The observed number of objects shows that IR-[WC] stars are comparatively long-lived. Both nebulae and stars must evolve much slower during the IR-[WC] phase than their compact nebulae suggest.

A way to slow down evolutionary time scales of the nebulae is by assuming the existence of a slowly expanding torus or disk (e.g. Waters et al 1997). This would lead to bipolar morphologies, possibly with high velocities perpendicular to the torus. Similar bipolar morphologies are seen in 10–15% of PN (Manchado et al. 1996). The evidence is at present inconclusive, but bipolar, torus-like morphologies are possible for all known IR-[WC] nebulae.

There are few or no Bulge [WC] stars. However, there are [WC] stars known in nearby galaxies of which three have IRAS detections: He2-436 in the Sagittarius dwarf spheroidal at 25 kpc (Sgr: Zijlstra & Walsh 1996), and SMP 58 and SNP 61 in the LMC (Zijlstra et al. 1994). Sgr has a metallicity of [Fe/H]$= -0.55$ and a main-sequence turn-off mass of 1.3 $M_\odot$ (Dudziak et al. 2000). Its two [WC] stars are the only such stars where these parameters are known.

The IR colours and IR flux (when scaled to a typical distance of 2 kpc) are similar to those of the IR-[WC] stars. However, their [WC] subclass is much earlier than for Galactic stars with similar IR characteristics. This suggests that the subclass is affected by metal blanketing in an extended wind, in which case lower metallicity causes the $\tau = 1$ surface to be closer to the real stellar surface. This would however imply that the late IR-[WC] stars have much hotter stars than the wind emission lines suggest.

4. THE PROBLEM OF EVOLUTION

There are three main problems regarding the IR-[WC] stars: (1) A unique nebular double chemistry seen *only* around IR-[WC] stars; (2) The conflict between the very short time scales implied both by the nebulae and by the low envelope mass expected for a hydrogen-poor Post-AGB star, and the much longer timescales implied by other considerations; (3) The fact that the dense nebulae favour a fatal-thermal pulse (FTP) scenario whereas stellar model calcuations favour a VLTP scenario to obtain hydrogen-poor stars.

Long time scales require slow post-AGB evolution of the star. Slow evolution is expected for stars with very low core mass. However, their luminosity would also be very low, which is not observed. The only other way to ensure slower

Post-AGB evolution is by replenishing the envelope of the star, to compensate for the effects of nuclear burning and wind mass loss. This would require accretion of mass from the envelope back onto the star. An accretion scenario could indeed explain the link between dense, compact nebulae and the peculiar evolution. Slow accretion could also trigger a VLTP on the cooling track when otherwise this might not occur.

An accretion model is highly speculative. It is however clear that the FTP and VLTP scenarios, while adequate for the other [WC] stars, do not as yet provide a good explanation for the IR-[WC] stars.

References

Cohen M., Barlow M.J., Sylvester R.J., Liu X.-W., Cox P., Lim T., Schmitt B., Speck A.K., 1998, ApJ, 513, L135

Dudziak G., Pequignot P., Zijlstra A.A., Walsh J.R., 2001, A&A, in press

Gorny S.K., Stasinska G., 1995, A&A, 303, 893

Herwig F., Blöcker T., Langer N., Driebe T., 1999, A&A, 349, 15

Manchado A., Guerrero M.A., Stanghellini L., Serra-Ricart M., 1996, The IAC Morphological Catalog of Nortern Galactic Planetary Nebulae

Waters L.B.F.M., Beintema D.A., Zijlstra A.A., de Koter A., Molster F.J., Bouwman J., de Jong T. Pottasch S.R., de Graauw Th., , 1998, A&A, 331, L61

Zijlstra A.A., Gaylard M.J., te Lintel Hekkert P., Menzies J.W., Nyman L.-A., Schwarz, H. E. 1991, A&A, 243, L9

Zijlstra A.A., van Hoof P.A.M., Chapman J.M., Loup C. 1994, A&A, 290, 228

ANALYSIS OF 5 PN WITH [WC8-11] NUCLEI

Krzysztof Gesicki
CA UMK, ul.Gagarina 11, PL-87-100 Torun, Poland
gesicki@astri.uni.torun.pl

Agnes Acker
Observatoire de Strasbourg, 11, rue de l'universite, F-67000 Strasbourg, France
acker@astro.u-strasbg.fr

Abstract We analyzed five planetary nebulae (PN) with [WCL]-type nuclei. These central stars are placed soon after the departure from AGB. The parameters obtained in our modeling show some discrepancies with the literature values, especially for He 2–142 and He 3–1333.

1. INTRODUCTION

The central stars of planetary nebulae (CSPN) are not a uniform group. There are among them objects classified as:
– type [WC] with subgroups [WCL] and [WCE]
– PG 1159
– something intermediate, sometimes variable, [WC]–PG 1159
– type Of i.e. with absorptions and emission lines
– type O i.e. with absorption lines, without emission
– continuous spectrum only

Out of this multitude of groups the class [WC]-type seems to be the easiest to identify by the spectral characteristics. Many publications have been devoted to this class, one interesting to us conclusion follows: the planetary nebulae around such CSPNs are not distinguished from other nebulae.

We are continuing our analysis of [WC]-type CSPN, see Gesicki & Acker (1996). The presented sample is limited to central stars which are youngest and therefore soon after the departure from AGB: PN G 001.5-06.7 = SwSt 1, PN G 027.6-06.7 = M 2–43, PN G 327.1-02.2 = He 2–142, PN G 332.9-09.9 = He 3–1333, PN G 352.9+11.4 = K 2–16.

R. Szczerba and S.K. Górny (eds.), Post-AGB Objects as a Phase of Stellar Evolution, 163–166.

2. THE ANALYZED NEBULAE

The five nebulae were observed by one of us (A.A.) at ESO with CAT telescope and with long slit. The spectral region covered Hα and neighboring [NII] lines only.

A detailed description of computer codes applied for the analysis is presented in our earlier paper (Gesicki et al., 1996). For each nebula we construct a model, which is a spherical shell defined by inner and outer radii, density distribution and velocity field. Then the photoionization structure is calculated and the emission lines are integrated. Simultaneously the free parameters are changed until the flux in Hβ, line ratios and plasma parameters agree with the observed values. Then the line shapes are compared with the observed profiles and the corrections to the velocity profiles are introduced and calculations repeated.

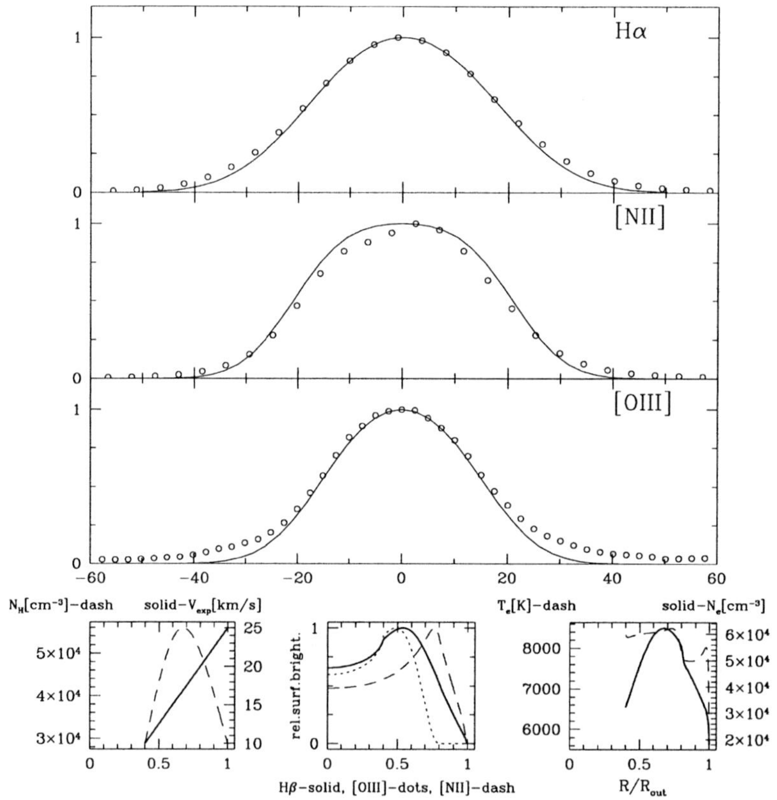

Figure 1 The nebula PN G 001.5-06.7 = SwSt 1. The observed spectral line (circles) and the fitted modelled line (continuous line) for hydrogen, [NII] and [OIII]. In the lowest boxes we present the adopted density and velocity field (left box), surface brightness profiles (middle) and electron temperature and density (right).

As four from our objects are unresolved and lacking the surface brightness images we adopted the same density radial structure for all PN. The inner nebular radius is assumed to be 0.4 of the outer radius, except of the one spatially resolved PN. The same average chemical composition is applied for all calculations.

For one object (SwSt 1) we present the results in a graphic form. For this PN the observations were supplemented by [OIII] 500.7nm line taken from the paper Gesicki & Zijlstra (2000). Therefore the modelling is more reliable than for the other objects. The small differences in width of the lines indicate a velocity gradient, but the broad Gaussian shape of lines can be best reproduced with the assumption of additional turbulent motions.

Table 1 The nebular parameters

PN G	001.5-06.7	027.6+04.2	327.1-02.2	332.9-09.9	352.9+11.4
name	SwSt 1	M 2–43	He 2–142	He 3–1333	K 2–16
		parameters found in literature			
type	[WC 10]	[WC 8]	[WC 9]	[WC 11]	[WC 11]
T_*	38000	65000	35000	17000	30000
T_{eb}	26600	73400	25900		19600
log L_*	4.1	3.24	4.0	3.0	
dist	4.0	1.4	3.5	1.5	3.4
log Ne		4.0	4.6		
ext C	0.7	2.8	1.3	1.3	
logF(Hβ)	-10.3	-13.1	-11.84	-12.15	
		parameters adopted for models			
T_{eff}	38000	65000	26000	26000	29000
log L/Lo	4.1	3.2	4.0	3.5	3.6
dist	4.0	1.4	3.5	1.5	3.4
R_{out}	0.02	0.005	0.03	0.02	0.18
M_{neb}	0.05	0.003	0.05	0.015	0.5
V_{exp}	10-25	con 20	10-30	20-80	con 40
V_{turb}	10	10	0	0	10
logF(Hβ)	-9.8	-9.6	-10.2	-9.9	-10.4
M_*	0.62	0.63	0.61	0.63	0.58

The Table presents some data found in literature concerning the objects from our sample. The parameters adopted for making the models often differ from their literature values because it was impossible to make the model consistent with all values.

In the last row we present the CSPN masses estimated according to the recipe discussed in the contribution of Gesicki & Zijlstra in this volume. Four of our PN have rather high nuclei masses.

3. DISCUSSION

We analyzed five PN with [WC8-11]- nuclei, placed soon after the departure from AGB. The parameters calculated in our modeling show some difference with those published in the literature, especially the stellar effective temperatures of He 2−142 and He 3−1333.

The presence of turbulent motions has been established for three PN from our sample. It can be seen that the two objects without turbulent motions are simultaneously the coolest ones. However one should remember that these two PN show some structures in their spectral lines which cannot be explained by the applied spherically symmetric model. These structures are present only in the line cores and the HWHM of the profiles is comparable for both [NII] and Hα lines what may still indicate turbulence.

The four CSPN with compact nebulae seem to be more massive than the "canonical" value of 0.6 $M_{\odot}$. The one central star with a very big nebula K 2-16 is less massive.

To be conclusive this work certainly needs to be extended on a larger sample. A synthetic paper about turbulence in PN with [WC] nuclei will appear soon (Acker, Gesicki, Durand, Grosdidier, 2001, in preparation).

Acknowledgments

This project was partially supported by the CNRS in the frames of the JUMELAGE "Astronomie Pologne" programme. K.Gesicki acknowledges partial support from Polish KBN grant No. 2.P03D.020.17.

References

Gesicki K., Acker A., 1996, ApSpSci 238, 101
Gesicki K., Acker A., Szczerba R., 1996, A&A 309, 907
Gesicki K., Zijlstra A.A., 2000, A&A 358, 1058

MASSES OF CENTRAL STARS OF PLANETARY NEBULAE

Krzysztof Gesicki
CA UMK, ul.Gagarina 11, PL-87-100 Torun, Poland
gesicki@astri.uni.torun.pl

Albert A. Zijlstra
Astrophysics Group, Phys. Dept., UMIST, PO Box 88, Manchester M60 1QD, United Kingdom
aaz@iapetus.phy.umist.ac.uk

Abstract We present the masses of central stars of 44 planetary nebulae (PN), obtained from measured [OIII] expansion velocities and derived from the dynamical ages. The stellar core mass distribution is narrow, peaking at 0.61 $M_{\odot}$. We discuss the correlations between central stars of PN (CSPN) masses and other stellar and nebular parameters.

1. INTRODUCTION

Recently we presented [OIII] expansion velocities for planetary nebulae in the Galactic Bulge and Halo, and in the Sagittarius Dwarf spheroidal galaxy (Gesicki & Zijlstra, 2000).

The observations were carried out with the ESO 1.4m CAT telescope in July 1993. The so-called short camera was used in the blue arm of the echelle spectrograph, with a resolution of 30000, corresponding to 6 km s^{-1}. The [OIII] 500.7nm line was observed.

Photo-ionization models are used to derive internal velocity fields, showing how the expansion velocity varies with radius. The velocities increase with the distance from the star, in agreement with hydrodynamical models. Dynamical ages are derived from these velocities and are corrected for the effects of post-AGB acceleration and non-uniform velocities.

Masses for the central stars are obtained from relations between dynamical ages and stellar temperatures. We used the fact that in such a plot the evolutionary tracks (parameterized by stellar mass) are almost parallel. The resulted stellar core mass distribution is narrow, peaking at 0.61 $M_{\odot}$. This is higher than

R. Szczerba and S.K. Górny (eds.), Post-AGB Objects as a Phase of Stellar Evolution, 167–170.

predicted for the Bulge by initial-final mass relations, but consistent with the local white dwarf mass distribution.

In this contribution we discuss in more detail the dependencies of the CSPN masses on some stellar and nebular parameters.

This technique for obtaining stellar masses can be used to compare different populations. As an example see the contribution of Gesicki & Acker in these proceedings, who suggest that [WC]-type stars have higher masses than average for CSPN.

1.1. CSPN MASSES ON THE H-R DIAGRAM

In Figure 1 we show schematically the CSPN masses on the H-R diagram, the size of each symbol corresponds to the mass. The smallest value is 0.58, and the largest 0.63. The lines show predicted luminosities versus temperature for three post-AGB models with core masses of 0.565, 0.6 and 0.625 $M_\odot$ (adopted from Schönberner 1983; Gorny et al. 1997 and Blöcker 1995 respectively). The masses correlate very poorly with the theoretical sequences in the H-R diagram. This indicates that the observational uncertainty in the luminosity is too large to be a useful indicator of stellar mass.

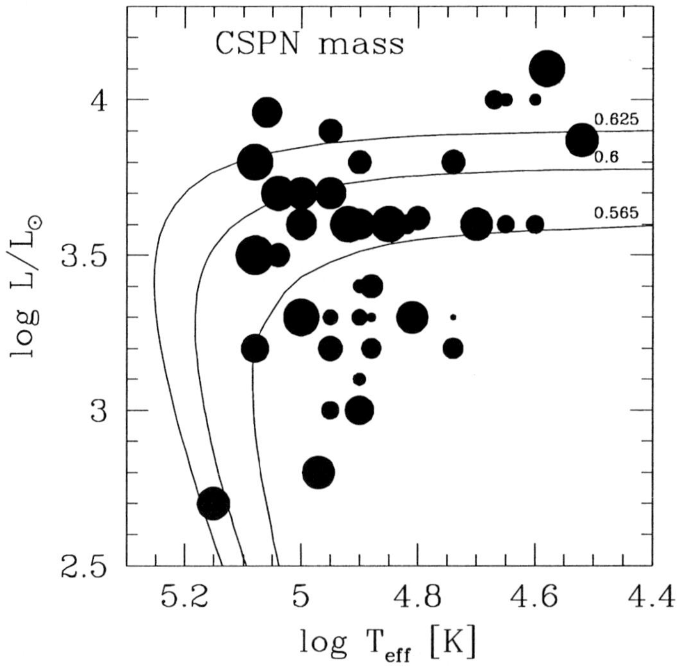

Figure 1 HR diagram for the PNe. The tracks correspond to theoretical models with core masses of 0.565, 0.6 and 0.625 $M_\odot$.

2. CSPN MASS VS. EFFECTIVE TEMPERATURE

Figure 2 shows a tendency for lower-mass central stars to have lower effective temperature. The upper boundary is probably where the PNe become optically thin and quickly diminish in brightness: this will happen at lower effective temperature for lower-mass stars because these stars evolve much slower. The lack of (Bulge) objects in the lower-right corner is unexplained. There is no observational bias against such objects. A possible explanation could be that the stellar evolution at higher core mass is accelerated due to post-AGB mass loss, up to a stellar temperature of 60000 K.

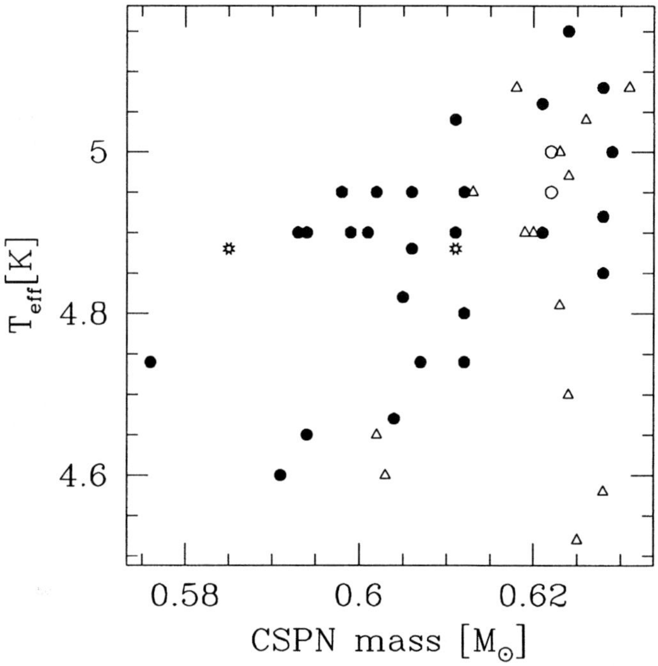

Figure 2 PN core mass versus effective temperature. Different stellar groups are marked with different symbols: filled circles: Galactic Bulge PNe; triangles: Galactic Disk PNe; stars: two Galactic Halo PNe; open circles: two Sagittarius Dwarf Galaxy PNe.

3. CSPN MASS VS. NEBULAR EXPANSION VELOCITY

Stasinska et al. (1997) showed that more massive CSPN tend to have nebulae with higher expansion velocities. Our data (Figure 3) however do not clearly support this trend. The only evident trend is that Bulge CSPN with mass below 0.6 $M_\odot$ have a (much) lower expansion velocity.

We note that the derivation of the CSPN mass depends on the dynamical age and therefore on the expansion velocity. Although there is no simple dependence, typically an error in V_{exp} of 5 km s^{-1} results in an error in the mass of 0.01; this corresponds to a roughly diagonal displacement in Figure 3. The absence of a trend at higher core masses cannot be explained with uncertainties in the expansion velocities.

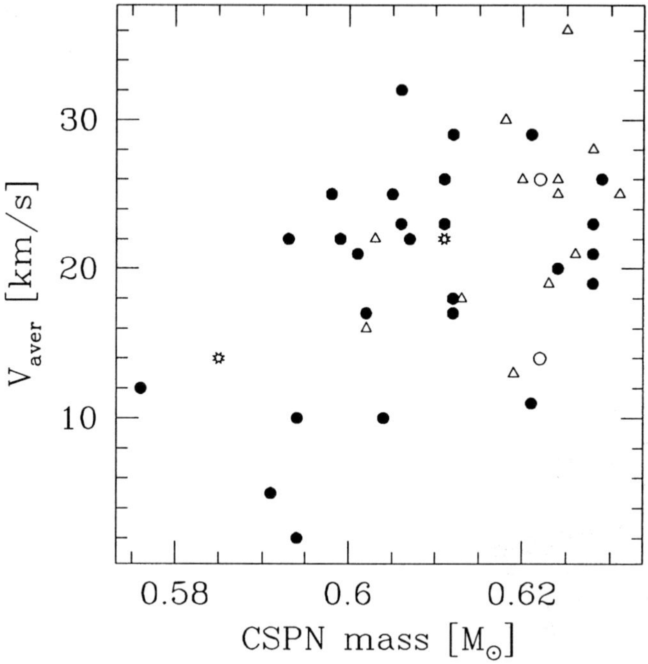

Figure 3 PN core mass versus the expansion velocity. Meaning of the symbols is the same as in previous figure.

Acknowledgments

This project was partially supported by the British-Polish Joint Research Collaboration Programme. K.Gesicki acknowledges partial support from Polish KBN grant No. 2.P03D.020.17.

References

Blöcker T., 1995, A&A 299, 755
Gesicki K., Zijlstra A.A., 2000, A&A 358, 1058
Gorny S.K., Stasinska G., Tylenda R., 1997, A&A 318, 256
Schönberner D., 1983, ApJ 272, 708
Stasinska G., Gorny S.K., Tylenda R., 1997, A&A 327, 736

III

BINARITY AND ONSET OF ASYMMETRICAL STRUCTURES

AGB EVOLUTION IN BINARY SYSTEMS

Adam Frankowski

N. Copernicus Astronomical Center, Toruń, Poland

torque@ncac.torun.pl

Abstract Results from numerical modelling of the AGB phase in binaries are presented. The method is to use single star evolutionary tracks and modify them by introducing binary interactions. Considered effects include enhancement of mass loss from the AGB primary, accretion of matter from the AGB wind onto the secondary, angular momentum loss from the system and angular momentum transfer between the binary components. Two different mass loss enhancement prescriptions are tested. An attempt to include magnetic effects is also made. The influence of various model parameters on the system evolution is briefly discussed.

1. INTRODUCTION

The connection between AGB evolution and binarity is not a new issue. First, we just *see* AGB stars in binaries – ca. 20% of symbiotic systems contain AGB giants, some of the so-called intrinsic S stars (which are believed to be AGB stars) also have a companion, and even the prototypic Mira, *o* Cet itself, is a binary star. Second, we have a bit more indirect indications – although 80% of symbiotics now contain 'plain' red giants, white dwarfs in these systems must have passed through the AGB stage. Similar systems are extrinsic S stars and also Ba and CH stars, enriched in s-process elements when their companions were wind blowing AGB giants. Further examples of systems that have passed an AGB episode include cataclysmic variables and binary central stars of planetary nebulae. Also, it is often suggested, that many of the amazing shapes of planetary nebulae result from interactions with a – stellar or substellar – companion (e.g. Soker 1995).

It is then interesting to pose a question if having a companion can noticeably influence the evolution of an AGB star. Such effects are common in close binaries, where the evolution of the system components may be significantly different than in the single star case. Apparently an AGB star – with its large radius of hundreds of $R_\odot$ – has problems with being a member of a *close* binary system. But all what counts is the relative size. An AGB giant, inflating on its

R. Szczerba and S.K. Górny (eds.), Post-AGB Objects as a Phase of Stellar Evolution, 173–180.

way to the tip of the asymptotic branch, can fill the Roche lobe even in a quite wide system. Because of the giant's deep convective envelope, this is likely to trigger a common envelope (CE) event, leading to substantial shrinkage of the system orbit and ejection of the envelope. It is also postulated (e.g. by Han et al. 1995) that a stable Roche lobe overflow (RLOF) is possible, when the mass of the AGB star is sufficiently reduced at the moment it overfills the Roche lobe.

Moreover, it has been hypothesized (Tout & Eggleton 1988, hereafter TE), that even when the giant is a member of a relatively wide, detached binary, but fills a significant fraction of its Roche lobe, the tidal interaction with a companion can noticeably enhance the giant's wind. However, the exact mechanism and magnitude of this possible enhancement remains unknown.

With or without such an enhancement, the AGB wind carries away mass and angular momentum from the system. A fraction of this mass and angular momentum can be transferred to the companion. As a result, the evolution of the AGB giant is accompanied by the evolution of the system parameters.

2. THE THEORETICAL MODEL

The model presented here has been constructed to explore this coupling between the stellar and orbital evolution. Full treatment of the problem should involve simultaneous time-dependent calculations of both evolutions – stellar and orbital. Given that full evolutionary calculations – even in the single star case – are complicated and time-consuming, the procedure used in the present study is based on a number of simplifying assumptions that make the problem more accessible and allow to circumvent some troublesome issues.

As a starting point for the model, the published single star evolutionary tracks of Blöcker (1995a, 1995b) were adopted. These tracks were then modified by applying enhanced mass loss rates. It was assumed, that the mass loss rate is the only stellar characteristic directly affected by binary interactions. The assumption that the luminosity evolution remains unchanged is justified by the effective decoupling of the giant's core (that dictates the luminosity) from the hydrogen-rich envelope (that is subject to mass loss). However, adopting the effective temperature (and thus the radius) from an original track is an acceptable approximation only to some point. This is so, because eventually the envelope mass reduction causes the convection to retreat in mass coordinate. The envelope mass at which this retreat occurs was estimated, and below this limit the effective temperature was derived from an appropriate relation between the envelope mass and the effective temperature.

A formula for the enhanced mass loss, $\dot{M}$, in the most general form can be written as:

$$\dot{M} = \dot{M}_s(1 + E),$$

where $\dot{M}_s$ is the mass loss rate of a single star of the same intrinsic parameters, and E is the enhancement term. There is no general agreement as to the nature (and even the very existence) of the tidally induced enhanced wind. Thus, three different prescriptions for the presumed mass loss enhancement have been implemented: (a) no enhancement, (b) the enhancement proportional to $(R/R_{RL})^6$, where R is the mass losing star radius and R_{RL} is the Roche lobe radius (as suggested by TE), (c) the enhancement proportional to $(R/R_{RL})^3$ (as derived in Frankowski & Tylenda 2001, hereafter FT).

Besides the modified evolution of the primary (by which I mean the AGB star, even if it is actually less massive than its companion), the evolution of the system parameters is followed. The most important of these parameters are the mass ratio, the orbital separation and the orbital period. Assumptions made as to the binary evolution are the following:

1 the orbits are circular,

2 the angular momentum considerations include the orbital motions and the rotation of the primary's envelope,

3 the rotation of the primary's core and of the secondary can be neglected.

Processes concerned include mass and angular momentum loss from the system, accretion of the matter from the AGB wind onto the secondary (described as in Han et al. 1995), and angular momentum transfer between the binary components via tidal forces. The tidal interactions leading to the exchange of the angular momentum between the orbital motion and the rotation are modelled according to concepts of Zahn 1977 and Keppens 1997.

The rate of the systemic angular momentum loss, that plays a key role in the orbital evolution, depends on the specific angular momentum of the wind material. In the simplest approach, this can be set to be equal to the specific orbital angular momentum pertaining to the primary. This is equivalent to assuming that the primary is not rotating or that the wind is emanating from a point source at the giant's center of mass. This, however, is not the case, and when the primary's radius is comparable to the distance from its center of mass to the center of mass of the system (this may be the case, especially when the inflated primary is more massive than the secondary), the correction may be significant, provided that the system is nearly at corotation. So, the fact that the wind is blown from the rotating surface of the giant should be accounted for in the calculations.

Much higher rates of the angular momentum loss can be expected if the wind corotates with the star up to a certain radius, due to viscosity (as suggested by Asida & Tuchman 1995), or magnetic fields (e.g. Mestel & Spruit 1987; Schrijver & Pols 1993; Lery et al. 1998). A rather conservative estimate of Soker & Harpaz 1992 allows for magnetic fields not exceeding ~ 10 G on the surface of an AGB star. This value, plus the dipole magnetic field geometry, has been adopted in the present calculations when the magnetic effects were taken into account.

Effects of the Roche lobe overflow and the common envelope phase are also investigated, when required, using a simple approach of de Kool 1990. It is assumed, that the common envelope is ejected at a cost of a fraction $\alpha_{\rm CE}$ of the gravitational binding energy released in the spiral-in process of the secondary and the giant core.

3. RESULTS AND DISCUSSION

With the assumptions and prescriptions described above, sequences of the evolutionary calculations have been performed, using the Blöcker's single-star AGB tracks of initial (ZAMS) mass 3, 4 and $5M_{\odot}$, for different values of input parameters. Here I present some preliminary results for the $3M_{\odot}$ track, which, in the Blöcker's calculations, ends up as an $0.625M_{\odot}$ white dwarf.

In most calculations the secondary initial mass, M_2, was set to $1M_{\odot}$, and the CE ejection parameter $\alpha_{\rm CE}$ was 1.0. Other model parameters, as the initial system separation, $A_{\rm i}$, the wind enhancement formula used and the value of the magnetic field, B, are indicated in the figure captions and in the descriptions below. The radius of the AGB star, R_{AGB}, is used as an additional convenient unit of measure of the initial orbital separation.

Fig. 1 presents three typical cases of the orbital evolution:

1 accelerating orbital shrinkage, leading to a CE (solid line),

2 accelerating widening of the orbit (dashed line),

3 accelerating shrinkage followed by accelerating and sometimes almost abrupt widening (dotted line).

The figure shows the results for a specific set of input parameters (the secondary initial mass $M_2 = 1M_{\odot}$, the wind enhancement according to TE, the magnetic field $B = 10$ G, and the initial separations of $723R_{\odot}$, $963R_{\odot}$, and $1445R_{\odot}$), but the indicated types of the orbital evolution apply in general. Steps or bumps visible in the plot are caused by thermal pulses of the AGB star.

The first behavior results when the mass transfer from the primary to the less massive companion dominates over the systemic mass loss, or when there are additional angular momentum sinks, as tidal lock of the AGB envelope, effective especially if the magnetic braking is present. The second case takes place when the mass loss from the system dominates over the angular momentum loss. The simplest example of such a situation is that of negligible tidal coupling and no accretion by the companion. Then, the orbital evolution is described by the Jeans theorem: if the mass lost from the system carries off the same specific angular momentum as the specific angular momentum of the mass losing star,

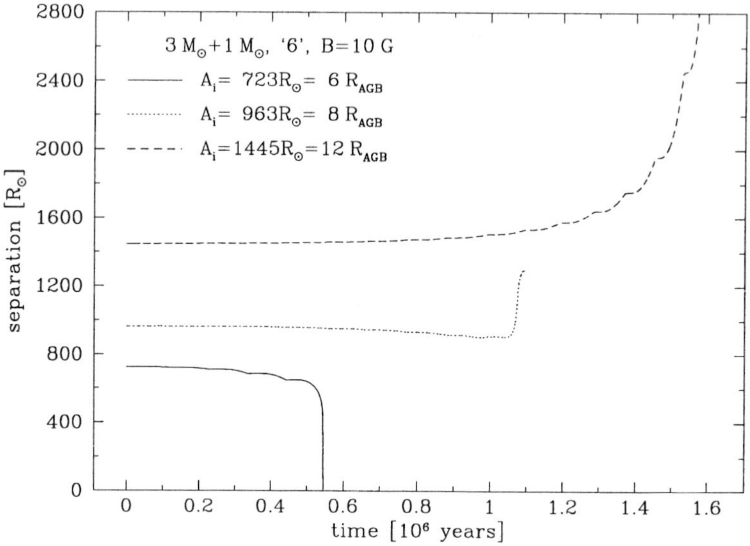

Figure 1 Three types of orbital separation evolutionary behavior (see text), shown for a $3M_\odot$ AGB star with a $1M_\odot$ secondary, with TE wind enhancement formula (denoted by '6' in the figure legend), and with the dipole magnetic field of 10 G. Initial separations are: solid line – $723R_\odot$ $(6R_{AGB})$, dotted line – $963R_\odot$ $(8R_{AGB})$, dashed line – $1445R_\odot$ $(12R_{AGB})$. Time is measured from the beginning of the original single star evolutionary track.

then the product of the separation and the total mass of the system remains constant. The third type of evolution is caused by an interplay of the processes of mass and angular momentum loss and transfer. The system loses orbital angular momentum quite efficiently, but the secondary manages not to dive into the primary's envelope until the mass ratio is reversed. The subsequent action of the mass loss from the system and the mass transfer from the now *less* massive primary to the now *more* massive companion results in an expansion of the system. The situation is similar as in the case of a stable RLOF.

The effect on the mass loss evolution of the three different tidal wind enhancement descriptions (see Sect. 2) is shown in Fig. 2. Other input model parameters are fixed: the secondary initial mass is $1M_\odot$, the initial separation is $963R_\odot$ $(8R_{AGB})$, and the magnetic field is absent ($B = 0$ G). The value of the proportionality constant in the enhancement formula of TE is taken from Han et al. 1995. The formula of FT is scaled to cause at $R/R_{RL} = 0.5$ an effect equal to that of the TE prescription. From Fig. 2 it can be seen that both prescriptions result in mass loss up to one magnitude higher than in the case of single star evolution.

Fig. 3 displays final ($P_{\rm f}$) vs. initial ($P_{\rm i}$) orbital periods for models calculated with various input parameters. All the models plotted have the secondary of

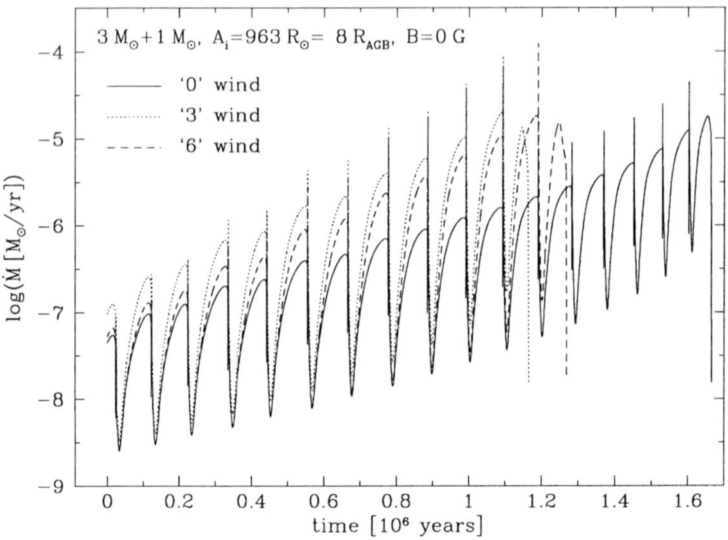

Figure 2 Mass loss evolution with various formulae for the tidal wind enhancement. Solid line – no tidal enhancement ('0' wind), dotted line – enhancement according to FT ('3' wind), dashed line – according to TE ('6' wind). Results for a $3M_{\odot}$ AGB star with a $1M_{\odot}$ secondary and initial separation of $963R_{\odot}$ ($8R_{AGB}$), in absence of magnetic effects ($B = 0$ G), are shown.

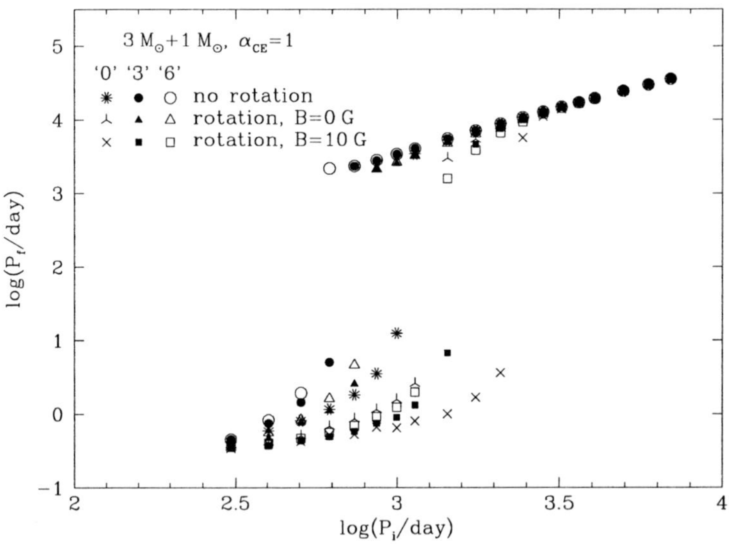

Figure 3 Final vs. initial orbital periods for models with various input parameters. In all the models $M_2 = 1M_{\odot}$ and $\alpha_{CE} = 1$. Circles denote models neglecting primary's rotation, triangles – models including rotation but without magnetic effects, and squares – models with rotation and magnetic field of 10 G. Skeletal symbols indicate no wind enhancement ('0'), filled symbols – the FT enhancement ('3'), open ones – the formula of TE ('6').

initial mass $M_2 = 1M_\odot$, and if CE occurs, the fraction of the orbital binding energy used to eject the common envelope is $\alpha_{\rm CE} = 1.0$. It can be seen that the enhanced mass loss leads to a shorter "critical" initial period, delimiting CE and non-CE cases. The increased angular momentum loss (caused by inclusion of the primary's finite dimensions, rotation and magnetic effects) has the opposite effect – the more effective is this loss, the wider systems end up in a CE.

The most striking feature of this figure is a gap in final periods, spanning from $\sim$10 days to $\sim$3 years. This is an effect of the competition between the tendencies to shrink and to widen the orbit analyzed above. When one of these wins, the system components are quite suddenly pushed away or together.

Interestingly, the position of the short-period part of the $P_{\rm f}$ distribution in Fig. 3 is consistent with observations of close binary nuclei of planetary nebulae (PNN), that all have periods ranging from 0.1 day to 16 days (see e.g. Bond 2000). Although observational selection effects may play some undetermined role in shaping the observed distribution, this coincidence seems promising. Moreover, the position of this gap was found to be the same also in sequences calculated with a less massive secondary of $0.5M_\odot$. However, the lower edge of the gap in $P_{\rm f}$ depends on the value of the CE ejection efficiency, $\alpha_{\rm CE}$. The effective $\alpha_{\rm CE}$ may be greater than 1 when there are additional energy sources available (e.g. recombination energy in the ionization zones). This energy, if deposited into the mass motion, can facilitate the ejection of the envelope (see e.g. Iben & Livio, 1993). Increasing $\alpha_{\rm CE}$ from 1 to, say, 10, causes an increase in $P_{\rm f}$ of the CE systems by a similar factor. This way the period gap is partially filled. But at the same time, this empties the region of the final periods below several days and is inconsistent with the cited observations of binary PNN.

On the other hand, the existence of systems, such as symbiotic stars, that have periods in the range avoided in the results presented here, seems to be in contradiction with these results. There are several possible explanations to this apparent conflict. First, this observational evidence could be interpreted in favor of the hypothesis that a binary companion causes strong wind enhancement. The enhanced mass loss could eject the giant's hydrogen-rich envelope before the star reached the gargantuan dimensions of the late AGB phase (perhaps even before the onset of thermal pulses, i.e. on the early AGB (E-AGB)), thus allowing for detached evolution of a system that otherwise would form a common envelope. The system period and separation would increase from their initial values, but not enough to exclude the final periods of the order of $\sim$100 days. However, this scenario can have some trouble with the extrinsic S stars and Ba/CH stars, which have similar parameters as symbiotic systems, but their peculiar surface abundances are believed to require accretion of s-process enriched matter from a thermally pulsing companion that has undergone the 'third dredge-up'.

Another possibility has been suggested by Iben 2000. He argues, that for some reason the Roche lobe filling leads to a common envelope only for E-AGB stars, and not in the case of thermally pulsing AGB (TP-AGB) stars. Then, systems that formally should enter a CE at the TP-AGB stage, evolve without any drastic shrinkage of the orbit. Also Jorissen et al. 1998 suggest that systems with an AGB stars can avoid the CE orbital decay. They list a number of possible reasons, from reduced effective binding energy of the giant's envelope, through the mass ratio reversal (due to a tidally enhanced wind), to degeneration of the critical Roche surfaces in the presence of the wind-driving force.

More theoretical work and calculations with different sets of input parameters are needed before these speculations could be replaced by more firm conclusions.

Acknowledgments

This work has been supported from the grant No. 2.P03D.020.17 of the Polish State Committee for Scientific Research.

References

Asida S.M., Tuchman Y., 1995, ApJ 455, 286

Blöcker T., 1995, A&A 297, 727

Blöcker T., 1995, A&A 299, 755

Bond H., 2000, in: Asymmetrical Planetary Nebulae II: From Origins to Microstructures, Kastner J.H., Soker N., Rappaport S. (eds.), ASP Conf. Series 199, 115

de Kool M., 1990, ApJ 358, 189

Frankowski A., Tylenda R., 2001, A&A 367, 513 (FT)

Han Z., Eggleton P.P., Podsiadlowski P., Tout Ch.A., 1995, MNRAS 277, 1443

Iben I. Jr., 2000, in: Asymmetrical Planetary Nebulae II: From Origins to Microstructures, Kastner J.H., Soker N., Rappaport S. (eds.), ASP Conf. Series 199, 107

Iben I. Jr., Livio M., 1993, PASP 105, 1373

Jorissen A., Van Eck S., Mayor M., Udry S., 1998, A&A 332, 877

Keppens R., 1997, A&A 318, 275

Lery T., Heyvaerts J., Appl S., Norman C.A., 1998, A&A 337, 603

Mestel L., Spruit H.C., 1987, MNRAS 226, 57

Schrijver C.J., Pols O.R., 1993, A&A 278, 51

Soker N., 1997, ApJS 112, 487

Soker N., Harpaz A., 1992, PASP 104, 923

Tout Ch.A., Eggleton P.P., 1988, MNRAS 231, 823 (TE)

Zahn J.-P., 1977, A&A 57, 383

PLANETS AND AXISYMMETRIC MASS LOSS

Noam Soker

Dept. of Physics, University of Haifa at Oranim, Tivon 26006, Israel

soker@physics.technion.ac.il

Abstract Bipolar planetary nebulae (PNe), as well as extreme elliptical PNe are formed through the influence of a stellar companion. But half of all PN progenitors are not influenced by any stellar companion, and, as I show here, are expected to rotate very slowly on reaching the upper asymptotic giant branch; hence they expect to form spherical PNe, unless they are spun-up. But since most PNe are not spherical, I argue that $\sim$ 50% of AGB stars are spun-up by planets, even planets having a mass as low as 0.01 times the mass of Jupiter, so they form elliptical PNe. The rotation by itself will not deform the AGB wind, but may trigger another process that will lead to axisymmetric mass loss, e.g., weak magnetic activity, as in the cool magnetic spots model. This model also explains the transition from spherical to axisymmetric mass loss on the upper AGB. For such low mass planets to substantially spin-up the stellar envelope, they should enter the envelope when the star reaches the upper AGB. This "fine-tuning" can be avoided if there are several planets on average around each star, as is the case in the solar system, so that one of them is engulfed when the star reaches the upper AGB. Therefore I retain earlier predictions (Soker 1996) that on average several planets are present around $\sim$ 50% of progenitors of PNe.

1. INTRODUCTION

1.1. BIPOLAR PNE: BINARY SYSTEM PROGENITORS (CLOSED CASE)

I distinguish two main groups of nonspherical planetary nebulae (PNe): bipolar and elliptical. Bipolar (also called "bilobal" and "butterfly") PNe are defined (Schwarz, Corradi & Stanghellini 1992) as axially symmetric PNe having two lobes with an 'equatorial' waist between them; while elliptical PNe have a general elliptical shape. Bipolar PNe amount to $\sim 10-15\%$ of all PNe (Corradi & Schwarz 1995). The only physical property for which the difference between elliptical and bipolar PNe is larger than the dispersion within each group is the expansion velocity, which is much faster for bipolars. It is almost certain that bipolar PNe are formed from binary systems. In most cases the stellar companion stays outside the mass losing star for the entire evolution, and in

R. Szczerba and S.K. Górny (eds.), Post-AGB Objects as a Phase of Stellar Evolution, 181–188.

the rest it forms a common envelope late in the evolution (Soker 1998a). I think that the observations, e.g., the similarity of bipolar PNe to many symbiotic nebulae, and the theoretical arguments, e.g., mechanism for blowing winds at several$\times$100 km s^{-1}, which support the binary model for the formation of bipolar PNe, are extremely strong (see summary in Table 1 by Soker 1998a). Some of the arguments can be found in Corradi (1995), Corradi & Schwarz (1995), Corradi *et al.* (1999, 2000), Morris (1987, 1990), Mastrodemos & Morris (1999), Soker (1997, 1998a), Soker & Rappaport (2000), and Miranda *et al.* (2000). Several scenarios were proposed for the formation of bipolar PNe from the evolution of single stars. I criticized these scenarios in different papers in the past, and showed that they fail to explain basic facts, and/or they suffer from unphysical arguments. I find the new paper by Matt *et al.* (2000) to fail on the same main points. Some problems with these single star models are: (1) These models do not distinguish elliptical and bipolar PNe, e.g., they can't account for the similarity of bipolar PNe and symbiotic nebulae. (2) The amount of angular momentum they require is much too large for a single star, whether angular momentum directly to influence the wind, or angular momentum required to amplify strong magnetic activity. (3) The single star models contradict the finding that most of the 16 known PNe with central binary systems (Bond 2000) possess an extreme elliptical shape, rather than a bipolar shape. Namely, if single stars can form bipolar PNe, how come AGB stars which go through a common envelope phase, and hence rotate much faster, form "only" elliptical PNe? These 16 PNe by themselves constitute very strong support for the binary model for the formation of both bipolar and extreme elliptical PNe (Bond & Livio 1990; Soker 1997).

1.2. ELLIPTICAL PNE: OPEN QUESTIONS

In addition to the $\sim 10 - 15\%$ of all PNe which are bipolar, and formed from binary systems, there are $\sim 20 - 30\%$ elliptical PNe formed from binary systems (Yungelson, Tutukov & Livio 1993; Han, Podsiadlowski & Eggleton 1995; Soker 1997). Most of these went through a common envelope phase and have extreme structures (Bond & Livio 1990; Bond 2000). By extreme structure I refer to a large concentration of mass in the equatorial plane, i.e., a torus, but there are no polar lobes, hence the PN is not a bipolar PN. The AGB progenitors of the PNe were spun up by their stellar companions to very high velocities, which led to high mass concentration in the equatorial plane. The relevant axisymmetric mass loss mechanisms for a common envelope evolution are summarized by Iben & Livio (1993) and Rasio & Livio (1996), while axisymmetric mass loss mechanisms for stellar companions outside the AGB envelope are listed in Soker (1998a).

But what about the rest of the axisymmetric PNe, which amount to $\sim 50\%$ of all PNe? They were not spun up by a stellar companion, either because the progenitor did not have a stellar companion, or the companion is at a very large orbital separation. The open questions are therefore:
(1) What is the mechanism by which slowly rotating AGB stars can blow axisymmetric winds? Clearly centrifugal forces are unimportant.
(2) Why do many of the elliptical PNe have an outer spherical halo while the inner region is elliptical? Put another way, why does the mass loss geometry change from spherical to axisymmetric only during the very end of the AGB and/or during the early post-AGB phase? (This question, and the answer given later, is the connection of my paper to the title of the meeting.)
(3) Can a single star blow an axisymmetric wind during its final AGB phase and/or post AGB phase, hence forming an elliptical PN?

2. SLOWLY ROTATING AGB STARS

I now give my answers to the questions raised in the previous section.

The axisymmetric mass loss mechanism, was proposed several years ago (Soker 1998b) and was further developed by Soker & Clayton (1999) and Soker & Harpaz (1999). It is assumed that a weak magnetic field forms cool stellar spots, which facilitate the formation of dust closer to the stellar surface, hence enhancing the mass loss rate there. If spots due to the dynamo activity are formed mainly near the equatorial plane, then the degree of deviation from sphericity increases. Based on a crude estimate I claimed (Soker 1998b) that this mechanism operates for slowly rotating AGB stars, having angular velocities of $\omega \gtrsim 10^{-4} \omega_{\rm Kep}$, where $\omega_{\rm Kep}$ is the equatorial Keplerian angular velocity. I would like to stress that I do not propose a new mass loss mechanism. I accept that pulsations coupled with radiation pressure on dust is the mechanism for mass loss (e.g., Bowen 1988), and that the luminosity, radius, and mass of the AGB star are the main factors which determine the mass loss rate (e.g., Höfner & Dorfi 1997). I only suggest that cool magnetic spots facilitate the formation of dust, and that their concentration near the equator causes the mass loss geometry to deviate from sphericity (Soker 1998b; Soker & Clayton 1999). It should also be noted that the required magnetic field is very weak, has no direct dynamic effect, and is expected to form only a very weak X-ray emission. Models based on strong magnetic fields (e.g., Matt *et al.* 2000) are in contradiction with observations that AGB stars are weak X-ray sources.

The transition to axisymmetric mass loss geometry, in the cool magnetic spots model, is attributed to the shielding of radiation by dust during the superwind phase (Soker 2000). Soker (2000) proposed that dust which is formed very close to the surface of a cool spot, practically at its surface, during a high mass loss rate phase (superwind), has a large optical depth, and it shields the region above it from the stellar radiation. As a result the temperature in the shaded region decreases rapidly relative to the surrounding temperature. This leads to further dust formation in the shaded region. Without the formation of dust close to the surface of the spot and the shielding, only large spots, with radii $b_s \gtrsim 0.3 R_*$, allow enhanced dust formation (Frank 1995). This process is effective for small cool spots, but only when mass loss rate is high, as in the superwind phase, hence optical depth is large. Therefore, the equatorial enhanced mass loss rate occurs mainly during the superwind phase at the end of the AGB. Another mechanism for the transition to axisymmetric mass loss, which can operate in parallel to the the shielding of radiation, is a more effective magnetic activity at the end of the AGB (Soker & Harpaz 1999). In addition to the formation of elliptical PNe, the local enhanced dust formation may lead to the formation of filaments, loops, and arcs, as observed in many PNe.

Most single stars rotate too slowly for the amplification of even the weak magnetic field required by the cool magnetic spots model. I argue that most are spun up by planets. This is the subject of the next section.

3. THE ROLE OF PLANETS

This section summarizes my recent paper (Soker 2001) in which I examine the implications of the recently found extrasolar planets on the planet-induced axisymmetric mass loss model for the formation of elliptical PNe. I first show that single stars rotate very slowly as they reach the upper AGB. I concentrate on stars with main sequence mass in the range of $1.3M_\odot < M_{\rm ms} < 2.4M_\odot$. In this mass range the transition from slow main sequence rotators to fast rotators occurs (e.g., Wolff & Simon 1997), hence these stars will clearly demonstrate the evolution of angular momentum, while avoiding some uncertainties with lower mass stars, e.g., the total mass they lose prior to the upper AGB is not well known. I assume that the star rotates as a solid body (i.e., the angular velocity is constant with radius inside the star) along its entire evolution, and that the wind carries specific angular momentum equal to that on the surface of the star. The average initial angular momentum on the main sequence is taken from Wolf & Simon (1997). For the stars considered here, most of the mass loss occurs on the AGB, when the mass of the core is $\sim 0.6M_\odot$. Under these assumptions an analytical expression can be obtained for the angular momentum (and angular velocity) on the upper AGB as a function of the envelope mass retained by the star as it loses mass (Soker 2001). As an example I present the result for a single star evolving on the upper AGB, and which had a mass of $M_{\rm ms} = 1.8M_\odot$ on the main sequence. The figure shows the evolution of the angular velocity (solid line), in units of $\omega_{\rm Kep}$, and the angular momentum (dashed line), in units of the orbital angular momentum of Jupiter J_J, as a function of the envelope mass left in the envelope. We note the fast decrease of the angular velocity as envelope mass decreases due to mass loss.

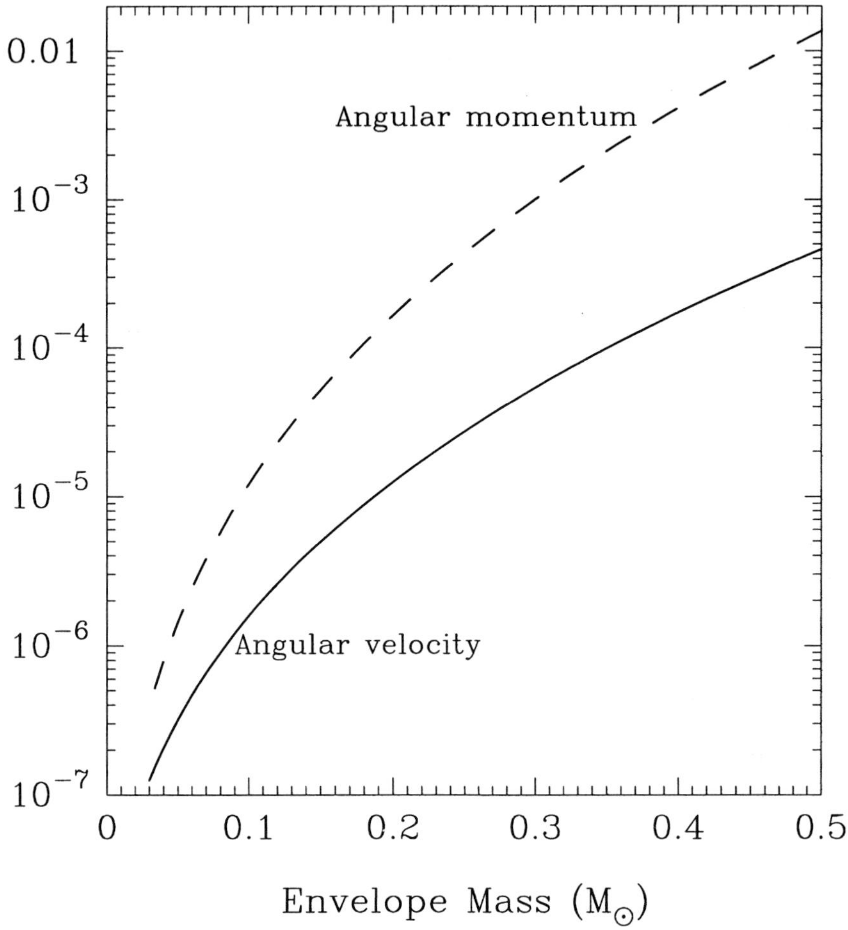

Figure 1 Evolution of the angular momentum and angular velocity as a function of the mass left in the AGB stellar envelope of a star which had a mass of $1.8M_\odot$ on the main sequence (for details see Soker 2001). Angular momentum is in units of Jupiter's orbital angular momentum, and angular velocity in units of the Keplerian angular velocity on the stellar equator.

In the cool magnetic spots model the role of the rotation is mainly to shape the magnetic field into an axisymmetric configuration (on average), and it may operate efficiently even for an envelope rotating as slowly as $\omega \sim 10^{-4}\omega_{\rm Kep}$ (Soker & Harpaz 1999). From the figure we see that single stars will not possess the required angular velocity when the envelope mass decreases below $\sim 0.3M_\odot$. However, very low mass planets, down to $\sim 0.01M_J$, where M_J is Jupiter's mass, are sufficient, if they enter the AGB envelope at late stages.

For example, a planet of mass $0.01M_J$ at an orbital separation of 2 AU has an angular momentum about equal to that of an AGB star with envelope mass of $M_{\rm env} = 0.4M_\odot$ which had a main sequence mass of $M_{\rm ms} = 1.8M_\odot$. If such a planet enters the envelope when $M_{\rm env} = 0.2M_\odot$, for example, it will increase the AGB envelope angular momentum by a factor of ~ 30. Taking the solar evolution to the AGB (see Soker 2001) I find that when the envelope mass becomes $0.15M_\odot$, the angular momentum of the AGB sun is $\sim 10^{-4}J_J$, or ~ 0.1 the angular momentum of Earth. By that time the orbital separation will be 1.33 AU (or $290R_\odot$). If the sun at this stage goes through a helium shell flash, so that the radius increases, say, to ~ 1.3 AU then another $\sim 10\%$ increase during the maximum radius in the pulsation cycles may reach the location of Earth, causing the Earth to spiral inside the solar envelope. A detailed analysis of the evolution of the Earth-sun system, until the sun leaves the AGB, for different assumptions and models, is given by Rybicki & Denis (2000). As a result of the deposition of the Earth's orbital momentum, the solar envelope will rotate ~ 10 times faster, or at $\omega \simeq 10^{-4}\omega_{\rm Kep}$. If this occurs indeed in about 7 billion years, then the Earth may be responsible for the PN of the sun being elliptical rather than spherical. However, it is not clear that the sun will engulf the Earth, or that it will form a PN at all (Rybicki & Denis 2000).

For a high probability that a planet will enter the AGB envelope at late stages, i.e., for it to occur in many stars, two things should happen. First, on average there should be several planets around each star (as is the case in the solar system), and second, there should be a fast and significant increase of the stellar radius on the upper AGB. Numerical simulations of AGB stars show that after thermal pulses (helium shell flashes) on the upper AGB, the envelope increases by $\sim 20-30\%$. This is in addition to the increase in the average AGB stellar radius as the core mass increases. So the second condition is fulfilled for upper AGB stars. The first condition is a requirement, hence a *prediction*, of the planet-induced axisymmetric mass loss model for the formation of elliptical PNe. The new addition of the present paper is the relaxation of the minimum mass demand on planets from $\sim 1M_J$ (Soker 1996) to $\sim 0.01M_J$. The motivations for reducing the lower mass limit are the new finding that only $\sim 5\%$ of sun-like stars have Jupiter-like planets around them, and a new model for axisymmetric mass loss, the cool magnetic spots model, which was constructed to work for very slowly rotating AGB stars, as discussed above.

Finally, note that many of the known sun-like stars that have planets around them will not form PNe at all. This is because their orbiting planet will spin-up the envelope and deposit energy already on the stellar red giant branch (RGB), hence mass loss on the RGB is expected to be high, and most of the stellar envelope will be lost already on the RGB. No observable nebula will be formed. So, while in most cases planet companions will lead to the formation of an elliptical rather than a spherical PN, in some cases Jupiter-like planets in close orbits around low mass stars will prevent the stars from forming a PN.

Acknowledgments

This research was supported in part by grants from the Israel Science Foundation and the US-Israel Binational Science Foundation.

References

Bond, H. 2000, in Asymmetrical Planetary Nebulae II: From Origins to Microstructures, ASP Conf. Ser. Vol. 199, eds. J. Kastner, S. Rappaport, and N. Soker, p. 115 (astro-ph/9909516).

Bond, H., & Livio, M. 1990, ApJ, **355**, 568.

Bowen, G. H. 1988, ApJ, 329, 299.

Corradi, R. L. M. 1995, MNRAS, 276, 521.

Corradi, R. L. M., Ferrer, O. E., Schwarz, H. E., Brandi, E., & Garcia, L. 1999, A&A, 348, 978.

Corradi, R. L. M., Livio, M., Schwarz, H. E., & Ulisse, M. 2000, in *Asymmetrical Planetary Nebulae II: From Origins to Microstructures*, eds. J. Kastner, S. Rappaport, and N. Soker, ASP Conference Series Vol. 199, p. 175.

Corradi, R. L. M. & Schwarz, H. E. 1995, A&A, 293, 871.

Frank, A. 1995, AJ, 110, 2457.

Han, Z., Podsiadlowski, P., & Eggleton, P. P. 1995, MNRAS, 272, 800.

Höfner, S., & Dorfi, E. A. 1997, A&A, 319, 648.

Iben, I. Jr., & Livio, M. 1993, PASP, 105, 1373.

Mastrodemos, N., & Morris, M. 1999, ApJ, 523, 357.

Matt, S., Balick, B., Winglee, R., & Goodson, A. 2000, ApJ, in press (astro-ph/0008129).

Miranda, L. F., Torrelles, J. M., Guerrero, M. A., Vazquez, R., & Gomez, Y. 2000, MNRAS, in press (astro-ph/0009396).

Morris, M. 1987, PASP, 99, 1115.

Morris, M. 1990, in *From Miras to Planetary Nebulae: Which Path for Stellar Evolution?* eds. M. O. Mennessier & A. Omont (Paris: Editions Frontieres), p. 520.

Rasio, F. A., & Livio, M. 1996, ApJ, 471, 366.

Rybicki, K. R., & Denis, C. 2000, preprint.

Schwarz, H. E., Corradi, R. L. M, & Stanghellini, L. 1992, A&AS 96, 23.

Soker, N. 1996, ApJL, 460, L53.

Soker, N. 1997, ApJSupp., 112, 487

Soker, N. 1998a, ApJ, 496, 833.

Soker, N. 1998b, MNRAS, 299, 1242.

Soker, N. 2000, MNRAS, 312, 217.

Soker, N. 2001, preprint (astro-ph/0006362).
Soker, N., & Clayton, G. C. 1999, MNRAS, 307, 993.
Soker, N., & Harpaz, A. 1999, MNRAS, 310, 1158.
Soker, N., & Rappaport, S. 2000, ApJ, 538, 241 (astro-ph/9911140).
Wolff, S. C., & Simon, T. 1997, PASP, 109, 759.
Yungelson, L. R., Tutukov, A. V., & Livio, M. 1993, ApJ, 418, 794.

BIPOLAR OUTFLOWS FROM OH/IR STARS

Albert A. Zijlstra
UMIST, department of Physics, Manchester M60 1QD, UK
a.zijlstra@umist.ac.uk

Abstract High-resolution OH data is used to study the onset of bipolarity in post-AGB stars. We find that high outflow velocities and irregular OH spectra are a reliable indicator for bipolar outflows. Velocity–radius diagrams show the presence of two components: a uniform, spherical wind and a collimated flow with velocity increasing linearly with radius. A momentum-driven wind–wind interaction model reproduces the observed features. We show evidence for enhanced post-AGB life times, which may indicate the effect of envelope re-accretion.

1. INTRODUCTION

The majority of planetary nebulae (PNe) show deviations from spherical symmetry. 10–20% show bipolar morphologies, and a few per cent show highly collimated, strong bipolar geometries. The range of morphologies are well explained by the interacting stellar wind model (e.g. Frank & Mellema 1994), where the fast wind from the PN central star star sweeps up the earlier slow AGB wind, amplifying pre-existing smaller asymmetries.

The amplification should not occur until the onset of the hot wind, during the PN phase. However, younger post-AGB objects, with cooler stars, show a large proportion of highly bipolar morphologies. Sahai & Trauger (1998) therefore propose that early in the post-AGB evolution, a fast jet-like wind operates. An alternative possiblity is that at the very end of the AGB, the mass loss suddenly becomes strongly equatorially enhanced, caused by e.g. the spiral-in of a Jupiter-like companion.

High-resolution OH maser observations are used here to study the onset of strong bipolarity. The results indicate wind–wind interactions already in the earliest post-AGB phase.

2. SAMPLE SELECTION AND OBSERVATIONS

A few per cent of all OH/IR stars have irregular 1612MHz OH spectra, sometimes with strong 1667MHz emission. These indicate a reduced velocity-coherence length, which may be an indication of asphericity. We selected ten

R. Szczerba and S.K. Górny (eds.), Post-AGB Objects as a Phase of Stellar Evolution, 189–196.

objects with an OH maser velocity range of at least 50 km s^{-1} in one or more of the OH lines (indicative of wind acceleration), and a spectrum with an emission plateau and/or more than two spectral features (indicative of asphericity). The prime object which shows such characteristics is OH 231.8+4.2, a classical proto-planetary nebula. These ten nebulae were observed with the VLA, the ATCA or MERLIN, at $0.3''$–$1.0''$ spatial resolution and 0.5–2.2 km/s velocity resolution. For each velocity channel, accurate positions were derived by fitting gaussians to the maser spots. The analysis was done using the resulting position–velocity components.

3. THE MODEL

During the earliest post-AGB evolution, stellar wind velocities will increase as the stellar escape velocities increase. This slightly faster wind will act as a snowplough on the earlier AGB wind. If either the AGB or early post-AGB wind has an asphericity, the interface will also be aspherical. OH maser emission is expected from the outer, unperturbed wind and from the dense interface.

While the velocity difference between the two winds is less than about 100 km/s, the interface will be momentum driven. If the two winds have constant velocity, the intershell will move with a time-independent, but direction-dependent velocity. For every direction from the star, the distance to the intershell will be proportional to the velocity: $r(\theta) = v(\theta)t$, so that a linear relation between v and r is observed. Shu et al. (1991) have found a similar result for CO outflows from young stellar objects. There are many observational similarities between the two classes of objects, and the physical basis of the bipolar outflows in old and young objects are likely related.

Fig. 1 shows the prediced velocity–radius relation for the OH masers, assuming the interface is located along two bipolar cones puncturing a spherical AGB shell.

4. OH 231.8+4.2

The best-resolved source in the sample is OH 231.8+4.2, which has a large bipolar nebula around a Mira central star. Although often called a proto-planetary nebula because of its morphology, the presence of a Mira shows it is still on the AGB. It is a likely member of the open cluster M46 (Jura & Morris 1985) with turn-off mass of 3 $M_{\odot}$. With this progenitor mass, a star at the tip of the AGB would be a carbon star. The strong OH emisison therefore indicates that OH 231.8+4.2 has not evolved as a typical OH/IR star. A binary companion has been suggested (Cohen et al. 1985).

The top panel in Fig.2 shows the location of the OH masers which seem to trace a ring or disk, with additional emission located along the position angle of the bipolar nebula (Reipurth 1987). The middle panel shows the velocity-

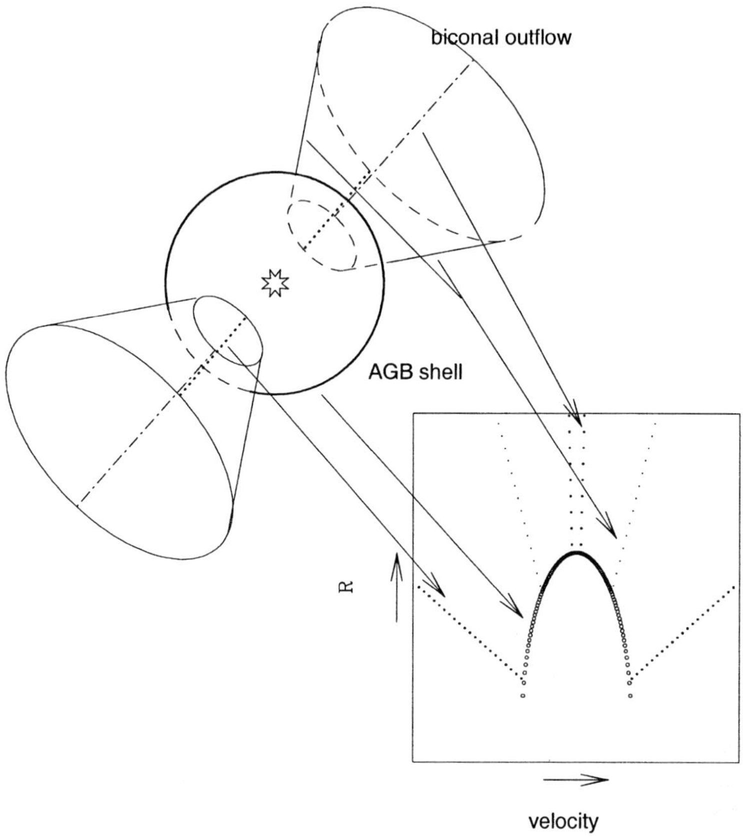

Figure 1 Predicted velocity–radius relations for a geometric model consisting of an AGB shell and a biconal, symmetric outflow. The arrows indicating the contributions from the different locations in the model.

radius diagram, and the bottom panel the total flux at each velocity. The diagram fits the model of Fig. 1 very well. This indicates that both components, the torus or shell with constant expansion velocity, and the intershell with velocity incerasing with distance from the star, are present. We interpret OH 231.8+4.2 as an interaction between two winds of limited ($\sim$ 100 km/s for the fastest) expansion velocity.

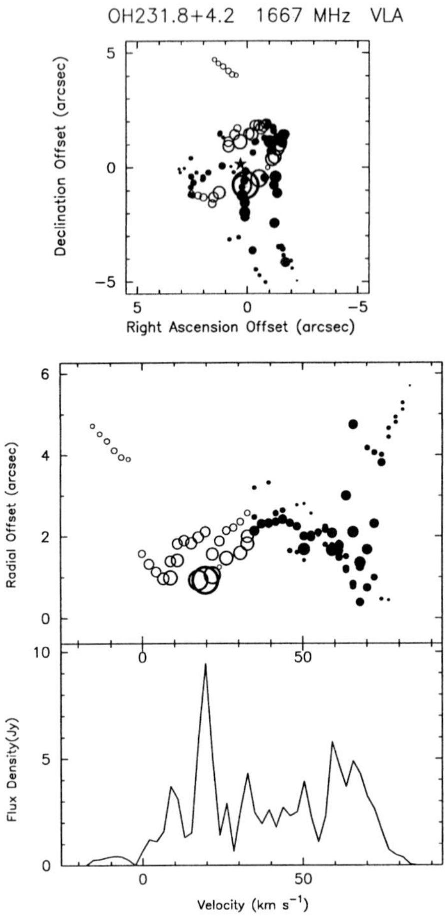

Figure 2 Diagnostic x-y, R-V and I-V diagrams for OH 231.8+4.2. The top panel shows the relative position of each emission component plotted relative to the unweighted emission centroid. The open and filled symbols show blue-shifted and red-shifted emission components respectively. The (estimated) stellar position is indicated by the filled star. The middle panel shows the projected radial offsets of the maser components from the stellar position, plotted against observed LSR velocity. The bottom panel shows the total flux density, I, seen in the channel maps at each velocity.

5. RESULTS

The results for the ten sources are summarised in Table 1. Six objects show a shell/torus component, and all but two show evidence for a 'linear outflow'. The expansion velocities for the shell components are 20–25 km s^{-1} which are typical values for OH/IR stars. Two objects have velocities too high for normal AGB winds and the shell-like region must already have undergone acceleration. Where both components are seen, the linear outflow velocity reaches about twice

Table 1 Component analysis for all observed OH/IR stars. V_* is the stellar velocity (LSR). V_{exp} is the expansion velocity of the shell/torus component (half the total velocity width) and R_0 the radius of this component. For the linear outflow we give the maximum distance from the star at which it is seen ('length') and its maximum velocity with respect to the star (V_l).

common name	V_* km s^{-1}	$V_{\rm exp}$ km s^{-1}	R_0 arcsec	length arcsec	V_l km s^{-1}	star class
OH 231.8+4.3	35	35	2	5.5	55	M9I/III +B?
IRAS 08005−2356	50	40	3.2	-	-	F5I
Roberts 22	3	20	0.8	1.5	35	A2I
HD 101584	40	-	-	2.2	40	B9II binary
IRAS 15405−4945	60	-	-	1	80	
IRAS 16342−3814	50	-	-	1	70	B?
IRAS 17253−2831	−62	9	0.25	0.15	16	
He 3−1475	50	25	0.5	-	-	B[e]
IRAS 18491−0207	75	-	-	1	70	
IRAS 22036+5306	−45	25	0.7	1	30	

the velocity of the shell component. The observed velocity ratio between shell and linear outflow is in good agreement with the predicted ratio of about a factor of 2 from the slow wind–wind model (section 3). The spatial size of the linear jet tends to be smaller than the shell (except for OH 231.8+4.2 where the sizes are similar), which suggests that the linear outflow is seen close to the line of sight, so that velocity deprojection effects are minor. This may be related to our selection criterium of high outflow velocities in the OH masers.

Where known or suspected, spectral types are indicated. Indications for central ionization exists for Roberts 22 and IRAS 08005−2356, from the OH profiles. This would require hotter stars than indicated by their spectral types. If confirmed, this would locate the bipolar OH/IR stars in the hotter part of the post-AGB evolution. The stellar wind velocities indicated by the OH agree with late B-type stars. This might imply that the observed intershell only forms shortly before the PN phase is reached. OH 231.8+4.2 is the strong exception and we suggest its interacting wind structure is due to a wind from a binary companion.

6. EVOLUTION

Fig. 3 shows the IRAS colour–colour diagram for the observed sources. They are seen to fall in two different groups. One group has colours similar to those of bipolar outflow soures (both young and old). Images for most of these confirm the presence of strongly bipolar reflection nebulae. None of the other observed OH/IR stars show this.

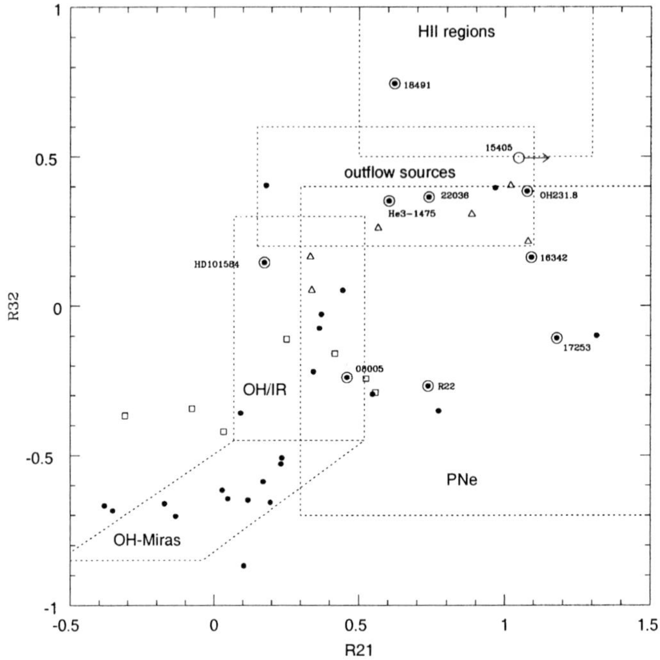

Figure 3 IRAS colour–colour plot. R21 = $\log F_{25}/F_{12}$ and R32 = $\log F_{60}/F_{25}$. Filled dots indicate OH/IR stars satisfying the selection criteria. Objects discussed in this paper are encircled. The open triangles show the extreme bipolar planetary nebulae, NGC 6302, M 2−9, Hb 4, He 2−437, 19W32, K 3−35. The open squares show the [WC] stars with mixed C/O chemistry: BD+30°3639, IRAS 07027−7934, He 2−113, CPD−56°8032 .

Fig. 4 shows an HST image of one the bipolar nebulae, IRAS 16342−3814, to show the typical morphology. The objects located outside the outflow box may have dense tori (e.g. Roberts 22) but do not show well-collimated outflows. As the (very rare) extreme bipolar PNe also fall in this region of the colour-colour diagram, we propose that there is an evolutionary connection, and that the collimated bipolar OH/IR stars will evolve into peculiar PNe, such as, M 2−9, K 3−35, and NGC 6302. It is noteworthy that the last two still show OH emission.

The remaining bipolar OH/IR star, without collimated outflows, have colours intermediate between the AGB stars and PNe, consistent with 'normal' early post-AGB evolution. However, both IRAS 08005−2356 and Roberts 22 show evidence for the presence of both carbon-rich and oxygen-rich dust. Among PNe, this is only found among a few objects with [WC] stars (Waters et al.

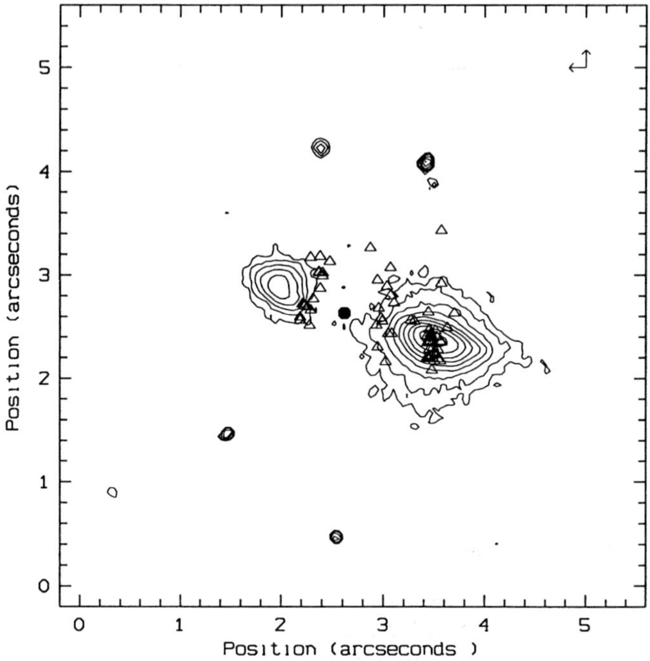

Figure 4 An F555 HST image. North is up and east is to the left. The pixel scale is 0.045 arcsec. The image has been de-rotated and corrected for field distortions. The contour levels are in steps which increase by a factor of 2. The triangles indicate the positions of the OH components. The filled star shows the estimated position of the central star. The error bars in the corner indicate the uncertainty in alignment of radio and optical coordinates.

1998; Zijlstra et al. 1991; Cohen et al. 1999), indicated by the open squares in Fig. 3. Again we propose a connection, although it is not clear whether our objects will actually evolve into [WC] stars and they may represent a closely related evolutionary track instead. The mixed chemistry does indicate that our OH/IR stars will not evolve into typical PNe. The indications are that the bipolar OH/IR stars represent a minority evolution only.

Within our sample, at least four objects are at a distance of 4 kpc or less, based on Hipparcos, cluster membership, and IR brightness. A lower limit to the observable life time can be obtained from the local birth rate of white dwarfs (e.g. Zijlstra & Pottasch 1991). The bipolar PNe account for $< 10\%$ of all PNe. We obtain that the post-AGB transition time for the bipolar OH/IR stars is $t_{\rm tr} \geq 7 \times 10^3$ yr, with a probable value in excess of 10^4 yr. This age is longer than normal post-AGB time scales, longer than the time over which OH remains observable, and longer than the dynamical time scales of the nebulae. The bipolar OH/IR stars appear to be affected by delayed evolution. We suggest that dense, very slowly expanding shells could lead to re-accretion of nebular gas unto the star. This could slow down the post-AGB evolution.

Acknowledgments

The author wishes to thank Peter te Lintel Hekkert and Jessica Chapman who are responsible for a large part of this result.

References

Cohen M., Dopita M.A., Schwartz R.D., Tielens A.G.G.M., 1985, ApJ, 297, 702

Cohen M., Barlow M.J., Sylvester R.J.. Liu X.-W., Cox P., Lim, T., Schmitt B., Speck A.K, 1999, ApJ, 513, L135

Frank A., Mellema G., 1994, A&A, 289, 937

Jura M., Kahane C., 1999, ApJ, 521, 302

Reipurth B., 1987, Nature, 325 787

Sahai, R., Trauger, J.T., 1998, AJ, 116, 1357

Shu F.H., Ruden S.P., Lada C.J., Lizano S, 1991, ApJ, 370, L31

Waters L.B.F.M., Beintema D., Zijlstra A.A., de Koter A., Molster F.J., Bouwman J., de Jong T., Pottasch S.R., de Graauw Th., 1998, A&A, 331, L61

Zijlstra A.A., Pottasch S.R., 1991, A&A, 243, 478

Zijlstra A.A., Gaylard M.J., te Lintel Hekkert P., Menzies J.W., Nyman L.-A., Schwarz H.E., 1991, A&A, 243, L9

THE SHAPE AND SHAPING OF PROTO-PLANETARY NEBULAE

Bruce J. Hrivnak
Dept. of Physics & Astronomy, Valparaiso University, U.S.A.
bruce.hrivnak@valpo.edu

Sun Kwok & Kate Y.L. Su
Dept. of Physics & Astronomy, University of Calgary, Canada

Abstract While AGB stars display spherical shapes and winds, planetary nebulae appear asymmetric. By studying proto-planetary nebulae (PPNs), the transitional objects between these two stages of stellar evolution, we can learn about when and how these asymmetric shapes develop. We describe in particular our HST imaging studies of PPNs in the visible and near-IR. They commonly reveal a bipolar morphology, in some cases with circumstellar, concentric arcs and in one case with a visible collimating disk. Mechanisms for shaping the nebula are reviewed, and the negative results of a radial velocity study to search for binary central stars of PPNs are discussed.

1. INTRODUCTION: THE SHAPE OF PPNS

While asymptotic giant branch (AGB) stars are found to display very spherical shapes and mass outflows, planetary nebulae (PNs), on the other hand, appear asymmetric, often with a bipolar or a point-symmetric morphology. Proto-planetary nebulae (PPNs) are the transitional objects between these two stages of stellar evolution, and as such they hold the clues to understanding the shape and shaping of PNs. Thus it is important to begin by answering the question, *what is the shape of PPNs?*

The first PPNs discovered in the 1970's, AFGL 2688 (Egg Nebula) and AFGL 618, both displayed a bipolar morphology. Following the IRAS mission in 1983, many new PPN candidates were identified based upon their mid-IR excesses. With typical ground-based imaging (R$\sim$1.5$''$), these objects were generally not resolved. However, the results of a higher-resolution survey of 15 PPNs with the Canada-France-Hawaii telescope (R$\sim$0.75$''$) found two to be bipolar, four elliptical, four round, and five unresolved/marginally resolved

R. Szczerba and S.K. Górny (eds.), Post-AGB Objects as a Phase of Stellar Evolution, 197–204.

(Kwok et al. 1996; Hrivnak et al. 1999a). Polarization measurements were also made that indicated an asymmetric distribution of dust around PPNs (Trammell et al. 1994). While these observations were of scattered visible light, some mid-IR imaging was also carried out that directly measured the thermal emission of the dust. Axial symmetry was observed in several of the brightest (nearest) sources (R$\sim$1.0$''$; Meixner et al. 1997; Dayal et al. 1998). *Thus evidence was found from ground-based observations that the circumstellar envelopes of PPNs were not spherical, but commonly possessed axial symmetry.*

2. PREVIOUS HST IMAGING

While ground-based imaging gave us hints of the shapes of PPNs, the HST (R$\sim$0.1$''$) has given us stunning, detailed images. Visible-band (WFPC2) images of the Egg Nebula show details in its bipolar lobes and display $\sim$20 pairs of circumstellar arcs and a pair of intersecting "searchlight beams" (Sahai et al. 1998a). IRAS 17150−3224 and IRAS 17441−2411 are both edge-on bipolar nebulae, and display eight and three pairs of circumstellar arcs, respectively; IRAS 17150−3224 also displays a pair of "searchlight beams" (Kwok et al. 1998; Su et al. 1998). IRAS 17245−3951 is an edge-on bipolar, while IRAS 16594−4656 is a bipolar seen at an intermediate orientation (Hrivnak et al. 1999b). A 2-D bipolar model with a disk has been computed to fit the images and also the spectral energy distributions (Su et al. 1998). Viewing the model from different orientations gives a good qualitative fit to the bipolar PPN images observed with HST. A recent snapshot survey resolved 17 of 18 PPNs with a variety of morphologies, some of which are elliptical with a bright central star and others clearly bipolar (Ueta et al. 2000). Our near-IR (NICMOS) images reveal the central star that is obscured in visible light in the edge-on bipolars (except in the Egg Nebula, which is still optically thick in the near-IR; Sahai et al. 1998b). Additional structures can also be seen, and the near-IR images nicely complement the visible (see Fig. 1). Polarization studies show in general a centro-symmetric pattern with high polarization values; these are consistent with a central source of light and single scattering. However, in the obscured lane between the lobes there exists a departure from central symmetry, where second and multiple scattering are not negligible (see Fig. 2).

Thus the evidence from higher-resolution HST imaging observations of PPNs is that bipolar morphologies are common.

3. RECENT HST IMAGING (CYCLE 8)

We were awarded HST WFPC2 time for a program to search for bipolar lobes and circumstellar arcs in a sample of PPNs (derived from our CHFT survey). Deeper (t=1000−1800 s) images were obtained. The goals were to see (a) how

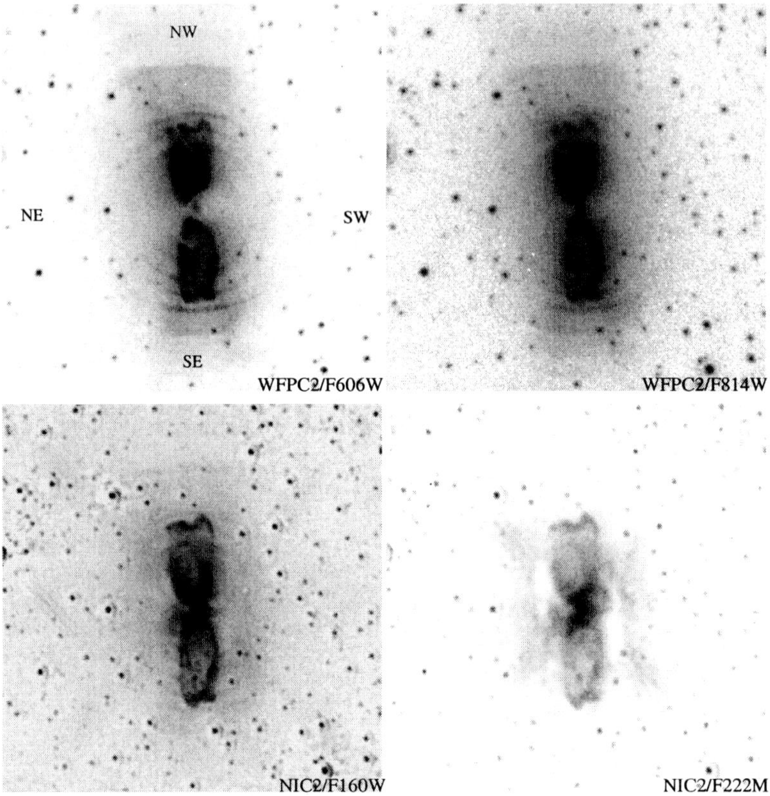

Figure 1 HST V,I,H,K images of IRAS 17150−3224. The arcs are seen more clearly in visible light, and the equatorial features in the near-IR.

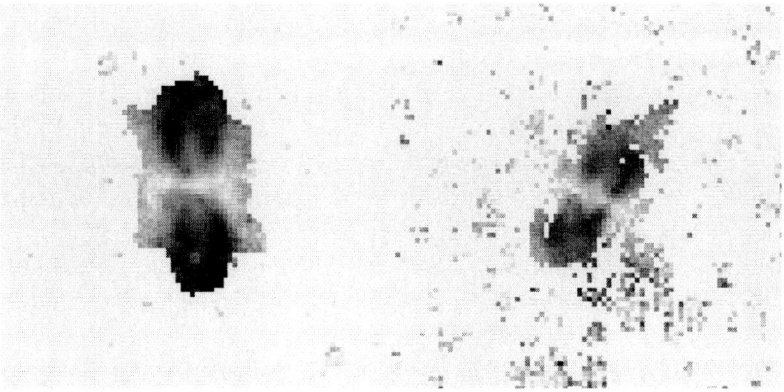

Figure 2 HST K-band polarization intensity maps for the bipolar PPNs IRAS 17150−3224 (left) and IRAS 17441−2411 (right). A high intensity is seen in the lobes and a low intensity in the obscuring lane between.

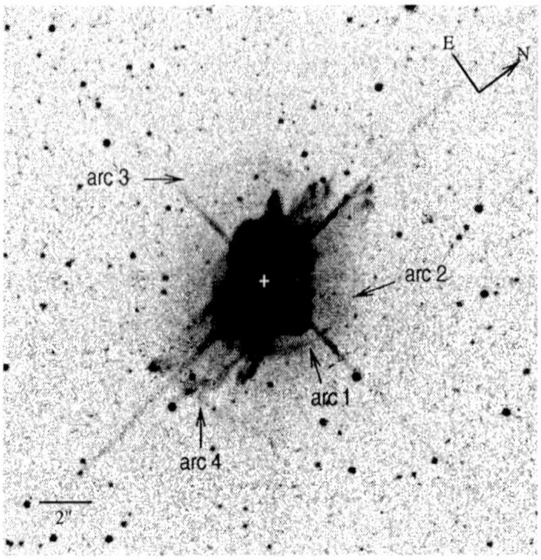

Figure 3 HST I-band image of IRAS 16594−4656, a bipolar PPN viewed at an intermediate orientation. Parts of four concentric arcs are seen.

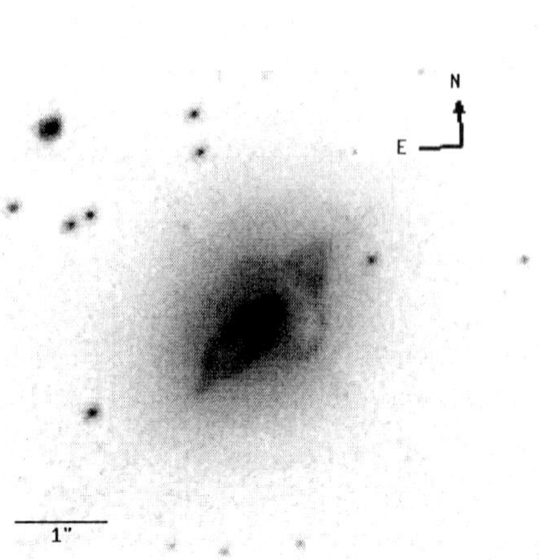

Figure 4 HST V-band image of IRAS 17106−3046, also a bipolar PPN viewed at an intermediate orientation. A visible disk can be seen confining the outflow, which forms two prolate lobes.

common is a bipolar structure – is it a universal structure in PPNs?; and (b) how common are circumstellar arcs.

Concentric circumstellar arcs were detected in two new PPNs, both of which are viewed at an intermediate orientation. In IRAS 16594−4656, four arcs are seen (see Fig. 3), while in IRAS 20028+3910, three short pairs of arcs are seen, in addition to a pair of intersecting "searchlight beams" (Hrivnak et al. 2000). These make a total of five PPNs in which circumstellar arcs are seen; in addition, they have been seen in three PNs and one AGB star. The timescales between arcs appear to be 100s to a few 1000 yr. Their presence in PPNs at a variety of viewing angles indicates that they are portions of spherical shells and suggests that they represent perturbations of spherical mass loss on the AGB by some periodic phenomonon. *Arcs may be very common in PPNs; the ones we detected were in bipolar PPNs that were the brightest and presumably the nearest.* A relatively deeper survey is needed to determine more quantitatively how common are circumstellar arcs around PPNs.

Bipolar lobes were detected in three PPNs. Both IRAS 19477+2401 and IRAS 22574+6609 are found to show edge-on lobes with additional "blobs" of light, while IRAS 17106−3046 is seen at an intermediate orientation. *Evidence continues to accumulate that a bipolar morphology is common and may be a universal model for PPN structure.* Variations in the morphology may be due to (a) viewing orientation and (b) optical depth of the CSE.

Models suggest that the collimation of the bipolar lobes can be accomplished by a torus or disk (see next section). Evidence for such a disk is found in the obscuring region seen between the lobes in edge-on cases (in visible light) and the suggestions of axial symmetry seen in some of the mid-IR images. In IRAS 17106−3046 we finally see the disk in scattered light (see Fig. 4). A very symmetric disk is seen that appears to collimate the outflow contained in two prolate lobes (Kwok et al. 2000).

4. MECHANISMS FOR SHAPING PPNS

The original interacting stellar winds model of Kwok (Kwok et al. 1978) explained the formation of PNs by the interaction of a fast wind from the central star with the slower AGB wind. This was later generalized by Balick (1987) with the addition of a density enhancement to produce the variety of basic morphologies seen in PNs (round, elliptical, bipolar). Detailed hydrodynamical models confirm that the interacting winds model can produce these morphologies (see review by Frank 1999, and references therein). In these models, an equatorial density enhancement provides the constraint for producing the collimated lobes. An alternate mechanism for constraining the mass outflow to collimated lobes is a magnetic field (García-Segura et al. 1999).

This raises the question as to the cause of such an equatorial density enhancement. One commonly suggested mechanism is a binary companion which, as it orbits close to or within the extended atmosphere of the AGB star, could divert some of the mass loss into the plane of the orbit (Livio & Soker 1988). Such a companion could be a star or a planet. Another mechanism is rapid rotation of the AGB mass loser, which would then yield a preferred plane for mass loss (see García-Segura et al. 1999, and references therein). This high rotation could perhaps be induced by a close binary companion (Livio 1993).

What about a mechanism to produce the concentric arcs and shells? These appear to be formed during the AGB phase. The timescale for these arcs, 10^2-10^3 yr, is longer than that of stellar pulsations, $\sim$1 yr, but shorter than that of thermal pulses, 10^4-10^5 yr. Several suggestions have been made, including density waves from the instability in gas−dust coupling (Deguchi 1997) and two-fluid (gas−dust) hydrodynamic flows (Simis et al., these proceedings). Binary companions are also invoked, that either divert the mass outflow periodically (due to their eccentric orbits) producing gaps in the CSE (Harpaz et al. 1997), or that lead to spiral shocks in the wind producing density enhancements (Mastrodemos & Morris 1999).

5. SEARCH FOR BINARY PPNS

Since several mechanisms suggested to shape the PPNs involve binary companions, it seemed worthwhile to search for companions to PPNs. In principle, there are several possible methods of detection. (1) Visual companions can be imaged with the HST; however resolvable companions would be too far appart to influence the mass loss. (2) Light variations could be searched for; a close binary companion would produce a periodic sinusoidal light variation due to the tidally-induced non-spherical shape. (3) Velocity variations could be measured caused by orbital motion. The last of these seems to be the most promising.

We have undertaken a radial velocity study to search for binaries among PPN central stars. Bond (2000) has found that $\sim$10% of PNs have a binary central star with a short period ($\leq$ 3 d). Note that a radial velocity search among PPNs has an advantage over PNs since the F and G central stars of PPNs have narrower and more lines than found in PNs. This leads to better precision and the ability to search for longer periods.

We carried out a five-year (1991–1995) radial velocity monitoring program of nine PPNs. The 1.2-m Coudé telescope at the Dominion Astrophysical Observatory was used, which achieved a typical precision of 0.65 km s^{-1}. Approximately 35 observations were made of each of the objects. All nine were found to vary in velocity, with a typical range of 10 km s^{-1}; however, in most cases the variation was not periodic. Three showed a periodicity, but with varying amplitude. The variations were determined to be due to pulsation (Hrivnak & Lu 2000; Hrivnak et al., these proceedings). No binaries were found; ones with P $>$ 1 yr could have escaped detection. *Why were no binaries found? Does this mean that the central stars of PPNs are not binaries, and that the shaping is caused by another mechanism?* Two possible caveats come to mind. Short-period companions would be within the extended atmospheres of the central stars and thus not detected. There may also be a viewing-angle selection effect. Since the brightest PPNs were observed, it may be that they are objects in which we are observing along the less-obscured polar direction of a bipolar nebula and the binary may thus be oriented near the plane of the sky. *Although we did not detect any binaries among these PPNs, there may be selection effects in the sample which limit our ability to see such binaries.*

6. SUMMARY AND CONCLUSIONS

Shapes of PPNs:

- Bipolar structure is common (universal model?).
- Arcs are common; this implies a spherical AGB mass loss, which was then interrupted or modified.
- An example of a disk-collimated outflow is seen (IRAS 17106–3046).

Shaping of PPNs:

- Shaping occurs early in the PPN transition.

- Models of wind expanding into an equatorial density enhancement successfully produce the observed basic shapes.
- A search for binary central stars of PPNs has not revealed any; however, there may be reasons and biases for this, so the models invoking a binary to shape the nebula may still be appropriate.

Acknowledgments

Support for this research has been provided by NASA through the STScI (GO-07840.02-96A, GO-08210.01-97A), by the NSF (AST-9900346), and by the Natural Sciences and Engineering Research Council of Canada.

References

Balick, B. 1987, AJ, 94, 671

Bond, H. 2000, in Asymmetrical Planetary Nebulae II: From Origins to Microstructures, eds. J.H. Kastner, N. Soker, & S. Rappaport (San Francisco: ASP), 115

Dayal, A., Hoffmann, W.F., Bieging, J.H., Hora, J.L., Deutsch, L.K., & Fazio, G.G. 1998, ApJ, 492, 603

Deguchi, S. 1997, in IAU Symp 180: Planetary Nebulae, eds. H.J. Habing & H.J.G.L.M. Lamers (Dordrecht: Kluwer), 151

Frank, A. 1999, New Astro. Rev., 43, 31

García-Segura, G., Langer, N., Różycka, M., & Franco, J. 1999, ApJ, 517, 767

Harpaz, A, Rappaport, S., & Soker, N. 1997, ApJ, 487, 809

Hrivnak, B.J., Kwok, S., & Su, K.Y.L. 1999b, ApJ, 524, 849

Hrivnak, B.J., Kwok, S., & Su, K.Y.L. 2000, submitted

Hrivnak, B.J., Langill, P.P., Su, K.Y.L., & Kwok, S. 1999a, ApJ, 513, 421

Hrivnak, B.J., & Lu, W. 2000, in IAU Symp 177, The Carbon Star Phenomenon, ed. R.F. Wing (Dordrecht: Kluwer), 293

Kwok, S., Hrivnak, B.J., & Su, K.Y.L. 2000, ApJ, in press

Kwok, S., Hrivnak, B.J., Zhang, C.Y., & Langill, P.L. 1996, ApJ, 472, 287

Kwok, S., Purton, C.R., & FitzGerald, M.P. 1978, ApJ, 219, L125

Kwok, S., Su, K.Y. L., & Hrivnak, B. J. 1998, ApJ, 501, L117

Livio, M. 1993, in IAU Symp 155, Planetary Nebulae, ed. R. Weinberger & A. Acker (Dordrecht: Kluwer), 279

Livio, M., & Soker, N. 1988, ApJ, 329, 764

Mastrodemos, N., & Morris, M. 1999, ApJ, 523, 357

Meixner, M., Skinner, C.J., Graham, J.R., Keto, E., Jernigan, J.G., & Arens, J.G. 1997, ApJ, 482, 897

Sahai, R. et al. 1998a, ApJ, 493, 301

Sahai, R. et al. 1998b, ApJ, 492, L163

Su, K.Y.L., Volk, K., Kwok, S., & Hrivnak, B. J. 1998, ApJ, 508, 744

Trammell, S.R., Dinerstein, H.L., & Goodrich, R.W. 1994, AJ, 108, 984
Ueta, T., Meixner, M., & Bobrowsky, M. 2000, ApJ, 528, 861

SPH MODELS OF PROTO-PLANETARY NEBULAE

Artur J. Gawryszczak[1,2], Joanna Mikołajewska[1], Michał Różyczka[1]
[1]*N. Copernicus Astronomical Center, Bartycka 18, 00-716 Warszawa, Poland*
[2]*Warsaw University Observatory, Al. Ujazdowskie 4, 00-478 Warszawa, Poland*

Abstract We present preliminary models of proto-planetary nebulae in binary systems. The term *proto-planetary nebula* denotes here an object several tens of AU across, resulting from the interaction between primary's slow and dense wind, and the gravitational field of the secondary. A significant flattening of the originally spherical density distribution of the wind is easily achieved in rather close binaries (a of a few AU). A fast and rarefied wind propagating through such a medium should naturally develop a bipolar structure. We find that it is also possible to obtain a flattened proto-planetary nebula in a wide binary, provided that the dense wind is either very efficiently cooled or its original velocity is very low.

1. MOTIVATION, METHOD AND ASSUMPTIONS

According to the widely accepted two-wind scenario, a planetary nebula (PN) is formed when a high-speed, low-density wind emitted by the central star interacts with a slow, high-density wind emitted by the same star in its AGB evolutionary phase (see e.g. [1] and references therein). Elliptical or bipolar nebulae are easily produced when the density distribution of the slow wind has a maximum in a well-defined plane. A viable mechanism for the production of such a distribution is gravitional deflection of the wind by a companion to the mass-losing star. Our report presents a set of preliminary models of dense winds emitted from a red giant in a binary system.

The wind is represented by Lagrangian particles launched at a constant rate from the surface of the red giant. To follow its evolution the SPH method is employed (see e.g. [2]). Whenever the pre-set limit of $\sim 4 \cdot 10^4$ particles is exceeded, those most distant from the red giant are removed from the domain. The distance at which removals occur stabilizes within several P_{orb}, and an almost stationary, rotating density distribution is obtained.

To facilitate the calculations we made the following assumptions: The red giant does not fill its Roche lobe, the wind is spherically symmetric, the wind gas is dynamically coupled to the dust, and the gravity of the red giant is balanced by

R. Szczerba and S.K. Górny (eds.), Post-AGB Objects as a Phase of Stellar Evolution, 205–208.

radiation pressure on the dust. Radiation pressure from the secondary, radiative cooling of the wind, and self-gravity of the wind are neglected. The equation of state of a perfect gas ($\gamma = 1.66$) is used. The initial temperature of the wind is 3000 K. The particles are injected at constant intervals of time, from points distributed randomly on the surface of the red giant. The wind generation procedure was tested on a freely expanding wind from a single star, and it gave entirely satisfactory results.

2. MODELS

The distance of particle removal does not exceed 50 AU, i.e. our models represent an early evolutionary stage of the dense wind (or the central part of an evolutionary advanced wind). We shall refer to them as protoplanetary nebulae (PPNe). Four PPNe models are presented here. Their parameters are listed in table 1, with M_{rg}, M_c, a, V_i and R_{rg} standing, respectively, for mass of the red giant, mass of the companion, orbital separation of the stars, initial velocity of wind particles at the surface of the red giant, and radius of the red giant.

Table 1 Parameters of the models.

	Model	M_{rg} $[M_\odot]$	M_c $[M_\odot]$	a [AU]	V_i [km/s]	R_{rg} [AU]
1	close binary, heavy companion	1.20	0.80	2.4	12.0	0.65
2	close binary, light companion	1.20	0.12	2.4	12.0	0.65
3	wide binary, slow wind	1.00	0.60	13.6	0.6	2.00
4	wide binary, pressureless wind	1.00	0.60	13.6	12.0	2.00

For each model, the plots in figures 1 and 2 show density distribution in the orbital plane, and azimuthally averaged density distribution in the plane perpendicular to the orbit. The interior of the red giant, where no wind is present, is black. Double white circles in azimuthally averaged plots for models 3 and 4 are an artifact of the averaging routine (the red giant is smeared into a torus, whose cross-section is seen in the plot).

The general effect of the companion is to deflect the wind toward the orbital plane. The effect is strong when the velocity of the wind does not exceed companion's orbital velocity, and it weakens with increasing a. In reality, the wind would accelerate due to radiation pressure from the primary, and it might achieve its terminal speed beyond the orbit of the companion. Models 1,2 and 4 overestimate the velocity of the wind approaching the companion, and underestimate the deflection. Allowing for a gradual increase of wind velocity, one may observe the deflection even in wide binaries (e.g. in model 3, where

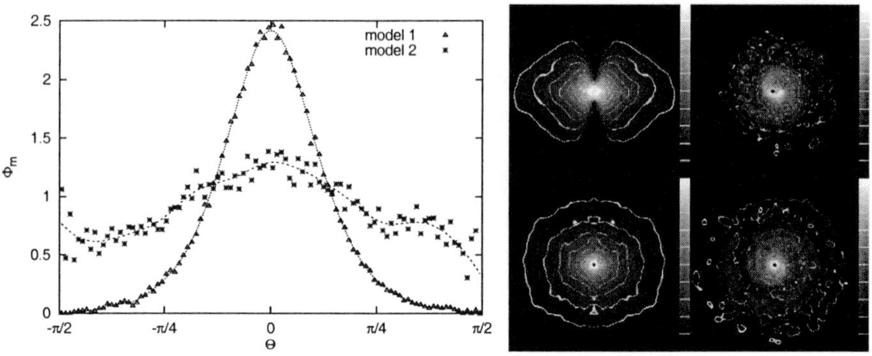

Figure 1 (left) Azimuthally averaged mass flux through the domain boundary as a function of the vertical angle, Θ. The orbital plane is at $\Theta = 0$. The flux is scaled to the initial mass flux of the wind. The continuous lines are spline-smoothed fits to numerical values shown with respective symbols. (right) Results for models 1 (top) and 2 (bottom). Left column: azimuthally averaged densities in r-z coordinates, right column: densities in the orbital plane. The black circle represents the interior of the red giant. Constant density lines (white) are logarithmically spaced with $\Delta log = 0.3$. The density range spans 3 decades, with maximum densities shown in white. The size of the region shown in each plot is 50 AU.

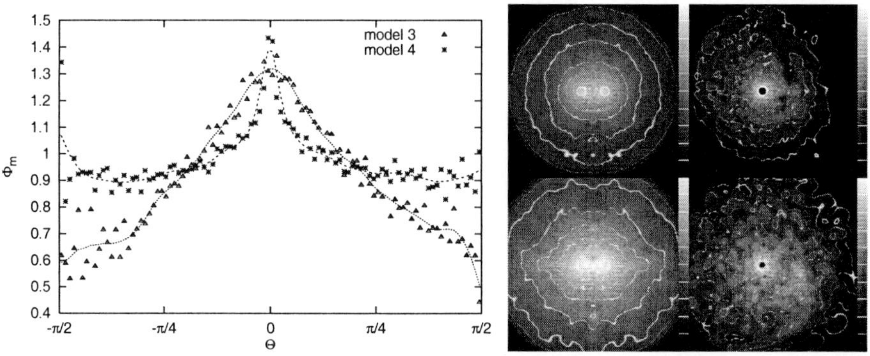

Figure 2 Same as above, but for models 3 and 4 with one exception: the size of the region shown in each plot is 75 AU.

an initially slow wind accelerates due to the thermal pressure gradient). With radiative cooling neglected, we overestimate temperature and pressure of the wind. To emphasize this effect, we calculated the pressureless model 4, which simulates a dust-dominated wind able to rapidly radiate all its heat away. In this model, a rather thin disk is formed in the orbital plane.

In all models a spiral pattern is visible in the orbital plane. However, the amplitude of azimuthal density variations seems to decrease with increasing distance from the binary. The latter observation is not entirely certain because of degraded numerical resolution at the edge of the computational domain. In any case, basic conclusions regarding the shape of the mature PN can be drawn from azimuthally averaged plots. In model 1, the fast wind would preferentially expand along the low-density funnel at the symmetry axis, and a bipolar nebula could form. In the remaining three models, an elliptical PN would be a more likely result.

At the outer boundary of the computational domain the wind is highly supersonic in all cases (reaching in the orbital plane 26, 14, 11, and 12 km/s for models 1-4, respectively), and the nonradial velocities are smaller than a few per cent of the radial velocity. Thus, beyond the computational domain our models would expand homologously.

3. COMPETITION

While this work was in progress, two papers on the same subject were published [3; 4]. In many aspects the work they report is superior to ours: a more sophisticated physics is employed, and the numerical resolution is better by a factor of 10. We show that, as far as the stationary global structure of PPNe is concerned, essentially the same conclusions can be drawn from very simple models. In the future we intend to focus on radiative effects and acceleration mechanism of the wind, trying to fit our models to observed objects. The models will also serve as initial conditions for simulations of fast wind phase and evolution of PNe.

Acknowledgments

This research was supported by the Committee for Scientific Research through grants 2P03D 014 19 and 2P03D 021 12. The original version of the SPH code was kindly provided by Dr. Gregory Laughlin.

References

1. García-Segura, G., Langer, N. & Rózyczka, M.: 1999, *ApJ* **517**, 767
2. Hernquist L. & Katz, N.: 1989, *ApJ Suppl.* **70**, 419
3. Mastrodemos, N. & Morris, M.: 1998, *ApJ* **497**, 303
4. Mastrodemos, N. & Morris, M.: 1999, *ApJ* **523**, 357

DYNAMICS OF THE COLLIMATED OUTFLOWS OF THE PPN HEN 3−1475

Angels Riera

Departament de Física i Enginyeria Nuclear. Universitat Politècnica de Catalunya. Av. Víctor Balaguer s/n. E-08800 Vilanova i la Geltrú, Spain

angels.riera@upc.es

Pedro García-Lario

ISO Data Centre. Astrophysics Division. Space Science Department of ESA. Villafranca del Castillo, Ap. de Correos 50727. E-28080 Madrid, Spain

Arturo Manchado

Instituto de Astrofísica de Canarias. E-38200 La Laguna, Spain

Matthew Bobrowsky

Challenger Center for Space Science Education. 1250 North Pitt Street, Alexandria, VA 22314, USA

Abstract Hen 3−1475 has recently been identified via ground-based and HST optical images as a bipolar proto-planetary nebula displaying a remarkable highly–collimated optical jet and a spectacular S-shaped string of knots along its bipolar axis, associated to extremely high velocities which decrease with the distance to the central star.

We present new long-slit high resolution spectroscopy of the highly collimated regions of Hen 3−1475. The high-dispersion spectra are used together with the high angular resolution images obtained with HST WFPC2 to provide a comprehensive picture of the complex structure and kinematics of the shocked–excited emission and the dust–scattered emission in this proto-planetary nebula. The overall picture supports a temporal and angular variability of the jet ejection.

R. Szczerba and S.K. Górny (eds.), Post-AGB Objects as a Phase of Stellar Evolution, 209–214.

1. INTRODUCTION

The kinematic study of the striking Hen 3–1475 is crucial for understanding the formation of jets in proto-planetary nebulae (hereafter PPNe). Hen 3–1475 displays a spectacular S-shaped string of point-symmetric knots extending over 17″ (Riera et al. 1995; Bobrowsky et al. 1995). With the help of HST WFPC2 observations by Borkowski et al. (1997), it is now possible to resolve the subarcsecond structure of the outer knots and of the inner regions.

The strong far infrared excess detected by IRAS is interpreted as the remmant emission from the AGB circumstellar shell while the known presence of strong P-Cygni profiles in the Balmer lines indicates that mass loss is still on-going. This is confirmed by the detection of thermal emission from hot dust in the near infrared (García-Lario et al. 1997).

The molecular gas in the envelope has been detected in CO as an expanding torus of material (Knapp et al. 1995). In addition, Hen 3–1475 also exhibits strong OH maser emission (te Lintel Hekkert 1991, Bobrowsky et al. 1995). Both CO and OH emission show a broad (60 km s^{-1}) component indicative of a fast molecular outflow which appears to be extended. The redshifted and blueshifted OH peaks, which are displaced 1" SE and NW, show a reversed behaviour with respect to the velocity field of the optical jets. It is remarkable that the OH radial velocity increases linearly with distance to the source, a situation that has been observed in other PPNe, and can be explained in terms of polar dependent winds (Zijlstra, these proceedings).

2. OBSERVATIONS

Long slit optical spectra of Hen 3–1475 were obtained in June 1998 using an Intermediate Dispersion Spectrograph (IDS) equipped with a 500 mm camera and a TEK CCD installed at the Cassegrain focus of the 2.5 m (f/15) Isaac Newton Telescope at the Observatorio del Roque de los Muchachos (La Palma, Spain) during service time. The spectra were taken along the major axis of the bipolar nebula covering the central star at a P.A. of 135° with a slit width of 2". Two exposures centered at 6300 and 6640 Å were performed with the H1800V grating yielding an spectral resolution of 0.23 Å/pixel, covering important diagnostic lines like [O I] 6300Å, [N II] 6548,6584Å, Hα and [S II] 6717, 6731Å with a spatial resolution of 0.33"/pixel.

High spatial resolution (0.0455"/pixel) optical WFPC2 images taken with HST using the Planetary Camera were also retrieved from the HST Data archive. An image of Hen 3–1475 taken through the narrow filter F658N ($\lambda_c = 6590$ Å; $\delta\lambda$=28.5 Å) was originally part of proposal 6347 (P.I. Borkowski) while those taken through the wide filters F555W ($\lambda_c = 5252$ Å; $\delta\lambda$= 1222.5 Å) and F814W (λ_c = 8269 Å; $\delta\lambda$= 1758.0 Å) were part of proposal 6364 (P.I. Bobrowsky).

3. MORPHOLOGY

The image taken with HST using the narrow F658N filter esentially traces the ionized gas in Hen 3−1475 through the detection of the nebular emission coming from the [NII] 6584 Å line. However, it should be noticed that small losses are expected in the detection of this line from the more extremely blueshifted regions of the nebula, not covered by the filter. In addition, extremely redshifted Hα emission is also expected to contribute to the emission detected in the inner regions of the redshifted lobe. This image shows clearly the jet-like structure and three pairs of symmetric emission knots. The pairs of knots are located at different position angles, which is the signature of point-symmetry, while the innermost knots are connected with the central source by converging thin lines (see Borkowski et al. 1997).

The F555W (shown in Figure 1) and F814W images trace better the morphology of the scattered light. They show that the brightest emission occurs at the limb-brightened bipolar reflection nebula, with a symmetry axis oriented in the same direction as the optical jet. The presence of a thick torus of dust perpendicular to the optical bipolar axis is inferred independently from our optical spectroscopy (see below) and from the CO observations, and explains the difference in brightness between NW and SE emission.

Figure 1 F555W WFPC2 HST image. North is up and East to the left.

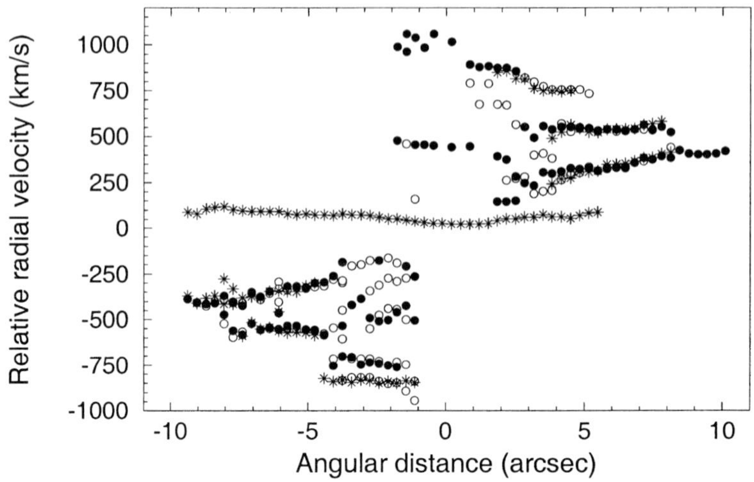

Figure 2 Position-velocity diagram for [NII] 6548 Å (open dots), [NII] 6584 Å (filled dots) and Hα (stars). Relative velocities and positions are quoted with respect to the systemic radial velocity (= 44 km s^{-1}) and the position of the central star. The NW lobe is at negative angular distances.

4. KINEMATICS

Figure 2 displays the relative radial velocity measured along the bipolar axis as a function of the distance to the central source. The velocity of the gas in the SE jet shows a symmetric behaviour with respect to the velocity detected in the NW gas when compared to the systemic velocity. In addition, there is a considerable deceleration in the gas as we move outwards from the central star to the outer knots in both lobes and both show an abrupt change in velocity at an intermediate angular distance.

The middle knots (NW2 and SE2) show spectra with double-peaked profiles indicative of two distinct velocity components, moving at 550 and 365 km s^{-1} away from the central source while, in contrast, at further distances from the centre, the emission lines show single-peaked broad profiles, at radial velocities $\pm$ 405 km s^{-1} relative to the systemic radial velocity.

The highest velocity gas is found near the center of the nebula at the position of the converging lines. Along SE1 (which shows three elongated condensations) the radial velocity decreases smoothly from +926 to +780 km s^{-1} in only 1.5″. A similar (but symmetric) behaviour is observed in NW1.

Figure 3 shows the relative radial velocities of the reflected Hα emission originated in the central core and being scattered by dust particles.

Several conclusions can be drawn from Figure 3: first, the scattered Hα emission is redshifted in both lobes; second, the scattered emission is redshifted

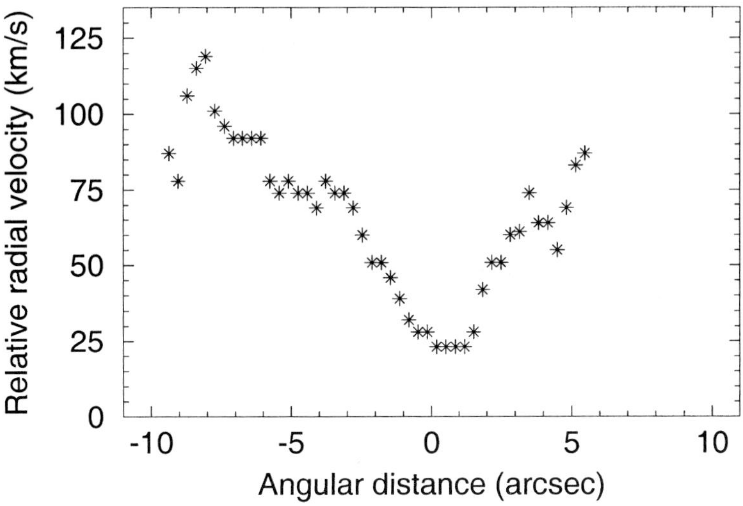

Figure 3 Expansion of the Hα position-velocity diagram showing the velocities of the scattered Hα emission.

by different amounts in both lobes, being the NW lobe redshifted with respect to the SE lobe. And finally, the radial velocity linearly increases with distance to the source.

For axial outflows from an object inclined with respect to the plane of the sky, one expects a red and blueshifted lobes as a consequence of its inclination. This is the case if the emission comes from the outflowing gas. However, if the observed emission is the light from the central object which is being scattered by dust, it should show the projected outflow velocity of the dust redshifted in both lobes since the dust particles are moving away from the central object. Assuming the NW and SE dust outflows have the same velocities, one should have a higher redshift than the other because of the projection effects of the inclination of the object. The observed velocities show that the reflected emission from the NW lobe is redshifted with respect to the SE lobe velocities, which is also the pattern shown by the OH radial velocities, indicating the NW flow is receding while the SE flow is approaching the observer.

In summary, the dust outflow velocities show common features with the OH radial velocities: a common inclination to the plane of the sky and a linear increase of the radial velocities with distance to the source. Thus, we conclude that the dust outflow is confined to the molecular expanding torus.

5. RESULTS AND DISCUSSION

Hen 3−1475 shows the combination of a fast bipolar post-AGB outflow oriented along the main bipolar axis with a low expanding optically thick torus of molecular gas and dust (the remnant from the AGB shell, still preserved from destruction as a consequence of the shielding provided by the dense material surrounding the central source). Due to the presence of this equatorial optically thick torus, photons are scattered off towards the polar direction giving its bipolar appearance to the refletion nebula.

Radial velocity decreases with increasing distance from the source in a step-like fashion. The decrease of the radial velocity with distance to the source might be due to: (i) precession, (ii) slowing down as a result of the drag due to its interaction with surrounding material, (iii) intrinsic variability. It is possible that some of the variations are due to energy losses along the flow and precession, but the abrupt decrease is the most compelling evidence for variability. Variability can also explain the formation of knots within the collimated outflow.

The overall picture supports the description of the system as the result of episodic ejections, with timescales $\sim 70/\tan(\phi)$ years, where ϕ is the inclination angle, in a precessing system with a period of few hundred years. The models proposed to explain the formation of high velocity precessing jets require the existence of a binary central system (see García-Segura et al. 2000 and Livio 2000, and references herein).

References

Bobrowsky, M. et al. 1995, ApJ 446, L89

Borkowski, K.J., Blondin, J.M., Harrington, J.P. 1997, ApJ 482, L97

García-Lario, P., Manchado A., Pych, W., Pottasch, S.R. 1997, A&AS, 126, 479

García-Segura, G., Franco, J., López, J.A., Langer, N., Rózyczka, M. 2000, ASP Conference Series 199, 235

Knapp,G.R., Bowers,P.F., Young,K., Phillips,T.G. 1995, ApJ 455, 293

Livio, M. 2000, ASP Conference Series 199, 243

Riera, A., García-Lario, P., Manchado, A., Pottasch, S.R., Raga, A.C. 1995, A&A 302, 137

te Lintel Hekkert. P. 1991, A&A 248, 209

STRONG ASYMMETRIES IN THE NEUTRAL ENVELOPE OF MIRA

E. Josselin, N. Mauron

Groupe d'Astrophysique, CC72, CNRS/U. Montpellier II, Place E. Bataillon, F-34095 Montpellier Cedex, France

josselin@graal.univ-montp2.fr

P. Planesas, R. Bachiller

Observatorio Astronómico Nacional (OAN), IGN, Apartado 1143, E-28800 Alcalá de Henares, Spain

Abstract We present high-resolution observations of the neutral envelope of *o* Ceti. The molecular component has been mapped in the CO(2–1) radio line and the atomic component is studied through long-slit spectroscopy of the optical K I line. These observations reveal strong asymmetries in the gas distribution, which can be interpreted as a spherical envelope disrupted by a bipolar outflow.

These data, combined with other observations of this object as well as similar observations of other AGB stars, suggest that asymmetrical mass loss processes occur early during the AGB stage, well before the proto-planetary nebula phase.

1. INTRODUCTION

Stars on the asymptotic giant branch (AGB) phase experiment strong mass loss, which generates circumstellar envelopes (CsE) made of atomic and molecular gas, and dust. Observations of these CsEs are generally interpreted with the assumption of spherical symmetry and a steady mass loss. However, recent observations have put in evidence that many CsE display departures from spherical symmetry (e.g. Lopez 1999). Moreover, many PNe are known to be elliptical or bipolar. Outstanding questions are therefore how and when such asymmetry appears during the AGB evolution.

Mira (*o* Ceti), the prototype of long period variable stars, is particularly well-suited to tackle with these problems, since it is one of the closest oxygen-rich AGB stars. This object is in fact a binary, maybe even a triple system (see Karovska et al. 1993). The atmosphere of the AGB star of this system (Mira

R. Szczerba and S.K. Górny (eds.), Post-AGB Objects as a Phase of Stellar Evolution, 215–219.

A) appears elliptical (Haniff et al. 1992, Karovska et al. 1997). Its companion (Mira B) is still of uncertain nature, being a main-sequence star or a white dwarf. It is relatively hot (about 10000 K), maybe also asymmetric and maybe surrounded by an accretion disk (Karovska et al. 1997). The first detailed study of the geometry of the molecular CsE of *o* Ceti has been carried out by Planesas et al. (1990a and 1990b). They showed that the CO spectra contain three velocity components, corresponding to a dominant spherical envelope with an expansion velocity of 7 km s^{-1} and a poorly collimated, low velocity (V_{exp} $\sim$4 km s^{-1}) bipolar outflow. The atomic component of the envelope also displays indications of asymmetry, as shown by Mauron and Caux (1992) from observations of the circumstellar atomic gas in the KI and NaI lines.

This paper reports new high angular resolution observations of the neutral envelope of *o* Ceti, in the millimeter CO(2–1) line and the optical KI lines. More details can be found in Josselin et al. (2000).

2. OBSERVATIONS

Interferometric observations, centered on *o* Ceti, of the millimeter CO(2-1) line were made with the IRAM interferometer (Plateau de Bure, near Grenoble, France; Guilloteau et al. 1992) from December 1995 to March 1996. We used as zero-spacing observations the single-dish data obtained at the IRAM 30m telescope by Planesas et al. (1990a). The restored gaussian clean beam for these observations is $2.5'' \times 2.4''$ at a position angle of 6°.

Long-slit spectroscopy of the envelope of *o* Ceti in the KI 7664.911 and 7698.974 Å lines was done at the Canada-France-Hawaii Telescope and the Haute-Provence Observatory. The slits were placed at perpendicular distances between 5″ and 20″ from the star. The spectral resolution was chosen to be ~1 Å. This is of the order of the width of the KI lines which consequently remain unresolved.

3. RESULTS

3.1. THE MOLECULAR GAS

The overall distribution of the molecular gas as observed in the CO(2–1) line is shown in the velocity-integrated map in Fig. 1. The emission extends over $250''^2$, which corresponds to a sphere of radius 1.7 10^{16} cm for a distance of 130 pc (HIPPARCOS). Regarding the low mass-loss rate (10^{-7} $M_\odot$ yr^{-1}, Mauron and Caux 1992 and references therein), this extent is mainly determined by photodissociation by interstellar UV radiation (Mamon et al. 1988).

It is obvious that the circumstellar envelope significantly deviates from sphericity. The inner region is much more extended in the East-West direction than the North-South one. The most striking characteristic is the presence

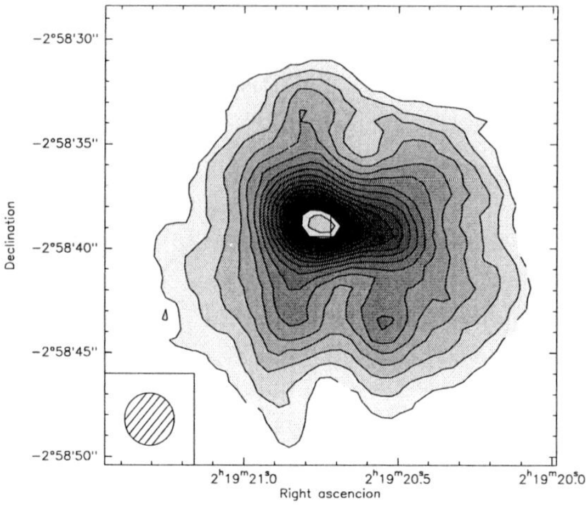

Figure 1 Contour map of the integrated ^{12}CO(2–1) emission from *o* Ceti over the whole velocity range of emission. The lower level contour and the contour spacing are 2 Jy beam^{-1} km s^{-1}. The beam after cleaning is shown in the box in the lower left corner.

of two "holes" located at $\sim 4''$ North and South of the star position. These holes generate a butterfly-like structure.

Emission extends over a velocity range of $\sim$ 7 km s^{-1}. A bipolar structure appears on channel maps (Fig. 3 in Josselin et al. 2000). In most velocity channels, the blue emission is dominated by a kind of equatorial bar, which in Fig. 1 appears as an asymmetric "peanut". On the contrary, this bar is hardly visible in the red emission, which displays a butterfly-like shape, with brighter South-East wing.

Altogether, we thus see a rather spherical circumstellar envelope being disrupted by a low-velocity bipolar outflow.

3.2. THE ATOMIC GAS

The brightness distribution of the $\lambda\lambda$ 7665, 7699 Å K I lines informs on the scattering of stellar light by circumstellar atomic potassium. An example of the obtained distribution of circumstellar atomic gas is shown in Fig. 2.

A striking characteristics of the K I spatial distributions along the slits is their asymmetries. These asymmetries can be explained by a structure in the north-west region, such as the bipolar outflow invoked to explain the radio map, superimposed on a spherically symmetric component. The positions of the "secondary" maxima are roughly aligned, together with the star position and the two holes in the molecular envelope. This reinforces the hypothesis of a bipolar outflow.

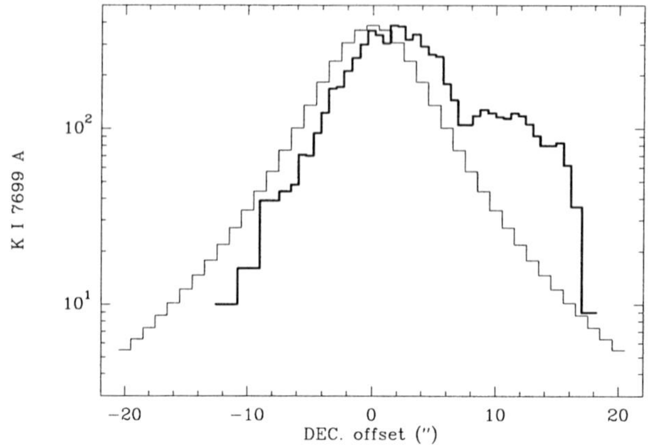

Figure 2 Comparison between an optically thin model of K I emission in a spherical envelope (thin line) and the observations from the 5″W slit (thick line).

4. DISCUSSION

The *o* Ceti system (star(s) + envelope) displays asymmetries on all scales. First of all, the primary star appears elliptical on HST images (Karovska et al. 1997). Infrared interferometry at 11 μm also reveals the complexity of the extended atmosphere and the inner dust envelope (R $\lesssim$ 10 $R_\star$; Lopez et al. 1997).

The coupling of all these data with our observations allows to draw out some conclusions about the origin of the asymmetries. First of all, it seems clear that mass loss is not isotropic. Indeed, the two kinematic components in the CsE probably prevent from explaining the geometry by external effects such as asymmetric photodissociation of CO and photoionization of K I by UV radiation of interstellar origin or emitted by the companion star. The presence of asymmetries on all scales, ranging from the star itself to the atomic envelope, suggests that they originate in the stellar atmosphere. An asymmetric mass-loss process, due to e.g. non-radial pulsations or giant convective cells and/or magnetic spots on the stellar surface, is thus the most natural explanation.

5. CONCLUSIONS

We presented new high-resolution observations of the neutral envelope of *o* Ceti envelope. The molecular envelope, observed in the CO(2–1) radio line with a 2.5″ resolution, appears very complex and may be composed of a rather spherical envelope being disrupted by a low-velocity bipolar atomic outflow. This outflow, oriented North-South, is inclined by about 30° with respect to the plane of the sky. The extent of the molecular envelope is essentially limited by photodissociation by interstellar UV radiation. The atomic envelope, observed in the optical K I lines, is about 4 times more extended. The bipolar outflow is also clearly visible in these lines.

o Ceti is probably a young AGB star. Our data bring new evidence that asymmetric mass loss is a common phenomenon, which occurs from the beginning of the AGB phase to the superwind and post-AGB phase. This may be particularly important to explain the shaping of bipolar Planetary Nebulae.

Acknowledgments

This paper is based on observations carried out with the IRAM Plateau de Bure Interferometer, the Canada-France-Hawaii Telescope, and at the Haute-Provence Observatory. IRAM is supported by INSU/CNRS (France), MPG (Germany) and IGN (Spain). We acknowledge the IRAM staff from the Plateau de Bure and from Grenoble for carrying the observations and help provided during the data reduction. E.J., P.P. and R.B. acknowledge partial support from spanish grants PB 96-104 and SB 97-22043, and N.M. support from the French CNRS "Physico-Chimie du Milieu Interstellaire" program.

References

Guilloteau S., Delannoy J., Downes D., et al., 1992, A&A 262, 624.

Haniff C.A., Ghez A.M., Gorham P.W., et al., 1992, AJ 103, 1662.

Josselin E., Mauron N., Planesas P., Bachiller R., 2000, A&A, in press.

Karovska M., Nisenson P., Beletic J., 1993, ApJ 402, 311

Karovska M., Hack W., Raymond J., Guinan E., 1997, ApJ 482, L175

Knapp G.R., Young K., Lee E., Jorissen A., 1998, ApJS 117, 209.

Lopez B., Danchi W.C., Bester M., 1997, ApJ 488, 807.

Lopez B., 1999, in: "Asymptotic Giant Branch Stars", eds T. Le Bertre, A. Lèbre, C. Walekens, IAU Symp. 191, p. 409.

Mauron N., Caux E., 1992, A&A 265, 711.

Planesas P., Bachiller R., Martin-Pintado J., Bujarrabal V., 1990a, ApJ 351, 263.

Planesas P., Kenney J.D.P., Bachiller R., 1990b, ApJ 364, L9.

MASS LOSS GEOMETRY IN SYMBIOTIC BINARIES

Romano L.M. Corradi
Isaac Newton Group of Telescopes,
Apartado de Correos 321, E–38700 Santa Cruz de la Palma, Spain
rcorradi@ing.iac.es

Abstract I briefly revise the properties of the extended ionised nebulae around symbiotic stars. These are key objects in the discussion of the effects produced by a detached stellar companion on the AGB mass loss, of the outflows from accreting post–AGB stars, and of the mutual interaction of the slow and fast winds produced in these binary systems.

The phenomenology is illustrated by new results on three of the best known symbiotics: He 2–104 (the Southern Crab), RX Pup and CH Cyg.

1. SYMBIOTICS, MIRAS AND PNE

Symbiotic stars are interacting binary systems containing a cool giant and an accreting compact companion, in most cases a hot white dwarf (e.g. Mikolajewska 1997). About 20% of symbiotic binaries contain a Mira variable as the cool component, i.e. a star on the way of loosing its envelope and form a planetary nebula (PN). This subgroup of symbiotics were originally distinguished by their near-IR emission indicating the presence of warm dust, and were therefore called as *D*, or "dusty", types. Their orbital periods are usually not known, but there is a general consensus that they are of the order of several tens, or even hundreds of years. Thus they are *detached* binary stars that certainly avoid a common envelope evolution during the whole AGB phase of their cool component.

In these systems, only a small fraction (1% or less) of the Mira wind is accreted onto the white dwarf (WD) of the system, while the vast majority forms a slowly expanding circumbinary nebula. As the Mira wind is partly ionised (and thus "illuminated") by the radiation of the hot WD, its evolution is generally easier to study than in single AGB stars. For this reason, symbiotic Miras are excellent laboratories where to study the effects of mass loss prior to the formation of a PN due to the presence of a detached stellar companion.

R. Szczerba and S.K. Górny (eds.), Post-AGB Objects as a Phase of Stellar Evolution, 221–226.

In addition, the hot WD companion of symbiotic Miras behaves as a central star of a PN, and many physical processes relevant to the discussion of the shaping of PNe can be investigated, such as winds interactions, the formation of accretion discs and associated collimated winds, or the mass release in thermonuclear runaways caused by accretion onto the WD. In this respect, jet formation by the WD + accretion disc is a most interesting topic to investigate in symbiotic Miras.

Table 1 Extended nebulae around symbiotic stars (cf. Corradi et al. 1999a for references on individual objects).

Name	IR Type	Size [″]	Shape
Nebulae resolved at optical wavelengths			
RX Pup	D (Mira)	5.6	irregular
AS 201	yellow D′	13	circular
H 2–2	S	1.4	elliptical?
BI Cru	D (Mira)	150	bipolar
V417 Cen	yellow D′	100	irregular
He 2–104	D (Mira)	95	bipolar,jet
He 2–147	D (Mira)	5	ring
H 1–36	D (Mira)	$\leq$1.5	–
CH Cyg	S	32	jet? irregular
HM Sge	D (Mira)	30	jet? irregular
V1016 Cyg	D (Mira)	20	elliptical
HBV 475	S	0.4	irregular
AG Peg	S	8	irregular
R Aqr	D (Mira)	120	bipolar
Nebulae resolved at radio wavelengths			
RX Pup	D (Mira)	1	spherical wind
Hen 3–1383	symbiotic?	4	jet?
SS 96	S	0.17	axisymmetrical
H 1–36	D (Mira)	5	knotty
RS Oph	S	0.2	bipolar
CH Cyg	S	1.5	jet
HM Sge	D (Mira)	4	bipolar?
V1016 Cyg	D (Mira)	4	irregular
AG Peg	S	60	elliptical, bipolar
R Aqr	D (Mira)	120	bipolar,jet

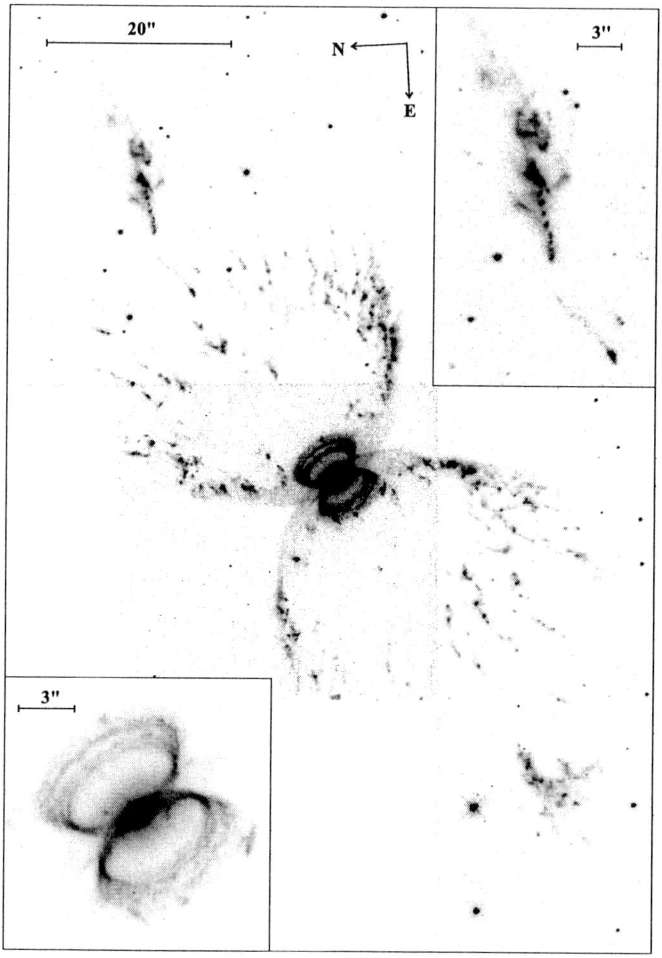

Figure 1 [NII] HST image of He 2−104 , on a logarithmic intensity scale. The inset boxes show details of the inner nebula and NW jet (adapted from Corradi et al. 2000).

2. IONISED OUTFLOWS IN SYMBIOTIC MIRAS

A complete list of the extended optical and radio nebulae known around symbiotic stars is given in Corradi et al. (1999a). An adapted version of that list is presented in Table 1. Optical ionised nebulae with sizes typical of PNe appear to be a common component of symbiotics Miras. On the contrary, very few nebulae are detected around the systems containing normal red giants. One obvious explanation is that Miras loose mass at a much higher rate than normal giants, and are therefore surrounded by denser nebulae which are easily observable when ionised by the WDs.

These nebulae show a variety of shapes, including elliptical or ring nebulae, jets, marked bipolar morphologies and irregular geometries. All the nebulae are highly aspherical, indicating that the kind of binary interactions occurring in symbiotic stars favour the formation of collimated outflows. In the following, I will illustrate the phenomenon by presenting some new results on symbiotic Miras. I will also mention the main physical processes which might play a role in shaping their outflows.

3. HENIZE 2-104 – THE SOUTHERN CRAB

He 2−104 is a symbiotic Mira whose WD companion has been undergoing a powerful thermonuclear outburst for $\geq$50 yr. As its nebular spectrum has not changed in the last 20 yr, stable thermonuclear burning on the surface of the WD is occurring, and the outburst might last for several hundred years. The extended nebula around He 2−104 consists of two nested pairs of bipolar lobes and a polar jet. An HST [NII] image is shown in Figure 1. The presence of two similar bipolar nebulae, but on a different scale, led us to interpret their formation as the result of two distinct (recurrent) nova–like outbursts on the WD component of the system (e.g. Corradi et al. 1999b). Fast winds setup up during the high luminosity phase of the outbursting WD would shape the circumstellar material into the bipolar lobes, the collimation being provided by a strongly flattened density distribution of the slower Mira wind, in a similar way as predicted by the standard Interacting-Winds theory of formation of bipolar PNe (e.g. Mellema 1997). Recurrent outbursts, separated by quiescence periods, would then result in multiple bipolar nebulae.

A recent kinematical study (Corradi et al. 2000b), however, reveals a different scenario. The inner pair of lobes, the outer one, and the polar jets are found to have the same kinematical age! Thus the different sizes of the two systems of lobes are just the result of different space velocities, the inner pair having expansion velocities of a standard Mira wind, while the outer pair and the polar jets being expanding with velocities up to ten times larger. (Three?) different collimation mechanism working in the system at about the same time are therefore needed. Speculatively, we associate the slowly expanding inner lobes with enhanced mass loss from the Mira, collimated by the density enhancement in the orbital plane create by the gravitational effects of the WD companion (Mastrodemos & Morris 1999). The higher velocity, outer bipolar lobes and jets would instead be produced by the hot component. Again, the collimation of the lobes might be purely hydrodynamical-inertial (e.g. caused by the equatorial density constraint), while jets might be collimated magnetically by an accretion disc wind (Livio 2000), or by means of a magnetized fast wind from the WD itself as proposed by García-Segura et al. (1999).

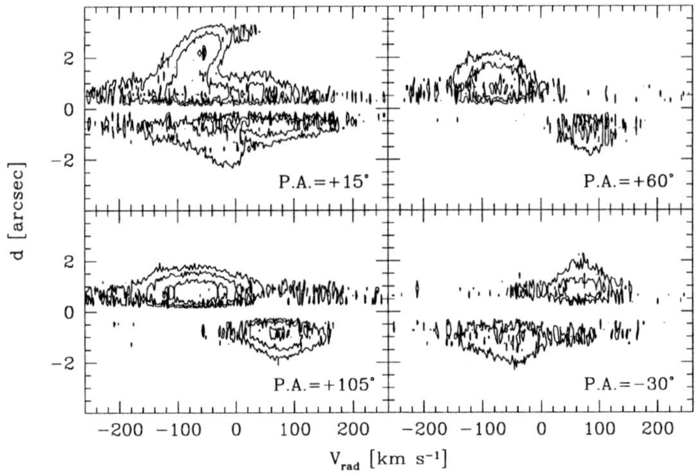

Figure 2 The [NII]658.3 nm spectra of RX Pup at various position angles, after subtraction of the unresolved core spectrum. Observed radial velocities V_{rad} are corrected to the systemic velocity. Successive levels of the contour plots increment of a factor $\sqrt{3}$.

4. RX PUPPIS

RX Pup is another symbiotic binary containing a long-period Mira variable. Recently, Mikolajewska et al. (1999) have presented a thorough study of its photometric and spectroscopic properties, and concluded that the the companion of the Mira is a hot WD that underwent a nova-like eruption around 1894, as well as a more recent one during the last three decades.

Paresce (1990) presented a coronographic [NII] image which revealed an elongated structure extending toward P.A.=+15° up to about $3''.7$ from the centre, which was interpreted as being a one-sided jet similar to the one of the other symbiotic Mira R Aqr (Hollis & Koupelis 2000).

The presence of this elongated feature is confirmed by recent echelle observations (Corradi & Schwarz 2000a, see also Figure 2), but the low radial velocities observed and their decrease with distance from the centre do not support the conclusion that it is a jet. In addition, a new nebular structure is found at ~90° to the main feature. The split lines indicate a nebula expanding with velocities $\geq$ 80 km s^{-1}, probably with a bipolar shape. This shows that outflows from symbiotic Miras are in some cases complex, with multiple directions of mass ejection, and with different symmetry axes appearing at different observing wavelengths. A similar case is that of HM Sge (Corradi et al. 1999b).

5. CH CYGNI

CH Cyg, located at about 250 parsec from the Sun, is one of the closest symbiotic binaries. It is composed of an early AGB star and a hot WD. The binary system is eclipsing, with an orbital period of 15 yr.

A spectacular activity of CH Cyg started on 1963, showing alternate periods of activity and quiescence reflecting in an irregular photometric and spectroscopic variability. A most notable event was the ejection in 1984 of a powerful radio jet (Taylor et al. 1986), whose proper motions indicated an expansion velocity of 400 km s^{-1}. We have imaged CH Cyg with HST in 1999, 15 yr after the jet ejection. The [OII] nebula is elongated roughly in the direction of the radio jet, and extends up to about 8 arcsec (2000 a.u.) from the central stars, slightly less than the distance at which the radio jet would be expected adopting the proper motions measured in 1984. A preliminary analysis of both the HST imaging and new ground-based echelle spectra suggest that the optical outflow lie very close to the plane of the sky. As the binary system is eclipsing, this is consistent with the general idea that the jet occurs along the polar direction of the orbit. What we observe in the [OII] near-UV light might be the remnant of the matter contained in the 1984 radio jet, or regions of a pre-existing nebula (shock–)excited by the passage of the jet. Owing to its vicinity and large expansion velocities, CH Cygis a very attractive environment where to follow in real time the evolution of a stellar jet.

References

Corradi, R.L.M., Brandi, E., Ferrer, O., Schwarz, H.E. 1999a, A&A, 343, 841

Corradi, R.L.M., Ferrer, O., Schwarz , H.E., Brandi, E., Garcia, L. 1999b, A&A, 348, 978

Corradi, R.L.M., Schwarz, H.E. 2000a, A&A, in press

Corradi, R.L.M., Livio, M., Balick, B., Munari, U., Schwarz, H.E. 2000b, ApJ Submitted

García-Segura, G., Langer, N., Rózyczka, M., & Franco, J. 1999, ApJ, 517, 767

Hollis, J.M., Koupelis, T.: 2000, ApJ 528, 418

Livio, M. 2000, in 'Asymmetrical Planetary Nebulae II: From Origins to Microstructures', ASP Conf. Series, Vol. 199, J. H. Kastner, N. Soker, and S. Rappaport eds, p. 243

Mastrodemos, N., Morris, M. 1999, ApJ, 523, 357

Mellema, G.: 1997, A&A 312, L29

Mikolajewska, J.:1997 in *Physical processes in Symbiotic binaries and related systems* ed. J. Mikolajewska, Copernicous Foundation for Polish Astronomy, Warsaw, p. 3.

Mikolajewska, J., Brandi, E., Hack, W., Whitelock, P.A., Barba, R., García, Marang, F. 1999, MNRAS, 305, 190

Paresce, F.: 1990, ApJ 357, 231

Taylor, A.R., Seaquist, E.R., Mattei, J.A.: 1986, Nature 319, 38

CIRCUMSTELLAR ENVIRONMENT OF RX PUPPIS

Joanna Mikołajewska
N. Copernicus Astronomical Center, Bartycka 18, 00716 Warsaw, Poland
mikolaj@camk.edu.pl

E. Brandi, L. G. García, O. E. Ferrer
Facultad de Ciencias Astronómicas y Geofísicas, UNLP - CIC - CONICET, Argentina

P.A. Whitelock, F. Marang
South African Astronomical Observatory, PO Box 9, 7935, Observatory, South Africa

Abstract The symbiotic Mira, RX Pup, shows long-term variations in its mean light level due to variable obscuration by circumstellar dust. The last increase in extinction towards the Mira, between 1995 and 2000, has been accompanied by large changes in the degree of polarization in the optical and red spectral range. The lack of any obvious associated changes in the position angle may indicate the polarization variations are driven by changes in the properties of the dust grains (e.g. variable quantity of dust and variable particle size distribution, due to dust grain formation and growth) rather than changes in the viewing geometry of the scattering region(s), e.g. due to the binary rotation.

1. INTRODUCTION

RX Pup is a symbiotic binary composed of a long-period Mira variable pulsating with $P \approx 578$ days, surrounded by a thick dust shell, and a hot $\sim 0.8\,M_{\odot}$, white dwarf companion which has been undergoing a nova-like eruption during the last three decades. The binary separation could be as large as $a \geq 50$ a.u., and the corresponding $P_{\rm orb} \geq 200$ yr, as suggested by the permanent presence of a dust shell around the Mira component (Mikołajewska et al. 1999, hereafter M99). In particular, the Mira is never stripped of its dust envelope, and even during relatively unobscured phases the star resembles high-mass loss galactic Miras with thick dust shells.

R. Szczerba and S.K. Górny (eds.), Post-AGB Objects as a Phase of Stellar Evolution, 227–233.

Our recent analysis of multi-frequency observations shows that most, if not all, photometric and spectroscopic activity of RX Pup in the UV, optical and radio range is due to activity of the hot component, while the Mira variable and its circumstellar environment is responsible for practically all changes in the IR range (M99). In particular, we have found large changes in the reddening towards the Mira accompanied by fading of the near IR flux (Fig. 1). However, the reddening towards the hot component and emission line regions remained practically constant and is generally less than that towards the Mira. These changes do not seem to be related to the orbital configuration nor to the hot component activity. Similar dust obscuration events occur in many well observed symbiotic Miras (e.g. Whitelock 1998), as well as in some single Miras (e.g. Mattei 1997, Whitelock 1998), and they are best explained as intrinsic changes in the circumstellar environment of the Mira variable, possibly due to intensive and variable mass loss (see also discussion in M99, and Mikołajewska 1999).

In the following we discuss polarization measurements for RX Pup obtained during the most recent obscuration phase.

2. OBSERVATIONS

$UBVRI$ polarization measurements were obtained between May 1996 and March 2000 with the five channel photopolarimeter of Turin Observatory, attached at the 2.15-m CASLEO telescope at San Juan, Argentina. More details about the instrumentation and data reduction can be found in Brandi et al. (2000). In addition, optical spectra with the same telescope and REOSC echelle spectrograph were obtained (see M99 for details). Finally, new $JHKL$ broad-band photometry on the SAAO system was obtained with the Mk II infrared photometer on the 0.75-m telescope at Sutherland (Carter 1990). These new IR data together with data from M99 are displayed in Figure 1 where the dates of the polarimetric measurements are also marked.

3. RESULTS AND DISCUSSION

In the period covered by our polarimetric observations, the near IR flux of RX Pup was gradually decreasing and the $J-K$ colour was increasing, suggesting that the Mira had entered a new obscuration phase. These secular changes in near IR flux are generally not correlated with changes in the optical; in particular the visual magnitude estimates reported by the Variable Star Section of the Royal Society of new Zealand (RASNZ) show that RX Pup was brightening in the optical between 1995 and 1998, and declining later (see also Fig. 4 in M99).

The results of polarimetric observations are shown in Figure 2. The left panel presents the degree of polarization and the position angle as a function of wave-

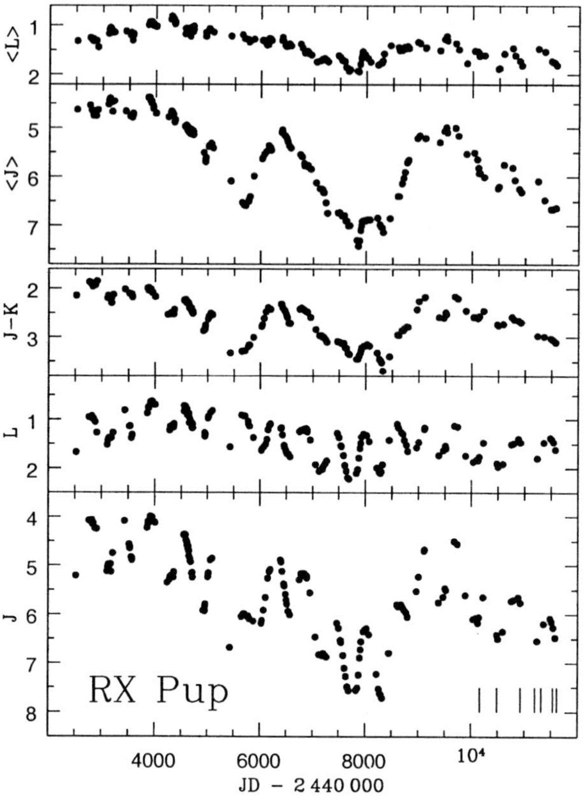

Figure 1 J and L light curves and $J - K$ colour of RX Pup. The upper panel show the same light curves after removal of the Mira pulsation. Bars in the bottom panel indicate dates of our polarimetric observations.

length for different observing runs. The observed wavelength dependence of p and the position angle could be accounted for by the presence of two components, one with p.a. $\sim 145-155°$, and a flat wavelength dependence resembling that of interstellar polarization which suggests a presence of interstellar-type large dust particles, and another component with p.a. $\sim 175 - 160°$ and p increasing steeply towards the ultraviolet, consistent with Rayleigh scattering by small particles. Both components show significant temporal changes which indicate circumstellar origin of the observed polarization.

Figure 2 also shows the polarization changes versus the visual magnitudes and $J - K$ colour. The visual magnitude is dominated by the light from the hot component, whereas the $J - K$ colour measures the amount of reddening towards the Mira component. The polarization degree seems to increase with both the visual magnitude and the $J - K$ colour. At the same time we do not find any correlation between the polarization and the Mira pulsation phase.

The polarized flux does not seem to be scattered light from the Mira, since in the optical the Mira is heavily obscured by the dust (M99). In particular, we do not see pulsational variations, nor can we detect the TiO bands in optical spectra at any epoch (see also M99). The $UVBR$ and probably also the I flux is dominated by the hot component with some contribution from the nebular line and continuum emission. M99 also found that the contribution from the nebular emission increases relative to other sources of emission as the optical brightness decreases, and as the hot component temperature and degree of ionization in the nebula, increase. In particular, the contribution from the nebular emission was negligible in 1995-1996. Then the relative contribution of the nebular emission increased from mid 1998 in-step with the declining visual brightness, although it never exceeded 0.1-0.15 mag. The polarized flux varied by a factor of ~ 2 in the visual range, whereas the strongest H I Balmer and Fe II emission-line fluxes remained constant within $\sim 20\%$. There was also no significant increase in polarization (although only $p_{\rm R}$ estimates are available) in 1987, when the nebular continuum and emission lines where much stronger than in the 1990s (M99). It is thus unlikely, that the polarized component corresponds to the nebular emission. Instead we believe that it is most likely to be radiation from the warm, $T \sim 6000$ K, component found in the 1995-96 optical spectra by M99 – scattered in the dust envelope surrounding the Mira. The warm component is responsible for the optical variability in the 1990s; however, there is no obvious, simple correlation between this activity and the reported polarization changes.

The changes in polarization are apparently correlated with the varying reddening towards the Mira. Between 1994/5 and 1999/2000 the Mira's average brightness weakened by ~ 1.5 mag at J and ~ 0.7 mag at K, respectively. The mean $J - K$ colour increased from ~ 2.2 in 1994/5 to ~ 2.6 in 1996–98, and to ~ 3 in 1999, respectively. These changes have been accompanied by an increase in p, by a factor of $\sim 2 - 3$, at all wavelengths, and a slow rotation in the position angle. The sense of this rotation, however, depends on the spectral range: the position angle decreased by $\sim 15°$ in UB and increased by $\sim 10°$ in the VRI range, respectively. Unfortunately, it is practically impossible to distinguish between a steady rotation of position angle with time and changes related to variable conditions in the circumstellar envelope of RX Pup because the reddening has been steadily increasing during the period covered by our observations.

There are very few previous polarization measurements for RX Pup. The UBV polarimetric measurements obtained during maximum of the hot component outburst and IR bright phase of the Mira in 1979-1980 show practically flat wavelength dependence with $p \sim 2\%$ and a constant position angle of $\sim 120°$ (Barbier & Swings 1982); however, these data were corrected, by an unknown amount, for interstellar polarization. In any case, the component with p steeply increasing towards ultraviolet was absent at this epoch. Schulte-Ladbeck et al. (1990) estimated $p_{\rm R} = 1.61$ and p.a. $= 121°$ on Mar 27, 1987, whereas Harries & Howarth (1996) measured $p_{\rm R} = 1.825$ and p.a. $= 146.7°$ on May 6-10, 1994.

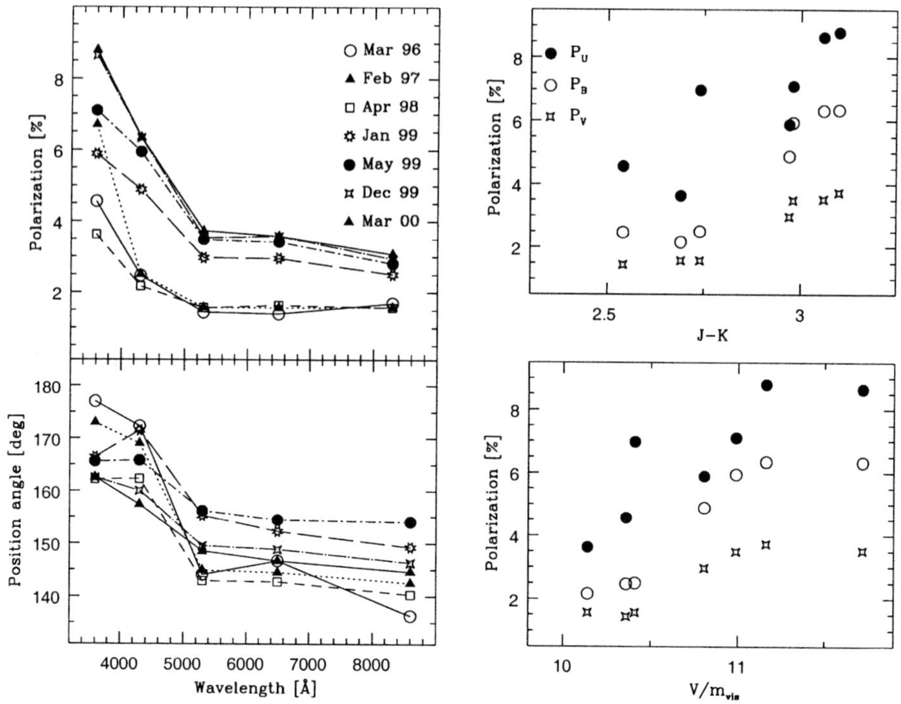

Figure 2 Left: Wavelength dependence of linear polarization, p [%] (top), and position angle for different epochs (bottom). The points correspond to weighted mean values calculated for each observing run. The error bars are not shown, however the relative mean errors in p [%] are < 0.06 in UB, < 0.008 in VRI, and < 0.01 in the position angle. Right: Dependence of the polarization on the reddening towards the Mira measured by the $J - K$ colour (top), and visual brightness (bottom).

On both of these dates, the visual magnitude of RX Pup was $V/m_{\rm vis} \sim 11.1$ and $J - K \sim 2.5$. We have observed similar values of $p_{\rm R} \sim 1.5 - 1.6$ and $J - K \sim 2.5 - 2.7$ in 1996-98; the system was then, however, much brighter, $V \sim 10.4 - 10.1$. This may suggest that the polarization changes are related predominantly to changes in the reddening towards the Mira, and not to the hot component brightness. The polarization changes are also not related to any particular changes in the emission line spectrum of RX Pup.

Interestingly, during 1996 March to 1998 April the degree of polarization at $BVRI$ was practically constant, while in U it varied. This may indicate that the 1996-98 changes were caused by variable conditions in the region where small particles dominate, while in 1999-2000 the contribution of large dust particles becomes comparably important. The lack of dramatic changes in the position angle associated with changes in p and the reddening towards the Mira component may indicate the polarization variations are driven by changes in the properties of the dust grains, for example variable amounts of dust with variable size distributions, due to dust grain formation and growth.

If the rotation in position angle with time is due to orbital motion, than the change in p.a. of $\sim 10 - 15^\circ$ in ~ 5 years corresponds to the binary period of $\sim 120 - 180$ years. Similarly, Δ p.a. ~ 30 between 1987 and the most recent observations is consistent with $P \sim 160$ yr. Note that an orbital period of ~ 200 yr is also implied by the properties of the dust envelope around the Mira component (M99).

In this context, it is interesting that a similar increase in p at $UBVRI$, together with a slow rotation of the position angle, increasing in UB and decreasing in RI, was observed during the dust obscuration event in R Aqr in the late 1970s (Nikitin & Khydyakova 1979). There is, however, a significant difference between R Aqr and RX Pup; in the former the optical and red light is dominated by the Mira, whereas in the latter the contribution from the Mira is negligible at these wavelengths. Moreover, in R Aqr, both the degree of polarization and the position angle (especially in the ultraviolet and blue spectral range) also vary with pulsation phase, which is not the case for RX Pup.

Since we do not know the orbital phase of RX Pup, it is impossible to drive any definite conclusions from the position angle of the polarized light. However, the fact that the hot component does not seem much affected when the reddening towards the Mira is strongly increasing (M99), suggests that it cannot be behind the dust cloud. Furthermore, the relatively large degree of polarization suggests that it is to one side of, rather than in front of, the dust. Assuming that the dust around the Mira is largely confined to a cone shaped region, the hot star is not far from the quadrature and is moving in our direction (somewhere between the quadrature and the inferior conjunction). According to the standard scattering model the position angle of the polarized light should be roughly orthogonal to the line connecting the source of light and the scattering region. The observed position angle short wavelengths, where the Rayleigh scattering by small particles dominates, is then roughly consistent with a scattering region close to the Mira, while the large particles are presumably dominating at larger distances from the Mira, which may account for the difference between p.a. at UB and VRI.

The polarization geometry of RX Pup can be compared with the geometry of its ionized nebula. There are two main directions for mass outflow distinguished in the nebula which are roughly perpendicular to each other (Corradi & Schwarz 2000): the previously known elongated optical/radio feature at p.a. $\sim 15^\circ$ with a velocity decreasing with distance from the central object, and a new EW compact component expanding at ≥ 80 km/s, probably with a bipolar shape. The elongated structure would then be roughly perpendicular to the deduced binary axis, whereas the compact bipolar flow would be aligned with the binary axis.

The orientation of the binary axis will of course change with the binary motion. The shape (geometry) of the neutral region where the dust can be effectively formed, and so the scattering geometry, will also change with the hot component properties, for example with temperature and luminosity - which determine the number of ionizing photons available, and with wind efficiency - which influences the position of the shock front between the components. Thus, further studies involving both high resolution imaging techniques together with spectropolarimetry are necessary to reveal much more detailed information about the geometry of the circumstellar environment of RX Pup.

Acknowledgments

This research was partly supported by KBN Research Grants No. 2 P03D 021 12 and No. 5 P03D 019 20 and makes use of observations obtained at the SAAO, Sutherland, South Africa.

References

Barbier, R., Swings, J.P., 1982, in Jaschek, M., Groth, H.-G., eds, Be Stars, IAU Symp. 98, D.Reidel, Dordrecht, p. 103

Brandi, E., Garcia, L.G., Piirola, V., Scaltriti, F., Quiroga, C., 2000, A&AS, 145, 197

Carter, B.S., 1990, MNRAS, 242, 1

Corradi, R.L.M., Schwarz, H., 2000, A&A, 363, 671

Harries, T.J., Howarth, I.D., 1996, A&AS, 119, 61

Mattei, J.A., 1997, JAAVSO, 25, 57

Mikołajewska, J., 1999, in Stecklum, B., Guenther, E., Klose, S., eds, Optical and Infrared Spectroscopy of Circumstellar Matter, ASP Conf. Ser., vol. 188, 291

Mikołajewska, J., Brandi, E., Hack, W., Whitelock, P.A., Barba, R., Garcia, L., Marang, F., 1999, MNRAS, 305, 190 (M99)

Nikitin, S.N., Khydyakova, T.N., 1979, Sov. A. Lett. 5, 327

Schulte-Ladbeck, R., Aspin, C., Magalhaes, A.M., Schwarz, H.E., 1990, A&AS, 86, 227

Whitelock, P.A., 1998, in Takeuri, M., Sasselov, D., eds, Pulsating Stars – Recent Developments in Theory and Observation, Universal Academy Press, Tokyo, 31

THE SUDDEN LOSS OF 0.01 $M_\odot$ FROM THE VV CEP'S PRIMARY AS A CHARACTERISTIC OF AGB EVOLUTION IN THE BINARIES.

Dariusz Graczyk & Maciej Mikołajewski

Center for Astronomy, Gagarina 11, Toruń 87-100

weganin@astri.uni.torun.pl, mamiko@astri.uni.torun.pl

Abstract We present the evidence of the large and sudden mass outflow from the primary star of VV Cep system. We are discussing the medium mass model for this system and propose that the VV Cep's primary is not a M supergiant but a massive AGB star. The timescale of the outflow is about 20 years and during the last 100 years there was no evidence of such a large mass loss event.

VV Cephei is a well-known eclipsing binary suspected to consist of a M2-type supergiant and a hot B type companion. Gaposchkin (1937) derived from photographic plates the orbital period $7430^{\rm d}$ (20.3 years) and published the ephemeris: MinI JD 2421070 + 7430 E. The hot component is surrounded by the accretion disk and the cool star itself is apparently a SRa variable with the period of about $110^{\rm d}$ (Saitō et al. 1980). We have observed this system using five-color photoelectric photometry during the last 1997/98 eclipse (Graczyk et al. 1999). The analysis of the data shows that the last ecplise was late by 68^d! This result was confirmed through spectroscopic observations of FeII, [FeII] emission lines and V/R ratio of Hα emission line (Leedjärv et al. 1999).

We have examined O$-$C diagram (Figure 1) for the past five eclipses observed since 1890, but we did not find any systematic deviation from the Gaposchkin's formula. This means that the period change is real and not "produced" by the presence of the third body. The period suddenly increased by almost 1% of the orbital period between the last two eclipses (20 years). We have discussed two reasons for the period change: 1) mass transfer between components and 2) ejection of mass from the system (Graczyk et al. 1999). The spectroscopic solution by Wrigh (1977) gives masses of both components equal to 20 $M_\odot$. Using these values and formulae for the period change as a function of the components' mass change:

$$\frac{\dot{P}}{P} = -\frac{3(1-q^2)}{q}\frac{\dot{M}_2}{M} + \frac{3}{q}\left(\frac{\dot{M}_2}{M}\right)^2$$

R. Szczerba and S.K. Górny (eds.), Post-AGB Objects as a Phase of Stellar Evolution, 235–238.

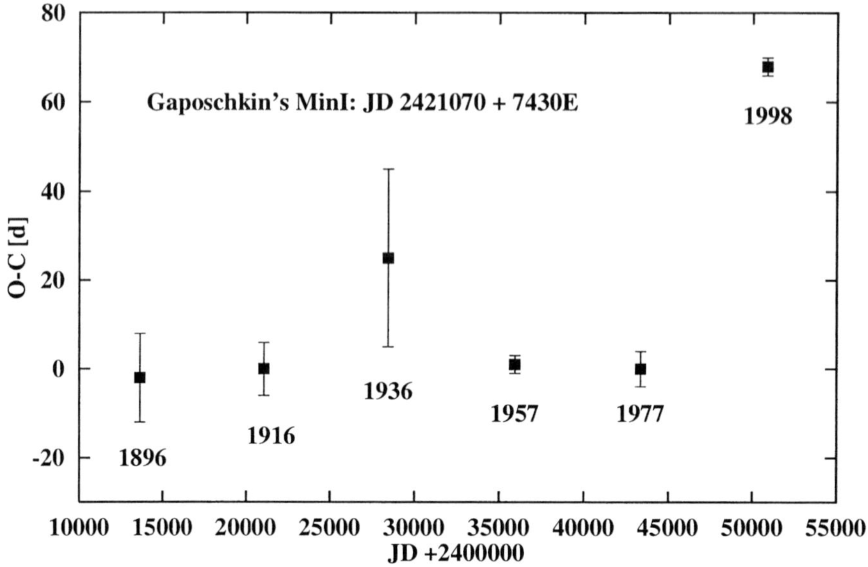

Figure 1 O-C diagram for VV Cep

in the case of mass transfer (accretion) and

$$\frac{\dot{P}}{P} = -\frac{(3+2q)}{q}\frac{\dot{M}_2}{M}$$

in the case of ejection of some mass from the system, we can calculate the total mass transferred (case 1) or ejected (case 2) during the last 20 years in VV Cep system. P is the orbital period, q is the mass ratio, M is the total mass of the system and M_2 denotes the mass of the losing star. For Wright solution we get 1 solar mass and 0.1 solar mass, respectively.

This enormously large amount of mass accreted onto hot component or ejected from the system (due to the strong wind from M star) should be observed as a considerable increase of VV Cep's activity visible in photometric and spectroscopic data between the last two eclipses. The lack of such observations entitles us to propose medium mass model for VV Cep instead of Wright's model. In this model the M component is the star with a mass of about 2.5 $M_\odot$ and radius 850 $R_\odot$ and B component is massive early type main sequence star with a mass of about 8 $M_\odot$. When the M star is near the periastron its radius is close to its Roche-lobe radius and some accretion and ejection events may occur. As the result, the total mass necessary to change the period of 1%, in the medium mass model, is 0.01 $M_\odot$ in the case of mass transfer and 0.008 $M_\odot$ in the case of mass outflow. We believe that both processes occured at the same time, but as the M star does not fil up exactly its Roche-lobe the more important process is mass loss from the system. Thus, during the 20 years we should expect the mass loss rate of $5 \times 10^{-4} M_{\odot} y^{-1}$ which is consistent with the estimates of the upper limits of mass loss rate from M star (Stencel et al. 1993). Figures 2 and 3 present comparison between Wright's dynamical solution (denotes by large filled square) and our solution (small squares) based on mass function for M star. Figure 2 presents the mass loss scenario and Figure 3 the mass accretion scenario. The dashed lines mark lines of constant mass loss rates in the plane: mass of the B component – mass ratio.

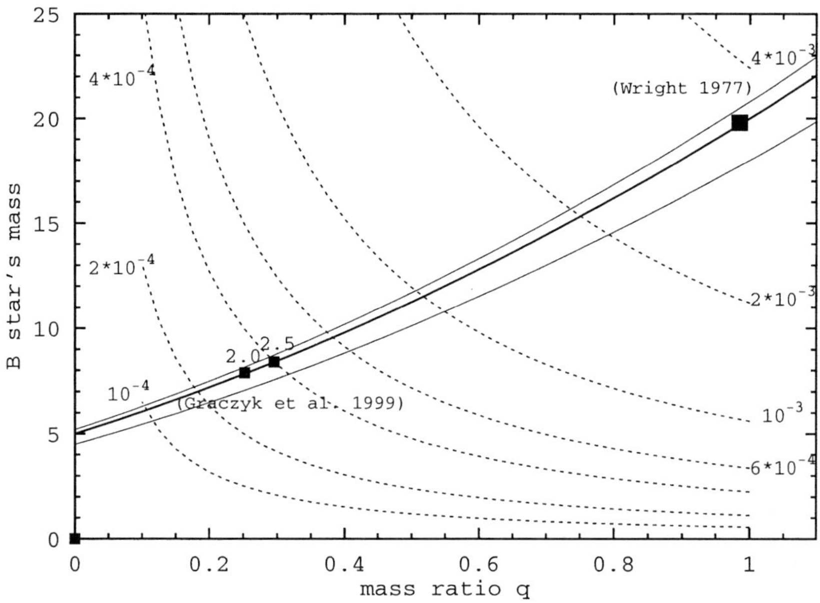

Figure 2 Dynamical solution for VV Cep in the case of the mass outflow (loss) scenario.

If we consider medium mass model we realise that the mass of the M component and its temparature 3100 K are fairly consistent with early AGB phase. But, there is no indication of Mira-variability, still the M star is semiregular variable. Its initial mass before mass loss and mass transfer phase should be around 10 $M_{\odot}$ and thus is larger than for typical star undergoing AGB phase. Numerical models suggest that binarity may lead to strong enhancement of mass loss during the time when the M primary almost fills up its Roche-lobe, especially in AGB phase (Frankowski & Tylenda 1998). The medium mass model for VV Cep does not answer to the question what processes induce very large, sudden mass loss in this binary. The timescale of the mass outflow/accretion is about 20 years and during last 100 years there was no evidence of the period change in VV Cep. We think that, maybe, there is a link between sudden processes of mass loss from AGB stars in binaries and the existence of tiny arcs around some

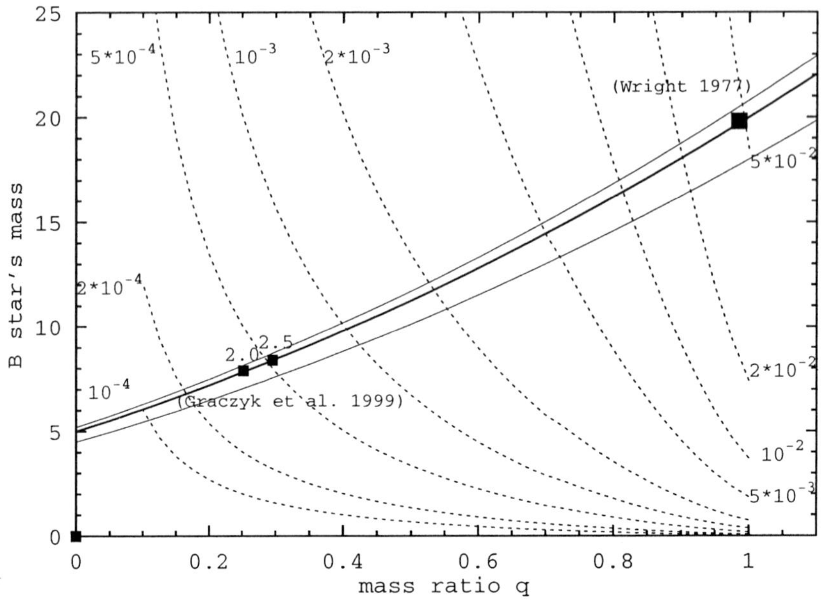

Figure 3 Dynamical solution for VV Cep in the case of the mass accretion scenario.

planetary nebulae. These arcs are concentric and circular and they origin from early phase of AGB evolution. However, their circumstellar symmetry may be difficult to explain by the mass outflow from binary systems which, in general, should have the asymmetrical spatial appearance.

Acknowledgments

We are much indebted to dr Laurits Leedjärv for correcting the English text. This work was supported by Polish KBN grants No. 2P03D 001 17 and No. 2 P03D 019 17.

References

Frankowski A., Tylenda R., 1998, Asymptotic Giant Branch Stars, IAU Symposium 191 Poster Session, P5-06

Gaposchkin S., 1937, Harvard Coll. Obs. Circ., No. 421

Graczyk D., Mikołajewski M., Janowski J.L., 1999, IBVS 4679

Leedjärv L., Graczyk D., Mikołajewski M., Puss A., 1999, A&A 349, 511

Saitō M., Sato H., Saijo K., Hayasaka T., 1980, PASJ, 32, 163

Stencel, R.E., Potter D.E., Bauer Hagen W., 1993, PASP, 105, 45

Wright K.O., 1977, J. Roy. Astron. Soc. Canada, 71, 152.

IV

EVOLUTION, SPECTROSCOPY AND CHEMICAL COMPOSITION

AGB AND POST-AGB EVOLUTION: STRUCTURAL AND CHEMICAL CHANGES

T. Blöcker, R. Osterbart, G. Weigelt

Max-Planck-Institut für Radioastronomie, 53121 Bonn, Germany

Y. Balega

Special Astrophysical Observatory, Nizhnij Arkhyz, 357147, Russia

A. Men'shchikov

Stockholm Observatory, 13336 Saltsjöbaden, Sweden

Abstract Structural and chemical changes during the AGB and post-AGB evolution are discussed with respect to two recent observational and theoretical findings. On the one hand, high-resolution infrared observations revealed details of the dynamical evolution of the fragmented, bipolar dust shell around the far-evolved carbon star IRC+10216 giving evidence for rapid changes of an already PPN-like structure during the very end of the AGB evolution. On the other hand, stellar evolution calculations considering convective overshoot have shown how thermal pulses during the post-AGB stage lead to the formation of hydrogen-deficient post-AGB stars with abundance patterns consistent with those observed for Wolf-Rayet central stars.

1. INTRODUCTION

During the evolution along the Asymptotic Giant Branch (AGB) and the subsequent post-AGB phases various structural and chemical changes of the stars and their surrounding dust-shells and nebulae take place. One important issue is the change of the dust-shell symmetry during transition from the AGB to the post-AGB stage. Unlike dust-shells around AGB stars, post-AGB objects as protoplanetary nebulae often expose prominent features of asphericities, in particular in axisymmetric geometry (e.g. Olofsson 1996), The shaping of planetary nebulae can be described by interacting stellar wind theories (see Kwok 2000) treating the interaction of a fast (spherical) central-star wind with the preceding slow (aspherical) AGB wind. The establishment of bipolar structures seems to begin already during the (very end of) AGB evolution. One example of

R. Szczerba and S.K. Górny (eds.), Post-AGB Objects as a Phase of Stellar Evolution, 241–248.

this stage of evolution is the carbon star IRC+10216. Its recent high-resolution observations indicate asymmetric mass-loss processes and even give insight into its dynamical dust-shell evolution (Osterbart et al. 2000).

Another matter of debate is the origin of hydrogen-deficient post-AGB stars. Although stars evolving through the AGB phase stay hydrogen-rich at their surfaces, a considerable fraction of their descendants, the central stars of planetary nebulae (CSPNe), show hydrogen-deficient compositions (Mendez 1991). Approximately 20% of the whole CSPNe population appears to be hydrogen deficient while the rest show solar-like compositions. Important constituents of the hydrogen-deficient population are the Wolf-Rayet (WR) central stars and the hot PG 1159 stars with typical surface abundances of (He,C,O)=(33,50,17) by mass (Dreizler & Heber 1998, Koesterke & Hamann 1997). Standard stellar evolution calculations failed to model these objects since they predict post-AGB stars either to have hydrogen-rich surfaces (e.g. Blöcker 1995) or, if hydrogen-deficient, to expose only a few percent of oxygen in their photospheres (Iben & McDonald 1995). However, if convective overshoot is considered, hydrogen-deficient post-AGB stars with abundance patterns as observed in WR central stars can be formed (Blöcker 2000, Herwig 2000).

2. THE DUST-SHELL OF IRC+10216

IRC+10216 (CW Leo) is the nearest ($d \sim 130$ pc, Groenewegen 1997) and best-studied carbon star and one of the brightest infrared sources in the sky. Due to strong stellar winds of $\dot{M} \approx 2-5 \times 10^{-5} \mathrm{M}_{\odot}\,\mathrm{yr}^{-1}$ (Loup et al. 1993) it is highly enshrouded by dust. The central star of IRC+10216 is a long-period variable with a period of $\sim$ 649 days (Le Bertre 1992). The bipolar appearance of the dust shell around this object was reported by Kastner & Weintraub (1994). The non-spherical structure is consistent with the conjecture that IRC+10216 is in a phase immediately before entering the protoplanetary nebula stage. High-resolution observations showed that the inner circumstellar dust shell is fragmented (Weigelt et al. 1998, Haniff & Buscher 1998). The results of Dyck et al. (1991) and Haniff & Buscher (1998) showed that the dust-shell structure of IRC+10216 is changing within a timescale of only several years. The dynamical dust-shell evolution was revealed in detail by recent observations of Osterbart et al. (2000) and Tuthill et al. (2000).

Figures 1-2 show J-, H-, and K-band observations of IRC+10216 (Osterbart et al. 2000). The images were reconstructed from 6 m telescope speckle interferograms using the bispectrum speckle interferometry method (Weigelt 1977, Lohmann et al. 1983, Weigelt 1991). The H and K images with resolutions between 70 mas and 92 mas consist of several compact components within a 0.2" radius and a fainter asymmetric nebula. A series of K-band images from five epochs between Oct. 1995 and Nov. 1998 shows the dynamical evolution

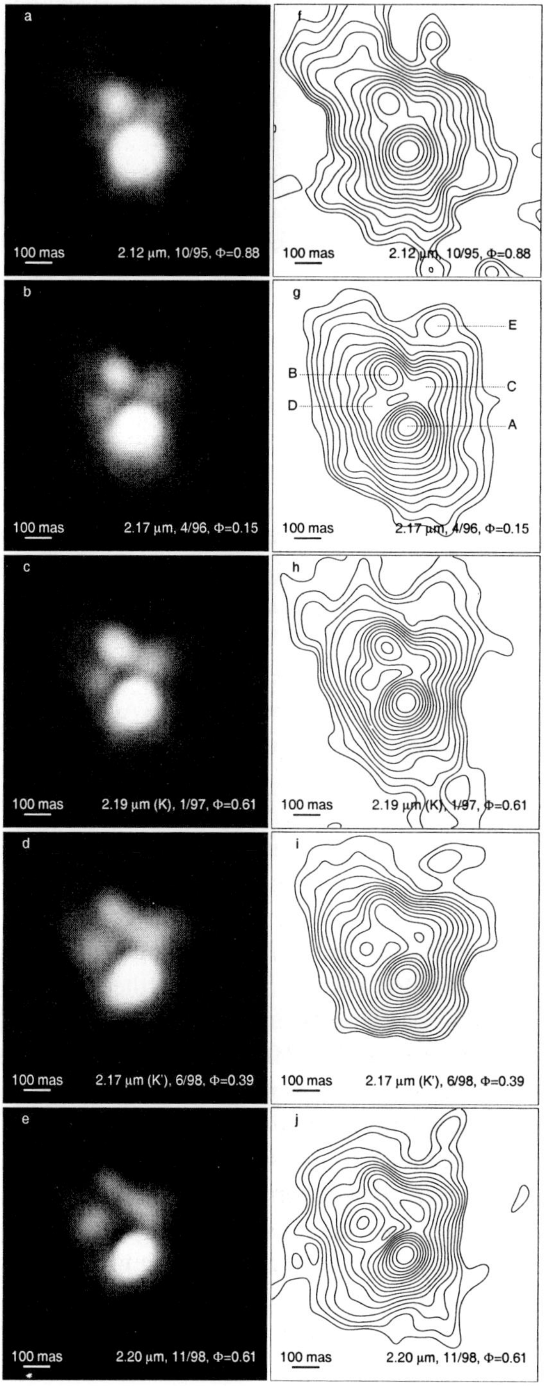

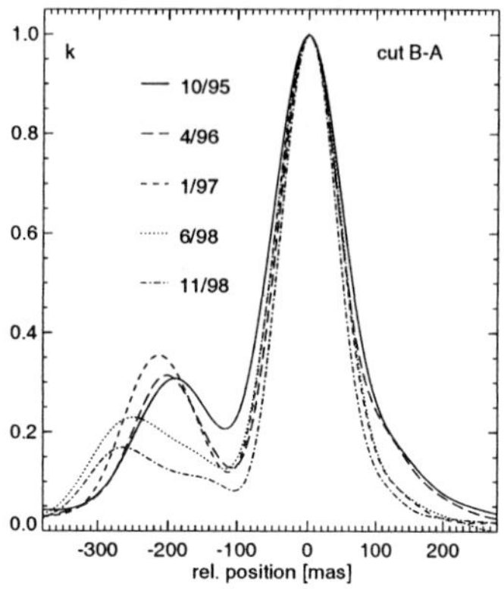

Figure 1 **a** to **e**: High–resolution bispectrum speckle interferometry images of IRC+10216. North is up and east to the left. The figures represent a time series showing the evolution of the subarcsecond structure of IRC+10216 from 1995 (top) to 1998 (bottom). In all figures the same gray level corresponds to the same relative intensity measured with respect to the peak. **f** to **j**: Same as a to e but as contour representation (contours at every 0.3 mag down to 4.8 mag relative to the respective peak). The resolutions of the images are 92 mas (a, f), 82 mas (b, g), 87 mas (c, h), 87 mas (d, i), and 75 mas (e, j). In all figures the filter wavelength, the epoch of the observation, and the photometric phase Φ are indicated. **k**: Cuts through the images a to e along the axis from component A to B (position angle 20°).

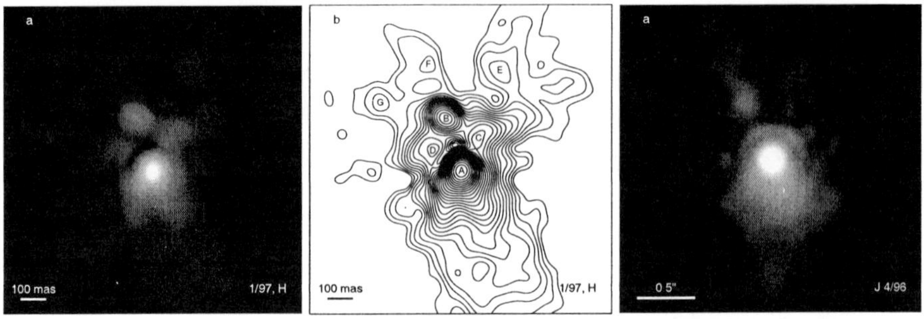

Figure 2 **a**: 70 mas resolution bispectrum speckle interferometry image of IRC+10216 in the H–band. North is up and east to the left. **b**: Same as (a) as a contour image with denotations (A to G) of compact structures. Contours are at every 0.2 mag down to 5.0 mag relative to the peak. **c**: J–band speckle reconstruction of IRC+10216 with 149 mas resolution.

of the inner nebula (Fig. 1). Denoting the brightest four components with A to D in order of decreasing brightness in the 1996 image, the separation of the two brightest components A and B increased from 191 mas in 1995 to 265 mas in 1998, i.e. by $\sim$ 35%, corresponding to a relative velocity of 23 mas/yr or 14 km/s within the plane of the sky at d = 130 pc. Within these 3 years the rather faint components C and D became brighter whereas component B faded. Note that the observations cover more than one pulsational cycle. Accordingly the apparent relative motions are not simply related to stellar variability. The general geometry of the nebula seems to be bipolar, most prominently present in the J-band image (Fig. 2) implying an asymmetric mass-loss.

The structures and changes in the inner nebula lead to the conclusion that the central star is close to or at the position of component B. Correspondingly, the initially brightest component A is the southern lobe of a bipolar structure. Then, the star is strongly but not totally obscured at H and K. Consistently, component B is very red in the $H - K$ color while A and the northern J-band components are relatively blue. This is supported by 1.1 μm archival HST polarimetry data (April 1997; see Osterbart et al. 2000). The polarization pattern is predominantly centro-symmetric with its center close to component B, whereas it shows strong polarization in the northern arms and a still significant polarization at component A. Finally, detailed two-dimensional radiative transfer calculations (Men'shchikov et al. 2000; this volume) show that the observed intensity ratio of A to B as well as the components' shapes clearly require the star to be at B. The inner nebula and the apparent motions seem to be rather symmetric around this position and the observed changes suggest an enhanced mass loss since 1997. A strongly variable mass loss operating on timescales of several stellar pulsation periods has, in fact, been predicted by dust-formation models of long-period carbon stars (Winters et al. 1995; this volume).

IRC+10216 is without doubt in a very advanced stage of its AGB evolution. The observed bipolarity of its dust shell even indicates that it has possibly entered the phase of transformation into a protoplanetary nebula.

3. H-DEFICIENT POST-AGB STARS

The most likely scenarios invoked to explain the H-deficient WR central stars are related to thermal pulses occuring either at the very end of the AGB evolution or during the post-AGB stage. During thermal pulses, recurrent instabilities of the He-burning shell, the luminosity of the He shell increases rapidly for a short time of 100 yr to $10^6 \ldots 10^8\,L_\odot$. The huge amount of energy produced forces the development of a pulse-driven convection zone which mixes products of He burning, i.e. carbon and oxygen, into the intershell region. Because the hydrogen shell is pushed concomitantly into cooler domains, hydrogen burning ceases temporarily allowing the envelope convection to proceed downwards after the pulse, to penetrate those intershell regions formerly enriched with carbon (and oxygen), and to mix this material to the surface (3rd dredge up). After the pulse hydrogen burning re-ignites and provides again the main source of energy (see Blöcker 1999 for a review).

Herwig et al. (1997) showed that the consideration of diffusive overshoot (Freytag et al. 1996) in all convective boundaries leads to important changes in AGB models, i.e. to (i) efficient dredge-up and formation of low-mass carbon stars; (ii) formation of ^{13}C as a neutron source to drive the *s*-process in these stars; and (iii) considerably changed intershell abundances. The overshoot efficiency was calibrated by the observed width of the main sequence. The latter finding (iii) turned out to be a key ingredient for the modelling of WR central stars. Overshoot leads to an enlargement of the pulse-driven convection zone and to enhanced mixing of core material from deep layers below the He shell to the intershell zone resulting in intershell abundances (mass fractions) of (He,C,O)=(40,40,16) instead of (70,25,2) as in non-overshoot sequences. These modified intershell abundances are already close to the observed surface abundances of Wolf-Rayet central stars. Finally, in contrast to standard evolutionary calculations, overshoot models do show dredge up for very low envelope masses and efficient dredge up was found even during the post-AGB stage (Blöcker 2000, Herwig 2000) leading to the mixture of the intershell abundances to the surface and to the dilution of hydrogen. Three thermal pulse scenarios for Wolf-Rayet central stars can now be distinguished:

AGB Final Thermal Pulse (AFTP), occurring immediately before the star moves off the AGB. In this case the envelope mass is already very small ($\sim 10^{-2} M_\odot$). During dredge-up a substantial fraction of the intershell region is mixed with the tiny envelope leading to the dilution of hydrogen and enrichment with carbon and oxygen. The resulting surface abundances depend on the actual envelope mass at which the AFTP occurs. For instance, Herwig (2000) found for $M_{env} = 4 \cdot 10^{-3} M_\odot$ (H,He,C,O)=(17,33,32,15) after an AFTP. The AFTP leads to a relatively high hydrogen abundance ($\gtrsim$ 15%) and predicts small kinematical ages for the PNe of WR central stars which emerge here directly from the AGB.

Late Thermal Pulse (LTP), occurring when the model evolves with roughly constant luminosity from the AGB towards the white dwarf domain. This kind of thermal pulse is similar to the one experienced by AGB stars but the envelope mass is even smaller than for the AFTP ($\sim 10^{-4} M_\odot$). Fig. 3 shows the evolutionary track of a $0.625\,M_\odot$ LTP model with diffusive overshoot (Blöcker 2000). After the flash the intershell abundances amounted to (He,C,O)=(45,40,13) and the model evolves towards the AGB domain. At minimum effective temperature ($\approx$ 6700 K) dredge up sets in and continues until the star has reheated to $\approx$ 12000 K. Hydrogen is diluted to 3% and the final surface abundances of He, C and O are close to those of the intershell region, viz. (45,38,12). Extension of convective regions and abundances are illustrated in Fig. 4 as function of age and effective temperature, resp. The kinematical age of the PN amount to a few thousand years .

Very Late Thermal Pulse (VLTP), occurring when the model is already on the white dwarf cooling track, i.e. after the cessation of H burning. Then, the pulse-driven convection zone can reach and penetrate the H-rich envelope and protons are ingested into the hot, carbon-rich intershell region raising a H flash (Iben & McDonald 1995). The energy released by this flash leads to a splitting of the convection into an upper zone powered by H burning and a lower one powered by He burning. The upper convection zone is, however, short-lived because the available hydrogen in the envelope is quickly consumed. Finally, the star becomes hydrogen-free and exposes its intershell abundances at the surface. Herwig et al. (1999) found for a $0.604\,M_\odot$ overshoot model surface abundances of (He,C,O)=(38,36,17). The kinematical age of the PN is relatively high since the star has first to fade along the cooling branch down to a few $100\,L_\odot$ before the flash. For $0.6\,M_\odot$ one obtains typically $t \gtrsim 20000$ yr.

All scenarios lead to hydrogen-deficient post-AGB stars with carbon and oxygen abundances as observed for Wolf-Rayet central stars. The variety of observations requires most likely all of these scenarios. Many objects have only very low hydrogen abundances, if any, favoring the LTP and VLTP. On the other hand, several Wolf-Rayet central stars are surrounded by young planetary nebulae (Tylenda 1996) and circumstellar shells (Waters et al. 1998) strengthening the AFTP and LTP. Within the current models roughly 20 to 25% of stars moving off the AGB can be expected to become hydrogen-deficient.

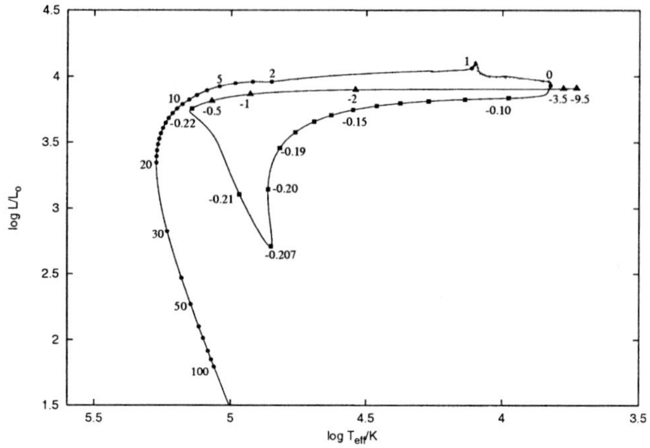

Figure 3 Evolutionary track for a 0.625 $M_\odot$ overshoot model suffering from an LTP (Blöcker 2000). Symbols refer to time marks given in units of 10^3 yr. Time is set to zero at minimum effective temperature after the flash. The age -3500 yr marks the begin of the central-star evolution and refers to a pulsational period of 50 d. The preceding AGB evolution (Blöcker 1995) did not include overshoot. Since the intershell abundances of carbon and oxygen in overshoot AGB models first increase and then level out after a series of thermal pulses, they can be accounted for by calculating the pulse-driven convection zone with an enhanced overshoot efficiency (f=0.064) leading to appropriate intershell abundances of (He,C,O)=(45,40,13). For the remaining evolution the standard efficiency (f=0.016; Herwig et al. 1997) was used.

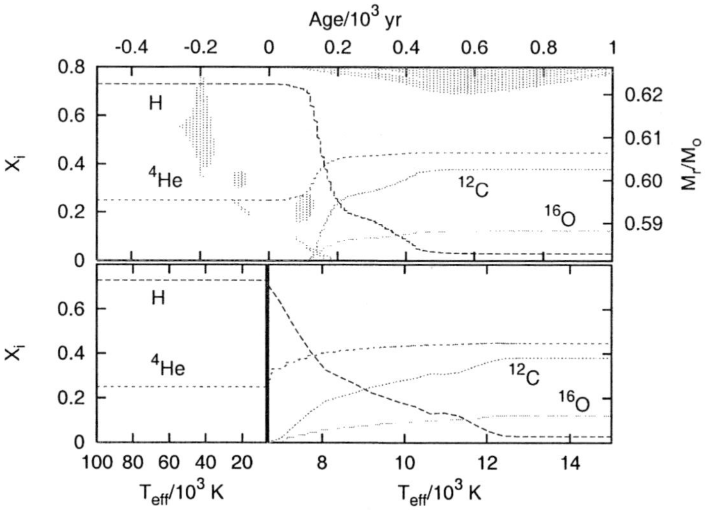

Figure 4 **Top**: Evolution of the surface abundances of H, He, C, and O (left scale) and extension of convective regions (right scale) for the 0.625 $M_\odot$ overshoot model in Fig. 3 during flash and dredge up. Shaded regions denote convective regions. Time is set to zero at minimum effective temperature after the flash. **Bottom**: Surface abundances of H, He, C, and O as function of the effective temperature. The line indicates the turn-around point at minimum effective temperature after the flash corresponding to age zero in the above panel.

References

Blöcker T., 1995, A&A 299, 755.

Blöcker T., 1999, AGB stars, IAU Symp. 191, T. Le Bertre, A. Lebre, C. Waelkens (eds.), p. 21

Blöcker T., 2000, ApSS, in press

Dreizler S., Heber U., 1998, A&A 334, 618.

Dyck H.M., Benson J.A., Howell R.R., Joyce R., Leinert C., 1991, AJ 102, 200

Freytag B., Ludwig H.-G., Steffen M. (1996), A&A 313, 497

Groenewegen M.A.T., 1997, A&A 317, 503

Haniff C.A., Buscher D.F., 1998, A&A 334, L5

Herwig F., 2000, ApSS, in press

Herwig F., Blöcker T., Schönberner D., El Eid, M., 1997, A&A 324, L81

Herwig F., Blöcker T., Langer N., Driebe T., 1999, A&A 349, L5

Iben I. Jr., MacDonald J., 1995, White dwarfs, Lecture Notes in Physics 443, D. Koester, K. Werner (eds.), Springer, p. 48.

Kastner J.H., Weintraub D.A., 1994, ApJ 434, 719

Koesterke L., Hamann W.R., 1997, A&A 320, 91.

Kwok S., 2000, Asymmetrical Planetary Nebulae II, Kastner J.H., Soker N., Rappaport S.A. (eds.), ASP Conf. Ser. 199, p. 9

Le Bertre T., 1992, A&AS 94, 377

Lohmann A.W., Weigelt G., Wirnitzer B., 1983, Appl. Opt. 22, 4028

Loup C., Forveille T., Omont A., Paul J.F., 1993, A&AS 99, 291

Mendez R.H, 1991, Evolution of Stars: The Photospheric Abundance Connection, IAU Symp. 145, G. Michaud, A. Tutokov (eds.), Kluwer, p. 375

Men'shchikov A.B., Balega, Y., Blöcker T., Osterbart R., Weigelt G., 2000, A&A, submitted

Olofsson H., 1996, ApSS 245, 169

Osterbart R., Balega, Y., Blöcker T., Men'shchikov A.B., Weigelt G., 2000, A&A 357, 169

Tuthill P.G., Monnier J.D., Danchi W.C., Lopez B., 2000, ApJ, in press

Tylenda R, 1996, Hydrogen-deficient stars, C.S. Jefferey, U. Heber (eds.), ASP Conf. Ser. 96, 267.

Waters L.B.F.M, Beintema D.A., Zijlstra A.A., de Koter A., Molster F.J., Bouwman J., de Jong T., Pottasch S.R., de Graauw T., 1998, A&A 331, L61.

Weigelt G., 1977, Optics Commun. 21, 55

Weigelt G., 1991, Progress in Optics 29, E. Wolf (ed.), North Holland, p. 293

Weigelt G., Balega Y., Blöcker T., Fleischer A.J., Osterbart R., Winters J.M., 1998, A&A 333, L51

Winters J.M., Fleischer A.J., Gauger A., Sedlmayr E., 1995, A&A 302, 483

STELLAR EVOLUTION AND NUCLEOSYNTHESIS OF POST-AGB STARS

Falk Herwig

Institut für Physik, Abteilung Astrophysik, Universität Potsdam, Potsdam, Germany

Institute for Physics and Astronomy, University of Victoria, Victoria BC, Canada

fherwig@uvastro.phys.uvic.ca

Abstract I discuss recent new models of post-Asymptotic Giant Branch stellar evolution. These models aim to clarify the evolutionary origin and status of a variety of hydrogen-deficient post-AGB stars such as central stars of planetary nebulae of Wolf-Rayet spectral type, PG 1159 stars or Sakurai's object. Starting with AGB models with overshoot such stars can evolve through one of four distinct channels. Each of these channels has typical abundance patterns depending on the relative timing of the departure from the AGB and the occurrence of the last thermal pulse. I discuss the responsible mechanisms and observational counterparts.

1. INTRODUCTION

Nuclear burning and mixing processes in post-Asymptotic Giant Branch (pAGB) stars are closely related to the final thermal pulse (Iben et al.1983; Iben1995). The pAGB stage will proceed undisturbed if the final thermal pulse (TP) occurs well before the departure from the AGB. However, if the final TP occurs immediately before the departure from the AGB (AFTP case) or during the early (LTP case) or the late post-AGB phase (VLTP case), then internal mixing and possibly as well nuclear burning processes will lead to distinctive alterations of the surface abundances (Herwig2000b; Blöcker2000). For example, the hydrogen-deficiency of central stars of planetary nebulae of Wolf-Rayet spectral type ([WC]-CSPN) is the result of such an abundance alteration as concluded from their large helium, carbon and oxygen abundance and the lack of hydrogen in their spectra (e.g. Koesterke and Hamann, 1997).

The final TP affects other elements as well. This is supported by the weird abundance pattern of Sakurai's object which apparently is a born-again AGB object (Asplund et al.1999). It shows a real-time evolution of lithium, hydrogen, s-process elements and others. Hybrid objects - with an intermediate hydrogen abundance between [WC]-CSPN and H-normal stars - are as well associated with a non-standard pAGB evolution (Napiwotzki et al.1991).

R. Szczerba and S.K. Górny (eds.), Post-AGB Objects as a Phase of Stellar Evolution, 249–256.

The TP is a thermonuclear instability of the He-shell and accordingly a TP can occur until He-burning ends at the latest at the pre-WD stage of pAGB evolution. On the AGB TPs are recurring mainly independently of mass loss on a timescale of a 10^4 to 10^5 yr. The AGB evolution is ended randomly by mass loss which strips the envelope off the core. If the envelope mass decreases below a critical value of about 0.001 $M_\odot$ contraction sets in and the star will begin its evolution along the horizontal part of pAGB evolution in the HRD. The last TP which occurs in the lifetime of a star may be past a few thousand years when the AGB star starts contraction to evolve into a central star. In that case this last TP does not interfere with an undisturbed and hydrogen-rich pAGB evolution finally leading to a white dwarf of spectral type DA.

However, if by coincidence the last TP occurs either immediately before or within a certain time interval after the transition towards the central star phase then a severely different evolution must be expected. In the following sections I will discuss in turn the three channels of evolution in which the final TP interferes with an undisturbed pAGB evolution.

2. THE AGB FINAL THERMAL PULSE

We found in our calculations of AGB models with overshoot that dredge-up is much more efficient if overshoot is not only applied to the bottom of the envelope convection zone but as well to the He-flash convection zone in the intershell during the TP (Herwig et al.1997; Herwig2000a). While the third dredge-up is usually difficult to find in stellar models of low core mass and low envelope mass we did find dredge-up after the last TP still on the AGB at a surprisingly low envelope mass of $4 \cdot 10^{-3}\,M_\odot$. This dredge-up was found under the assumption of an increased efficiency of convective mixing ($\alpha_{\rm MLT} = 3.0$) and overshoot with an efficiency parameter of $f = 0.016$. The amount of dredged-up material was of the order of $3 \cdot 10^{-3}\,M_\odot$.

In the AFTP case the final TP coincidently occurs when the envelope mass is already very close to the critical envelope mass which marks the departure of the star from the AGB regime. The remaining envelope mass and the amount of dredged-up mass are of comparable quantity and therefore the surface abundance change due to dredge-up is considerable. If the star is O-rich (C/O<1) before the final TP it will turn into a carbon star during the following final dredge-up episode, just at the onset of the proto-planetary transition phase. This carbon star is different from the C-stars known on the AGB in that its C abundance is much larger. In fact, after an AFTP dredge-up episode as described above the emerging central star is hydrogen-deficient. The extent of H-deficiency depends on the amount of envelope mass at the final TP and the efficiency of following dredge-up. As an example, the two AFTP calculations by Herwig (Herwig 2000b) produced central stars with mass fractions of (H/He/C/O) of (0.55,0.31,0.07,0.04) and (0.17,0.33,0.32,0.15).

Although the current models of the AFTP have systematically larger H-abundances than found in the majority of [WC] CSPN or PG 1159 stars this evolutionary scenario has otherwise observational support. Planetary nebulae (PNe) of [WC]-type central stars are on average not older than PNe surrounding chemically normal central stars (Gorny and Stasinska1995). This indicates that these objects are on their first evolutionary departure from the AGB. The alternative evolutionary scenario described below, in particular the very late TP, would rather require H-deficient central stars to have old PNe since they are supposed to be on their second departure from the AGB following a born-again evolution which might last more than 10^4 yr. Thus, the AFTP scenario naturally resolves the time scale problem of PNe around H-deficient CSPN.

Recent ISO observations support the AFTP scenario as well (Waters et al.1998; Cohen et al.1999). Surprisingly, both oxygen and carbon rich dust has been found around [WC]-late ([WC-L]) CSPN (called here the *C+O dust feature*). C-rich dust (PAH = Polycyclic Aromatic Hydrocarbon) and O-rich dust (crystalline silicates) usually exclude each other because of the initial formation of CO molecules. Consequently the simultaneous presence of O- and C-rich dust must be the result of a recent change of surface abundance pattern in the mass losing star.

The AFTP scenario does provide such a sudden abundance change immediately before the onset of the pAGB phase. In this picture the O-rich AGB star evolves towards the final TP with presumably high mass loss and experiences efficient dredge-up of intershell material which has a C/O ratio of about 3 . . . 4. Instantly the star will become C-rich. Possibly the now C-rich surface composition will once more enhance mass loss and accelerate the evolution towards the pAGB phase. The timescale of the abundance change and the following transition evolution off the AGB are around 1000 yr, in agreement with the timescale inferred for the abundance change from the ISO observations of the circumstellar dustshells.

If *all* [WC-L] CSPN show the signature of simultaneous presence of O- and C-rich dust (see the contribution by Hony et al., these proceedings) we may assume that *all* [WC-L] CSPN are originating from an AFTP because this is currently the only scenario we know for the C+O dust phenomenon. This raises the question why there are no [WC-L] type descendants of the LTP or VLTP evolution (see below) in their second departure from the AGB. This would imply that a star like Sakurai's object does not show up as a [WC-L] type CSPN, possibly because these objects camouflage themself by ejecting puffs of dust as suggested by another born-again candidate V605 Aql (Clayton and de Marco1997). However, there is at least one [WC-L] star, V348 Sgr, which is likely to be a born-again object because of its old PNe (Pollacco et al.1990) and thus, no O-rich dust signatures should be detected for this object.

There might be as well another clue for the progenitors of [WC-L] CSPN. If all or at least the overwhelming majority of [WC-L] show the C+O dust feature then an AFTP predominantly occurs in O-rich AGB stars (S- and M-giants). An AFTP event occurring in a C-giant could not host crystalline silicates in its dustshell. Possibly, the mass loss of S- and M-giants depends more strongly than that of C-giants on the stellar parameter variation during a TP and thereby correlates the occurence of the final TP with the immediately following departure from the AGB. Maybe the C-giants have a low fraction of immediate departures from the AGB after a final TP because their mass loss is more evenly distributed over the TP.

3. THE LATE THERMAL PULSE

While in the AFTP case the final TP on the AGB is immediately followed by the transition of the star towards the pAGB phase the order of events is reversed in the LTP case. The departure from the AGB begins up to about 5000 yr before the last TP. Accordingly the star will be disturbed by a TP during the pAGB evolution which leads to a born-again evolution back to the AGB. This evolutionary scenario is closely related to the VLTP described below with one main distinction. In the LTP case hydrogen burning has not yet ceased at the time of the TP. As a consequence the star will not experience the violent nuclear processes induced by the ingestion of the H-rich envelope material into the He-flash convection zone as found for VLTP models (see below). Instead, Blöcker (Blöcker 2000) and Herwig (Herwig 2000b) found that dredge-up during the born-again AGB stage after the LTP can lead to abundance alterations which are in accordance with abundances of [WC]-CSPN as well. If the time interval between the departure from the AGB and the TP is small this scenario could even account for the normal ages of PNe around [WC]-CSPN. Moreover this evolutionary channel more naturally leads to the very low H-abundances actually observed in the [WC]-CSPN. The envelope mass is by definition of the order of $10^{-4}\,M_\odot$ when the star has entered the CSPN stage and the TP occurs. In the LTP calculation by Herwig (Herwig 2000b) the amount of dredged-up material is of the order of a few $10^{-3}\,M_\odot$ and the resulting surface mass fractions are (H/He/C/O)=(0.02,0.37,0.40,0.18).

Taking into account the comparatively large mass loss of [WC]-CSPN of $\dot{M} \sim 10^{-6}\,M_\odot$ (Koesterke and Hamann1997), it is conceivable that a pAGB star after a LTP born-again event loses its entire top layer during the second pAGB evolution. This top layer represents the region well mixed during the dredge-up after the LTP and consists of the just mentioned amount of a few $10^{-3}\,M_\odot$. If this stripping of the outer dredge-up layer is possible for a LTP descendant star then the LTP evolution could even yield non-DA WDs which do not contain any hydrogen at all.

One of the unsolved problems with this scenario is the description of mass loss during the second, born-again AGB phase when the abundance suddenly and dramatically changes due to dredge-up. The large amounts of C and O brought to the surface could initiate a rapid increase in dust formation and thereby mass loss.

4. THE VERY LATE THERMAL PULSE

As in the LTP case the final TP occurs after the departure from the AGB. However, here the time interval between the AGB departure and the TP is longer, typically exceeding $\sim$ 5000 yr. At this time hydrogen burning has already stopped and the entire envelope material will be included in the convective He-burning region (see Herwig et al. (herwig et al.1999) and Herwig (Herwig 2000b) for details and references). The star takes a deep loop through the HRD in order to return to the AGB again, similar to the LTP case. While the VLTP model abundances agree with the observations of [WC]-CSPN, the PNe timescale problem is most severe for this case. Before the potential of the LTP and the AFTP to provide scenarios for the origin of the H-deficient [WC]-CSPN and PG 1159 was realized the VLTP was thought to be the only possible evolutionary origin of these objects (Iben and MacDonald1995).

The VLTP provides unique conditions of convective nucleosynthesis originating from the ingestion of protons from the unprocessed envelope into the He-flash convection zone. The nuclear burning and mixing occur both on the time scale of about one hour once the protons have reached sufficient temperature while being transported inward by convection. Temporarily and locally the luminosity of hydrogen burning exceeds the peak-flash He-burning luminosity and the He-flash convection zone is fragmented. (Asplund et al.1999).

The nucleosynthesis during the VLTP is highly dependent on the details of convective mixing because here both processes are closely related. While the protons from the envelope enter a hotter environment on their way down into the He-flash convection zone their nuclear burning time scale diminishes. At the position where it matches the convective turnover time scale the greatest part of the hydrogen-burning luminosity will be created. Therefore the position in the star where the peak H-burning luminosity can be found will depend both on the assumptions of convective mixing efficiency and the numerical treatment of the simultaneous burning and mixing.

For the calculation by Herwig et al. (Herwig et al.1999) we developed a fully coupled numerical scheme for convective nucleosynthesis. The equations of

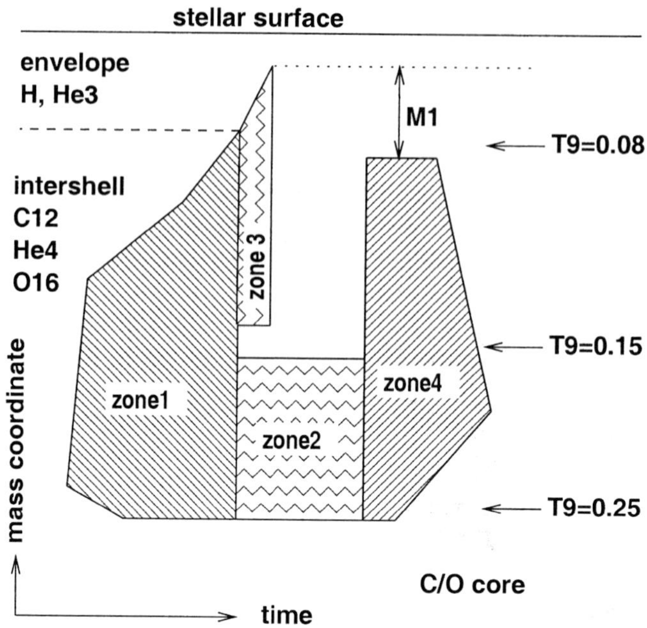

Figure 1 Schematic (not to scale) of the time-evolution of convection zones in the top 0.01 $M_\odot$ of a post-AGB star of 0.6 $M_\odot$ during a VLTP. All shaded areas indicate convectively unstable zones. The two solid horizontal lines in the upper part of the diagram indicate the stellar surface and the mass coordinate of the envelope-intershell (=core) transition.

material transport (one diffusion-like equation for each isotope) and the nuclear network equations at each depth mass grid are solved altogether fully implicit in one scheme (Herwig2000b). This treatment returns consistent abundance profiles within the convective region. Thus, the energy generation rates calculated as a function of position in the stellar burning region reflect the rapid consumption and simultaneous convective mixing of protons.

The evolution of the pulse-driven convection zone during a VLTP model sequence is sketched in Fig. 1. It shows the sequence of events found in the models of Herwig et al. (Herwig et al.1999). The convective instability can be divided in four different regions. Zone 1 represents the onset of the He-flash. During this first phase the He-burning luminosity increases by several orders of magnitude. It is essentially the same evolution as during the onset of any AGB TP or the LTP. The difference occurs when the upper boundary of the He-flash convection zone approaches the H-rich envelope (second horizontal line from top in Fig. 1). In the VLTP case the convective instability reaches out of the intershell and spreads into the H-rich envelope as represented by zone 3 in Fig. 1. This is not possible in the LTP case or during typical AGB TP because there the H-burning shell is still active at the time of the TP (Iben1976). The fresh nuclear fuel which is ingested into the deeper and hotter region burns on ever shorter timescales while at the bottom of zone 2 the unstable He-burning continues mainly undisturbed.

Due to the locally large energy generation of up to $10^8\,L_\odot$ by H-burning a small radiative layer between zone 2 and 3 establishes. This radiative layer affects the permeability for particle transport from zone 2 to 3. If overshoot is very efficient a considerable fraction of freshly produced particles, e.g. ^{13}C will be burned at the bottom of zone 3 instead of the bottom of zone 2. Thus, the assumed overshoot efficiency has an important influence on the model predictions. Moreover the position of the split between zone 2 and 3 depends on the assumption of convective efficiency. For a larger convective efficiency the mass coordinate of peak H-burning luminosity and thereby of the split shifts inwards and accordingly the maximum temperature ruling the nucleosynthesis in zone 2 increases.

The additional energy supply by proton captures causes zone 2 to stretch out and engulf the entire envelope. The moment of abundance change at the surface due to the mixing of zone 2 depends on the initial conditions of the VLTP. On one extreme is the immediate abundance change after the VLTP while the star still resides in the blue and faint part of the born-again loop in the HRD. This will only occur for the most vigorous cases of proton burning during the VLTP. In other cases a tiny layer of original envelope material of less than $10^{-6}\,M_\odot$ may cover the changed abundances close to the surface until during the evolution back to the AGB either mass loss or the onset of convection will reveal the H-poor abundance.

In any case, zone 3 is only short-lived due to the restricted amount of hydrogen available in the envelope. If that amount is consumed the related convective region will die away. After some cooling of that layer the original He-flash convection zone resumes control (zone 4). Note, that the most outward reach of zone 4 is smaller than that of zone 3. In the layer M1 swept over by zone 3 but not by zone 4 an element distribution formed only in zone 3 can be preserved close enough to the stellar surface in order to show up later on. It is here where we suspect lithium to form according to the mechanism of *hot hydrogen-deficient* ^{3}He *burning* (Herwig and Langer2000). This lithium could explain the observed abundance in Sakurai's object (Asplund et al.1999) which is believed to be a VLTP star caught in its born-again evolution.

The VLTP does also provide a rich neutron capture nucleosynthesis initiated by the production of ^{13}C in zone 3 (Malaney1986). The neutrons are partly released immediately by the well known $^{13}C(\alpha, n)^{16}O$ reaction. Alternatively ^{13}C is stored in the region covered by zone 3 until the He-flash convection zone recovers from the disturbation of H-burning. Then ^{13}C burns at higher temperatures in zone 4. Future models of the s-process during the VLTP should predict a distinctive heavy-element signature to be compared with that of Sakurai's object.

Acknowledgments

This work has been supported by the *Deutsche Forschungsgemeinschaft, DFG* (La 587/16). I would like to thank T. Blöcker, W.-R. Hamann, L. Koesterke, N. Langer, D. Schönberner and K. Werner for very useful discussions.

References

Asplund, M., Lambert, D. L., Kipper, T., Pollacco, D., and Shetrone, M. D., 1999, A&A 343, 507

Blöcker, T., 2000, in T. Blöcker, R. Waters, and B. Zijlstra (eds.), Low mass Wolf-Rayet Stars: origin and evolution, Ap&SS, in press

Clayton, G. C. and de Marco, O., 1997, AJ 114, 2679

Cohen, M., Barlow, M. J., Sylvester, R. J., Liu, X. ., Cox, P., Lim, T., Schmitt, B., and Speck, A. K., 1999, ApJ Lett. 513, L135

Gorny, S. K. and Stasinska, G., 1995, A&A 303, 893

Herwig, F., 2000a, A&A 360, 952

Herwig, F., 2000b, in T. Blöcker, R. Waters, and B. Zijlstra (eds.), Low mass Wolf-Rayet Stars: origin and evolution, Ap&SS, in press, astro-ph/9912353

Herwig, F., Blöcker, T., Langer, N., and Driebe, T., 1999, A&A 349, L5

Herwig, F., Blöcker, T., Schönberner, D., and El Eid, M. F., 1997, A&A 324, L81

Herwig, F. and Langer, N., 2000, Nucl. Phys. A, in press, astro-ph/0010120

Iben, Jr., I., 1976, ApJ 208, 165

Iben, Jr., I., 1995, Physics Reports 250, 1

Iben, Jr., I., Kaler, J. B., Truran, J. W., and Renzini, A., 1983, ApJ 264, 605

Iben, Jr., I. and MacDonald, J., 1995, in D. Koester and K. Werner (eds.), White Dwarfs, No. 443 in LNP, p. 48, Springer, Heidelberg

Koesterke, L. and Hamann, W. R., 1997, A&A 320, 91

Malaney, R. A., 1986, MNRAS 223, 683

Napiwotzki, R., Schönberner, D., and Weidemann, V., 1991, A&A 243, L5

Pollacco, D. L., Hill, P. W., and Tadhunter, C. N., 1990, MNRAS 245, 204

Waters, L. B. F. M., Beintema, D. A., Zijlstra, A. A., and et al., 1998, A&A 331, L61

WHAT CAN WE LEARN FROM POST-AGB CHEMICAL STUDIES ON THE AGB 3RD DREDGE-UP PHENOMENA?

Hans Van Winckel, Maarten Reyniers

Instituut voor Sterrenkunde, K.U.Leuven, Celestijnenlaan 200B, 3001 Heverlee, Belgium

Hans.VanWinckel@ster.kuleuven.ac.be

Abstract In this contribution we focus on the remarkably divers detailed chemical patterns observed in post-AGB stars and we address the question whether AGB (nucleosynthetic) evolutionary models can be further constrained by the results on post-AGB stars.

1. INTRODUCTION

There is general agreement that the stellar evolution on the AGB is accompanied by a *chemical* evolution in which mainly He-burning products and yields from the slow neutron capture process (s-process) are mixed to the photosphere. The neutrons needed for the s-process nucleosynthesis come from the $^{13}C(\alpha,n)^{16}O$ reaction which takes place in a radiative environment during the intershell period (Straniero et al., 1995). The primary ^{13}C pocket is probably built up after the mixing of protons into the ^{12}C-rich intershell during a thermal pulse. It is only when the remaining ^{14}N is brought to hotter layers in the next thermal pulse, the $^{22}Ne(\alpha,n)^{25}Mg$ will be activated. The proton profile engulfed in the intershell determines therefore the richness of the s-process nucleosynthesis.

The models on the photospherical chemical enrichment during the AGB evolution are very dependent on not well constrained or even free parameters like the details of the 3rd dredge-up extra mixing, the engulfed proton profile, the numerical treatment of convection and convective overshoot, possible rotationally induced mixing, etc. (see e.g. Blöcker and Herwig this volume, Busso et al. 1999 and references therein).

The observational data guiding the theoretical models come from several types of objects. First of all, of course, the intrinsic AGB stars of the M-MS-S-SC-C stellar type sequence. This sequence is thought to be mainly defined by an

R. Szczerba and S.K. Górny (eds.), Post-AGB Objects as a Phase of Stellar Evolution, 257–264.

increasing C/O ratio. The enrichment in ^{12}C seems to correlate well with an increase in s-process enrichment along this sequence (e.g. Smith & Lambert 1990 and references therein). The strong molecular opacity and sometimes unstable photospheres make a quantitative analysis, however, difficult and some species even untraceable. Especially the spectra of Carbon stars are very difficult to analyse quantitatively.

Another source of data is the chemical information obtained by spectra of objects of the Ba-star group (e.g. Jorissen 1999 and references therein). This is the generic term for all the stars for which the excess is acquired by mass-transfer from the late AGB companion, which is now evolved into a white dwarf. The interpretation on these data in the framework of theoretical AGB chemical evolutionary models is not straightforward since the abundances may have been diluted by mixing of the unprocessed material during the evolution of the gainer. Certainly for the binaries with shorter periods, the evolution (especially the mass-loss) of the system might have been influenced by binary interaction.

A very important source of data concerning different nucleosynthetic processes and possible yields of a completed stellar life-cycle is also coming from the precise laboratory measurements of the isotopic ratios in stellar grains from meteorites (Busso et al. 1999 and references therein).

In this contribution we focus on the diverse chemical patterns observed in post-AGB stars. With their F-G spectral type and high luminosity, post-AGB stars are in principle ideal to study the chemical evolution of stars on the AGB since the atmospheric parameters allow the use of standard model atmospheres; they show a large spread in metallicity; the molecular veiling is often negligible; and a wide range of elements (including many s-process elements) can be studied by the analysis of atomic lines. The evolutionary phase is, however, fast so the significance of results on individual objects is difficult to extrapolate to the whole AGB stellar evolution theory.

2. CHEMICAL PATTERNS OBSERVED IN POST-AGB STARS

In recent years many studies of the detailed chemical composition of post-AGB stars appeared in the literature (for a review see Van Winckel 1999 and references therein) and these showed that the chemical patterns are remarkably divers with only a fraction of the objects displaying evidence for an efficient enrichment during the 3rd dredge-up.

We will give a small overview of the variety of patterns and focus first on the results on stars that indeed do show a large overabundance of s-process elements accompanied by a large C/O photospheric ratio. Next we discuss some examples of very similar stars where no chemical signatures of dredge-up

Figure 1 H_γ spectra of two post-AGB stars of very similar temperature, gravity and metallicity. HD 187885 clearly shows the presence of lines of the s-process elements Zr and La while in SAO 239853 no trace of these lines is observed.

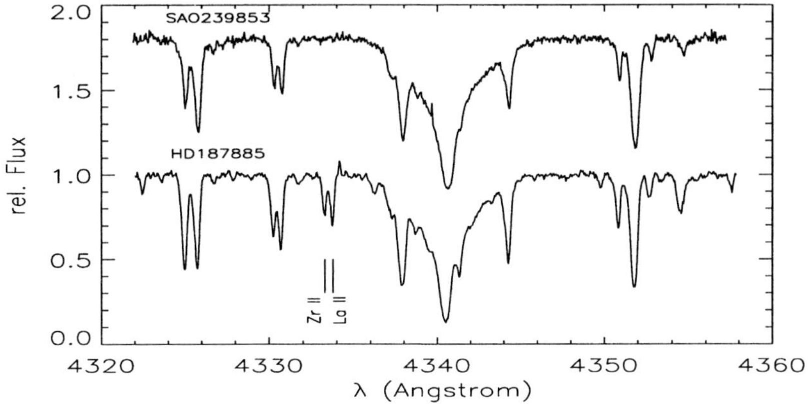

enrichment were found at all, giving useful information on the occurrence of the 3rd dredge-up.

Since it became clear that some very peculiar chemical patters are observed in objects that turned out to be binaries with orbital periods between 100 and 2500 days, we will also discuss the chemical influence of this binarity.

3. THE HAVES : 21 MICRON STARS

An important separate chemical class seems to be stars that display in their IR-spectra a feature at 21μm (Kwok et al. 1989 ; Kwok these proceedings), the carrier of which was recently suggested to be TiC (Von Helden et al. 2000; Hony et al. these proceedings).

Several studies of the photospherical chemical composition of individual sources appeared in the literature (Klochkova 1995; Začs et al. 1995; Van Winckel et al. 1996; Reddy et al. 1997; Decin et al. 1998; Klochkova et al. 1999; Reddy et al. 1999; Van Winckel & Reyniers 2000) and indicate that they show clear signatures of very strong enrichment: a C/O ratio larger than one and, most interestingly, they are among the most s-process enriched objects known till now (see Fig.1). The chemical composition together with their double-peaked energy distribution, galactic latitude, often low metallicity and high radial velocities and the carbon rich nature of the circumstellar dust around these objects are indeed the prototypical characteristics of what a post 3rd dredge-up object should look like.

Since different authors use different atomic data, spectral resolution, signal-to-noise, line-strength limits and method of analysis it is often very difficult to

Figure 2 Left: The observed strong correlation between the total enrichment of s-process elements ([s/Fe]) and the [hs/ls]-ratio which is a measure for the integrated neutron irradiation (Van Winckel & Reyniers 2000). Right: The [Zr/Fe] abundance ratio versus the metallicity for post-AGB stars showing a double peaked energy distribution. The enriched objects are all confirmed 21 μm stars (except one for which no ISO spectrum is available). The none enriched objects have very similar SED, metallicity, kinematics and global characteristics of the 21 μm stars but do not show any sign of an s-process enrichment. The non-enriched objects plotted here are SAO 239853, IRAS 18025−3906, HD 161796, HD 112374, HD 107369.

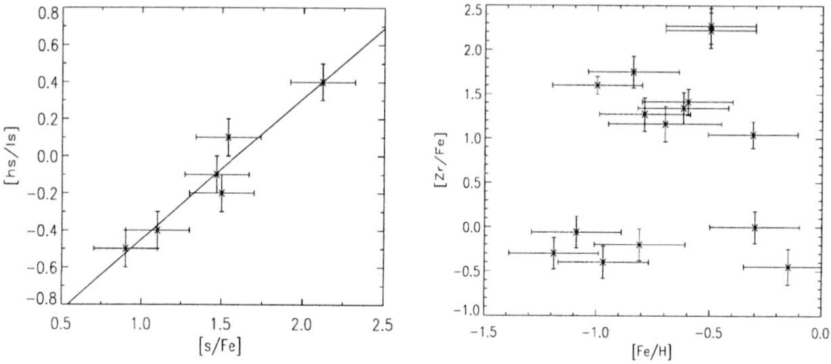

compare the results quantitatively. Van Winckel & Reyniers (2000) therefore initiated a strictly homogeneous study of the chemical composition of six 21 μm objects. The differential abundances should limit the possible systematic effects. The main results of this study are that the enrichment observed in the 21 μm stars is indeed very high, also when compared to other intrinsically enriched objects (see fig. 4 and 9 of Van Winckel & Reyniers 2000). Moreover, since the objects have a considerable spread in metallicity, the anti-correlation of the s-process nucleosynthesis on the metallicity could be studied. A trend is indeed observed but the spread is large and certainly intrinsic. This large spread is confirmed when also the extrinsically enriched objects are sampled (see Fig. 10 of Van Winckel & Reyniers 2000).

Another interesting and much stronger correlation was determined (Fig. 2) : the most enriched objects show the strongest integrated neutron irradiation. The latter is often parameterised by [hs/ls]-index which is the abundance ratio of the heavy s-process elements of the Ba peak to the light s-process elements of the Sr-peak. This probably means that the dredge-up efficiency is strongly linked with the neutron production in the intershell and thus the proton engulfment.

4. THE HAVE-NOTS

The 21 μm stars are till now, however, the only single post-AGB stars showing overabundances of s-process elements and the results on other objects with a double-peaked energy distribution are much less easy to interpret (see e.g. Luck et al. 1990; Van Winckel 1997). In none of these other objects studied so far s-process overabundances were found larger than the [s/Fe] $\sim$ 0.2 error level and often even underabundances were reported. The post-AGB stars, even with O-rich circumstellar dust, are expected to have already undergone several dredge-up episodes after each thermal pulse and we therefore expect a gradual increase in the s-process enrichment assuming the transition from O-rich to C-rich occurs over many thermal pulses. According to the still rather limited data available now, there seems to be no such gradual increase : or an object is strongly enriched in s-process elements or it is not. Moreover, the s-process enriched objects all belong to the 21 μm class of stars (except IRAS 07430+1115 for which no ISO spectrum is available). This is illustrated in Fig 2 where we plotted the Zr overabundance relative to the metallicity of post-AGB stars with a double-peaked energy distribution and which were analysed using the same method and atomic data. It is clear that the metallicity alone does not seem to be a key ingredient to determine whether an object will undergo an efficient 3rd dredge-up or not.

Our poor understanding of the 3rd dredge-up phenomena was recently emphasized by the results on the chemical composition of J-type carbon stars (Abia & Isern 2000). These authors obtained accurate chemical abundances of 12 J-type Galactic carbon stars and found surprising patterns which are not anticipated in the current theoretical predictions: these Carbon stars show very high Li abundances, high $^{13}C/^{12}C$ ratios but most surprisingly they do not show enhancement of any observed s-process element. The nature of the J-type carbon stars is still a matter of debate but CO measurements (Jura & Kahane 1999 and references therein) and ISO results (Yamamura et al. 2000) do indicate, at least for some objects, a binary nature. The best model (see Yamamura et al. 2000), however, suggests the stars to be in wide binaries for which direct interaction seems less plausible. The relation between the special photospherical chemical patterns and the eventual wide binary nature of these objects is certainly not clear at the moment.

5. DEPLETION

There is now general consensus that the extremely low iron abundances down to [Fe/H]=$-$4.8 observed in some post-AGB stars do not reflect the initial conditions but are acquired during stellar evolution by a depletion process in which gas, devoid of refractory elements, was separated from the circumstellar dust and was re-accreted on the photosphere (Venn & Lambert 1990; Bond 1991; Van Winckel et al. 1992). The extreme stars were found to be all binaries with a period around one to a few years (Van Winckel et al. 1995) and surrounded by a stable circumstellar dusty disc. This led Waters et al. (1992) to suggest that such a long-lived dusty disc is needed to guarantee a low density and long timescale for the gas-dust separation process to occur.

In recent years it became clear that the photospheric depletion pattern is surprisingly common : many RV Tauri stars turned out to show similar, albeit less extreme, patterns (Giridhar et al. 1994, 1998, 2000; Gonzalez & Wallerstein 1996; Gonzalez et al. 1997a, b; Van Winckel et al. 1998). The often large amplitude and semi-regular pulsations of RV Tau stars render direct detection of binary motion cumbersome although for regular pulsators like AC Her, a large monitoring programme of the radial velocity did prove the star to be a binary with a period of 1200 days (Van Winckel et al. 1998). If indeed binarity is a necessary condition for the depletion process to be efficient, the binary nature of all these depleted stars should be explored.

The presence of a stable dusty disc is for some of the extremely depleted objects well established (see Van Winckel 1999 and references therein). Alternative tracers to look for binaries should be explored and the special observational characteristics of dusty reservoirs given in the next paragraph, can be exploited to suspect binarity in new objects for which direct detection by radial velocity monitoring is difficult.

6. OBSERVATIONAL EVIDENCE FOR A STABLE DUSTY DISC

SED: For optically bright post-AGB objects there is a very good correlation between the SED characteristics and the binary nature : stars showing a broad IR excess revealing the presence of both hot and cool dust turned out to be binaries (Waters et al., 1997; Van Winckel 1999) while for objects with a double peaked energy distribution (like the 21 μm sources) where there is only cool dust, binarity is commonly not observed (see also Hrivnak these proceedings).

Photometric monitoring: Several objects show clear signs for variable circumstellar reddening during orbital motion (see Van Winckel 1999 and references therein). The photometric extremes occur at conjunction and the objects are faintest when reddest during inferior conjunction passage. This behaviour is best explained assuming a geometrically rather thick dust torus surrounds the binary system (Waelkens et al. 1991). For the famous Red Rectangle nebula, the disc is resolved at high angular resolution images (Roddier et al. 1995; Bond et al. 1997; Osterbart et al. 1997).

CO kinematics and depletion: The CO microwave lines in orbiting reservoirs around evolved stars are, on general, very narrow and weak (e.g. Jura & Kahane 1999). The best documented example is the Red Rectangle where the narrow microwave emission feature is 100 times weaker than anticipated from the 60μm flux indicating a very low gas to dust ratio. The low kinematics point to keplerian motion rather than gas expansion.

Crystallisation process: One of the spectacular results of the spectrographs on ISO was the discovery of a rich mineralogy in circumstellar disks, not only around young stellar objects but also around evolved objects (see e.g. Waters et al. 1999 and references therein). In the long-lived discs around evolved binaries, crystalline silicates and even possible planetisimals evolve, shedding new and unexpected light on the universality of the structure, formation and evolution of circumstellar disks (Waters et al. 1998; Molster et al. 1999; Jura & Turner 1998)

Large grains: Here again the Red Rectangle is the best documented object and the large continuum fluxes on mm and cm wavelengths reveal the presence of large grains (Jura et al., 1997). These large grains must have formed by coagulation during the long storage time in the disc and in the Red Rectangle even macro-structures might have been formed (Jura & Turner 1998).

The circumstellar dust discovered around RV Tauri stars is also peculiar and many objects share (some) of the above mentioned observational characteristics of a stable orbiting dusty disk. This led Van Winckel et al. (1999) to conclude that also among the RV Tau stars, binarity may be widespread.

7. CONCLUDING REMARKS

The photospherical chemical properties of 21μm stars show that they are the prototypical examples of post-3rd dredge-up stars and as such yield very important information on the nucleosynthesis and dredge-up phenomena associated with the late AGB evolution. The identification of the feature as TiC has, however, very important consequences on the mass-loss history of the systems (see Hony these proceedings and references therein) which are, till now, theoretically not well understood.

Also interesting objects are those which show *no* evidence for chemical enrichment although they seem to have very similar metallicity, kinematics and total IR excesses as the 21μm stars. At present there seems to be no gradual increase in s-process overabundances observed in post-AGB stars: or an object is strongly enriched or it is not.

There is growing observational evidence that binaries with a large range in periods are able to form a long-lived orbiting dusty disc during the heavy mass-loss episodes of their final evolutionary stage. The presence of such a disc has far reaching consequences on the evolution and the chemistry of the system. With their prolonged IR lifetime, any sample of post-AGB stars selected by their IR properties, will be biased towards binarity.

Acknowledgments

HVW is post-doctoral fellow of the Fund for Scientific Research, Flanders (Belgium). MR is researcher of the same Fund.

References

Abia C., Isern J., 2000, ApJ 536, 438

Bond, H.E. 1991, in Michaud G., Tutukov A.V. (eds.) IAU Symposium 145, *Evolution of stars: the photospheric abundance connection*, Kluwer, Dordrecht, p. 341

Bond H.E., Fullton L.K., Schaefer K. et al., 1997, in Habing H.J., Lamers H.J.G.L.M. (eds.) IAU Symp. 180, *Planetary Nebulae*, 211

Busso M., Gallino R., Wasserburg G.J., 1999, ARA&A 37, 239

Decin L., Van Winckel H., Waelkens C., Bakker E.J., 1998, A&A 332, 928

Giridhar S., Rao N.K., Lambert D.L., 1994, ApJ 437, 476

Giridhar S., Lambert D.L., Gonzalez G., 1998, ApJ 509, 366

Giridhar S., Lambert D.L., Gonzalez G., 2000, ApJ 531, 521

Gonzalez G., Wallerstein G., 1996, MNRAS 280, 515

Gonzalez G., Lambert D.L., Giridhar S., 1997a, ApJ 479, 427

Gonzalez G., Lambert D.L., Giridhar S., 1997b, ApJ 481, 481

Jorissen A., 1999, in : IAU Symp. 191 "AGB stars", Le Bertre T., Lèbre A., Waelkens C. (eds.), 437

Jura M., Turner J., Balm S.P., 1997, ApJ 474, 741

Jura M., Turner J., 1998, Nature 395, 144

Jura M., Kahane C., 1999, ApJ 521, 302

Klochkova V.G., 1995, MNRAS 272, 710

Klochkova V.G., Szczerba R., Panchuk V.E., Volk K., 1999, A&A 345 , 905

Kwok S., Volk, K.M., Hrivnak, B.J., 1989, ApJ 345, 51

Lloyd-Evans., 1999, in : IAU Symp. 191 "AGB stars", Le Bertre T., Lèbre A., Waelkens C. (eds.), 453

Luck R.E., Bond H.E., Lambert D.L., 1990, ApJ 357, 188

Molster F.J., Yamamura I., Waters L.B.F.M., et al., 1999, Nature 401, 563

Osterbart R., Langer N., Weigelt G., 1997, A&A 325, 609

Reddy B.E., Parthasarathy M., Gonzalez G., Bakker E.J., 1997, A&A 328, 331

Reddy B.E., Bakker E.J., Hrivnak B.J., 1999, ApJ 524, 831

Roddier F., Roddier C., Graves J.E., Northcott M.J., 1995, ApJ 443, 249

Smith V.V., Lambert D.L., 1990, ApJSS 72, 387

Straniero O.,Gallino R., Busso M., et al., 1995, ApJ 440, L85

Van Winckel H., 1997, A&A 319 , 561

Van Winckel H., 1999, in : IAU Symp. 191 "AGB stars", Le Bertre T., Lèbre A., Waelkens C. (eds.), 465

Van Winckel H., Reyniers M., 2000, A&A 354, 135

Van Winckel H., Mathis J.S., Waelkens C. 1992, Nature 356, 500

Van Winckel H., Waelkens C., Waters L.B.F.M., 1995, A&A 293, L25

Van Winckel H., Waelkens C., Waters L.B.F.M., 1996, A&A 306, L37

Van Winckel H., Waelkens C., Waters L.B.F.M., et al., 1998, A&A 336, L17

Van Winckel H., Waelkens C., Fernie J.D., Waters L.B.F.M., 1999, A&A 343, 202

Venn K.A., Lambert D.L. 1990, ApJ 363, 234

Von Helden G., Tielens A.G.G.M., Van Heijensbergen D., et al., 2000, Science 288, 313

Waelkens C., Lamers H.J.G.L.M., Waters L.B.F.M., et al., 1991, A&A 242, 433

Waters L.B.F.M., Trams N.R., Waelkens C., 1992, A&A 262, L37

Waters L.B.F.M., Waelkens C., Van Winckel H., 1997, in Habing H.J., Lamers H.J.G.L.M. (eds.), IAU Symp. 180,*Planetary Nebulae*, p. 313

Waters L.B.F.M., Cami J., de Jong T., et al., 1998, Nature 391, 868

Waters L.B.F.M., Beintema D.A., Cami J. et al., 1999, in Cox P. & Kessler M.F. *The universe as seen by ISO*, ESA-SP 427, 219

Yamamura I, Dominik C., de Jong T., Waters L.B.F.M., Molster F.J., 2000, A&A in press

Začs L., Klochkova V.G., Panchuk V.E., 1995, MNRAS 275, 764

OPTICAL SPECTROSCOPY OF PECULIAR SUPERGIANTS

V.G. Klochkova, V.E. Panchuk
Special Astrophysical Observatory, RAS, Nizhnij Arkhyz, Russia

R. Szczerba
Nicolaus Copernicus Astronomical Center, PAS, Toruń, Poland

Abstract Based on optical echelle spectra we have determined the main parameters, the atmospheric chemical abundances and the evolutionary status of a sample of peculiar supergiants associated with infrared sources. In the spectra of some post–AGB objects we have revealed strong **emission** C_2 Swan bands and absorption features at the positions of well–known DIBs. An analysis of radial–velocities and intensities led us to the conclusion that these spectral features originate predominantly in the expanding circumstellar envelopes.

1. INTRODUCTION

The surface abundance determination of optically bright stars is a powerful tool to study their evolutionary status, but also to investigate physical processes which lead to the chemical diversities. Among peculiar supergiants, the post–Asymptotic Giant Branch (post–AGB) objects are ideally suited as tracers of the chemical evolution of low–mass stars since their spectral types (K–A) make that a large number of elements (including s–process ones) can be studied by their atomic lines. We report in this paper the results of high resolution spectroscopic observations and detailed photospheric abundance analysis for a group of peculiar supergiants which includes mostly post–AGB candidates.

2. OBSERVATIONS AND ANALYSIS

The observations were performed in 1994–1999 with the 6 m BTA telescope, at the Nasmyth–2 focus using the echelle spectrometer LYNX (Panchuk et al. 1999) equipped with a CCD of 1040 × 1170 elements. LYNX provides spectral resolution of R = 25000 and a signal–to–noise ratio higher than 100 in the wavelength range 4700–7200 or 5200–8800 Å. Besides, the echelle spectrom-

R. Szczerba and S.K. Górny (eds.), Post-AGB Objects as a Phase of Stellar Evolution, 265–270.

eter PFES (Panchuk et al. 1998) placed at the BTA prime focus has been used to observe the faintest programme stars ($V > 12^m$). For the objects as bright as $V \leq 14.5^m$ the device ensures a $S/N \approx 100$ desired for our purpose with a spectral resolution $R \approx 15000$.

For the determination of the main parameters (effective temperature, $T_{\rm eff}$, and surface gravity, log g) and for the computation of the chemical composition the grids of Kurucz (1993) and Gustafsson et al. (1975) models have been used. The effective temperatures derived spectroscopically are in good agreement with the values obtained from the spectral energy distribution modeling.

3. MAIN RESULTS

Practically all post–AGB objects studied by us have reduced (relative to the Sun) metallicity (see $[Fe/H]_\odot$ in Table 1), which indicates that they belong rather to the intermediate and old population of the Galaxy. Our investigations (see Table 1 and references therein) revealed also significant excesses of carbon, oxygen and heavy s–process elements in the atmosphere of most supergiants.

The Classical post–AGB object IRAS 07134+1005. The first success of our programme was to show a large excess (about 2 dex relative to metallicity) of heavy metals synthesized in the processes of neutronization (Y, Zr, Ba, La, Nd) for the metal–poor ($[Fe/H]_\odot = -1.0$) post–AGB supergiant IRAS 07134+1005 (Klochkova 1995). Along with excess of heavy metals, very large (over an order of magnitude) excesses of CNO–group elements were found.

The Bipolar nebula AFGL 2688. For the first time we have determined the chemical composition of AFGL 2688 (Klochkova et al. 2000c). The iron abundance of AFGL 2688 ($[Fe/H]_\odot = -0.59$) indicates that this object belongs rather to the intermediate population of the Galaxy. High overabundances of carbon and nitrogen ($[C/Fe] = +0.73$, $[N/Fe] = +2.00$, $C/O > 1$) are found. The revealed overabundance of s–process elements (Y and Ba) relative to iron is not large: $[X/Fe] = +0.55$. The abundances of lanthanoides are even less increased: for La, Ce, Pr and Nd the average value is $[X/Fe] = +0.26$. Such an abundance pattern for heavy metals seems to be in agreement with a low intensity of the 21 μm band observed in AFGL 2688. The intensity of this emission band is higher in post–AGB objects with larger s–process elements excess.

The analysis of radial velocities of the spectral features formed both in the photosphere and in the circumstellar envelope of AFGL 2688 permitted us to reveal a high velocity (about 60 km s^{-1}) component of the stellar wind.

The Fast–evolving object IRC +10420 (IRAS 19244+1115). An analysis of the IRC +10420 spectra taken with BTA permits us to conclude that the effective temperature of this object has increased significantly over the last 20 years. The spectral class has changed from F8 (Humphreys et al. 1973) to A5 (Klochkova et

Table 1 Peculiar supergiants studied at the 6 m BTA telescope.

name	*sp type*	*V*	$T_{\rm eff}$ [K]	log g	$[{\rm Fe/H}]_\odot$	*IR-exc*	*remarks*
04296+3429	G0Ia	14.2	6300[a]	0.0	−0.86	Y	21 μm
04440+2605	G2Iae	9.2	5600[b]	1.0	+0.07	Y	puls
07134+1005	F5I	8.4	7000[c]	0.1	−1.00	Y	bin, 21 μm
07284−0940	F8Ibe	5.7-6.6	4950[b]	0.0	−0.69	Y	bin?, puls
07331+0021	G5Iab	8.0-9.3	4500[d]	0.0	−1.13	Y	puls
09276+4454	K7II	8.1	3400[e]	1.0	−0.34	Y	
UU Her	F5Ib	9.0	6000[f]	0.7	−1.32	N	bin? puls
17436+5003	F3Ib	7.0	7100[g]	0.5	−0.25	Y	
Sakurai's ob	F2Ia	11.4	7250[h]	1.0	−1.54		
18062+2410 [1]	B5I	11.4				Y	var
18095+2704	F3Ib	10.4	6700[c]	1.0	−0.78	Y	
18281+2149	F2Ipe	6.8-9.0	6100[b]	1.5	−0.82	Y	bin, puls
19102+0329	F8-G2	9.1	5700[i]	2.0	+0.17		puls
19114+0002	G5Ia	8.0	5000[g]	−0.75	−0.98	Y	bin?
			6800[k]	1.3	−0.10		
19244+1115	A5Ia	11.2	8500[l]	1.0	−0.12	Y	
20004+2955	G7Iab	8.9	5000[m]	1.0	−0.2	Y	
22223+4327 [1]	G0Ia	13.3				Y	21 μm
22272+5435	G5Ia	9.3	5600[n]	0.5	−0.49	Y	21 μm
23304+6147	G0Ia	13.5	5900[o]	0.0	−0.64	Y	21 μm
AFGL 2688	F5Iae	14.0	6500[p]	0.0	−0.59	Y	bipolar neb.

[1] – in preparation; bin := binary star; *IR–exc* := IR–excess; puls := pulsating star.
references: [a] Klochkova et al. (1998), (1999); [b] Klochkova & Panchuk (1998); [c] Klochkova (1995); [d] Klochkova & Panchuk (1996); [e] Klochkova & Mishenina (1998); [f] Klochkova et al. (1997a); [g] Klochkova (1998); [h] Kipper & Klochkova (1997); [i] Klochkova et al. (2000d); [k] Začs et al. (1996); [l] Klochkova et al. (1997b); [m] Klochkova et al. (2000b); [n] Začs et al. (1995); [o] Klochkova et al. (2000a); [p] Klochkova et al. (2000c).

al. 1997b). The abundance of the iron–group elements is solar while carbon has been found to be underabundant and nitrogen overabundant. Such an abundance pattern is typical for massive supergiants of the disk. Calculation of abundances for elements synthesized in the processes of slow neutronization has shown them to be normal (solar). These results combined with the high luminosity (close to the limit of Humphreys–Davidson) confirm the hypothesis that most likely IRC +10420 is a massive star (up to 40 $M_\odot$) at a short and therefore rarely observed evolutionary stage: transition from OH/IR star to a Luminous Blue Variable (LBV) or Wolf–Rayet (WR) star.

Molecular emission features. We have identified the emission bands of the Swan's system of C_2 molecule in the optical spectra of 4 sources (IRAS 22223+4327, IRAS 23304+6147, IRAS 04296+3429 and AFGL 2688 see Fig. 1) which belong to the small group of post–AGB objects with 21 μm feature in their IR–

spectra. Comparison between the Hale-Bopp comet spectrum and spectra of these four objects allowed us to propose that in both cases the same mechanism (resonance fluorescence – Zanstra 1928, Swings 1941) is responsible for the emission in the C_2 molecular bands.

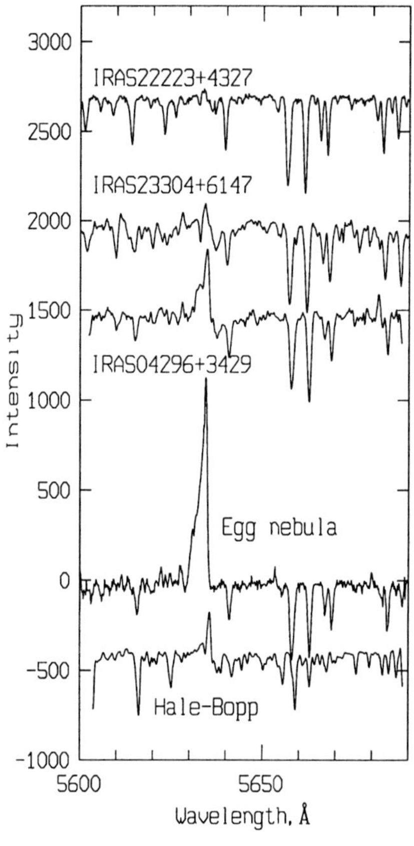

Figure 1 Emission in the Swan band λ 5635 Å of C_2 molecule in the spectra of some post–AGB objects.

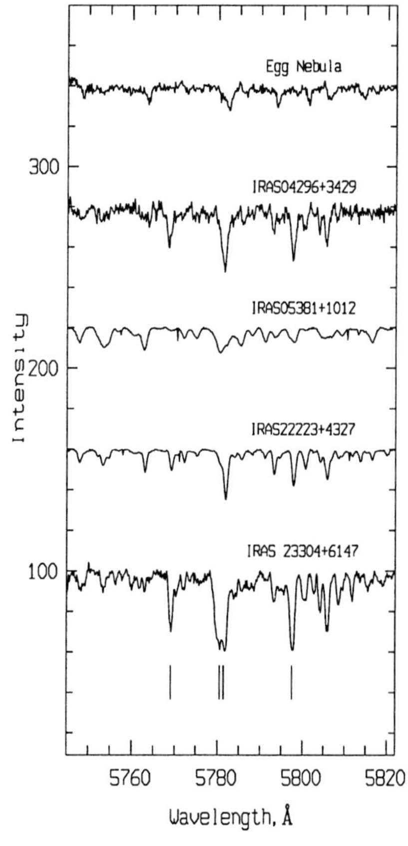

Figure 2 Spectra of selected post–AGB objects around diffuse band at λ 5780 Å.

Circumstellar bands. By modeling the spectra of some post–AGB objects we have revealed absorption features identified with the known but still unidentified diffuse interstellar bands (DIBs). The strongest absorptions have been found in the spectra of IRAS 23304+6147, IRAS 04296+3429, IRAS 22223+4327 and the Egg Nebula (see Fig. 2). Radial velocities of these spectral details coincide with those from the emission bands of C_2 and differ significantly from the star systemic velocity. It allows to estimate the expansion velocity (V_{exp}) of the shell in which the formation of these molecular features takes place. The values of V_{exp} derived from our optical spectra agree very well with V_{exp} determined from the CO measurements in the radio range. It means that diffuse bands are formed (mainly) in the expanding circumstellar envelopes around some post–AGB sources.

4. SUMMARY

The results of optical spectroscopy of peculiar supergiants with the 6 m BTA telescope in 1994–1999 are shortly reviewed. The investigated sample contains mostly low–mass post–AGB objects but also includes stellar objects of different nature. For example, IRC+10420 which, most likely, is a member of the not so numerous population of massive stars evolving towards LBV or WR–stage.

All post–AGB objects with large excesses of heavy metals are associated with an IR–source having the spectral feature at 21 μm (Decin et al.1998, Klochkova 1998). In the optical spectra of some (relatively faint) post–AGB candidates which belong to the 21 μm emitters we have found the *emission* Swan's bands of C_2 molecule and strong features which we identified with diffuse *circumstellar* bands (DCBs).

Acknowledgments

The authors acknowledge support from the Russian Foundation of Basic Research (project 99-02-18339), from the U.S. Civilian Research and Development Foundation (award No. RP1–2264) and from the grant No.2.P03D.002.13 and No.2.P03D.020.17 of the Polish State Committee for Scientific Research.

References

Decin L., van Winckel H., Waelkens C., Bakker E.J., 1998, A&A 332, 928
Gustafsson B., Bell R.A., Eriksson K., et al., 1975, A&A 42, 407
Humphreys R.M., Strecker D.M., Murdock T.L., Low F.J., 1973, ApJ 179, L4
Kipper T., Klochkova V.G., 1997, A&A 324, L65
Klochkova V., 1995, MNRAS 272, 210
Klochkova V.G., Panchuk V.E, 1996, Bull. Spec. Astrophys. Observ. 41, 5
Klochkova V.G., Panchuk V.E., Chentsov E.L., 1997a, A&A 323, 789
Klochkova V.G., Chentsov E.L., Panchuk V.E., 1997b, MNRAS 292, 19
Klochkova V.G., 1998, Bull. Spec. Astrophys. Observ. 44, 5
Klochkova V.G., Mishenina T.V., 1998, Bull. Spec. Astrophys. Observ. 44, 83
Klochkova V.G., Panchuk V.E., 1998, Astron. Letters 24, 754
Klochkova V.G., Panchuk V.E., Szczerba R., 1998, Ap&SS 255, 485
Klochkova V.G., Szczerba R., Panchuk V.E., Volk K, 1999, A&A 345, 905
Klochkova V.G., Szczerba R., Panchuk V.E., 2000a, Astron. Letters, 26 N2, 115

Klochkova V.G., Mishenina T.V., Panchuk V.E., 2000b, Astron. Letters, 26 N6, 465

Klochkova V.G., Szczerba R., Panchuk V.E., 2000c, Astron. Letters, 26 N7, 510

Klochkova V.G., Andrievsky S.M., Kovtyukh V.V., 2000d, A&A submitted

Kurucz R., 1993, CDROMs

Panchuk V.E., Najdenov I.D. Klochkova V.G., et al., 1998, Bull. Spec. Astrophys. Observ. 44, 127

Panchuk V.E., Klochkova V.G., Najdenov I.D., 1999, Preprint SAO, N135

Swings P., 1941, Lick Observ. Bull. 19, 31

Začs L., Klochkova V.G., Panchuk V.E., 1995, MNRAS 275, 764

Začs L., Klochkova V.G., Panchuk V.E., Spelmanis R, 1996, MNRAS 282, 1171

Zanstra H., 1928, MNRAS 89, 178

SPECTROSCOPICAL MASSES OF POST-AGB STARS

M.R. Schmidt
N. Copernicus Astronomical Center, Toruń
schmidt@ncac.torun.pl

Abstract The literature has been reviewed in search of available spectroscopic parameters for the post-AGB objects. The surface gravity and the effective temperature have been used for determination of spectroscopic masses from comparison with theoretical evolutionary tracks. The derived distribution of masses is relatively narrow and concentrates around 0.55 $M_\odot$ in agreement with analogous distributions for central stars of planetary nebulae and white dwarfs.

Systematic searches for post-AGB stars conducted in past 10 years resulted in an increasing number of candidates to this phase of evolution. According to the theory, stars at this phase move horizontally at a constant luminosity, and the time they need to reach the phase of planetary nebulae depends strongly on the mass of the post-AGB remnant. Hence, the mass of the object provides an unique information on the consistency of the evolution theory with observations.

We present here preliminary results of an extensive review of spectroscopical parameters based on literature. The starting point of this review is database of post-AGB objects presented in these proceedings (Szczerba et al.). Most of the objects in this catalogue have been appointed to the spectral type - luminosity class, only about 25 percent have been analysed spectroscopically.

The results of the review are summarized in the Table 1. The usual names of the objects are given in column (1), their effective temperatures in column (2), the surface gravities in column (3) and references in column (4). In the cases where analyses from different authors were available for the same object, we have chosen the values derived from the highest quality observations (high spectral resolution, high S/N). In the case of RV Tauri type stars observations from different phases by the same author have been analysed, the effective temperature and the surface gravity have been averaged. This assumes that during pulsations the star moves along evolutionary track.

In general, the effective temperature and gravity in the cool end of the sample are determined from analyses of the FeI and FeI/FeII ionization equilibrium. In the hot end, however, the effective temperature is determined from

R. Szczerba and S.K. Górny (eds.), Post-AGB Objects as a Phase of Stellar Evolution, 271–276.

Table 1 Spectroscopic parameters for the post-AGB objects

Name	T_{eff}[K]	log g	Source
BPS CS 22946−005	20000.	2.70	1
LB 3193	13000.	2.20	2
LB 3219	21250.	2.80	2
IRAS Z02229+6208	5500.	0.50	3
TW Cam	4800.	0.00	4
IRAS 04296+3429	7000.	1.00	5
RV Tau	4500.	0.00	4
Barnard 29	20000.	3.00	6
CPD−61°455	25000.	3.60	7
IRAS 05341+0852	6500.	1.00	8
DY Ori	6000.	1.50	9
SS Gem[a]	5600.	0.20	11
CT Ori[a]	5750.	1.00	11
HD 46703	6000.	0.40	12
HD 52961	6000.	0.50	10
HD 56126	7250.	0.50	8
U Mon	5000.	0.00	4
AI CMi	4500.	0.00	13
IRAS 07430+1115	6000.	1.00	3
AR Pup	6000.	1.50	9
IW Car	6700.	2.00	14
IRAS 10301+0933	24200.	3.09	15
HD 95767	7300.	1.50	16
HD 101584	8500.	1.50	17
SAO 239853	7350.	0.75	16
HD 107369	7600.	1.50	16
CE Vir[a]	4375.	0.30	11
PG 1323−086	16600.	2.50	18
ROA 24	6950.	1.15	34
HD 119608	24000.	3.00	19
EN TrA	6150.	1.25	16
BT Lib	5800.	1.40	4
BD+33°2642	20200.	2.90	20
SAO 243754	9300.	1.70	21
LS 3591	8500.	0.90	21
UU Her	6000.	0.70	33
TT Oph	4800.	0.50	4

Table 1 cont.

Name	T_{eff}[K]	log g	Source
LS IV −4 01	11000.	2.00	2
PG 1704+222	17600.	2.78	18
LS 4039	22000.	3.40	21
UY Ara	5500.	0.30	4
V453 Oph	5800.	0.75	26
IRAS 17279−1119	7400.	0.50	15
IRAS 17436+5003	7250.	1.30	22
89 Her	6550.	0.65	23
IRAS 18062+2410	22000.	3.00	24
IRAS 18095+2704	6700.	1.00	25
LS 4825	22000.	2.90	21
AC Her[a]	5900.	1.12	26
R Sct	4500.	0.00	4
AD Aql	6300.	1.25	26
HD 177566	30000.	3.80	27
EP Lyr[a]	6100.	1.40	9
LB 3116	16000.	2.70	32
DY Aql	4250.	0.60	11
IRAS 19500−1709	8000.	1.00	8
IRAS 19590−1249	23750.	2.70	2
IRAS 20004+2955	5000.	1.00	28
R Sge[a]	5100.	-0.25	9
AFGL 2688	6500.	0.00	29
V360 Cyg[a]	5375.	0.37	26
PHL 1580	24000.	3.60	30
BPS CS 29493−046	20000.	3.00	1
PHL 174	18000.	2.70	30
IRAS 22223+4327	6500.	1.00	5
IRAS 22272+5435	5600.	0.50	31
DS Aqr[a]	6500.	1.50	26, 4
IRAS 23304+6147	6750.	0.50	8

[a]Straight means of log g and T_{eff} from few observations.

References: (1) Kendall et al. 1997, (2) McCausland et al. 1992, (3) Reddy et al. 1999, (4) Giridhar et al. 2000, (5) Decin et al. 1998, (6) Conlon et al. 1994, (7) Hambly et al. 1996, (8) Van Winckel and Reyniers 2000, (9) Gonzalez et al. 1997a, (10) Waelkens et al. 1991 (11) Giridhar et al. 1997b, (12) Luck and Bond 1984 (13) Klochkova and Panchuk 1998, (14) Giridhar et al 1994, (15) Gies and Lambert 1992, (16) VanWinckel 1997, (17) Sivarani et al. 1999 (18) Moehler and Heber 1998, (19) Conlon 1992, (20) Napiwotzki et al. 1994, (21) Venn et al. 1998, (22) Yermakov, Schmidt 1998, (23) Luck et al. 1990, (24) Parthasarthy et al. 2000, (25) Klochkova 1995, (26) Giridhar et al. 1998, (27) Kendall et al. 1994, (28) Klochkova et al. 2000a, (29) Klochkova et al. 2000b, (30) Conlon et al. 1991, (31) Začs et al. 1995, (32) Kilkenny and Lydon 1986, (33) Klochkova et al. 1997, (34) Gonzalez and Wallerstein 1992

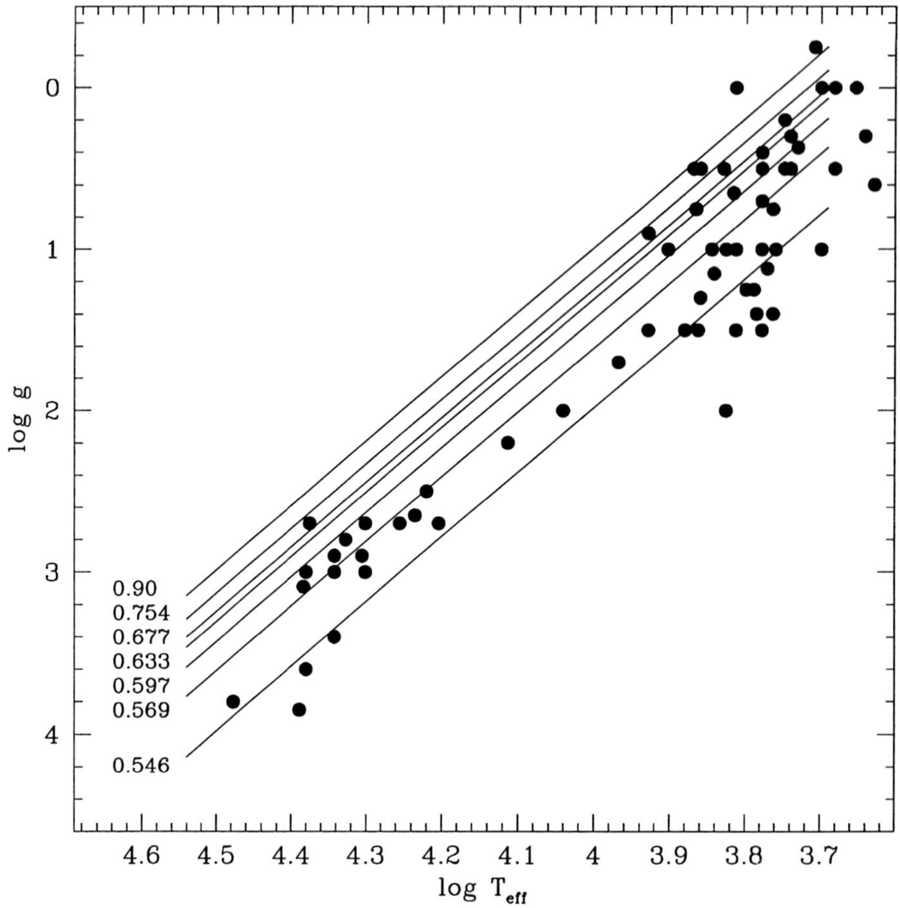

Figure 1 Diagram log g - T_{eff} for the objects from Table 1. Full lines show the theoretical evolutionary tracks for different stellar masses given in $M_\odot$.

SiII/SiIII/SiIV or MgII/MgIII ionization equilibrium, or from Strömgen photometric colours while the gravity is deduced from the profiles of Balmer lines.

The errors of the determination of log g are usually declared between 0.3 and 0.5. The smaller value is more characteristic for hotter objects. In most cases, they are usually estimated from a single spectrum. At the cooler edge of the post-AGB sample the errors are closer to 0.5. The differences between determinations from different studies, if available, seem to confirm this estimation.

In summary a sample of 68 objects has been collected for which spectroscopic determinations of the effective temperature and gravity have been found.

The mass of a post-AGB star can be estimated from a comparison of the observed atmospheric parameters (T_{eff}, log g) with those predicted from theoretical evolutionary tracks. This method has been succesfully applied to the central stars of planetary nebulae (e.g. Méndez et al. 1992). Fig. 1 compares the observed parameters for the objects from our sample with the theoretical tracks for different stellar masses. These are H-burning models for the population I composition taken from Vasiliadis and Wood (1994) and Schönberner (1983). Small differences between both calculations are of negligible importance for mass determination if they are to compared with the accuracy of the spectroscopical parameters.

The mass distribution for the post-AGB objects displayed in Fig. 1 is shown in Fig. 2. It is rather narrow with a well defined peak around 0.55 $M_{\odot}$. It can be compared to those for the planetary nebula central stars (Stasińska et al. 1997) and the white dwarfs (Bergeron et al. 1992, Bragaglia et al. 1995). Our distribution for the post-AGB stars resembles quite well that for the white dwarfs. When compared with the distribution for the planetary nebula nuclei we see that the objects with masses smaller that 0.55 $M_{\odot}$ are missing in the latter. Stasińska et al. (1997) explains this by observational selection. Low mass stars evolve too slowly in the post-AGB phase to produce observable nebulae.

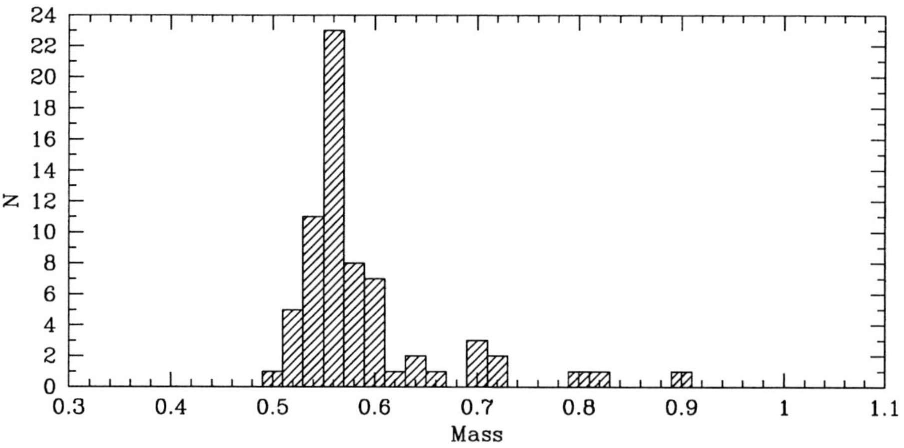

Figure 2 The post-AGB mass distribution.

The accuracy of the applied method is practically limited by the precision of determination of log g. As can be seen in Fig. 1 the width of the evolutionary tracks is 1 in log g. Hence, the spectroscopical accuracy of 0.5 in log g is hardly enough to make any conclusion about the mass distribution. This uncertainty is further enhanced by the limitation of the derivation of gravity to the nearest atmospheric model from the Kurucz's grid listed with step of 0.5 in log g. Taking into account inadequacy of static atmospheric models for analysis of mostly pulsating stars, further progress in the determination of spectral masses may be achieved only by better understanding of the line formation in pulsating atmospheres.

Acknowledgments

This research was supported by the Polish State Committee for Scinetific Research (grant No. 2 P03D 020 17).

References

Conlon E.S., 1992, Proc. of the IAU Symposium 155, eds. Weinberger R. and Acker A., 360
Conlon E.S., Dufton P.L., Keenan F.P., McCausland R.J.H., 1991, MNRAS 248, 820
Conlon E.S., Dufton P.L., Keenan F.P., 1994, A&A 290, 897
Decin L., Winckel, Waelkens, Bakker E., 1998, A&A 332, 928
Gies D.R., Lambert D.L., 1992, ApJ 387, 673
Giridhar S., Rao N.K., Lambert D.L., 1994, ApJ 437, 476
Giridhar S., Lambert D.L., Gonzalez G., 1998, ApJ 509, 366
Giridhar S., Lambert D.L., Gonzalez G., 2000, ApJ 531, 521
Gonzalez G., Wallerstein G., 1992, 254, 343
Gonzalez G., Lambert D.L., Giridhar S., 1997a, ApJ 479, 427
Gonzalez G., Lambert D.L., Giridhar S., 1997b, ApJ 481, 452
Hambly N.C., Dufton P.L., Keenan F.P., Lumsden S.L., 1996, MNRAS 278, 811
Kendall T.R., Dufton P.L., Keenan F.P., Beers T.C., Hambly N.C., 1997, A&A 317, 82
Kendall T.R., Brown P.J.F., Conlon E.S., Dufton P.L., Keenan F.P., 1994, A&A 291, 851
Kilkenny D., Lydon J. 1986, MNRAS 218, 279
Klochkova V.G., 1995, MNRAS 272, 210
Klochkova V.G., Panchuk V.E., Chentsov E.L., 1997, A&A 323, 789
Klochkova V.G., Panchuk V.E., 1998 Astron. Lett. 24, 650
Klochkova V.G., Mishenina T.V., Panchuk V.E., 2000a, Astron. Lett. 26, 465
Klochkova V.G., Szczerba R., Panchuk V.E., 2000b, Astron. Lett. 26, 510
Luck R.E., Bond H.E., 1984, ApJ 279, 729
Luck R.E., Bond H.E., Lambert D.L., 1990, ApJ 357, 188
McCausland R.J.H., Conlon E.S., Dufton P.L., Keenan F.P., 1992, ApJ 394, 298
Méndez R., Kudritzki R.P., Herrero A., 1992, A&A 260, 329
Moehler S. & Heber U., 1998, A&A 335, 985
Napiwotzki R, Heber U., Koppen J., 1994, A&A 292, 239
Parthasarathy M., Garcia-Lario, P., Sivarani T., Manchado A., Sanz Fernández de Córdoba L., 2000, A&A 357, 241
Reddy B.E., Bakker E.J., Hrivnak B.J., 1999, ApJ 524, 831
Sivarani T., Parthasarathy M., Garcia-Lario, Manchado A., Pottasch S.R. 1999, A&A Suppl. 137, 505
Schönberner D., 1983, ApJ 272, 708
Veen K.A., Smartt S.J., Lennon D.J., Dufton P.L. 1998, AA 334, 987
Van Winckel H., 1997, A&A 319, 561
Van Winckel H., Reyniers M., 2000, A&A 354, 135
Vassiliadis E., Wood P.R., 1994, ApJS 92, 125
Waelkens C., Waters. L.B.F.M., Casatella A., Le Bertre T., Lamers H.J.G.L.M., 1987, A&A 181, L5
Waelkens C., Van Winckel H., Bogaert E., Trams N.R., 1991, AA 251, 495
Yermakov S., Schmidt M.R., 1998, private comm.
Začs L., Klochkova V.G., Panchuk V.E., 1995, MNRAS 275, 764

BD+33°2642: ABUNDANCE PATTERNS IN THE CENTRAL STAR OF A HALO PN

Ralf Napiwotzki, Markus Herrmann, Ulrich Heber
Dr. Remeis-Sternwarte, Sternwartstr. 7, 96049 Bamberg, Germany
(napiwotzki,herrmann,heber)@sternwarte.uni-erlangen.de

Martin Altmann
Sternwarte, Universität Bonn, Auf dem Hügel 71, 53121 Bonn, Germany
maltmann@astro.uni-bonn.de

Abstract We present a model atmosphere analysis of high resolution UV and optical spectra of the PN central star BD+33°2642. The derived abundance pattern is best explained by a dust-gas fractionation scenario. Since it it suspected that this process is related to binarity we checked the available spectroscopic material and found evidence for radial velocity variations, indicating a binary nature of BD+33°2642. The measured mean radial velocity is combined with a HIPPARCOS proper motion measurement and the galactic orbit of BD+33°2642 is calculated. The derived orbit provides strong evidence that BD+33°2642 is a member of the galactic halo population.

1. INTRODUCTION

BD+33°2642 is a low mass central star of a planetary nebula (PN). It is the coolest central star known ($T_{\rm eff} = 20200$ K; Napiwotzki et al. 1994) and marks the beginning of the PN phase. A spectroscopic analysis based on medium resolution optical and high resolution IUE spectra (Napiwotzki et al. 1994) revealed that the atmosphere of BD+33°2642 is metal poor. C, N, O, Mg, Si turned out to be depleted by about 1 dex, and Fe by 2 dex relative to the solar values. We proposed that this chemical mixture is best explained by a gas-dust-separation, which seems to be at work in some peculiar cool post-AGB stars as well (Bond et al. 1991 and references therein).

We obtained new high resolution spectra with the Keck I, the Calar Alto 2.2 m telescope, and the STIS spectrograph of HST and performed a LTE analysis. The high quality of our new spectra allowed the detection and measurement of lines of several additional elements (S, Al, Mn, Cr, Ar, P). We will use this

R. Szczerba and S.K. Górny (eds.), Post-AGB Objects as a Phase of Stellar Evolution, 277–282.

improved data base to rediscuss the gas-dust separation scenario. Since Van Winckel et al. (1995) presented evidence that the extremely iron-deficient post-AGB stars are members of binary systems, we use our high-resolution spectra to check for radial velocity variations. Finally, we compute a galactic orbit of BD+33°2642 and determine the population membership of this star.

2. MODEL ATMOSPHERE ANALYSIS

We performed a standard LTE model atmosphere analysis of BD+33°2642. Model atmospheres were calculated with the ATLAS 9 program of Kurucz (1993) and theoretical equivalent widths were computed with an updated version of the spectrum synthesis code LINFOR[1]. The basic parameters $T_{\rm eff} = 20200$ K and $\log g = 2.9$ were adopted from Napiwotzki et al. (1994). We used the B2IV star γ Peg analyzed by Peters (1976) to check our results, in particular the validity of the LTE assumption for certain ionization stages.

The spectral range from 4270 Å to 6760 Å is covered by a high-resolution spectrum (0.14 Å FWHM) taken with the HIRES spectrograph of the Keck I telescope. This spectrum is complemented by a spectrum of moderately lower resolution (0.4 Å FWHM) observed with the Coudé spectrograph of the Calar Alto 2.2 m telescope covering the range 4050 Å to 5000 Å. A very high resolution UV spectrum (0.01 Å FWHM) of the HST-STIS spectrograph taken for calibration purposes was extracted from the HST archive (spectral coverage: 1200Å to 1400 Å). Equivalent widths were determined by fitting Gaussian profiles (cf. Napiwotzki et al. 1994). We derived a microturbulence velocity of 10 km/s from the iron lines.

The resulting abundances are displayed in Fig. 1. Our new analysis confirms our previous results (Napiwotzki et al. 1994), but with higher accuracy and an increased number of investigated elements. BD+33°2642 is metal-poor, the iron content is only 1/30 of the solar value. Since this indicates a population II membership of BD+33°2642, we added in Fig. 1 the abundances expected for a pop. II star with the iron abundance of BD+33°2642 (data taken from Caretta et al. 2000; McWilliam 1997; Wheeler et al. 1986). While predicted and measured abundances are in good agreement for the iron group elements Cr and Mn, the values measured for C, O and the α process elements S and Ar are much higher than expected. However, the abundances of the α process elements Mg and Si are well explained by a pop. II composition and the measured value is even below the expected one for Al.

How to explain this strange abundance pattern? Objects probably related to BD+33°2642 are the metal-poor B-type post-AGB stars discussed by McCausland et al. (1992). McCausland et al. proposed an enhancement of α process

[1]www.sternwarte.uni-erlangen.de/ãi26/linfit/linfor.html

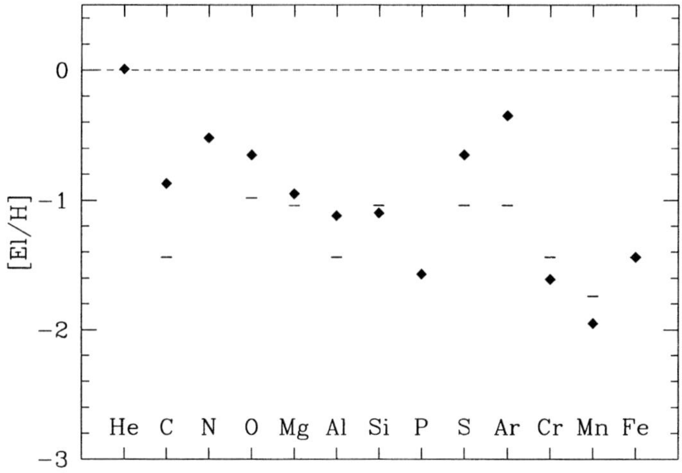

Figure 1 Abundance determinations for BD+33°2642 (diamonds) compared with abundances expected for a pop. II star with the iron content of BD+33°2642 (dashes).

elements due to the third dredge-up. However, this scenario can be ruled out, because formation of α process elements beyond ^{16}O is inhibited in low mass stars by the extremely low cross section of the $^{16}\mathrm{O} \rightarrow {}^{20}\mathrm{Ne}$ reaction (see e.g. Rolfs & Rodney 1988). Other possibly related objects are the members of a small class of very iron-deficient post-AGB stars of spectral type F. Most extreme examples are HR 4049 and HD 52961 (Waelkens et al. 1991) with an atmospheric iron abundance $[\mathrm{Fe/H}] = -4.6$. However, the oxygen abundance is close to the solar value: $[\mathrm{O/H}] = -0.3$ (HR 4049) and -0.5 (HD 52961). Bond (1991) proposed dust-gas fractionation to explain these abundance patterns. At the end of the AGB phase dust forms in the extended envelope of the red giant. Elements with a high condensation temperature (like iron) condensate onto dust grains, while other elements with a lower condensation temperature (like C,N,O,S,Ar) remain in the gas phase. If the dust grains are removed from the circumstellar enviroment by radiation pressure and the gas falls back to the stellar surface this can explain the observed abundances.

Since S and Ar are not modified by dredge-up processes and remain in the gas phase due to their low condensation temperatures, both elements are valuable tracers of the initial stellar abundance. Our spectral analysis indicates S and Ar abundances of $\approx$1/3 of the solar value. This would be consistent with BD+33°2642 being a metal-poor disk object or a relatively metal-rich halo star. A further discussion of the population membership will be given below.

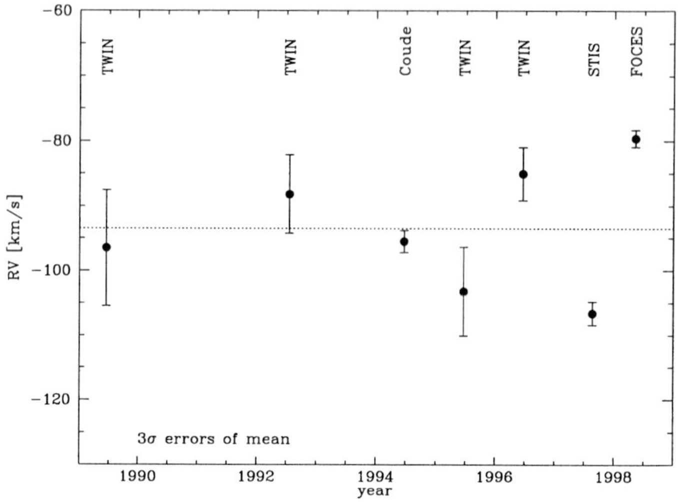

Figure 2 Radial velocity measurements of BD+33°2642. The origin of the spectra is indicated (HST-STIS spectrograph; COUDE and FOCES spectrograph at the Calar Alto 2.2 m; TWIN spectrograph at the Calar Alto 3.5 m).

3. RADIAL VELOCITY

Van Winckel et al. (1995) have shown that all known extremely iron-deficient post-AGB stars reside within a binary system. Orbital periods are of the order of one year and the measured radial velocity varies by typical 30 km/s. This provides strong support that binarity is an essential ingredient for the gas-dust fractionation. In the binary scenario the interaction with the companion during the AGB phase produces a circum-system disk of matter lost by the AGB star. After the primary left the AGB the mass loss drops and the gas can be reaccreted. However, the accretion of dust is prevented by radiation pressure.

If we interprete the abundance pattern of BD+33°2642 as the result of a dust-gas fractionation analogous to that observed in the extremely iron-deficient post-AGB stars, we would expect that BD+33°2642 is the member of a binary system. We used a collection of high and medium resolution spectra taken during the last ten years to check this hypothesis. Our radial velocity measurements (Fig. 2) show variations up to 30 km/s. Although our data is too coarse to detect a periodicity, this indicates that BD+33°2642 is a member of a binary system with properties similar to the extremely iron-deficient stars. We conclude that BD+33°2642 displays a mild form of the dust-gas fractionation observed in these stars.

Since BD+33°2642 is reasonably bright, it's proper motion was measured by the HIPPARCOS satellite ($\mu_\alpha = -14.24$ mas, $\mu_\delta = +0.45$ mas). The star is too far away for a direct parallax measurement and we used the distance

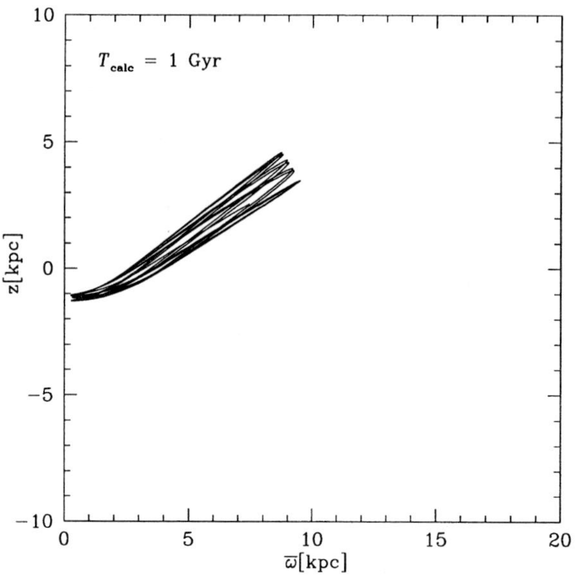

Figure 3 Meridional projection of the galactic orbit of BD+33°2642 during the last 10^9 years.

estimate from our determination of the fundamental parameters ($d = 3.3$ kpc; Napiwotzki et al. 1994). We combined the proper motion with the mean radial velocity (−94 km/s) and calculated a galactic orbit of BD+33°2642 with Allen & Santillen's (1991) model of the galactic mass distribution. A detailed description of our method is provided in Altmann & de Boer (2000). A meridional projection of the BD+33°2642 orbit is shown in Fig. 3. The orbit is highly eccentric ($e = 0.95$) and reaches a maximum height above the galactic disk of 4.8 kpc. The orbital velocity $\Theta = 16$ km/s is much lower than typical values of disk stars (sun: $\Theta = 235$ km/s). All this is characteristic for population II stars and provides evidence that BD+33°2642 is a member of the galactic halo population. Thus the PN surrounding BD+33°2642 is one of the very few known PNe of type IV.

4. CONCLUSIONS

BD+33°2642 is a metal poor star with an iron content typical of a pop. II star: $[\mathrm{Fe/H}] = -1.4$. However, the abundances of C and O and most measured α process elements is higher than expected for an pop. II star of this iron content. While C and O abundances could have been altered by dredge-up processes there is no way to change the abundances of the α process elements in general by nuclear processing in the low mass progenitor of BD+33°2642. This abundance is best explained by a dust-gas fractionation as it was shown to be at work in extremely iron-deficient post-AGB stars. Since van Winckel et al. (1995) demonstrated that all known extremely iron-deficient post-AGB stars are members of binary systems, we checked BD+33°2642 for binarity. Our analysis of a number of available high- and medium-resolution spectra revealed radial velocity variations of up to 30 km/s indicating binarity. Although the data is not sufficient to determine a period or radial velocity curve, the variations are compatible to the radial velocity amplitudes measured in iron deficient post-AGB stars, indicating a relationship between these stars and BD+33°2642. Finally we combined our radial velocity measurement with a proper motion measurement and determined a galactic orbit of BD+33°2642. It turned out that BD+33°2642 does have a highly excentric orbit typical for a member of the galactic halo population. Thus we conclude that BD+33°2642 started it's career as metal poor pop. II star, but dust-gas fractionation caused an additional depletion of elements with a high condensation temperature, like Fe, Cr, Mn, Si, and Mg. Best indicators of the initial metallicity are probably the elements Ar and S, which are not modified during nuclear processing in low mass stars and have low condensation temperatures. Both elements indicate that BD+33°2642 was initially a relatively metal-rich pop. II star.

Acknowledgments

We thank Neil Reid for providing his spectrum of BD+33°2642 and Sabine Moehler for sharing her knowledge of the spectral analysis software.

References

Allen C., Santillan A. 1991, RMxAA 22, 255

Altmann M., de Boer K.S. 2000, A&A 353, 135

Bond H.E. 1991, in: IAU Symp. No. 145, Evolution of stars: the photospheric abundance connection, eds. G. Michaud & A. Tutukov, Kluwer, Dordrecht, p. 341

Caretta E., Gratton R.G., Sneden C. 2000, A&A 356, 238

Kurucz R.L. 1993, ATLAS9 stellar atmosphere programs and 2km/s grid, CD-ROM No. 13

McCausland R.J.H., Conlon E.S., Dufton P.L., Keenan F.P. 1992, ApJ 394, 298

McWilliam A. 1997, ARA&A 35, 503

Napiwotzki R. 1993, Acta Astron. 43, 415

Napiwotzki R., Heber U., Köppen J. 1994, A&A 292, 239

Rolfs L.E., Rodney W.S. 1988, Cauldrons in the cosmos, University of Chicago Press

Van Winckel H., Waelkens C., Waters L.B.F.M. 1995, A&A 293, L25

Waelkens C., Van Winckel H., Bogaert E., Trams N.R. 1991, A&A 251, 495

Wheeler J.C., Sneden C., Truran J.W.Jr. 1989, ARA&A 27, 279

THE VERY LITHIUM RICH POST-AGB SB2 BINARY HD 172481

Maarten Reyniers, Hans Van Winckel

Instituut voor Sterrenkunde, Celestijnenlaan 200B, 3001 Heverlee, Belgium

Maarten.Reyniers@ster.kuleuven.ac.be

Abstract Double lined spectroscopic binaries in an evolved stage of evolution are expected to be extremely rare since they must consist of equally luminous and thus almost equally evolved objects, which requires an extremely similar initial mass. In this contribution we discuss such rare double evolved SB2 system: HD 172481. This binary includes an F-type post-AGB object and an M-type AGB companion. The spectrum shows a surprisingly strong Li I 670.8 nm line with an equivalent width of $W_\lambda = 54$ mÅ, yielding a lithium abundance of $\log \epsilon(\mathrm{Li}) = 3.6$. Several explanations for this huge lithium content are explored.

1. HD 172481 IN THE POST-AGB SAMPLE

In the course of our ongoing program to study the chemical composition of post-AGB stars (see Van Winckel, this volume), high resolution, high signal-to-noise spectra were taken of the F2Ia supergiant HD 172481. This star was previously selected as a candidate post-AGB star by Oudmaijer et al. (1992) because of its far infrared excess and its high galactic latitude ($b = -10.37°$).

2. OBSERVATIONAL DATA

Our **photometric** data consist of 54 data points in the Geneva photometric system between 1989 and 1996, near-IR JHKLM photometry taken with the ESO 1m telescope in 1992 and the IRAS-fluxes. 62 **radial velocity** measurements were obtained with the CORAVEL radial velocity spectrometer (Danish 1.5m telescope, La Silla, Chile) between 1983 and 1997 and 4 recent measurements with the CORALIE spectrograph (Euler telescope, La Silla, Chile). 5 **high resolution, high signal-to-noise spectra** were taken with several spectrographs and telescopes: 2 spectra with the UES (WHT, La Palma, Spain) in 1995, 2 with the EMMI spectrograph (NTT, La Silla, Chile) in 1997 and 1998 and 1 with FEROS (ESO 1.5m, La Silla, Chile) in 2000.

R. Szczerba and S.K. Górny (eds.), Post-AGB Objects as a Phase of Stellar Evolution, 283–288.

Figure 1 The 1998 EMMI (top) and FEROS (bottom) spectra in the 700-715 nm TiO band head region. The top spectrum is offset by 0.2 units. The TiO bands in the red spectrum of HD 172481are a clear signature of its cool companion. Moreover, these bands vary in time as evidenced by our 1998 EMMI spectrum. Band head wavelengths are from Valenti et al. (1998).

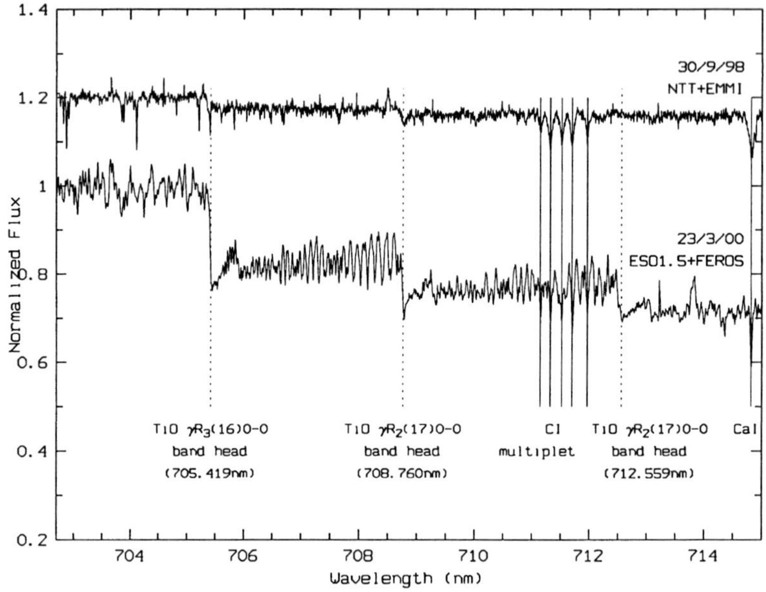

3. HD 172481 AS A BINARY

A comparison of the blue and green regions of the spectra of this object with spectra of the normal massive F0Ib supergiant HR 1865 confirms the F2Ia0 spectral type given by the SIMBAD database. However, our 1997 EMMI and FEROS spectra show clear and strong discontinuous jumps at certain wavelengths in the red region which we identified as **TiO band heads** (Fig. 1). Since the pulsational amplitude is very low ($\Delta V \sim 0.2$), we conclude that these bands are caused by a cool M-type companion. The red region of the spectrum of HD 172481 is therefore a mixture of two spectral types: atomic lines induced by a hot component (~7250 K) and TiO bands caused by a cool component (~3500 K). Moreover, the strength of these bands is clearly variable in time (also Fig. 1).

Other evidence confirming the binary nature of HD 172481 was found in its **spectral energy distribution** (SED hereafter), which is constructed from the collected photometry in Fig. 2 with an adopted reddening of E(B-V)=0.44. The fit of only the F-type (Kurucz) model atmosphere clearly did not satisfy from about $\lambda \sim 10^3$ nm redwards. Therefore, we had to invoke a second, cool component with $T_{eff} = 3500$ K. The latter temperature is, however, highly uncertain because it is derived by only a few photometric points.

Figure 2 The dereddened SED of HD 172481 with a E(B-V) = 0.44. The dotted line represents the Kurucz model with T_{eff} = 7250 K and log(g) = 1.5; the dashed line the model with T_{eff} = 3500 K and log(g) = 0.5; the full line is the combined model.

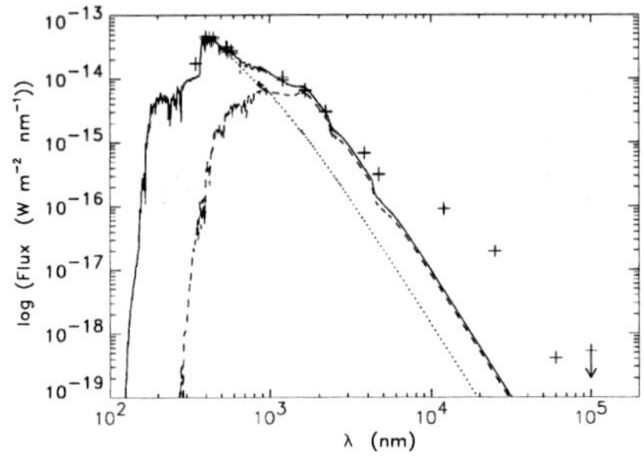

Figure 3 The abundances of HD 172481 relative to iron [el/Fe]. Errorbars are given if the number of lines is more than 3. The horizontal line represents [el/Fe]=0. Solar abundances are mainly taken from Grevesse (1989).

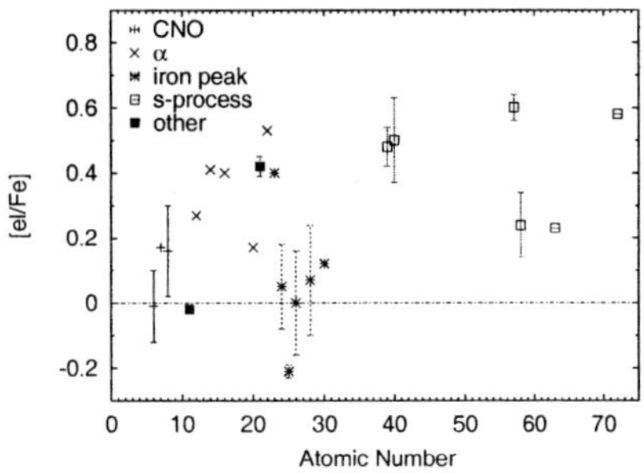

Two more interesting remarks can be made concerning the SED. The *luminosity ratio* of both components L_{F-type}/L_{M-type} is found to be 1.8. This ratio is rather dependent on the adopted reddening: using E(B-V) = 0.2, it equals to 1. In any case, the two components must be of very similar luminosity. Consequently, also the M-type component must be very luminous and probably in its AGB phase of stellar evolution. A second remark concerns the IRAS-fluxes (the 4 most redward points on the SED) pointing to a cool infrared excess which we interpret as caused by a circumstellar dust shell or disk.

The binary nature of HD 172481 could, however, not be resolved by our **radial velocity monitoring**: although a peak-to-peak amplitude of 17 km s^{-1} is observed, we could not discover a long-term trend in the radial velocity data. A possible explanation for this fact can be an unfavourable inclination angle, but most likely it is due to a long orbital period. The variation induced by the orbital motion is then completely washed out by the (low amplitude) pulsation. This (semi-regular) low-amplitude pulsation is also seen in our Geneva photometry.

4. CHEMICAL ANALYSIS

A detailed chemical abundance analysis was carried out based on the high resolution spectra. Atmospheric parameters were obtained with the commonly used spectroscopic method: excitation equilibrium for the effective temperature (T_{eff}), ionisation equilibrium for the gravity ($\log(g)$) and the microturbulent velocity (ξ_t) is obtained by demanding that the iron abundance is independent of the reduced equivalent widths (W_λ/λ) of the Fe I-lines. This method resulted in T_{eff}= 7250 (±250) K, $\log(g)$ = 1.5 (±0.5) (cgs) and ξ_t = 4 (±1) km s^{-1}. 48 Fe I and 14 Fe II lines were involved in this determination. We used the atmospheric models of Kurucz (CD-ROM set, 1993) in combination with his abundance calculation program WIDTH9. Our linelist contains only lines with well determined oscillator strengths in detail described in Van Winckel & Reyniers (2000).

209 lines of 28 different ions were measured. We obtained a metallicity of [Fe/H] = −0.55. The results of our analysis are graphically presented in Fig. 3. On this figure, one can clearly see that
– the CNO elements are clearly *not* enhanced;
– the α elements are enhanced by ~0.3 to 0.4 dex, which is not an intrinsic enhancement, but a galactic evolution effect for this metallicity range;
– the other iron peak elements do follow the iron deficiency of [Fe/H] = −0.55;
– the light s-process elements Y and Zr and the heavy s-process element La are enhanced by ~0.5 dex.

5. POST-AGB STATUS OF HD 172481

In this section we collect our arguments for the proposed post-AGB status of HD 172481.
1. The high galactic latitude ($b = -10.37°$),
2. the large radial velocities ($<v> \sim -85$ km s^{-1}) and
3. the moderate metal deficiency ([Fe/H]=−0.55)
point to a population II membership. Furthermore,
4. the SED (Fig. 2) shows the presence of dust, as a result of previous mass loss, probably during the AGB phase.
5. The photometry and the Hα profiles on our 4 spectra that include Hα show a variability very similar to other post-AGB stars (*cfr.* Lèbre, this volume).
6. Finally, the slight s-process overabundances are probably the result of the third dredge-up, when He burning products are brought to the surface. This last argument is, however, not straightforward. If we expect the signature of a 3rd dredge-up, also the CNO elements should be enhanced, which is *not* detected: the total CNO abundance is equal to the initial value.

6. LITHIUM IN HD 172481

The most surprising discovery during our spectral analysis was the detection of a strong Li I resonance line at 670.8 nm (Fig. 4). Kurucz's synthesis program SYNTHE was used to account for the doublet structure of this line. A least-squares fit between the observed and the synthetic spectrum yielded an abundance of $\log(\mathrm{Li}) = 3.57 \pm 0.2$ (Fig. 4). It is far from clear how HD 172481 can be so Li rich, exceeding even the interstellar medium value of $\log \epsilon(\mathrm{Li}) = 3.3$. During the evolution of a star, lithium is destroyed, as it burns at relatively low temperatures. Lithium destruction and depletion is expected to occur during the main-sequence evolution (e.g. $\log(\mathrm{Li})_{\odot} = 1.16$) as well as during the RGB and AGB evolution, as the deeper convective envelope reaches into lithium-depleted interior layers.

Figure 4 The Li I resonance line in HD 172481. *left panel:* The UES and FEROS spectrum around 671 nm with the Li I doublet of 54 mÅ. The top spectrum is offset by 0.3 units. The contribution of the cool companion is larger for our FEROS run as evidenced by the detection of the (weak) TiO band head at 671.447 nm and the apparently lower s/n for this spectrum. *right panel:* Synthesis of the Li I resonance line using our UES spectrum. A velocity broadening (ζ) of 14 km s^{-1} is used in this synthesis, deduced by the synthesis of unblended lines with a smooth profile. The points are the observed spectrum, the lines synthetic spectra with log ϵ(Li) = 3.47, 3.57 and 3.67 resp. A least-squares fit resulted in log ϵ(Li) = 3.57 (full line).

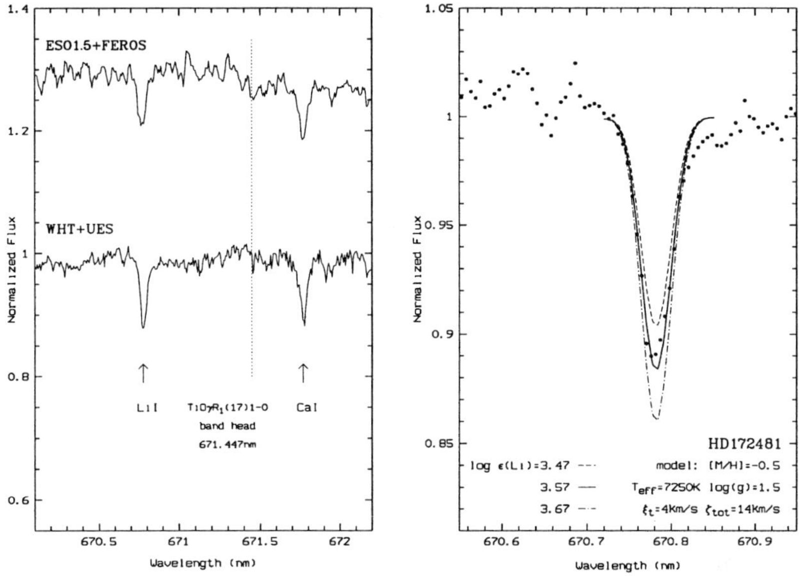

Nevertheless Li rich evolved stars exist. In general three classes of Li rich evolved (AGB and post-AGB) stars can be distinguished: (1) the luminous "hot bottom burning" (HBB) S-type AGB stars in the Clouds (Plez et al. 1993, Smith et al. 1995) and the Galaxy (García-Lario et al. 1999); (2) galactic low-mass C stars (Abia et al. 1993; Abia & Isern 1997, 2000) which are mostly J-type and (3) 4 metal-poor low-mass post-3rd dredge-up objects being IRAS 22272+5435 (Začs et al. 1995), IRAS 05341+0852 (Reddy et al. 1997), IRAS Z02229+6208 and IRAS 07430+1115 (Reddy et al. 1999). Only for the HBB stars, Li production is supported by the models via the ^{7}Be-transport mechanism (Cameron & Fowler 1971). All models suggest that HBB and the associated Cameron-Fowler mechanism only works in intermediate mass stars (4-7 $M_\odot$; e.g. Sackmann & Boothroyd 1992).

For classes (2) and (3), which are most probably low-mass stars, no envelope burning is predicted in any AGB model. By its old nature, and hence low mass, a "hot bottom burning" scenario is not applicable to HD 172481. The most promising models to date, which could shed some light on the Li rich evolved low-mass stars are the "cool bottom processing" models by Wasserburg et al.

(1995) and Sackmann & Boothroyd (1999). These models are successfully developed to explain the Li rich K giants (e.g. de La Reza et al. 1997 and references therein) using a non-standard mixing mechanism. The same group of authors suggests that their models might also work in the TP-AGB phase of low-mass stars (Boothroyd & Sackmann 1999).

7. CONCLUSIONS

In this contribution we shortly described our study on the very peculiar supergiant HD 172481. The TiO band heads superimposed on a typical F-type spectrum reveal that it is a binary object, which is further strengthened by the SED. The high radial velocities, the high galactic latitude and the moderate metal deficiency found in our chemical analysis suggest that the F-type component is an old and hence low-mass object. It is probably in its post-AGB phase of evolution as evidenced by the dust, the nature of the (low amplitude) variability and the slight s-process overabundances. Above all, this star is found to be super rich in lithium, with a Li abundance of $\log \epsilon(\mathrm{Li}) = 3.6$. A more detailed description can be found in Reyniers & Van Winckel (2000).

Acknowledgments

Both authors acknowledge financial support of the Fund for Scientific Research - Flanders.

References

Abia C., Isern J., 1997, MNRAS 289, L11
Abia C., Isern J., 2000, ApJ 536, 438
Abia C., Boffin H.M.J., Isern J., Rebolo R., 1993, A&A 272, 455
Boothroyd A.I., Sackmann J.-I., 1999, ApJ 510, 232
Cameron A.G.W., Fowler W.A., 1971, ApJ 164, 111
de La Reza R., Drake N.A., da Silva L., Torres C.A.O., Martin E.L., 1997, ApJ 482, L77
García-Lario P., D'Antona F., Lub J., Plez B., Habing H.J., 1999. In: Le Bertre T., Lèbre A., Waelkens C. (eds.) Proc. IAU Symp. 191, AGB Stars. ASP Conference Series, p. 91
Grevesse N., 1989. In: Waddington C.J. (ed.) AIP Conferences Series 183, Cosmic Abundances of Matter. American Institute of Physics, New York, p. 9
Oudmaijer R.D., van der Veen W.E.C.J., Waters L.B.F.M., et al., 1992, A&ASS 96, 625
Plez B., Smith V.V., Lambert D.L., 1993, ApJ 418, 812
Reddy B.E., Parthasarathy M., Gonzalez G., Bakker E.J., 1997, A&A 328, 331
Reddy B.E., Bakker E.J., Hrivnak B.J., 1999, ApJ 524, 831
Reyniers M., Van Winckel H., 2000, A&A, in press
Sackmann I.-J., Boothroyd A.I., 1992, ApJ 392, L71
Sackmann I.-J., Boothroyd A.I., 1999, ApJ 510, 217
Smith V.V., Plez B., Lambert D.L., Lubowich D.A., 1995, ApJ 441, 735
Valenti J.A., Piskunov N., Johns-Krull C.M., 1998, ApJ 498, 851
Van Winckel H., Reyniers M., 2000, A&A 354, 135
Wasserburg G.J., Boothroyd A.I., Sackmann I.-J. 1995, ApJ 447, L37
Začs L., Klochkova V.G., Panchuk V.E., 1995, MNRAS 275, 764

CIT 6: THE EARLY PHASE OF PPN?

Laimons Začs
Ventspils International Radio Astronomy Center, Riga, Latvia
zacs@latnet.lv

Andrejs Alksnis
Institute of Astronomy, University of Latvia, Riga, Latvia

Raivis Spēlmanis
University of Latvia, Riga, Latvia

Abstract Photometric and spectroscopic studies for one of the early phase candidate for proto-planetary nebulae, the dust enshrouded extreme carbon star CIT 6, are presented.

1. INTRODUCTION

Only a small fraction of proto-planetary nebulae (PPNe) candidate studied shows clear evidence of chemical evolution as expected after the 3rd dredge-up: an enhancement of carbon and s-process elements in their atmospheres (see Van Winckel 2000; and references therein). What are the progenitors of such objects? Apparently, the carbon stars. However, among the intrinsic (AGB) carbon stars only N stars have an enhancement of s-process elements. On the other hand, masquerades within binary systems can produce carbon and s-process rich AGB (post-AGB) stars (Jorissen & Van Eck 2000). Here we present the results of long-term photometric and spectroscopic observations for the one fascinating carbon star CIT 6.

CIT 6 (= RW LMi = IRC +30219 = AFGL 1403 = IRAS 10131+3049), the dust enshrouded extreme carbon star, is the second brightest carbon star at 12 μm (3319 Jy), next to IRC+10216. CIT 6 lies at large distance from the Galactic plane (b = 56°). The optical spectrum of CIT 6 has been previously classified as C 4,3 (Cohen 1979), however, Trammell et al. (1994) suggest a spectral type SC. Alksnis (1995; and references therein) noted an unusual light and color variations of this source. According to the LRS classification carried out by

R. Szczerba and S.K. Górny (eds.), Post-AGB Objects as a Phase of Stellar Evolution, 289–294.

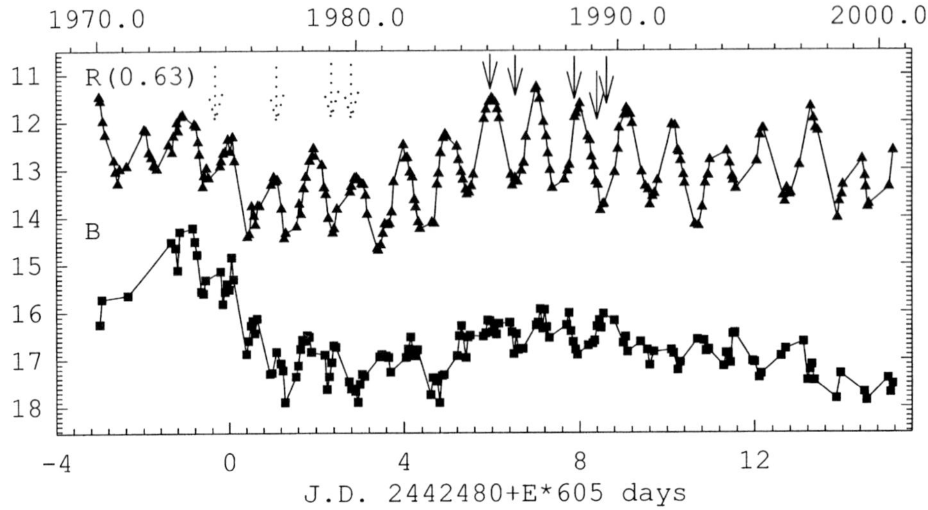

Figure 1 The averaged values (0.05P) of the photometric observations for CIT 6 in the red (R) and blue (B) light. The long period variability with a typical period of 605-628 days and the slow light variations with a possible cycle ≥15 years have been detected. The moments of spectroscopic observations are indicated by arrows.

Kwok, Volk, & Bidelman (1997) CIT 6 belong to the Group C with 11 μm SiC dust emission. Such sources are generally volved carbon stars undergoing mass loss. The mass loss rate was found $6 \times 10^{-6} M_{\odot}\, yr^{-1}$ during the 3000 yr (Meixner et al. 1998).

2. A PHOTOMETRIC MONITORING

The multicolor photometric observations have been made with the Baldone (nearly Riga) Schmidt telescope since 1970. All the observations are close to the standard UBV system or were reduced to R(0.63) and I(0.81). Instrumental details can be found in Alksnis (1995). Standard errors are typically 0.05-0.09 mag depending on the brightness of the star. Figure 1 shows the mean light curves for CIT 6 in R and B bands, based on observations carried out by Alksnis and collaborators.

In the red and infrared typical light curves of carbon Miras have been observed: 1) long period variability with a typical period (P) of 605 days until 1987 and 628 days later on; 2) slow light variations with a possible cycle ≥15 years. Sometimes fast light fluctuations were observed in the time scale of a few (maximum 30) days. The long period variability in the B colour was clearly observed only before 1976, when CIT 6 in B colour was brightest and had the lower colour index (B-R). Afterwards correlation between the long period variability in the red and blue sharply decreased, and sometimes (e.g. 1988

-1990) anticorrelation was observed. However, the fast fluctuations are always simultaneously in the red and blue.

3. OPTICAL SPECTROSCOPY

The observations reported here were carried out using two medium resolution spectrographs, the TV scanner and the echelle spectrometer ZEBRA equipped with a two dimensional photon-counting system, at the Nesmith focus of the 6-m telescope at Northern Causasus at different phases of light variations. The spectra cover 3800-6800 Å at a resolution of 1-3 Å; a few - in a wider region at a lower resolution.

An inspection of the spectra for CIT 6 in the blue region (Fig.2) shows similarity with the spectrum for PPN CRL 2688 (Alksnis et al. 1988; Začs 1997). The principal photospheric lines of hydrogen and Ca II H K, C_3 molecular absorption bands and [S II] emission features are indicated. Note that the C_3 band at 4050 Å was found also in the spectrum of the carbon and s-process rich PPN IRAS 22272+5435 (G 5Ia; variable with P $\simeq$ 133 days (see Hrivnak et al.; this volume)).

A qualitative comparison of the spectra for CIT 6 in the yellow region shows significant variability. Figure 3 shows the difference among the spectra at different light variations phases. The typical carbon (Y Tau) and SC stars (GP Ori) are provided for comparison purposes.

The analysis of our spectra (see Fig.1; solid arrows) for CIT 6 and the data published (Cohen 1980; dashed arrows) shows that the maximums of periodic (P) light variations are accompanied by the appearance of C_2 absorption; thus the normal carbon star spectrum can be observed. However, the minimums are accompanied by emission and drastic changes of all the spectrum. Therefore, spectral classification could be contradictory for CIT 6. Possibly, these phenomena could be best explained by the ejection of gas and the formation of dust around the long period variable (see, for example, Fleischer, Winters, & Sedlmayr 1999).

4. DISCUSSION AND CONCLUSIONS

The millimeter interferometry (Meixner et al. 1998) give evidence about the history of mass loss; although the clumpiness of the CO emission was detected for CIT 6, the circumstellar molecular envelope have characteristics of the spherical symmetry. However, the future of this object can be guessed through observations of the central part of its circumstellar environment and the object itself. The Michelson interferometry at 2.2 and 3.1 μm with the Keck I teleskope (Monnier et al., private communication) resolved the complex structure (elongated central source and evidences of bipolarity) of CIT 6 at the tens of milliarcsecond scale.

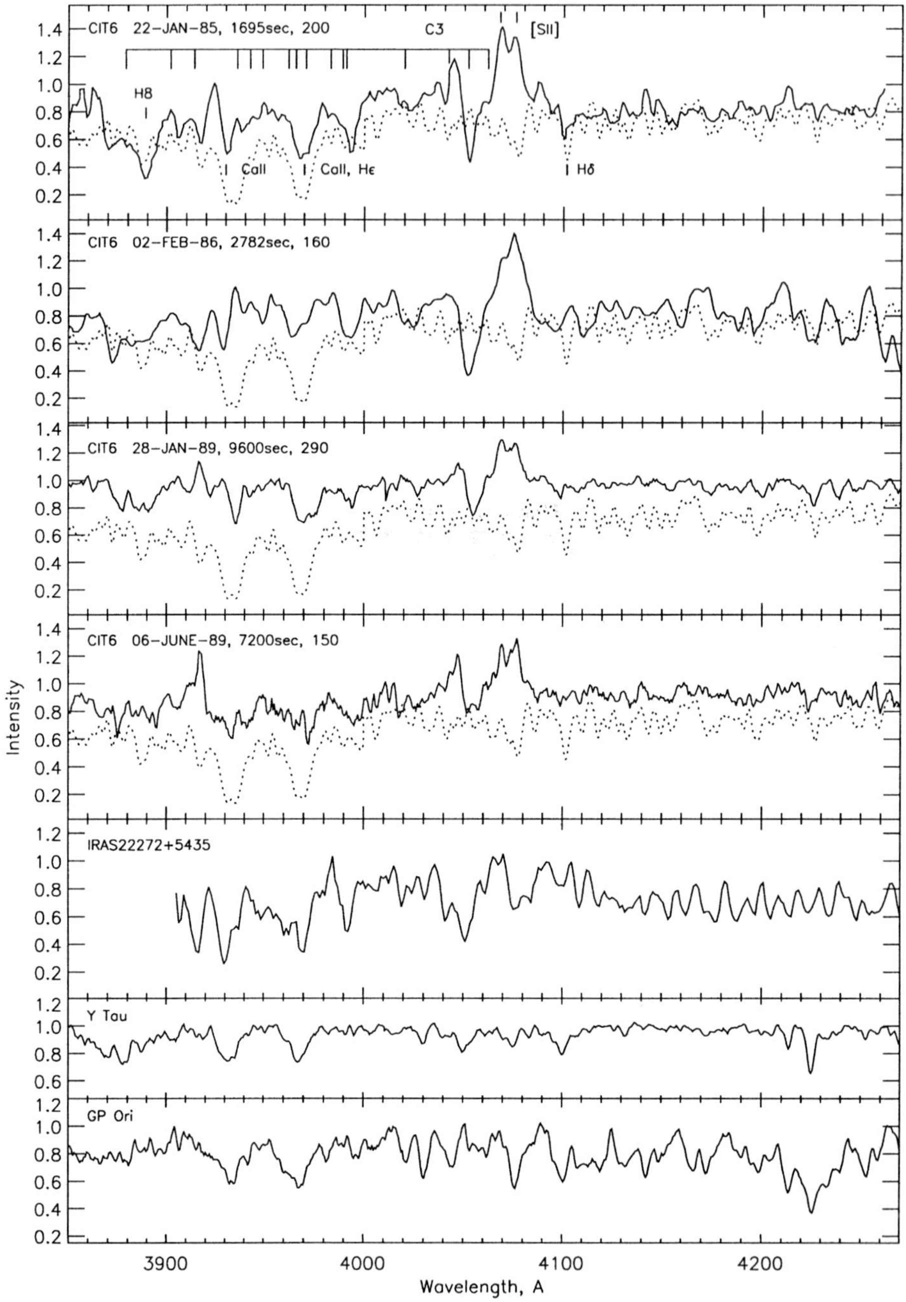

Figure 2 The spectra of CIT 6 in the blue spectral region along with those for the carbon and s-process rich PPN IRAS 22272+5435 the carbon star Y Tau (C 7,4e) and the SC star GP Ori (SC 7/8). On the left upper corner an observational and exposure time and an average photon count at continuum for the each spectrum have been indicated. Dashed line: a synthesized photospheric (atomic) spectrum for PPN central star with the typical $T_{\rm eff} = 5500$ K, $\log g = 0.5$, $[M] = -1.0$, and $[s] = +1.0$ convolved with an instrumental profile.

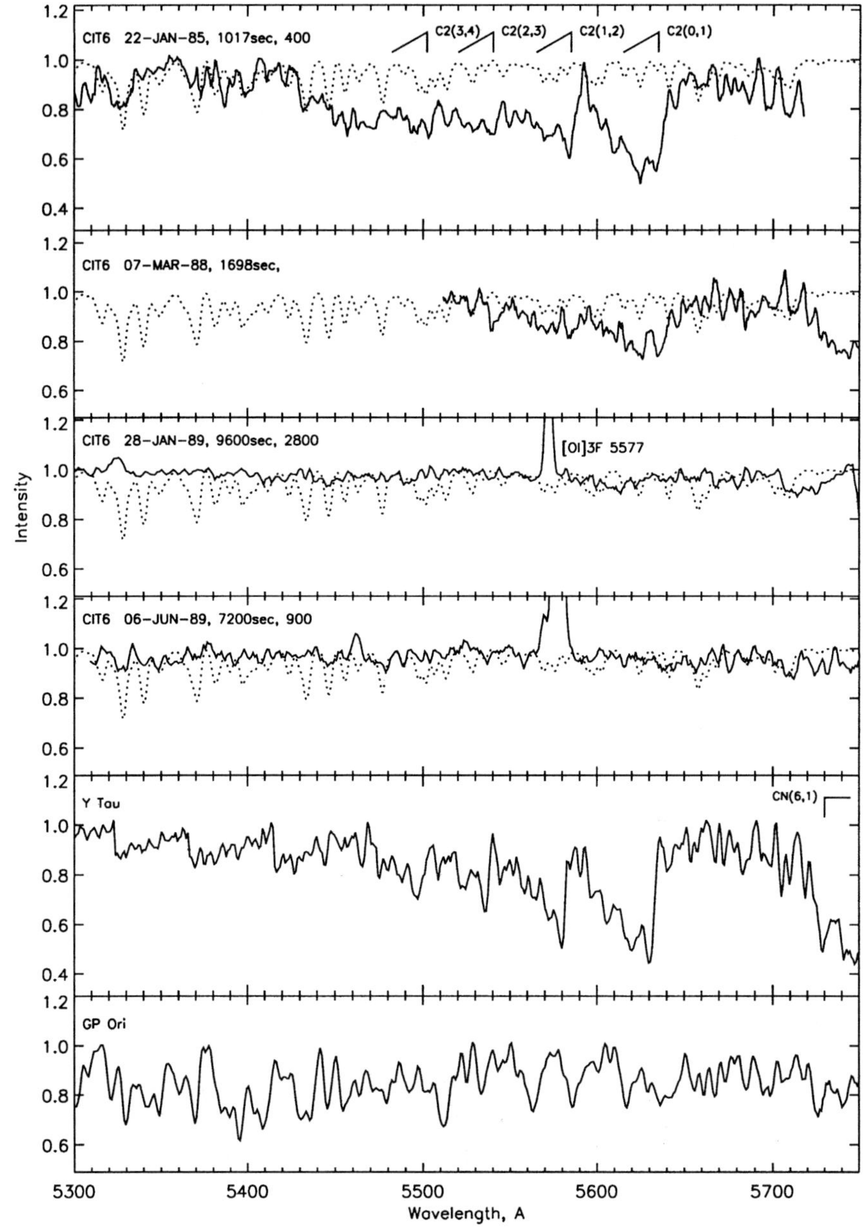

Figure 3 Same as Fig.2, but in the yellow, showing the difference among the spectra at the different light variations phases. The strong molecular (C_2) features have been observed nearly the maximum (1985) of the periodic light variations. Nearly the minimum (1989) the continuum is almost featureless, however, very strong [O I] emission at 5577 Å was detected.

First, in the infrared and red the typical light curves of carbon Miras have been observed for CIT 6, the long period variability with a period of 605-628 days and slow light variations with a possible cycle $\geq$15 years. The long period variability in the blue colour was clearly observed until 1976.

Second, the blue spectrum of CIT 6 is similar to that of PPN CRL 2688. The photospheric lines of hydrogen and Ca II HK identified encourage the stellar nature of the blue light.

Third, the long period light variations of CIT 6 are accompanied by the appearance (absence) of molecular (C_2) absorption and atomic (H_α, H_β, Na I D_{12}, [O I]) emission. Such changes are periodic and could be explained by the ejection of gas and the formation of dust around the long period variable.

Finally, CIT 6 seems to be the one of suitable candidates to investigate the formation of asymmetries detected among an important fraction of AGB and post-AGB sources, to understand better the role of (possible?) companion (see Soker, this volume), and to examine the formation of molecules and dust in the atmospheres and circumstellar envelopes of carbon-rich long period variables.

Acknowledgments

LZ thanks the LOC (chaired by R.Szczerba) of the Workshop for travel support. B.Hrivnak has been thanked for providing us the blue spectrum of IRAS 22272+5435. This research has been supported in part from the Latvian Council of Science (grant 96.0173.1).

References

Alksnis A., 1995, Baltic Astronomy 4, 79
Alksnis A., Začs L., Kopylov I.M., Somov N.N., 1988, SvA 14, L352
Cohen M., 1979, MNRAS 186, 837
Cohen M., 1980, ApJ 238, L81
Fleischer A.J., Winters J.M., Sedlmayr E., 1999, In Asymptotic Giant Branch Stars, ed. Le Bertre T., Lébre A., Waelkens C., IAU Symp. 191, ASP, 187
Kwok S., Volk K., Bidelman W.P., 1997, ApJS 112, 557
Meixner M., Campbell M.T., Welch W.J., Likkel L., 1998, ApJ 509, 414
Trammell S.R., Dinerstein H.L., Goodrich R.W., 1994, AJ 108, 984
Van Winckel H., 2000, A&A 354, 135
Začs L., 1997, Ap&SS 251, 289

SPECTROSCOPY OF POST-AGB A-SUPERGIANT HD 101584 (IRAS 11385−5517)

T. Sivarani[1], M. Parthasarathy[1], P. García-Lario[2], A. Manchado[3], S.R. Pottasch[4]
[1]*Indian Institute of Astrophysics, Bangalore 560 034, India.*
[2]*ISO Data Centre, Astrophysics Division, Space Science Department of ESA, Villafranca del Castillo, Apartado de Correos 50727,28080 Madrid, Spain.*
[3]*Instituto de Astrofisica de Canarias, E-38200 La Laguna, Tenerife, Spain.*
[4]*Kapteyn Astronomical Institute, Postbus 800, NL-9700 AV Groningen, The Netherlands.*

Abstract In the high resolution optical spectra we identified forbidden lines of [OI] and [CI], which indicate the presence of low excitation nebula. There are many emission lines due forbidden and permitted lines of neutral and singly ionised metals. The Hα, FeII 6383Å, NaI D_1, D_2 lines and the CaII IR triplet lines show P-Cygni profiles indicating a mass outflow. From the spectrum synthesis analysis of NI lines, we find the atmospheric parameters to be T_{eff}=8500K, log g=1.5, V_{turb}=13km s^{-1} and [Fe/H]=0.0. Carbon and nitrogen are found to be overabundant. From the analysis of Fe emission lines and forbidden lines, we derived T_{exi} = 6100K±200 for the emission line region.

1. INTRODUCTION

Humphreys and Ney (1974) found near-infrared excess in HD 101584 and suggested that it is a massive F-supergiant with an M-type binary companion star (Humphreys 1976). On the basis of its far-infrared colors, Parthasarathy and Pottasch (1986) suggested that it is a low mass star in the post-Asymptotic Giant Branch (post-AGB) stage of evolution.

CO molecular emission lines at millimeter wavelengths were detected by Trams et al. (1990). The complex structure of the CO emission shows large Doppler velocities of 130 km s^{-1} with respect to the central velocity of the feature indicating a very high outflow velocity. Te Lintel Hekkert et al. (1992) reported the discovery of OH 1667 MHz maser emission from the circumstellar envelope of HD 101584. The OH spectrum has a velocity range of 84 km s^{-1} and shows two unusually broad emission features. Te Lintel Hekkert et al. (1992) found from the images obtained from the Australian Telescope, that the OH masers are located along the bipolar outflow. The post-AGB nature of HD 101584 is also suggested by the space velocity of the star derived

R. Szczerba and S.K. Górny (eds.), Post-AGB Objects as a Phase of Stellar Evolution, 295–300.

from the central velocity of the CO and OH line emission. This velocity of V_{rad} = 50.3 ± 2.0 km s^{-1} does not agree with the galactic rotation curve assuming it to be a luminous massive population I F supergiant. However, Bakker et al. (1996a) based on the He I, CII and N II lines suggested a high temperature for HD 101584.

2. OBSERVATIONS AND ANALYSIS

High resolution and high signal to noise ratio spectra of HD 101584 were obtained with the European Southern Observatory (ESO) Coude Auxiliary Telescope (CAT) equipped with the Coude Echelle Spectrograph (CES) with a CCD as detector. We have also obtained 2.5Å resolution spectra of HD 101584 from 3900Å to 8600Å with the 1m telescope and UAGS spectrograph with a CCD at the Vainu Bappu Observatory (VBO), Kavalur, India. We have carried-out spectrum synthesis calculations using KURUCZ stellar models (1994). SYNSPEC code (Hubeny et al. 1985) was used for calculating the theoretical line profiles. The gf values were taken from Wiese et al. (1966), Wiese and Martin (1980), Hibbert et al.(1991), Parthasarathy et al. (1992) and Reddy et al. (1997 and references therein). For the analysis of forbidden lines we have used the IRAF software package NEBULAR under STSDAS.

2.1. RESULTS

From forbidden emission lines due to [OI] (5577Å, 6300Å and 6363Å) we obtained flux ratio, I(6300)+I(6363)/I(5577)=13.3. Using this flux ratio and from the analysis of FeI emission lines, we derived T_e=6300±1000K and N_e=1 x 10^7. The above given flux ratio of [OI] lines is not accurate because of very the weak 5577Å line and poor signal to noise ratio. So, we could not estimate the oxygen abundance.

There are very few absorption lines and most of these are affected by emission and or a shell component, therefore we derived the average radial velocity from the well defined emission lines. The average radial velocity from the emission lines is found to be 50 ± 2km s^{-1}.

The UV (IUE) low resolution spectrum of HD 101584 matches well with that of an A6Ia star (HD 97534) indicating a T_{eff} of 8400K (Lang 1992). From the analysis of several nitrogen lines around 7440Å and 8710Å we derived V_{turb}=13km s^{-1}, T_{eff} = 8500K, log g = 1.5, V_T=13km s^{-1} and [Fe/H] = 0.0.

The line at 5876Å was identified as a He I line by Bakker et al.(1996a). They also identified the lines at 5047Å and 5045Å as due to He I and N II respectively. However, we find that the 5047 and 5045 lines are in fact due to FeII. Except He I 5876Å, we have not found any other helium lines in the spectrum and nor have we found any N II or O II lines. In fact, Hibbert et al. (1991) indicate the

Table 1 Chemical composition of HD 101584.

Element	[element/H]		Lines
	T_{eff}=8500K log g=1.5	T_{eff}=9000K log g=1.5	
C	1.0±0.1	0.3±0.1	C II 6578,6582
N	0.5±0.1	0.6±0.1	N I 7423,7442,8703, 8711,8718,8728
O	0.1	0.18	O I 6158
Mg[1]	0.2	0.02	Mg II 4481.13
Ti[1]	0.0±0.4	0.4±0.4	Ti II 15 lines
Fe[1]	-0.1±0.1	0.5±0.1	Fe I 3 lines
	0.2±0.4	0.3±0.4	Fe II 6lines

[1] The equivalent widths were taken from Bakker (1996a)

presence of a CI line at 5876Å. It is likely that the line at 5876Å may be due to CI instead of He I.

If we assume that the 5876Å line is due to He I then for a solar helium abundance and log g = 1.5, T_{eff} = 9000K is found. Since we do not see any other helium lines, if 5876Å line is due to helium , it is likely that it may be formed in the stellar wind or in the chromosphere of the star. On the basis of the presence of the helium line and N II Bakker et al. (1996a) suggested that HD 101584 is a B9II star of T_{eff} 12000K. On the basis of the analysis of our spectra we have not found any evidence for such a high temperature. We have also analysed the equivalent widths of absorption lines in the spectrum of HD 101584 given by Bakker et al.(1996a). The final abundances of some of the elements are listed in Table 1. The abundances listed in Table 1 show that the star is overabundant in carbon and nitrogen. It appears that the material processed by the triple alpha C-N and O-N cycle has reached the surface.

3. DISCUSSION AND CONCLUSIONS

The optical spectrum of the post-AGB star HD 101584 is rather complex. We find several emission lines and P-Cygni profiles indicating an ongoing mass-loss and the presence of a circumstellar gaseous envelope. From the analysis of the absorption lines we find the atmospheric parameters to be T_{eff}=8500K, log g=1.5, V_t=13km s^{-1} and [Fe/H]=0.0.

Carbon and Nitrogen are found to be overabundant indicating that material processed by triple alpha C-N and O-N cycles has reached the surface. Since our blue spectra are of relatively low resolution and because of the presence of emission and shell components it is difficult to estimate reliable abundances of s-process elements. The OI line at 6156Å is blended with a weak FeI emission line. The OI triplet at 7777Å is very strong and affected by NLTE. Therefore the photospheric oxygen abundance is uncertain. A NLTE analysis of the high resolution OI 7777Å triplet may yield a more reliable oxygen abundance.

The nitrogen abundance is based on 6 lines in the 7440Å and 8710Å region. We have not used the strong nitrogen lines. Nitrogen seems to be clearly overabundant. The carbon abundance is based on two CII lines at 6578Å and 6582Å. There is a clear indication that carbon is overabundant. The abundance of Mg, Ti, and Fe are nearly solar. The Ti abundance is based on 15 lines and the Fe abundance is based on 6 lines. Many of the other atomic lines are affected by emission and shell components. In our opinion, the line at 5876Å might be due to CI (Hibbert et al. 1991) and not due to He I, as previously suggested by Bakker et al. (1996a). We have not found any other He I, N II or O II lines. Our analysis shows that the T_{eff} of HD 101584 is 8500±500K.

Bakker et al. (1996b) found small amplitude light and velocity variations and suggested that HD 101584 is a binary with an orbital period of 218 days. The radial velocity variations may be due to pulsation, macroturbulence motions or shock waves in the outer layers of the stellar atmosphere. Many post-AGB supergiants show small amplitude light and velocity variations (Hrivnak 1997). These variations may not be interpreted as due to the presence of a binary companion. Long term monitoring of the radial velocities is needed in order to understand the causes for these variations.

The spectrum and the brightness of HD 101584 appears to remain the same during last two or three decades. There is no evidence for significant variations in brightness similar to those observed in Luminous Blue Variables (LBVs). The chemical composition and all the available multiwavelength observational data collected during the last two decades by various observers indicates that HD 101584 is most likely a post-AGB star.

The presence of several P-Cygni lines with significant outflow velocities, the OH maser and CO emission profiles (Te Lintel Hekkert et al. 1992, Trams et al. 1990) and the IRAS infrared fluxes and colours (Parthasarathy and Pottasch 1986) indicates the possibility that HD 101584 is a post-AGB star with a bipolar outflow and with a dusty disk. Since HD 101584 shows a strong Hα emission line, high resolution imaging with the Hubble Space Telescope (HST) may reveal the bipolar nebula and the presence of a dusty disk similar to that observed in other post-AGB stars like IRAS 17150−3224 (Kwok et al. 1998) or IRAS 17441−2411 (Su et al. 1998). Recent CO observation (Olofsson and Neyman 1999) reveal a molecular gas envelope of 0.1 $M_{\odot}$, with very similar characteristics to those of well-known young post-AGB objects viewed edge-on. Also the C^{12}/C^{13} ratio derived by Olofsson and Nyman (1999) indicates that it may be a carbon rich post-AGB star.

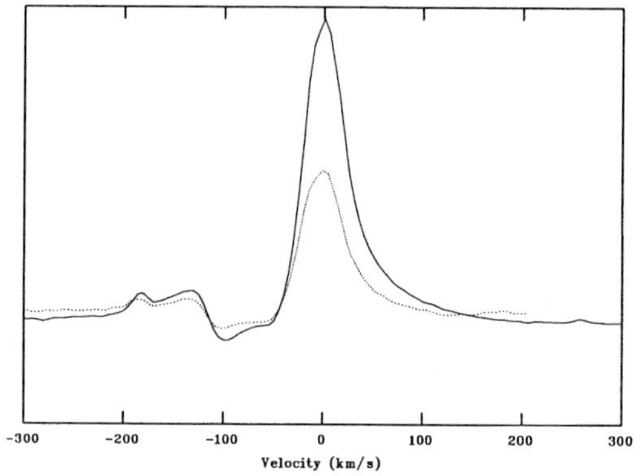

Figure 1 P-Cygni profiles of Hα and FeII(6383Å) lines showing similar velocity structures

References

Bakker E.J., 1994, A&AS 103, 189

Bakker E.J.,Lamers H.J.G.L.M., Waters L.B.F.M., Waelkens C.,Trams N,R., Van Winckel H., 1996a, A&A 307, 869

Bakker E.J.,Lamers H.J.G.L.M., Waters L.B.F.M., Waelkens C., 1996b, A&A 310, 861

Friedjung M., Muratorio G. 1987, A&A, 188, 100

García-Lario P., Parthasarathy M., de Martino D., Monier R., Manchado A., de Córdoba S.F., Pottasch S.R., 1997, A&A, 326, 1103

García-Lario P., Manchado A., Parthasarathy M., Pottasch S.R., 1994, A&A, 285, 179

Hibbert A., Biémont E., Godefroid M., Vaeck N., 1991, A&AS, 88, 505

Hrivnak B., 1997, in Planetary Nebulae, ed. H. J.Habing and H.J.G.M. Lamers, IAU symp. No. 180, 303

Hoffleit D., Saladya M., Wlasuk P., 1983, Supplement to the Bright star Catalogue, Yale University Observatory, USA

Hubeny I., Stefl S., Harmanec P., 1985, Bull. Astron. Inst. Czechosl. 36, 214

Humphreys R.M., 1976, ApJ 206, 122

Humphreys R.M., Ney E.P., 1974, ApJ 190, 339

Kurucz R.L., 1994, Solar Abundance Model Atmospheres, Kurucz CD-ROM No.19, Smithsonian Astrophysical Observatory

Kwok S., Su K.Y.L., Hrivnak B.J., 1998, ApJ, 501, L117

Lang K.R., 1992, Astrophysical data: Planets and stars, Springer Verlag

Morrison N.D., Zimba J.R., 1989, BAAS 21, 1022

Olofsson H., Nyman L.A., 1999, A&A 347,194

Osterbrock D.E., 1989, Astrophysics of Gaseous Nebulae and Active Galactic Nuclei, Oxford University Press, p.117

Parthasarathy M., García-Lario P., Pottasch S.R., 1992, A&A, 264, 159

Parthasarathy M., Pottasch S.R., 1986, A&A, 154, L16

Reddy B.E., Parthasarathy M., Gonzalez G., Bakker E.J., 1997, A&A 328, 331

Rosenzweig P., Reinoso E.G., Naranjo O., 1997, JRASC, 91, 255

Su K.Y.L., Volk K., Kwok S., Hrivnak B.J., 1998, ApJ 508, 744

Te Lintel Hekkert P., Chapman J.M., Zijlstra A.A., 1992, ApJ 390, L23

Trams N.R., Van der Veen W.E.C.J., Waelkens C., Waters L.B.F.M. 1990, A&A 233, 153

Viotti R., 1969, Astrophys. Space Sci. 5, 323

Wiese W.L., Smith M.W., Glennon B.M., Atomic Transition Probabilities, Vol. 1, NSRDS-NBS(U.S.) 4, 1966

Wiese W.L., Martin G.A., Wavelength and Transition Probabilities for Atoms and Atomic Ions, NSRDS-NBS(U.S.), 68, 1980

SPECTROSCOPY OF POST-AGB CANDIDATE HD 331319 (IRAS 19475+3119)

T. Sivarani[1], M. Parthasarathy[1], P. García-Lario[2] and A. Manchado[3]
[1]*Indian Institute of Astrophysics, Bangalore 560 034, India.*
[2]*ISO Data Centre, Astrophysics Division, Space Science Department of ESA, Villafranca del Castillo, Apartado de Correos 50727,28080 Madrid, Spain.*
[3]*Instituto de Astrofisica de Canarias, E-38200 La Laguna, Tenerife, Spain.*

Abstract HD 331319 is a F3 Ib star with a detached cold circumstellar dust shell with far-IR colours similar to planetary nebulae. From an analysis of high resolution spectra using LTE stellar atmospheric models, we have derived the atmospheric parameters and the chemical composition of HD 331319. By matching the theoretical and the observed spectrum we obtained T_{eff}=7500K, log g=0.5 and ξ_t=7 kms^{-1} and estimated [C/Fe] = 0.4±0.3, [N/Fe] =1.4±0.5, [O/H]=0.3±0.3, [Fe/H] = -0.2±0.3 We find that nitrogen is overabundant, oxygen and iron are nearly solar. The total C,N,O abundances with respect to iron indicate mixing due to the third dredge-up.

1. INTRODUCTION

HD 331319 (F3 Ib) is having IRAS fluxes similar to that of a PPN (Proto Planetary Nebula) (Parthasarathy 1993, Preite-Martinez, 1988) with a dust temperature of 136K. HD 331319 was found to be a coldest objects in the CO survey of very cold IRAS objects (Likkel et al. 1987,1991). The CO emission profile has symmetrical wings extending from -20km s^{-1} to +20km s^{-1} from the line center. There is no OH maser emission from this object (Likkel et al. 1991). So HD 331319 could be a carbon rich object or the OH column density may be less due to decreased mass loss. From the color temperature, IRAS variability index and from the CO observations Likkel et al. (1991) conclude that it is a post-AGB supergiant. The CO expansion velocities from CO 1-0 and 2-1 lines are typical of a post-AGB star. The infrared spectroscopy of HD 331319 (Hrivnak et al. 1994) shows no 21μm and 3.3μ features in emission, indicating that the circumstellar dust may not be carbon rich.

R. Szczerba and S.K. Górny (eds.), Post-AGB Objects as a Phase of Stellar Evolution, 301–304.

2. OBSERVATIONS AND ANALYSIS

The observations were made on 3rd August 1993 using the 2.5m Isaac Newton Telescope (INT) at La Palma with a 1800 grooves/mm grating (H1800V) and a GEC CCD as detector. The resolution is about 0.3 Å at 5000Å. The spectra covers wavelength regions 4350-5200Å, 5240-5550Å and 7000-7230Å. We have used Kurucz (1994) LTE stellar model atmospheres and the computer code MOOG, (Sneden 1973, which is recently modified C. Sneden 1997, private communication) for the theoretical spectrum synthesis. We have used the Atomic line database at Vienna (VALD-2, Ryabchikova et al. 1999).

3. ATMOSPHERIC PARAMETERS AND ABUNDANCES

The observed Hβ profile when compared with the profile computed using model atmospheres indicates T_{eff}=6000K and log g=0.0 (Fig. 1). However Balmer lines could be affected by the extended atmosphere and circumstellar matter.

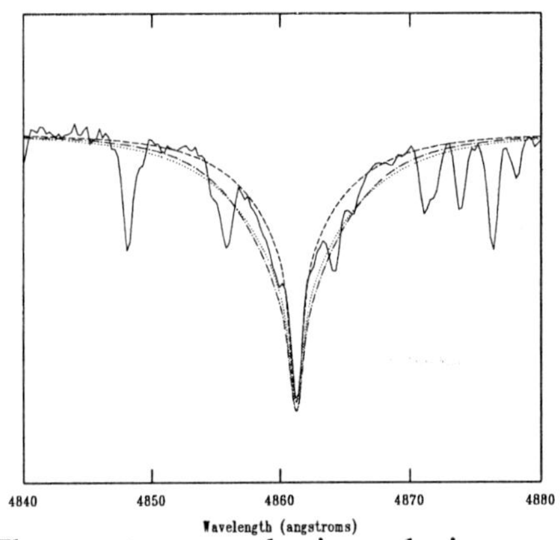

Figure 1 The observed Hβ line profile is compared with the theoretically calculated line profile for different stellar atmospheric models. The dashed line which matches well with the observations corresponds to T_{eff} =6000 K and log g=0.0, the dotted line corresponds to T_{eff} =6500K and log g=0.5 and the dashed dotted line corresponds to T_{eff} =7000K and log g=0.5

The spectrum synthesis analysis agrees for the stellar atmospheric model T_{eff}=7500K and log g=0.5. We found differences in the abundances between neutral and ionized lines. It has been noticed in several post-AGB supergiants that the atmosphere is far from ionization equilibrium. A difference of about 0.2-0.3 dex between the abundances derived from neutral and ionized lines has been reported in the literature (Van Winckel, 1997). For T_{eff} =7700K and log g=1.0 both the FeI and FeII lines give same abundances (Fig. 2). We have listed the abundance for various atmospheric models in Table.1.

4. DISCUSSIONS AND CONCLUSIONS

The IRAS colors and the variability indicates that AGB phase of evolution of HD 331319 was terminated recently . The expansion velocities obtained from the CO observations (15-20 kms^{-1}) is similar to that found in post-AGB stars.

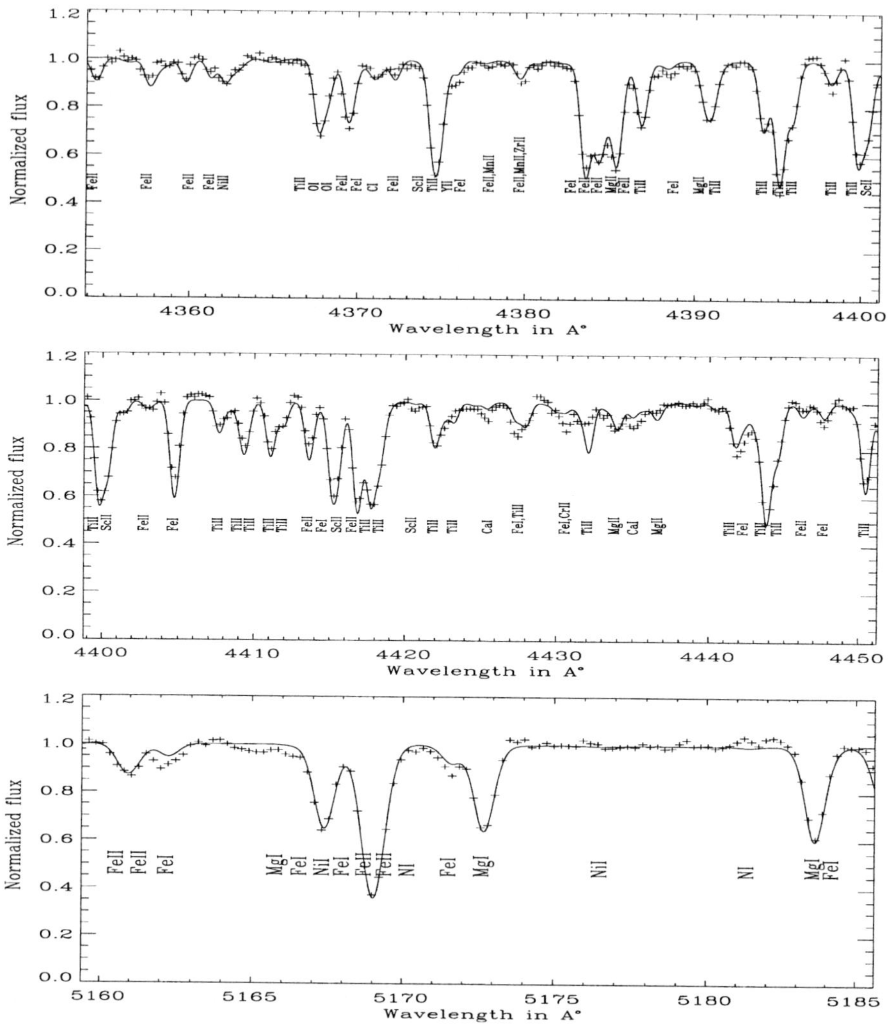

Figure 2 The observed spectrum of HD 331319 (resolution of 0.3Å) is compared with the theoretically synthesized spectra for a model with T_{eff}=7700K, log g=1.0, ξ_t=7 km s^{-1} and [M/H]=0.0

The near-IR imaging shows that the dust shell is very compact indicating that it is a very young PPN (Kastner et al. 1995). Nearly solar abundance of iron and oxygen indicates that there is depletion due to metals locked up in the dust grains. The overabundance of nitrogen and moderate overabundance of carbon indicates that HD 331319 is most likely an evolved post-AGB star and not a massive pop-I F supergiant.

Table 1 Chemical composition of HD 331319. The microturbulent velocity used is ξ_t=7kms^{-1}

Element	[element/H]		
	T_{eff}=6500K log g=0.5	T_{eff}=7500K log g=0.5	T_{eff}=7750K log g=1.0
C	-0.5	0.2	0.3
N	1.2	1.2	1.2
O	0.1	0.1	0.2
Mg	-0.55	-0.5	-0.5
Si	0.0	-0.3	-0.3
Ca	-1.2	-0.3	0.0
Sc	-1.2	-0.5	-0.3
Ti	-1.0	-0.5	-0.3
V	-1.0	-0.5	0.0
Cr	-1.0	-0.5	0.0
Mn	-1.0	-0.5	0.0
FeI	-1.4	-0.4	-0.1
FeII	-0.8	-0.1	-0.1
Co	-1.0	-0.5	0.0
Ni	-1.0	-0.5	0.0
Y	-1.5	-1.0	-0.1
Zr	-1.0	-0.3	0.0
Ba	-1.8	-0.3	0.0

References

Hrivnak B.J., Kwok S., Geballe T.R., 1994, ApJ 420, 783

Kastner J.H., Weintraub D.A., 1995, ApJ 452, 833

Kurucz R.L., 1994, Solar Abundance Model Atmospheres, Kurucz CD-ROM No.19, Smithsonian Astrophysical Observatory

Likkel L., Omont A., Morris M., Forveille T., 1987, A&A 173, L11-L14

Likkel L., Forveille T., Omont A., Morris M., 1991, A&A 246,153

Parthasarathy M., 1993, ASP Conf. ser. 45, 173

Preite-Martinez A., 1988, A&AS 76, 317

Ryabchikova T.A. Piskunov N.E., Stempels H.C., Kupka F., Weiss W.W. Proc. of the 6th International Colloquium on Atomic Spectra and Oscillator Strengths, Victoria BC, Canada, 1998, Physica Scripta T80, (1999), in press. (VALD-2)

Sneden C., 1973, PhD Dissertation, Univ. of Texas at Austin

Sneden C., 1998, Private communication

Van Winckel H., 1997, A&A 319, 561

SPECTROSCOPY OF POST-AGB CANDIDATE IRAS 10215−5916

T. Sivarani[1], M. Parthasarathy[1] P. García-Lario[2] and A. Manchado[3]

[1]*Indian Institute of Astrophysics, Bangalore 560 034, India.*

[2]*ISO Data Centre, Astrophysics Division, Space Science Department of ESA, Villafranca del Castillo, Apartado de Correos 50727,28080 Madrid, Spain.*

[3]*Instituto de Astrofisica de Canarias, E-38200 La Laguna, Tenerife, Spain.*

Abstract IRAS 10215−5916 was classified as a post-AGB star and also as a massive red supergiant. It was also found to be a spectroscopic binary. In our high resolution spectra we found molecular features from a cool star of T_{eff}=3750K. We also see the spectral signature of a relatively hotter star of T_{eff}=7250K. We found that both the components of the binary system are metal poor. From the theoretical spectrum synthesis we obtained [C/H]=-0.2±0.4, [O/H]=-1.0±0.4 and [Fe/H]=-2.0±0.5 for the hot component of the binary system. The enhanced carbon abundance and metal poor characteristics show that the hot component of the binary system could be a post-AGB star.

1. INTRODUCTION

The evolutionary status of IRAS 10215−5916 is not clear whether it is in post-AGB or post-RGB phase. García-Lario et al. (1994) discovered nebular emission from this IR object. From their high resolution spectra they found an expansion velocity of 17 kms^{-1}, which is similar to that observed in PPNs. The imaging in near-IR shows IRAS 10215−5916 is extended and asymmetric. The nebular lines and the extended morphology was explained by incorporating an unseen hot binary which ionizes the nebula and forms a disk. IRAS 10215−5916 was thought to be G type star in the transition between AGB and PN stage (Parthasarathy 1989). The expansion velocity of 40 kms^{-1} was measured from the CO(J=1-0) rotational line (Josselin et al. 1998), which is high for a low-mass AGB star. Molster et al.(1999) found that IRAS 10215−5916 is binary system with a cool component of T_{eff} 3750K and a hot component of T_{eff} 7250K from the flux distribution in the Infrared region (the ISO data). The hot component was found to be 1.8 times more luminous than the cooler one.

R. Szczerba and S.K. Górny (eds.), Post-AGB Objects as a Phase of Stellar Evolution, 305–308.

2. OBSERVATIONS AND ANALYSIS

The high resolution spectra were obtained from 1.4m CAT-CES, with a resolving power of R=50000. The spectra covers wavelength regions 5090-5130Å, 6135-6180Å, 6340-6380Å, 6535-6590Å, 7995-8060Å and 8396-8456Å. The spectra were wavelength calibrated and normalised using IRAF.

We have assumed the values given by Molster et al. (1999) for the temperatures of two components of the binary system and the relative luminosity and calculated the theoretical spectrum. We used the Kurucz stellar atmospheric models (Kurucz 1994). For calculating the theoretical spectrum. We used the SYNSPEC code by Hubeny (1985) and SPECTRUM code by R. O. Gray (1998). We used E(B-V)=1.2 given by Molster et al. (1999).

3. RESULTS

IRAS 10215−5916 shows a composite spectrum of a M type and F type star. In the red region there are strong lines of CaII IR triplet from the cool star (Fig. 3). In the blue region around 5100Å we see absorption lines due to molecules of CN, C_2 and CH from the cool star (Fig. 2). There are many absorption features due to the diffuse interstellar bands. There are nebular lines of [N II]. The Hα and FeII 6383Å shows a double peak emission (Fig.1). The velocities of the double peak emission is of the order of 50kms^{-1}, which is close to the value obtained from the CO observations. The CO line could be from rotating disk around the star or from a bipolar outflow. We found the E(B-V) values derived from the DIBs are different for lines at different wavelengths. We also found a slight correlation between the E(B-V) and the wavelengths of the lines, which could be due to the circumstellar origin of DIBs close to the hotter component of the binary system.

The similarity in the line profile of Hα and FeII 6383Å is also seen in HD 101584 (Sivarani et al. 1999). We found that even the DIBs in both the stars have same equivalent widths. So, the DIBs could be of circumstellar origin. DIBs are generally formed in carbon rich environment. HD 101584 is also found to be a carbon rich post-AGB star with bipolar OH and CO emission (Sivarani et al 1999).

The chemical composition of the hot component of IRAS 10215−5916, is listed in Table 1 and the observed and synthetic spectra are shown in Fig.2. From the abundances of O, Mg, Si, Y, Sr, Zr and Ba indicate that the metallicity, [Fe/H] is -1.0. Our preliminary analysis show that the hot star is more depleted of metals than the cool component. This could be due to dust formation around the hot component which has gone through the mass-loss phase on the AGB stage. The cool star was also found to be metal poor. Due to the uncertainty in the model, spectrum synthesis code and line lists for cool star, we did not try estimating the abundance of the cool component.

Element	C	O	Mg	Si	Fe-peak	Y	Sr	Zr	Ba
[element/H]	-0.2	-1.0	-1.0	-1.0	-2.0	-1.0	-1.0	-1.0	-1.0

Table 1 Chemical composition of hot component of the binary derived for the atmospheric parameters, T_{eff}=7250K,ξ_t=4km/s and log g=0.5 for the hot component and T_{eff}= 3750K,log g=0.0 and ξ_t=4km/s for the cool component

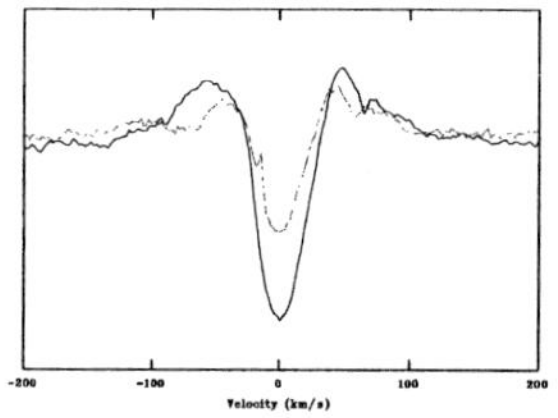

Figure 1 The Hα and FeII 6383Å lines with double peak emission

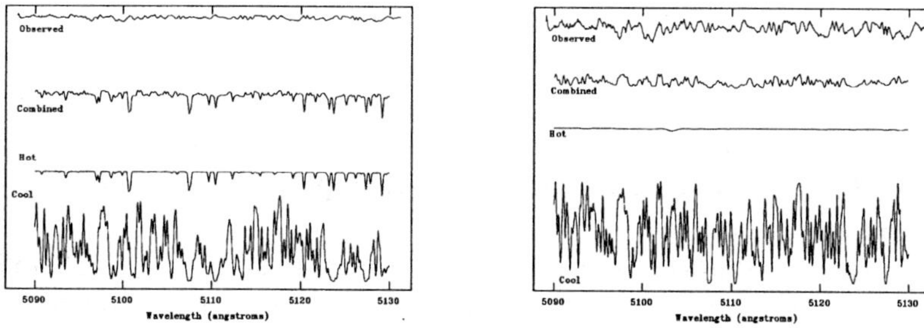

Figure 2 The theoretical spectra of a cool star of T_{eff} =3750K, log g=0.0 and a relatively hot star of T_{eff}=7250K, log g=0.5 is combined for L_{hot}/L_{cool}=1.8 and the combined spectrum is compared with observed spectra. The left panel is for [Fe/H]=0.0 and the right panel is for [Fe/H]=-1.0. The spectra in the right panel matches better with observations

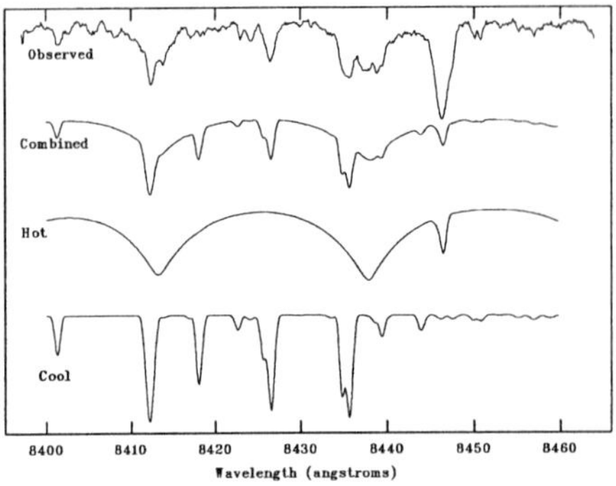

Figure 3 The observed spectrum shows paschen lines due to the hot component and Ca II triplet lines due to a cool component

The abundances of the hot star shows enhancement of carbon with respect to oxygen, indicating that IRAS 10215−5916 is most likely a carbon-rich post-AGB star. The presence of the DIBs also support carbon enrichment.

4. CONCLUSIONS

The double peak emission of Hα and FeII 6383Å could be from a bipolar emission which is collimated due to the circumbinary disk. The binary system is found to be metal poor and this cannot be attributed to depletion due to dust, because we find oxygen also underabundant. The enhancement of carbon with respect to iron shows that IRAS 10215-5916 is carbon rich post-AGB star.

References

García-Lario P., Manchado A., Parthasarathy M., Pottasch S.R., 19 94, A&A, 285, 179

Gray R.O., 1998, http://www1.appstate.edu/ dept/physics/ spectrum/spectrum.html

Hubeny I., Stefl S., Harmanec P., 1985, Bull. Astron. Inst. Czechosl. 36, 214

Josselin E., Loup C., Omont A., Barnbaum C., Nyman L., Sevre F., 1998, A&AS 129, 45

Kurucz R.L., 1994, Solar Abundance Model Atmospheres, Kurucz CD-ROM No.19, Smithsonian Astrophysical Observatory

Molster F.J., Waters L/B.F.M., Trams N.R., van Winckel H., Decin L., van Loon J. Th., Jäger C., Henning Th., Käufl H.-U., De Koter A., Bouwman J., 1999, A&A 350, 163

Parthasarathy M. 1989, IAU coll 106, "Evolution of peculiar redgiant stars", eds. H. R. Johnson and B. Zuckerman, Cambridge Univ. Press, p 384

Sivarani T., Parthasarathy M., García-Lario P., Manchado A., Pottasch S.R. 1999, A&AS 137, 505

SPECTROSCOPY OF HD 168625 (IRAS 18184−1623)

P. García-Lario[1], T. Sivarani[2], M. Parthasarathy[2] and A. Manchado[3]
[1]*ISO Data Centre, Astrophysics Division, Space Science Department of ESA, Villafra:.. a del Castillo, Apartado de Correos 50727,28080 Madrid, Spain.*
[2]*Indian Institute of Astrophysics, Bangalore 560 034, India.*
[3]*Instituto de Astrofisica de Canarias, E-38200 La Laguna, Tenerife, Spain.*

Abstract HD 168625 (B9-A0 Ia) was classified as a post-AGB star and also as a luminous blue variable. We have analysed the high resolution optical spectra of HD 168625 and derived the atmospheric parameters and the chemical composition. We obtained T_{eff}=15000K, log g=2.5, ξ_t=6km/s and [M/H]=0.0, and estimated the chemical composition. We find [C/H] = -0.3±0.1, [N/H] = 0.5±0.1, [O/H] = 0.0±0.1, [Mg/H] = -0.3±0.2, [S/H] = -0.4±0.1, [Si/H] = -0.5±0.1 and [Ne/H] = 0.8±0.2. HD 168625 has overabundance of nitrogen and underabundance of carbon. The oxygen abundance is solar. This indicates mixing of CNO cycle burning products to the surface.

1. INTRODUCTION

HD 168625 is classified as a B8-type hypergiant (Morgan et al 1955). Much later it is been classified as B5.6 by Chentsov and Lund (1989) (See also Hutsemekers et al 1994 and references therein). The photometric history over past several years(1973-1999) does not indicate any major spectrophotometric variations (Sterken et al 1999). But there is quasi periodic low amplitude variations seen over years (Sterken 1977, Van Genderen et al 1992). Sterken et al (1999) classify HD 168625 as as a variable hypergiant or marginally dormant LBV. On the basis of spatially resolved nebula around HD 168625 Hutsemekers et al (1994) classified it as an LBV. From the velocities and chemical composition derived from spatially resolved spectroscopy Nota et al (1996), confirms that the nebula is associated with the central star and it is nitrogen rich. HD 168625 was also thought to be a post-AGB candidate. It has IRAS far-IR colors similar to planetary nebulae and post-AGB stars (Parthasarathy 1989).

R. Szczerba and S.K. Górny (eds.), Post-AGB Objects as a Phase of Stellar Evolution, 309–312.

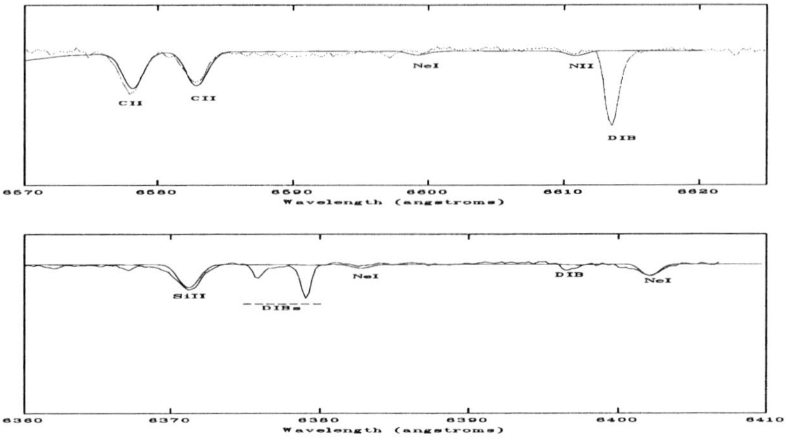

Figure 1 The observed spectrum is compared with the theoretical spectrum for T_{eff}=15000K, log g=2.5 and [Fe/H]=0.0

2. OBSERVATIONS AND ANALYSIS

High resolution spectra were obtained using the 2.5m Isaac Newton Telescope (INT) located at La Palma, which is equipped with a Intermediate Dispersion Spectrograph (IDS). The observations were made on 24th May 1991. The spectra cover wavelength regions 5325-5450Å, 6125-6225Å, 6300-6400Å, 6525-6625Å, 7060-7175Å, 7400-7500Å and 8680-8780Å with a resolution of 0.3Å at 6000Å. The theoretical spectrum is calculated using the SYNSPEC code by Hubeny (1986). We used the Kurucz (1994) model atmospheres. The linelist for the spectrum synthesis is also taken from Kurucz (1994).

3. RESULTS

The high resolution optical spectrum of HD 168625 show absorption lines of CII, N II, O II, Mg II, S II, Si II Ne II and He I. From the S II lines we estimated T_{eff}=15000K, log g=2.5 and ξ_t=6km/s. The long slit spectra shows an extended emission in Hα and also in [N II] 6548, 6583Å lines.The spectra also shows lot of strong diffuse interstellar bands (DIBs).

From the spectrum synthesis (Fig. 1), we found overabundance in nitrogen and underabundance of carbon. The oxygen abundance is solar. We see overabundance of neon which could be due to NLTE effect. The Mg, S and Si abundances are less than the solar value. The estimated chemical composition is listed in Table.1 We have obtained the IUE NEWSIPS data from Vilspa. From the UV spectrum we derived E(B-V)=1.4. From the Hipparcos parallax

Table 1 Chemical composition of HD 168625.

[element/H]	[C/H]	[N/H]	[O/H]	[Mg/H]	[S/H]	[Si/H]	[Ne/H]
T_{eff}=15000K	-0.3	0.5	0.0	-0.3	-0.4	-0.5	0.8
log g=2.5,ξ_t=6km/s							
no.lines	3	3	3	1	3	1	5

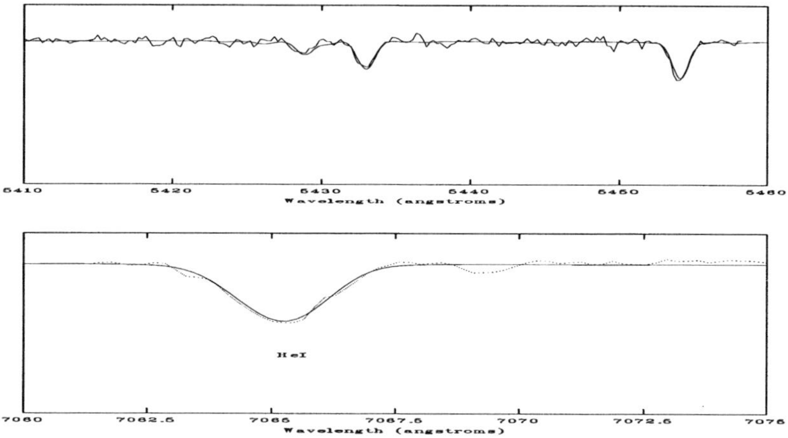

Figure 2 The top panel is shows the synthesis of S II lines

(2.49 mas) for HD 168625, and (B-V)=1.227, V=8.96 we derived, a distance of 401.6 pc and estimated M_v=-4.1, $(B-V)_o$=-0.173. This places the star to be B5II.

4. DISCUSSIONS

HD 168625 is a variable star and the equivalent widths of the lines are also seem to vary. So the derived abundance assuming a static atmosphere should be taken with a caution. The enhancement in the nitrogen abundance seen in HD168625 is consistent with the nitrogen rich nebula. This also gives an evidence that the nebula has formed from the material ejected from the star after it has gone through the CNO cycle and mixing.

The distances derived from the Hipparcos parallax seem to be inconsistent with the previous results. The distance 0.4kpc from the Hipparcos measurement is very less than the distance of 2.2kpc, assuming that HD 168625 belongs to the Omega nebula. The reddening derived from the nebular lines is found to be E(B-V)=0.75, which is less than that derived from the 2200Å feature in the UV(IUE) spectrum of the star.

The excess reddening of the star light compared to the nebula is probably due to circumstellar reddening due to a dusty disk.

The [N II] lines indicate the presence of a low excitation nebula. This is typical of planetary nebulae. The enrichment of nitrogen in the nebula as well as in the star shows that it is an evolved star.

There are lot of unidentified IR lines and also DIBs seen in the spectra. Generally the origin of these lines are thought to be the Poly Aromatic Hydrocarbons (PAHs). These are strong in the cases where the C/O ratio is very high. But we see enriched nitrogen. So more detailed study is needed to know about the origin of these features.

The evolutionary status of HD168625 is not clear. It may be a LBV or a post-AGB star with a core mass of the order 1.0 $M_\odot$. The overabundance of nitrogen is similar to that observed in type I planetary nebulae. Accurate distance is needed to estimate the absolute luminosity of HD 168625. The LBV star HR Caralso shows over abundance of nitrogen. In order to understand the evolutionary status of HD 168625, we need a detailed spectroscopic and photometric monitoring.

References

Chentsov E.L., Lund 1989, Astrofizika 31, 5

Hutsemekers D., Van Drom E., Gosset E., Melnick J., 1994, A&A 290, 906

Morgan W.W., Code A.D., Whitford A.E., 1955, ApJS 2, 41

Nota A., Pasquali A., Clampin M., Pollacco D., Scuderi S., Livio M., 1996, ApJ 473, 946

Parthsarathy M., 1989 IAU coll. 106 "Evolution of Peculiar redgiant stars", eds. H.R. Johnson and B. Zuckerman, Cambridge Univ. Press, p. 384

Sterken C., 1977, A&A 57, 361

Sterken C., Arentoft T., Duerbeck H.W., Brogt E., 1999, A&A 349, 532

Van Genderen A.M. et al 1992, A&A, 264, 88

THE EVOLUTIONARY STATUS OF HD 179821

Thévenin F.
Observatoire de la Côte d'Azur, France
thevenin@obs-nice.fr

Jasniewicz G.
Université de Montpellier, France

Parthasarathy M.
Indian Institute of Astrophysics, Bangalore, India

Abstract HD 179821 (IRAS 19114+0002) is classified as a moderately metal-poor ([Fe/H] = - 0.5) post-AGB star of Teff = 5660K and also as a very massive supergiant of Teff = 6750K. Recent detailed studies of this star are not able to establish whether HD 179821 is a low mass star in post-AGB stage of evolution or a massive (30 Solar Mass) yellow supergiant caught in between RGB and Wolf Rayet phase (Reddy and Hrivnak (1999), Jura and Werner (1999), Thévenin et.al (2000)). Our LTE analysis of the spectrum of HD 179821 and the recent radio study of circumstellar envelope by Josselin and Lébre (2000) favours the hypothesis that it is a low mass star in post-AGB stage of evolution.

1. INTRODUCTION

The chemical composition of bright post-AGB stars has been recently reviewed by Bond (1991) and Luck (1993). The observed abundance patterns of post-AGB supergiants indicates the following groups:

I) Post-AGB stars with nearly normal abundance of Fe, slightly overabundant in CNO with respect to normal Population I supergiants. These stars have oxygen-rich circumstellar dust shells. The high latitude F-supergiant HD 161796 (Parthasarathy and Pottasch 1986) is a good example of stars belonging to this group (Luck 1993).

II) Post -AGB stars showing extreme underabundance of Fe and other refractory elements, and nearly normal CNO, S and Zn. Stars belonging to this

R. Szczerba and S.K. Górny (eds.), Post-AGB Objects as a Phase of Stellar Evolution, 313–316.

group are HR 4049, HD 52961, HD 44179 etc (Waelkens et al. 1992). The underabundance of Fe and other refractory elements is due to the dust grain condensation. The abundance pattern is correlated with the condensation temperatures (Bond 1991; Van Winckel et al. 1992; Parthasarathy et al. 1992) indicating that the depleted refractory elements are locked up in circumstellar dust grains. According to Van Winckel et al. (1995), the gas-dust separation needed to account for the depletion pattern is likely to take place in a circumbinary dusty disc.

III) Post-AGB stars belonging to this group are carbon-rich and overabundant in s-process elements indicating that they have gone through the third dredge-up during their AGB stage of evolution. IRAS 05341+0852 is a good example of a post-AGB star belonging to this group (Reddy et al. 1997). The other two post-AGB F supergiants showing overabundance of carbon and s-process elements are HD 56126 (Parthasarathy et al. 1992; Klochkova 1995) and HD 187885 (Van Winckel et al. 1996).

IV) The high latitude low gravity hot post-AGB supergiants were found to show underabundance of Fe and significant underabundance of carbon indicating that these are old disc or halo stars in post-AGB stage of evolution (McCausland et al. 1992). The abundance pattern of these stars indicates that they left the AGB before undergoing the third dredge-up.

Detailed chemical composition analysis is available for only a few post-AGB stars. In order to further understand the abundance pattern of post-AGB F and G supergiants, we carried out chemical composition analysis of bright post-AGB candidate HD 179821 (Pottasch and Parthasarathy 1988).

1.1. HD 179821

Having nearly the coldest IRAS colours (12 μm=31 Jy, 25 μm=650 Jy, 60 μm=520 Jy and 100 μm=170 Jy) among evolved stars, HD 179821 has gained considerable attention in recent literature as it displays many unusual properties. In the CO survey of cold IRAS objects, Zuckerman and Dyck (1986) have detected CO molecular emission in HD 179821. It has large CO radial velocity (V_r=100 km s^{-1}) with respect to local standard of rest and also it has large outflow velocity (V_{exp}= 32 km s^{-1}). Based on its kinematical properties, large IRAS fluxes and outflow velocity, they suggested that the star might once have been a runaway O-type star. Pottasch and Parthasarathy (1988) were the first to discuss the post-AGB nature of this star based on its IRAS fluxes. The spectral energy distribution (SED) of this source closely resembles that of the high latitude post-AGB candidate HD 161796 (Parthasarathy & Pottasch 1986) and its IRAS colours are similar to planetary nebulae. The large far infrared excess of this star has been attributed to large scale mass-loss in the AGB phase of evolution. The studies of SED from 0.4 μm to 100 μm (Pottasch & Parthasarathy

1988; Hrivnak et al. 1989; van der Veen et al. 1989, 1994) points out that the star has a detached cold ($\approx$ 130 K) dust shell and no hot dust component indicating recent cessation of mass-loss. However very recently Hawkins et al. (1995) discovered an extended nebula around HD 179821 at near-IR wavelengths and they suggested that it is an extremely massive star which has recently undergone an enormous mass-loss (Jura and Werner 1999). They estimate mass of the circumstellar envelope to be around 8$M_\odot$. Polarimetric and coronagraphic near-IR images (Kastner and Weintraub 1995) revealed the presence of extended dust reflection nebula around HD 179821. Under the assumption that the star lies at its kinematical distance (6 kpc), Kastner and Weintraub (1995) suggested that its circumstellar envelope has a dynamical lifetime of 5000 yr, and a mass of $\approx$ 5 $M_\odot$ of gas and dust. Thus HD 179821 is either classified as a post-AGB star descended from a low-mass or intermediate-mass progenitor (Reddy and Hrivnak 1999) or as a very massive red supergiant (see : Jura and Werner 1999). We used spectra from the 1.52m telescope at Observatoire de Haute Provence

Adopted atmospheric parameters resulting from synthetic spectrum are : $T_{\rm eff} = 5660\,$K, $\log g = -1.00$, [Fe/H] = -0.5, $\xi_{turb} = 11.0$km.s^{-1}. Abundance fits were done using spectral regions where most of the lines are weak.

The chemical composition of HD 179821 clearly differs from the average chemical composition of yellow supergiants (Barbuy et al. 1996) and dwarfs stars (Wheeler et al. 1989). Yellow supergiants have elemental abundances [Fe/H]=0.0, [C/Fe]=$-$0.2, [N/Fe]=0.1, [O/Fe]=$-$0.2 (Barbuy et al. 1996). However HD 179821 is found to be moderately metal-deficient indicating that it is not a massive population I supergiant. The large radial velocity coupled with low-metallicity implies that HD 179821 is not a massive supergiant.

Recently Kastner and Weintraub (1995) and Hawkins et al. (1995) suggested that HD 179821 may be a massive yellow supergiant caught in between RGB and Wolf Rayet phase. They estimated mass of the circumstellar dust shell to be 5-8$M_\odot$. Another important point to be noted is the non-detection of HCN toward HD 179821 attributed to the recent cessation of mass-loss (Hrivnak et al. 1989) which is also suggested by its low IRAS variability index VAR=0. Presence of strong HCN is the characteristic of heavily mass losing supergiants (Bujarrabal et al. 1992; Nercessian et al. 1989).

Note that our derived abundances of HD 179821 are remarkably resembling to those deduced for the halo-PN DDDM 1 (see Clegg et al. 1987). Iron is moderately underabundant with carbon and s process elements underabondant. Same situation is found for the object Barnard 29(see Colon et al. 1994). As evidence these objects, HD 179821, DDDM-1 and Barnard 29 did not undergo a 3rd dredge-up. By these abundance aspects these objects have similar abundances as the HII regions of the Small Magellanic Cloud and the same iron abundances as supergiant stars also belonging to the SMC. Did they have the same material origin of a C-poor halo gas similar to the present SMC ?

Our abundances of HD 179821 shows that it has left the AGB before the third dredge-up. It is having chemical composition similar to that of high latitude hot post-AGB stars. Therefore HD 179821 is a cooler analogue of high latitude hot post-AGB star (group IV mentionned above). As HD 179821 evolves to higher temperatures it may become a hot post-AGB star with underabundance of carbon similar to that of high latitude hot post-AGB stars and Barnard 29.

Recently Josselin and Lébre (2000) carried out an analysis of high-sensitivity radio observations and favour the hopothesis that HD 179821 is a low mass post-AGB star.

References

Barbuy B., De Medeiros J.R., Maeder A., 1996, A&A 305, 911

Bond H.E., 1991, in "The evolution of stars: Photospheric abundance connection", IAU Symp. 145, ed. G. Michaud and A. Tutokov, p.341

Bujarrabal V., Alcolea J., Planesas P., 1992, A&A 257, 701

Clegg R.E.S., Peimbert M., Torres-Peimbert S., 1987, MNRAS 224, 761

Colon E.S., Dufton P.L., Keenan F.P., 1994, AA 290, 897

Josselin E., Lèbre A., 2000, AA in press

Jura M., Werner M.W, 1999, ApJ 525, L113

Hrivnak B.J., Kwok S., Volk K.M., 1989, ApJ 346, 265

Kastner J.H., Weintraub D.A., 1995, ApJ 452, 833

Klochkova V.G., 1995, MNRAS 272, 710

Luck R.E., 1993, in "Luminous High-Latitude Stars", ASP Conference

Mc Causland R.J.H., Conlon E.S., Dufton P.L., Keenan F.P., 1992, ApJ 394, 298

Nercessian E., Omont A., Benayoun J.J., Guilloteau S., 1989, A&A 210, 225

Parthasarathy M., Pottasch S.R, 1986, A&A 154, L16

Parthasarathy M., Garcia-Lario P., Pottasch S.R, 1992, A&A 264, 159

Pottasch S.R., Parthasarathy M., 1988, A&A 192, 182

Reddy B.E., Hrivank B.J, 1999, AJ 117, 1834

Reddy B.E., Parthasarathy M., Gonzalez G., Bakker E.J., 1997, A&A 328, 331

Thévenin F., Parthasarathy M., Jasniewicz G, 2000, A & A 359, 138

van der Veen W.E.C.J., Habing H.J., Geballe T.R, 1989, A&A 226, 108

van der Veen W.E.C.J., Waters L.B.F.M., Trams N.R., Matthews H.E., 1994, A&A 285, 551

Van Winckel H., Mathis J.S., Waelkens C., 1992, Nature 356, 500

Van Winckel H., Waelkens C., Waters L.B.F.M., 1995, A&A 293, L25

Van Winckel H., Waelkens C., Waters L.B.F.M., 1996, A&A 306, L37

Waelkens C., Van Winckel H., Bogaert E., Trams N.R., 1992, A&A 264, 159

Wheeler J.C., Sneden C., Truran J.W., 1989, ARAA 27, 279

Zuckerman B., Dyck H.M., 1986, ApJ 311, 345

IRAS 01005+7910, A HIGH GALACTIC LATITUDE POST-AGB STAR?

Jingyao Hu
Beijing Astronomical Observatory, Chinese Academy of Sciences
National Astronomical Observatory, Chinese Academy of Sciences
20A Datun Road, Beijing 100012, China
hjy@class1.bao.ac.cn

Abstract IRAS 01005+7910 is a cold IRAS source. In this paper we present the results of optical identification, photometric and spectroscopic observations. Its spectral type is classified as B2Ie, and V magnitude is 10.85. Hα shows P Cygni profile in the spectrum. According to its location in the Galaxy (b=16.6), we considered that it is a post-AGB star or a proto-planetary nebula.

1. INTRODUCTION

The IRAS mission give us a chance to find the objects in post-AGB phase, a number of high latitude F-, A-, and B-type supergiants found in high galactic latitude are in fact post-AGB stars. In the systematic search for proto-planetary nebula based on IRAS data (Hu et al. 1993, 1994), we found one object in the sample (IRAS 01005+7910), located at galactic latitude of b=16.6. Recently, Hrivnak et al. (2000) found that this source shows carbon-rich features in infrared. In this paper we present the results of optical identification, photometric and spectroscopic observations and discussions on its nature.

2. OPTICAL IDENTIFICATION AND OBSERVATIONS

2.1. OPTICAL IDENTIFICATION

The IRAS source was identified optically by using Digitized Sky Survey (DSS). A finding chart, showing the optically bright object within IRAS position error box, is shown in fig.1. In table 1 we present the optical position, derived from USNO A2.0 and IRAS position. As we will show later on, P Cygni profile of the Hα line from this optical source and large E(B-V)=0.39 (b=16.6) confirms our identification.

R. Szczerba and S.K. Górny (eds.), Post-AGB Objects as a Phase of Stellar Evolution, 317–320.

Figure 1 Finding chart (5' × 5') for IRAS 01005+7910, the north is up.

2.2. PHOTOMETRIC OBSERVATION

The optical photometric observation was carried out in October 1990, using a CCD detector on the 1.0 meter Kaptyen telescope at La Palma. The Johnson BVRI system and standard stars listed by Landort (1983) were used for brightness calibrations. The result of BVRI photometry and IRAS flux are listed in table 2.

2.3. SPECTROSCOPIC OBSERVATIONS

The low resolution optical spectra were taken in November 1990 using 2.5m Issac Newton Telescope (INT) at La Palma and the FOS with RCA CCD detector at a resolution of 10.7A/pixel. About ten years later we took another low

Table 1 Optical and IRAS positions of IRAS 01005+7910

	R.A.(2000.0)	Dec.(2000.0)
Optical	01 04 45.48	+79 26 46.3
IRAS	01 04 44.5	+79 26 43

Table 2 BVRI magnitude and IRAS flux of IRAS 01005+7910

magnitude	IRAS flux
B=11.08	3.90Jy (12 μm)
V=10.85	24.23Jy (25 μm)
R=10.68	10.07Jy (60 μm)
I=10.50	2.42Jy (100 μm)

resolution spectrum in January 2000 using 2.16m telescope at Xinglong Station, Beijing Astronomical Observatory and OMR spectrograph with TEK1024 CCD at similar resolution. We cannot find any difference between these two spectra. A follow-up intermediate resolution spectra were obtained immediately using IDS attached on the INT and same OMR spectroscopy but with higher resolution gratting. In both intermediate spectra P Cygni profile is clearly shown in Hα line and we could not find any evidence of variability. Fig.2 shows the intermediate spectra covered from 3600Å to 6900Å taken in 2000, the inserted panel in Fig.2c is the enlarged portion near Hα emission.

3. DISCUSSIONS

3.1. THE SPECTRAL TYPE OF OBJECT

The two-dimensional spectral type was classified based on the blue part of intermediate resolution spectrum taken in 2000 and criteria are followed from Jaschek and Jaschek (1987). In the spectrum HeI lines are strong but no HeII lines present, so it should be later than O-type. According to the existance of O II, Si III and Si IV lines, It should be earlier than or equal to B2 star. The ratio of 4552 Si III/4089 Si IV is small, so we can confirm that it is a B2 type star. The luminosity of this object determined by the ratio of 3995 N II/4009 He I, and 4121 He I/4144 He I, confirms that it is a supergiant. The narrow profile of Balmer lines are consist with our luminosity classification.

3.2. NATURE OF THE OBJECT

If this object is a normal B2I star, we have $(B\text{-}V)_0$ =-0.16 and M_V=-6.4. The observed color is B-V=0.23. So we can derive the redening as E(B-V)=0.39 and interstellar absorption A_V=1.20. The distance module m-M=16.05 and distance d=16.2 kpc. Due to the galactic latitude b=16.6, the object should be located at about 4.6 kpc above the galactic plan. It does not fit with normal B-type supergiant and we confirm that this object is a post AGB or proto-planetary nebula. Recently, IRAS 01005+7910 has been observed by ISO (Hrivnak et al., 2000) and found that it is a carbon-rich proto-planetary nebula. The galactic latitude of this object is high but not so high as well-known high galactic latitude post-AGB stars, and we do not know its luminosity, so we can not derive its distance. The observing data to-date are still not enough for determining its abundance, so we can not distinguish its population.

3.3. MASS-LOSS FROM IRAS 01005+7910

The Hα line of IRAS 01005+7910 shows P Cygni profile with velocity of 450 km/sec. This means that the mass-loss process in this object is still on-going. This is consistent with result given by Hrivnak et al.(2000).

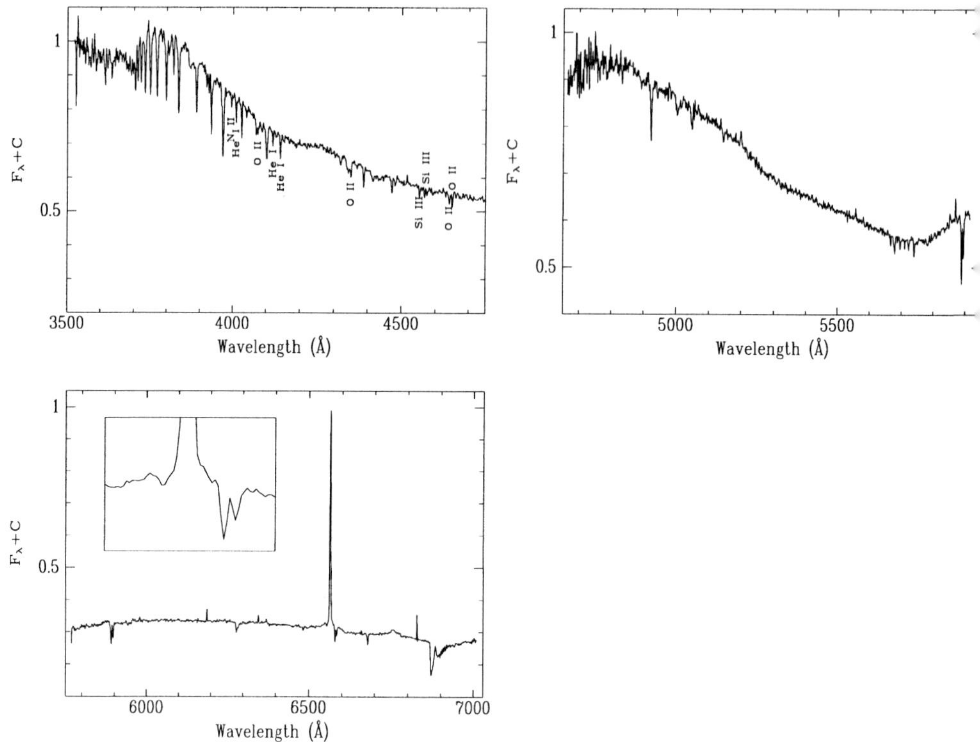

Figure 2 Optical spectra of IRAS 01005+7910

4. CONCLUSIONS

Finally we conclude that

1. the optical counterpart of IRAS 01005+7910 is a B2Ie star;
2. but it is a post-AGB star or proto-planetary nebula;
3. mass-loss process of this object is still on-going.

References

Hrivnak B.J.,Volk K.,Sun K. 2000, ApJ 535,275

Hu J.Y., Slijkhuis S., de Jong T., Jiang B.W. 1993, A&AS 100,413

Hu J.Y., Te Lintel Hekkert P., Slijkhuis S.,Baas F. 1994, A&AS 103,301

Jaschek C., Jaschek M. 1987, The Classification of Stars, Cambridge University Press

Landolt A.U. 1983, AJ 88,439

V

DUST AROUND POST-AGB SOURCES

ISO RESULTS FOR PROTOPLANETARY NEBULAE

Kevin Volk
University of Calgary
volk@iras.ucalgary.ca

Sun Kwok
University of Calgary

Bruce Hrivnak
Valparaiso University

Ryszard Szczerba
NCAC Torun

Abstract We discuss the results of analysis of *Infrared Space Observatory* spectra of PPN candidates. In particular for the carbon-rich PPNs in our sample we have analysis of the 21 and 30 μm features as well as the UIR features and plateaus. The 30 μm feature appears to have two sub-components, at about 27 and about 34 μm, which vary from source to source (Hrivnak, Volk, & Kwok, 2000). Based upon dust radiative transfer models we are able to fit the underlying continuum and so compare feature strengths and shapes for different objects The UIR features appear to vary somewhat from source to source, whereas the broad features seem to be the same in all sources. We also briefly discuss the galactic distribution of the PPNs

1. INTRODUCTION

The *IRAS* satellite was a milestone in protoplanetary nebulae (PPN) research because it allowed the identification of a large number of PPN candidates from an unbiased survey for the first time. Our approach to finding PPN candidates was to look for objects in the lower right corner of the *IRAS* 12/25/60 colour-colour diagram (i.e. Volk & Kwok, 1988). This region was chosen in part based

R. Szczerba and S.K. Górny (eds.), Post-AGB Objects as a Phase of Stellar Evolution, 323–328.

on dust shell models and in part to minimize confusion with other types of objects. Follow-up spectroscopy has demonstrated the PPN nature of a number of these objects. The sample is biased towards younger and dustier PPNs.

The *ISO* mission gave us the chance to get detailed observations of the dust shells of these PPNs. We have *ISO* SWS01, SWS06, and LWS01 spectroscopy of a number of our PPN candidates.

PPNs give us a better chance to understand the dust emission and dust features than do the more extreme AGB stars or PNs, since the dust shells are usually optically thin in the far-infrared and there is no ionized region to complicate things. We can see the intrinsic feature shapes and strengths reflected directly in the observed spectra.

We have observations of a number of oxygen-rich PPNs, including IRAS 18095+2704, IRAS 20004+2955, and IRAS 17150−3224. We have not yet analyzed these observations in detail. One thing that we can say is that none of our sources shows much, if any, in the way of crystalline silicate emission features. So there is some difference between our objects and PPN candidates such as HD 161796, which has very obvious crystalline silicate features.

2. CARBON-RICH SOURCES

The *ISO* observations of carbon-rich PPNs have shown that:

• The 21 μm feature is a bit more common than previously thought; it may be present in the Egg Nebula, and is observed in of order 10 to 15 sources including a few AGB stars (Volk, Xiong, & Kwok, 2000).

• The shape of the feature is the same in all sources within the limitations of the observations (see Volk, Kwok, & Hrivnak, 1999), and it has been tentatively identified as due to TiC (vonHelden, *et al.*, 2000). The identification seems quite impressive, but we need more follow-up laboratory work to make it conclusive.

• The 30 μm feature and the UIR plateaus are strong in many of these objects, and show signs of sub-features that vary from source to source. This is discussed in Hrivnak, Volk, & Kwok (2000) where we show that there are two components to the 30 μm feature.

• the C_2H_2 molecular band observed in carbon star spectra is not seen in post-AGB sources. This may imply chemical changes in the gas, perhaps forming PAH-type molecules from C_2H_2.

• The underlying dust continuum is well matched by plain amorphous carbon dust grains (the BE1 opacity from Rouleau and Martin (1991)). This allows us to explicitly study the features.

Our goal is use detailed study of these spectra, for PPNs, planetary nebulae, and the progenitor AGB objects, to understand the chemistry and the physical conditions in the circumstellar shells as the objects evolve.

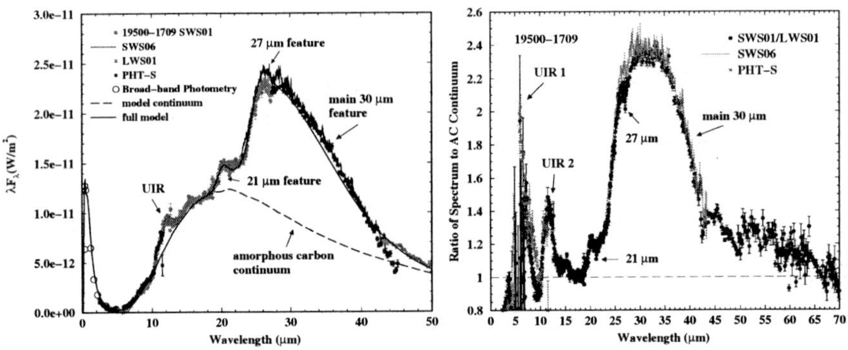

Figure 1 The left panel shows the *ISO* data for IRAS 19500–1709 along with the fit to the spectrum and the fit to the continuum. The right panel shows the resulting ratio data with the various features marked.

3. FITTING DUST FEATURES IN PPNS

To study the features we need to understand the continuum spectrum, since these PPNs at different stages of evolution can have very different dust temperatures and that strongly changes the spectral appearance in the infrared. We have carried out fitting of the continuum and of the features using a normal spherical dust radiative transfer program. The continuum is defined by the longer wavelength data–at worst, the *IRAS* photometry at 60 and 100 μm, at best a LWS01 spectrum–and by a small number of points at shorter wavelengths. It appears that the 15 to 18 μm region is unaffected by features, and there is usually a continuum region at roughly 10 μm between the UIR features, unless they are particularly strong. Since the dust emission typically peaks at about 20 μm in these objects, these continuum regions are sufficient to define the continuum uniquely.

With the continuum level deduced from the model, we can then divide out the continuum shape and display the spectrum in ratio form, whereupon the features can be explicitly compared for different objects. This requires assuming that the dust optical depth is small over the entire infrared wavelength region. Figure 1 shows the original continuum fit with the *ISO* data for IRAS 19500–1709, and the resulting ratioed data.

Figure 2 shows how the division of the 30 μm feature into its two subcomponents was carried out based upon the SWS06 spectrum of SAO 163075. The 27 μm feature was fit with a smooth curve up to the peak, which was (arbitrarily) assumed to be right at 27 μm, and the profile was reflected about the peak. This produced a reasonable shape for the longer wavelength component, which works for all the sources we have looked at.

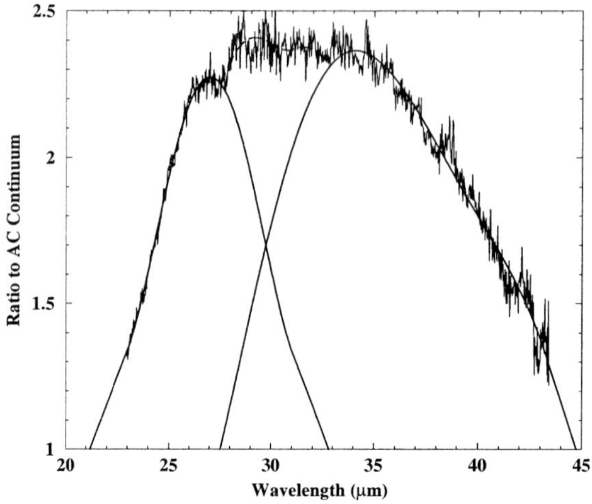

Figure 2 The component fits to the 30 μm feature. The dashed curve shows the combined shape of the two features. The data is from the SWS06 observation of IRAS 19500−1709.

Having the "real" feature shapes from the ratioed data, we can then derive standard feature shapes. For the components of the 30 μm feature we used the SWS06 *ISO* data for IRAS 19500−1709 = SAO 163075. The UIR plateau shapes were derived from the observations of IRAS 22272+5435 = SAO 34504. These template features were then applied to the data from other sources using a least-squares fit of the five feature components (the 21 μm feature, two UIR features, and the two components of the 30 μm feature) along with a baseline fit. This has been done for 8 sources in all. Figure 3 shows the normalized feature shapes we have derived.

To check the fit, we can recalculate the dust shell model with the same density and temperature distribution as in the continuum fit but with the opacity function that includes the features. This works rather well, except that in some cases the UIR features are too weak in the models. It seems likely that quantum heating effects in the dust grains are responsible for the observed UIR features being much stronger than expected from the models.

We have derived feature strengths for all the features in this way; a table of the results can be obtained from the WWW page
`http://www.iras.ucalgary.ca/~volk/ppn.html`
which also includes colour versions of the Figures. We have not been able to relate the feature strengths to each other or to obvious properties of the stars as yet.

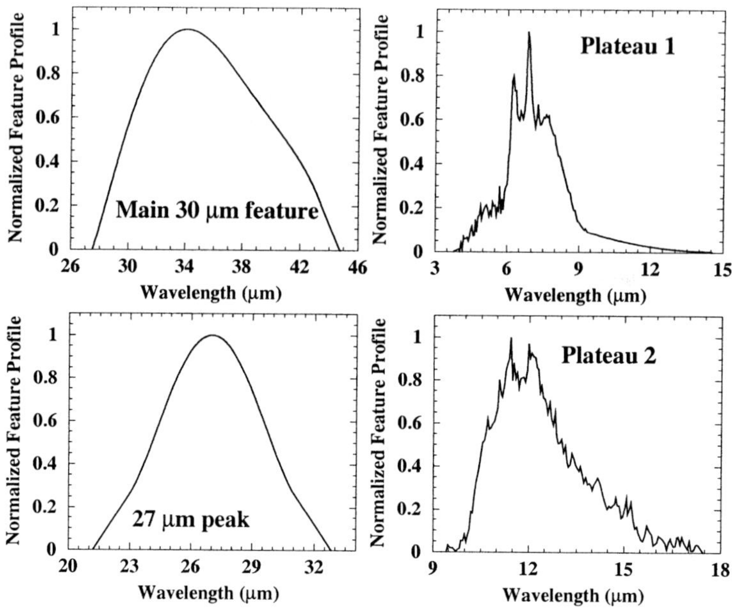

Figure 3 Normalized profiles for the UIR plateau features and the components of the 30 μm feature.

4. GALACTIC DISTRIBUTION OF PPNS

Having accurate models for the spectral emission of these objects, we can estimate their distances if their total luminosities are known. Of course this assumes that the dust shell is not decidedly asymmetric, as is seen for a number of PPNs (i.e. Hrivnak, Kwok, &Su, 1999). Thus we have used our dust shell models of 11 PPNs and an assumed luminosity of 8318 $L_\odot$, suitable for a 0.635 $M_\odot$ core mass, to plot the galactic positions of these objects. We have also included 6 extreme carbon stars and NGC 7027 for comparison; for NGC 7027 we assume a distance of 880 pc. We assume a solar circle radius of 8.5 kpc.

Figure 4 shows the results in "top" and "side" views. In the top view the spiral arm positions are marked. The objects show no change in surface density with galactic radius, but there is a very odd left/right asymmetry in the top view which we cannot account for. The distribution in z is consistent with moderate mass progenitors, say of order 2 $M_\odot$, although there may be a division between objects near the plane and those more than 400 pc off the plane. However the infrared properties of the sources show no difference between the "high" and "low" objects.

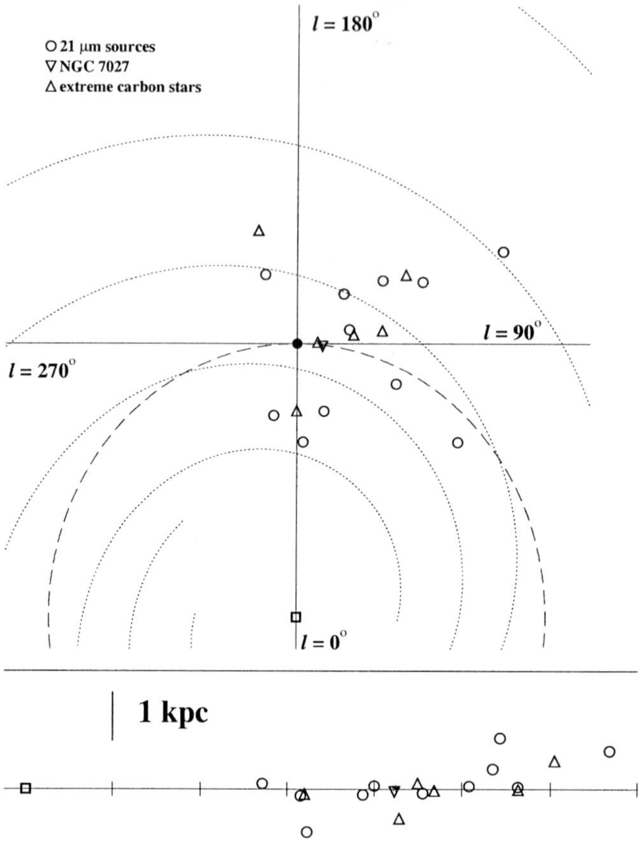

Figure 4 The derived galactic distribution of the carbon-rich PPNs, some extreme carbon stars, and NGC 7027. In the upper panel, the dotted lines trace the spiral arms and the dashed line traces the solar circle. In the lower panel, the tick marks are placed every 2 kpc out to 14 kpc.

Acknowledgments

This work was supported in part by the Natural Sciences and Engineering Research Council of Canada and grant No. 2.P03D.020.17 of the Polish State Committee for Scientific Research.

References

Hrivnak, B., Volk, K., & Kwok, S., 2000, ApJ, 535, 285
Hrivnak, B., Kwok, S., & Su, K. Y. L., 1999, ApJ, 524, 849
Rouleau, F., & Martin, P. G., 1991, ApJ, 377, 526
vonHelden, G., *et al.*, 2000, Science, 288, 313
Volk, K., Hrivnak, B., & Kwok, S., 1999, 516, L99
Volk, K., & Kwok, S., 1988, ApJ, 342, 345
Volk, K., Xiong, G.-Z., & Kwok, S., 2000, ApJ, 530, 408

AN ATLAS OF ISO SWS SPECTRA: THE TRANSITION FROM AGB STARS TO PNE

Pedro GARCÍA-LARIO

ISO Data Centre, Astrophysics Division, Space Science Department of ESA, Villafranca del Castillo, Apartado de Correos 50727, E-28080 Madrid, Spain

pgarcia@iso.vilspa.esa.es

José Vicente PEREA CALDERÓN

INSA S.A., Villafranca del Castillo, Apartado de Correos 50727, E-28080 Madrid, Spain

Abstract The ISO Data Archive has been searched for stars observed with the Short Wavelength Spectrograph (SWS) in the transition phase between the Asymptotic Giant Branch (AGB) and the Planetary Nebula (PN) stage. More than 350 sources have been identified with a reasonable signal-to-noise SWS01 spectrum corresponding to 77 original ISO proposals. The selected spectra have been classified according to their chemical characteristics (C-rich or O-rich) and overall spectral energy distribution from 2 to 45 μm. Based on the analysis made, an evolutionary scheme is proposed which reproduces the temporal evolution of the far infrared spectra of stars in this short and still poorly known transition phase.

1. INTRODUCTION

Stars in the transition phase from the AGB to the PN stage are characterised by severe mass loss (up to 10^{-4} $M_{\odot}yr^{-1}$) which produces the formation of circumstellar envelopes easily detectable with ISO in the SWS range.

It is also in this short transition phase when dramatic changes can also take place in the chemistry of some of these stars turning their O-rich atmospheres into C-rich as a consequence of the dredge-up of byproducts of helium burning to the surface of the star. Eventually, this change of chemistry can also be reflected in the composition of the dust grains formed in the circumstellar envelope. All these changes can easily be followed with spectroscopy in the range of SWS through the detection of prominent gas and solid phase features which are specific to each of the two possible chemistries.

R. Szczerba and S.K. Górny (eds.), Post-AGB Objects as a Phase of Stellar Evolution, 329–332.

With the analysis of the large number of SWS spectra contained in the ISO Data Archive it is now possible for the first time to draw a complete picture of this short and poorly understood evolutionary phase.

2. CLASSIFICATION OF STARS IN THE SAMPLE

The ISO Data Archive was systematically searched for spectra taken with the SWS in the low resolution mode (AOT 01) corresponding to stars classified in the literature as AGB, post-AGB stars or PNe. This resulted in the identification of more than 350 sources for which a reasonable signal-to-noise SWS01 spectrum was available corresponding to 77 original ISO proposals.

Esentially, the SWS spectra in the ISO Data Archive can be classified according to: 1) the dominant chemistry, through the detection of specific gas phase and/or solid phase features corresponding to an O-rich (70%) or a C-rich (23%) chemistry (in a few cases - 7% - a mixed chemistry was found); and 2) the overall spectral energy distribution, from which we can derive the temperature of the dust in the envelope as well as the relative strength of the emission from the central star with respect to the emission originated in the circumstellar shell.

Based on this, we propose an evolutionary scheme which is summarized in Figure 1 as two different sequences of spectral templates, one for each dominant chemistry.

A description is given below of the main changes observed in the infrared spectra of these stars. A detailed interpretation will be presented in a forthcoming paper.

The O-rich sequence

At the beginning of the AGB (phase O1 in Figure 1) the SWS spectrum is dominated by the cool stellar continuum over which O-rich gas phase photospheric features (H_2O at 2.6 μ, SiO at 4.0 μm, CO_2 at 4.3 μm or SO_2 at 7.3 μm) are clearly detected.

As a consequence of the increasing mass loss and the cooling of the outer layers of the stellar atmosphere dust grains are formed. At this stage (phase O2-O3) a broad band in the 9-13 μm region together with a narrow peak at $\sim$12.7 μm are detected, attributed to a combination of α-Al_2O_3 (corundum) with $MgAl_2O_4$ (spinel) and TiO_2 (rutile), suspected to be the core materials of O-rich dust grains (Posch et al. 1999). Later, broad amorphous silicate emission bands at 9.7 and 18 μm of increasing strength (O4-O5) are also visible. Amorphous silicates are known to be the main constituent of the mantles of O-rich dust grains.

The maximum strength of the amorphous silicate bands occurs at phase O6-O8. The thermal emission from the circumstellar dust at this stage becomes dominant in the SWS range while the central star gets completely obscured by the thick circumstellar envelope at the end of the AGB phase (O9), when the

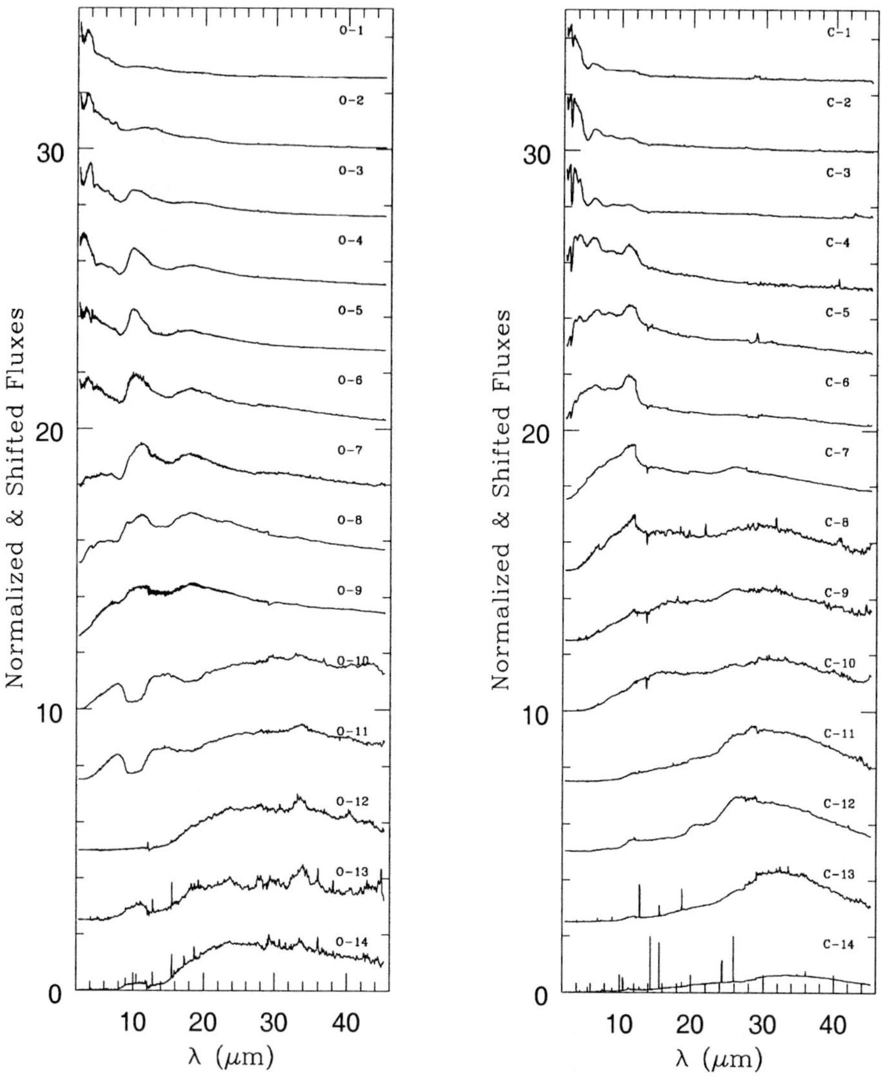

(a) O-rich sequence (b) C-rich sequence

Figure 1 Evolutionary scheme proposed.

maximum mass loss rate is reached. In the most extreme cases (large progenitor masses) the silicate bands can turn from emission into absorption (phase O10).

Just after the mass loss suddenly stops at the early post-AGB phase, solid state features corresponding to crystalline silicates become visible as broad and complex emission bands at various wavelengths in the SWS range. Specially prominent are the complex features found at 23, 28, 33 and 40 μm (Molster et al. 2001) Crystalline silicate emission at 10.5 μm contributes to the asymmetry observed at the base of the 9.7 μm absorption band (phase O11).

During the post-AGB phase the central star increases its effective temperature very rapidly while the thickness of the circumstellar envelope decreases. At this moment the emission in the SWS spectrum is characterised by the presence of strong crystalline silicate features (O12-O13).

Finally, a new (O-rich) PN is born when the effective temperature of the central star is hot enough to ionize the envelope. Nebular emission lines are usuallt detected while we can still see some crystalline silicate emission overimposed on a very cold dust continuum (O14).

The C-rich sequence

As in the O-rich case, at the beginning of the AGB, the SWS spectrum is dominated by the cool stellar continuum over which C-rich gas phase photospheric absorption features (like C_2H_2 at 3.0 and 13.7 μm, CS at 4.0 μm, C_3 at 5.1 μm, or HCN at 3.0 and 14.0 μm) are detected (C1).

As the mass loss increases and the outer layers of the stellar atmosphere cool down, a broad feature at 11.3 μm becomes detectable with increasing strength (C2-C4). This feature is attributed to amorphous SiC, expected to be the main constituent of C-rich dust grains, and reaches its maximum strength at phase C5-C6.

At the end of the AGB phase the strong 11.3 μm feature becomes sharp and asymmetric while the C_2H_2 absorption at 13.7 μm reaches its maximum strength (C7-C8). At this moment the dust continuum completely dominates the observed spectrum in the SWS range.

After the mass loss stops at the early post-AGB phase, strong (yet unidifentified) features become visible at 21, 26.5 and 29.5 μm (C9-C10) for a very short time. At the same time, the SiC feature at 11.3 μm and the C_2H_2 absorption at 13.7 μm decrease in strength while a broad plateau of emission in the 11-15 μm range attributed to large hydrogenated PAHs appears (C11-C12), followed by other PAH features at 3.3, 6.2, 7.7, 8.6 and 11.3 μm, which are identified as small, dehydrogenated PAH clusters.

A new (C-rich) PN is born when the effective temperature of the central star is hot enough to ionize the envelope. In some cases nebular emission lines are observed while we still see prominent emission from PAHs overimposed on a very cold continuum (C13-C14).

References

Posch, T., Kerschbaum, F., Mutschke, H., et al. 1999, A&A 352, 609

Molster, F.J., Waters, L.B.F.M., Tielens, A.G.G.M. 2001 (in preparation)

BIG, BUMPY DUST SHELLS AROUND PROTOPLANETARY NEBULAE

Angela Speck, Margaret Meixner & Toshiya Ueta
University of Illinois at Urbana-Champaign
akspeck@astro.uiuc.edu

Abstract We present evidence for parsec-sized dust shells around post-AGB objects, seen in ISOPHOT data. Furthermore, some of these data show evidence for enhanced mass-loss episodes on the timescales of thermal pulses. Preliminary studies suggest that the mass of the progenitors of these stars should be of the order of $\sim 3M_{\odot}$.

1. INTRODUCTION

In order to fully understand protoplanetary nebulae (PPNe) we need to study the stars from which they evolve. The direct precursors of the PPN are believed to be the asymptotic giant branch (AGB) stars, which suffer intensive mass-loss and become surrounded by circumstellar dust shells. These circumstellar shells form the nebulae seen as PPN. However, the mass-loss processes are not well understood. Several models exist which suggest that the mass-loss could be constant, varying, episodic or steadily increasing (e.g. van der Veen & Habing 1988; Vassiliadis & Wood 1993, hereafter VW), but there is little observational data to clarify the situation. The most promising theoretical models suggest that mass-loss is due to a combination of stellar pulsation and radiation pressure on dust grains (e.g Höfner & Dorfi 1997)

While there is evidence that the mass-loss from these objects becomes asymmetric towards the end of the AGB (e.g. Josselin et al. this volume; Ueta et al. 2000), studies of the dust (e.g. Mauron & Huggins 1999) and gas (Olofsson et al. 2000) around AGB stars shows that they initially have fairly symmetric mass-loss. Furthermore, the velocity of the outflowing material seems to be fairly constant (e.g. Huggins et al. 1988). Therefore, the PPN dust shells contain a fossil record of the mass-loss from the precursor AGB stars. We present an ISOPHOT far-infrared imaging study of three PPNe: oxygen-rich HD 161796, and carbon-rich AFGL 2688 and AFGL 618.

R. Szczerba and S.K. Górny (eds.), Post-AGB Objects as a Phase of Stellar Evolution, 333–338.

2. OBSERVATIONS

We have obtained linear scans of the thermal emission from three PPN using the ISOPHOT AOT PHT32. Details of the observations are listed in Table 1. A more detailed discussion of the the observations, as well as the data reduction techniques, can be found in Speck et al. (2000a; 2000b).

Table 1 Observations

Source	filter (μm)	scan length	image pixel size	θ^*	spectral type	CS[†] chem.
AFGL 618	120/180	53′	30″×92″	9°	B0	C
AFGL 2688	120/180	53′	30″×92″	8°	F5I	C
HD 161796	90/160	36′/53′	15″×46″/30″×92″	28°	F2-F5	O

* θ is the orientation of the scan, measured east of north.
† Chemistry of the circumstellar envelope.

3. RESULTS

HD 161796 is a low metallicity supergiant which is believed to have recently left the AGB on its way to becoming a planetary nebula (PN; Hrivnak et al. 1989). At first glance this object looks like a point source (Fig.1a). However, if we look at the ~1-2% flux level (Fig.1b) we can see that the observation deviates from a point source to show weak extended emission out to a radius of ~400″. We have simulated the extended emission using a mapping simulator (Gabriel & Hur 2000) in the ISOPHOT interactive analysis (PIA) package (Gabriel et al. 1997)[1]. The simulations shown in Fig.1b demonstrate that our 1-D scans have

[1]This simulator allows us to "build" an object out point sources and continuous extended sources, convolve it with a beam profiles and produce an image of how this object would appear if observed using PHT32. See Gabriel & Hur (2000) for details

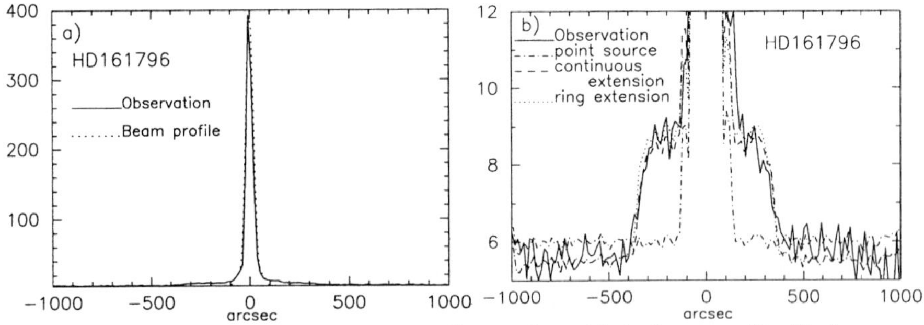

Figure 1 HD 161796 at 90μm: a) scan profile together with the beam profile; b) close up on the extended emission: scan profile together with the a simulated point source and two simulated extended sources. y-axes = brightness in MJy/sr.

insufficient resolution to distinguish between a point source with a continuous extended source and a point source surrounded by a detached ring with an inner radius of up to 50″. Therefore, for this object, we cannot determine whether mass-loss has been continuous or episodic.

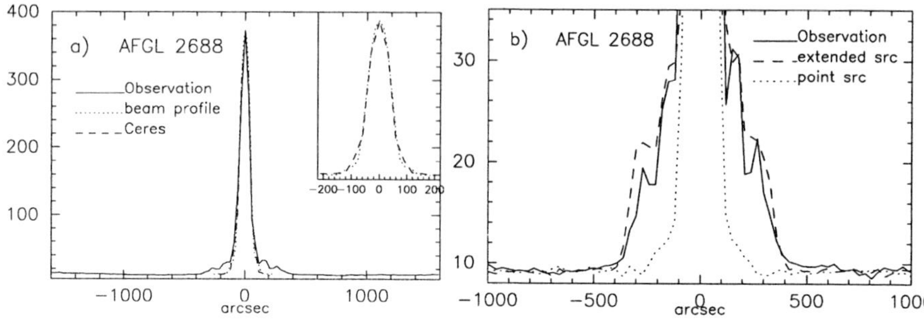

Figure 2 AFGL 2688 at 120μm: a) scan profile (solid line) together with the beam profile (dotted line) and the scan profile of Ceres (dashed line), a point source with similar peak flux to AFGL 2688 at 120μm and observed using the same AOT; b) close up on the extended emission: scan profile together with the simulated point and extended sources. y-axes = brightness in MJy/sr.

Like HD 161796, AFGL 2688 is believed to have left the AGB in the last few hundred years (e.g. Skinner et al. 1997). In Fig.2a the 120μm scan of AFGL 2688 shows a bright point source surrounded by less bright ($\sim$10%) extended emission out to $\sim$350″. Fig.2a shows that the distribution of the extended emission is not smooth, but rather shows enhancements at certain radii from the central point source. This is demonstrated more clearly in Fig.2b, an enlargement at the 10% flux level which also shows the simulated emission using the PIA map simulator. Enhanced emission occurs at radii of about 150″ and 300″.

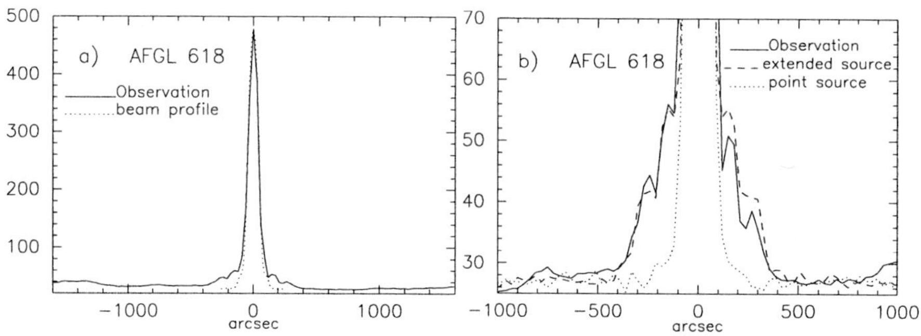

Figure 3 AFGL 618 at 120μm: a) scan profile (solid line) together with the beam profile (dotted line); b) close up on the extended emission: scan profile together with the simulated point and extended sources. y-axes = brightness in MJy/sr.

AFGL 618 is considered to be either an evolved PPN or a very young PN. This object has progressed further along the path from the tip of the AGB to

Table 2 Radii and timescales dervied from the observations

Source	Distance (kpc)	v_{exp} (km/s)	R_{max} (pc)	t_{AGB} (10^4yrs)	R_1 (pc)	t_1 (10^4yrs)	R_2 (pc)	t_2 (10^4yr)
AFGL 618	1.8	~20	~3.5	~17	~1.3	~6.4	~2.6	12.8
AFGL 2688	1.2	22.4	~2.0	~9	~0.87	~3.7	~1.7	7.4
HD 161796	1.2	12	~2.3	~19	...	...	...	...

v_{exp} = the expansion velocity from CO rotation lines; t_{AGB} = time since oldest part of dust shell was ejected*; R_1 = radius of inner emission enhancement; R_2 = radius of outer emission enhancement; t_1 = time since inner enhancement was ejected*; t_2 = time since outer enhancement was ejected*; * all ages assume constant expansion velocity;

the PN phase than the previous two objects. Fig.3 (a&b) shows that the thermal emission from AFGL 618 closely resembles that of AFGL 2688, showing a bright point source surrounded by lower level (~10% flux) "bumpy" extended emission out to ~400″. The enhanced emission bumps occur at radii of 160″ and 275″ in the simulated extended emission. Fig.4 shows that the emission bumps appear at slightly different radii for the two C-rich objects. Moreover, these emission bumps occur at the same position when the same object is observed at different wavelengths (120 and 180μm), indicating that they are not related to instrumental effects.

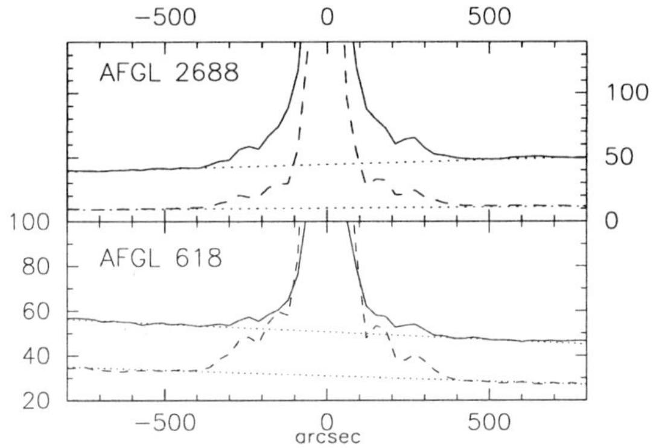

Figure 4 Comparison of the scans for AFGL 2688 and AFGL 618. solid lines = 120μm profiles; dashed lines = 180μm profiles; dotted lines = background level

4. DISCUSSION

Using our observations together with expansion velocities and distances from the literature, we can deduce the timescales associated with mass-loss for these objects. These are shown in Table 2. If we assume that we are observing the entire dust shell, then t_{AGB} represents the time these stars were in the mass-losing AGB phase. Furthermore, the enhanced emission seen in the observations of the two C-rich objects (AFGL 2688 and AFGL 618) is assumed to be due to enhanced mass-loss caused by thermal pulses (e.g VW; Steffen et al. 1998), which implies that the time between the ejection of the enhancements gives the time between thermal pulses ($t_{interpulse}$). We can compare these timescales with those computed through the stellar evolution models of VW. Table 3 lists $t_{interpulse}$, t_{AGB} and the average duration of pulses, t_{Swind}, for various progenitor masses, together with the timescales for AFGL 2688 and AFGL 618. The resolution of our scans only constrains the upper limit of t_{Swind} for these objects.

Table 3 Timescales for thermal pulses/enhanced mass-loss from VW

M_{MS} ($M_\odot$)	$t_{interpulse}$ (10^5yrs)	t_{AGB} (10^5yrs)	t_{Swind} (10^3yrs)
0.945	~1	5.704	0.4
1	~1	6.502	0.8
1.5	~1	9.385	0.8
2	~0.9	13.39	6
2.5	~0.7	18.27	4
AFGL 618	*~0.53*	*1.6*	*<10*
AFGL 2688	*~0.37*	*0.9*	*<10*
3.5	~0.3	3.509	9
5	~0.1	3.601	3

M_{MS} is the main sequence mass. Z=0.008

The outflows of both AFGL 2688 and AFGL 618 have been modeled to derive mass-loss rates based on CO observations (e.g Skinner et al. 1997; Meixner et al. 1998). These modeled mass-loss rates are of the order of a few $\times 10^{-5} M_\odot yr^{-1}$. Taking the duration of this mass-loss to be t_{AGB}, we find the total mass lost by the star to be approximately $3M_\odot$. If the remnant star has a mass in the range 0.6-1$M_\odot$, that gives a total mass for the progenitors of these objects of $<4M_\odot$, in reasonable agreement with the value of $\sim 3M_\odot$ from Table 3. Estimates of the mass-loss rate for HD 161796 suggest a value of $<10^{-5} M_\odot yr^{-1}$ (see Likkel et al. 1987). Using t_{AGB} from Table 2 gives a mass for the entire dust shell of $<2M_\odot$. Again, we assume that the mass of the remnant star is in the range 0.6-1$M_\odot$, giving a mass for the progenitor of $\lesssim$2.5-3$M_\odot$, slightly smaller than for the C-rich objects.

5. CONCLUSIONS

There is evidence for extremely extended dust emission (radius $\gtrsim$ 1pc) around the three PPNe presented here. The sizes of these dust shells can be used to calculate their ages ($1-2\times 10^5$yrs). For the C-rich objects there also appear to be periodic enhancements on a timescale of a few $\times 10^4$yrs, which may coincide with thermal pulses on the AGB. Comparison with models suggest that the C-rich stars evolved from progenitors with masses of $\sim$3-4$M_\odot$, while the O-rich object seems to have evolved from a slightly less massive progenitor, $\lesssim$2.5-3$M_\odot$.

References

Gabriel, C., Acosta-Pulido, J., Heinrichsen, I., Morris, H., Tai, W.-M., 1997, in ASP Conf. Ser. Vol. 125, Astronomical Data Analysis Software and Systems VI, ed. G. Hunt & H.E.Payne, (San Francisco: ASP), p.108.

Gabriel, C., Hur, M., 2000, in ISO Beyond Point Sources: Studies of Extended Infrared Emission, R.J. Laureijs, K. Leech and M.F. Kessler (eds), ESA-SP 455, 11.

Höfner, S., Dorfi, E.A., 1997, A&A,319, 648.

Hrivnak, B.J., Kwok, S., Volk, K.M., 1989, ApJ, 346, 265.

Huggins, P.J., Olofsson, H., Johansson, L.E.B., 1988, ApJ, 332, 1009.

Mauron, N., Huggins, P.J., 1999, A&A, 349, 203.

Meixner, M., Campbell, M.T., Welch, W.J., Likkel, L., 1998, ApJ, 509, 392.

Olofsson, H., Bergman, P., Lucas, R., Eriksson, K., Gustafsson, B., Bieging, J.H., 2000, A&A, 353, 583.

Skinner, C.J. et al. 1997, A&A, 328, 290.

Speck, A.K., Meixner, M., Knapp, G.R., 2000a, in ISO Beyond Point Sources: Studies of Extended Infrared Emission, R.J. Laureijs, K. Leech and M.F. Kessler (eds), ESA-SP 455, 83.

Speck, A.K., Meixner, M., Knapp, G.R., 2000b, ApJL, in review.

Steffen, M., Szczerba, R., Schönberner, D., 1998, A&A, 337, 149.

Ueta, T., Meixner, M., Bobrowsky, M., 2000, ApJ, 528, 861.

van der Veen, W.E.C.J., Habing, H.J., 1988, A&A, 194, 125.

Vassiliadis, E., Wood, P.R., 1993, ApJ, 413, 641 (VW).

Young, K., Phillips, T.G., Knapp, G.R. 1993, ApJ, 409, 725.

SPATIAL DISTRIBUTIONS OF MULTIPLE DUST COMPONENTS IN THE PPN/PN DUST SHELLS

Toshiya Ueta, Angela K. Speck, and Margaret Meixner
University of Illinois at Urbana-Champaign
ueta@astro.uiuc.edu, akspeck@astro.uiuc.edu, meixner@astro.uiuc.edu

Aditya Dayal
IPAC/JPL
adayal@ipac.caltech.edu

Joseph L. Hora and Giovanni Fazio
Harvard-Smithsonian Center for Astrophysics
jhora@cfa.harvard.edu, gfazio@cfa.harvard.edu

Lynne K. Deutsch
Boston University
deutschl@bebop.bu.edu

William F. Hoffmann
Steward Observatory/University of Arizona
whoffmann@as.arizona.edu

Abstract We investigate spatial distributions of specific dust components in the circumstellar shells of a proto-planetary nebula candidate, HD 179821, and a planetary nebula, BD+30°3639, by means of spectral imaging. With high-resolution ground-based images and *ISO* spectra in the mid-infrared, we can derive "dust feature only" maps by subtracting synthesized continuum maps from the observed images at the feature wavelength. Such spatially detailed information will help to develop models for the evolution of dust grains around evolved stars.

R. Szczerba and S.K. Górny (eds.), Post-AGB Objects as a Phase of Stellar Evolution, 339–342.

1. INTRODUCTION

With *ISO* discoveries of a wealth of infrared (IR) emission features from evolved stars (e.g., Waters et al. 1996; Waters et al. 1998) and laboratory measurements of IR spectra for candidate dust species (Speck 1998 for a compilation of references), we are now able to identify individual circumstellar dust components.

Using such detailed information to interpret high-resolution ground-based mid-IR images we can spatially separate dust emission due to different species by means of spectral imaging. This can be done by (1) identify an emission feature and nearby continuum bands, (2) construct a color temperature map from the continuum images, (3) synthesize a continuum image at the feature wavelength from the temperature map taking into account diffraction effects, and (4) subtract the synthesized continuum-only map from the observed feature band image.

2. DUST FEATURE MAP CONSTRUCTION

Fig. 1 shows mid-IR images and spectra of an oxygen-rich proto-planetary nebula candidate, HD 179821 (IRAS 19114+0002, AFGL 2343). The images indicate the presence of a dust torus surrounding the central star, and the spectra reveal both amorphous and crystalline silicate features (Waters et al. 1996). We can isolate dust emission at 10.3 and 18μm by subtracting continuum emission maps synthesized using 8.8, 12.5, and 20.6μm band images as reference continuum.

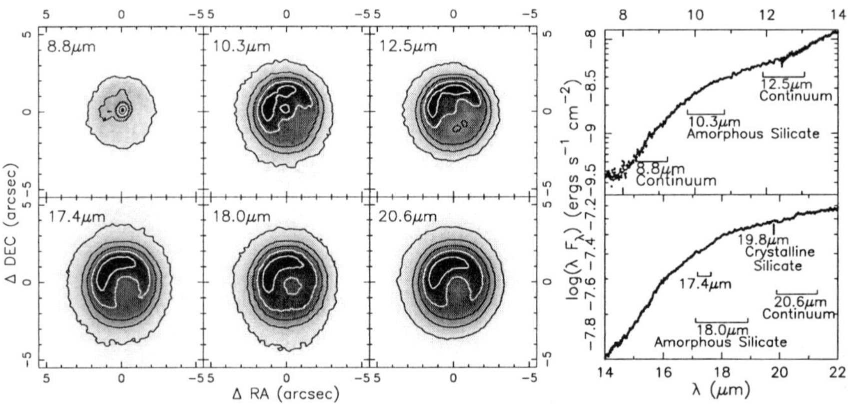

Figure 1 MIRAC3/IRTF images (left: contours are 90, 70, 50, 30, and 10% of the peak) and *ISO* spectra (right; Waters et al. 1996) of HD 179821. Band/filter locations are indicated in the spectra.

BD+30°3639 (IRAS 19327+3024) is a carbon-rich planetary nebula, in which crystalline silicates have been detected (Waters et al. 1998). Mid-IR images (Fig. 2) show a box-like elongation which is also seen in optical emission line images (e.g., Harrington et al. 1997).

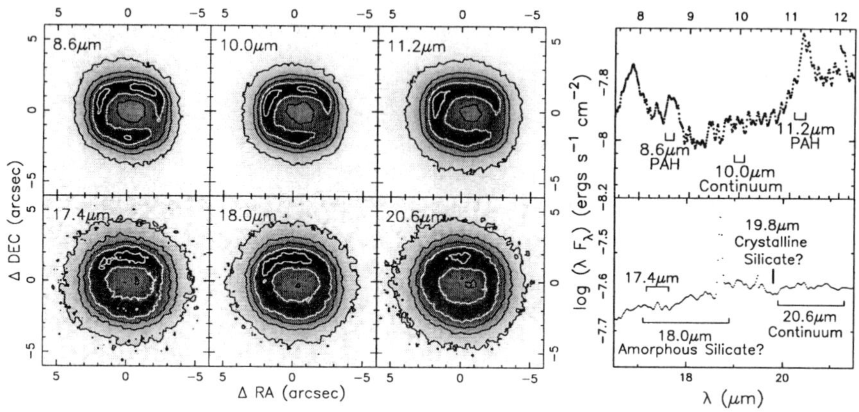

Figure 2 MIRAC3/IRTF images (left) and *ISO* spectra (right; Waters et al. 1996) of BD+30°3639 in the mid-IR. Band/filter locations are indicated in the spectra.

While polyaromatic hydrocarbon (PAH; e.g., Allamandola et al. 1999) emission at 8.6 and 11.2μm can be isolated by using 10 and 20.6μm continuum band images (Fig. 4), we were unable to isolate crystalline silicate at 19.8μm because of the weak feature and apparent lack of clean continuum around 18μm.

3. RESULTS AND IMPLICATIONS

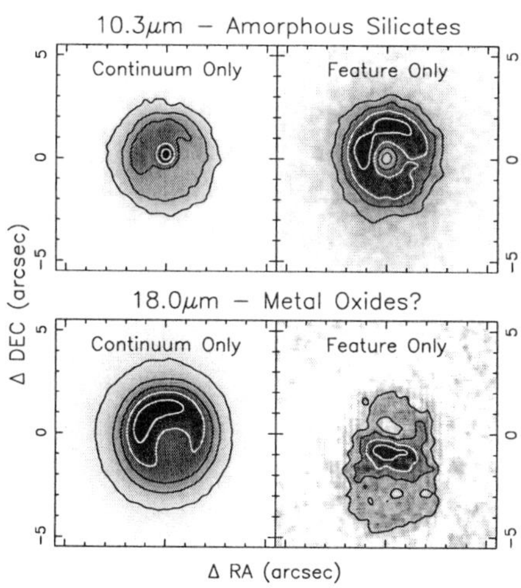

Figure 3 Continuum/feature only images at 10.3 and 18.0μm for HD 179821 .

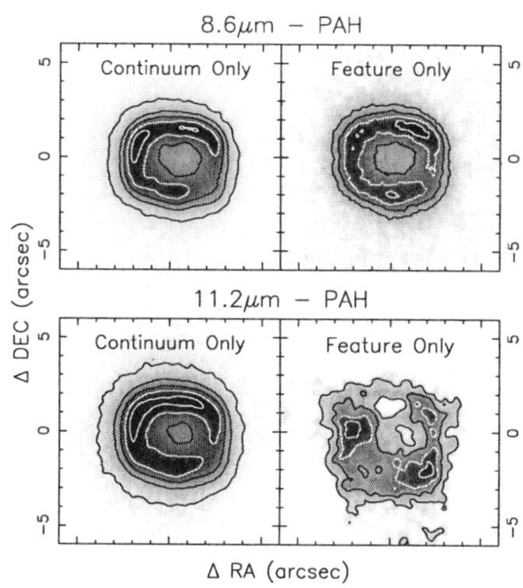

Figure 4 Continuum/feature only images at 8.6 and 11.2μm for BD+30°3639.

Fig. 3 shows that the 10.3 and 18.0μm bands have different spatial distributions, indicating that these emissions are due to different carrier species. We thus need to find a dust species with an emission feature at 18μm but not at 10.3μm (i.e., not silicates). Magnesium iron oxides are such species as shown in the spectrum (Henning et al. 1995).

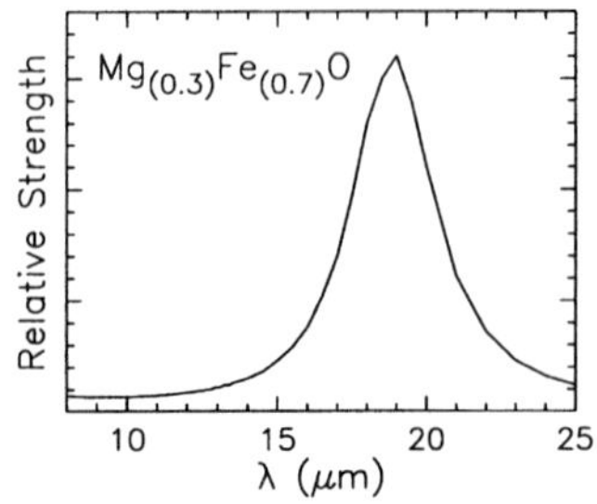

PAH emissions at 8.6 and 11.2μm show drastically different spatial distributions in Fig. 4, implying that these PAH bands may have been originated differently (e.g., Beintema et al. 1996, Molster et al. 1996).

The 8.6μm feature		**The 11.2μm feature**
• slightly compact		• slightly extended
• ionized PAHs	$\Longleftrightarrow$	• neutral PAHs
• in-plane C-H mode		• out-of-plane C-H mode

Observation of the differences listed above suggests that the difference between the two bands may arise from the availability of hydrogen, either in the form of an atom or an out-of-plane C-H stretch. The former implies the difference in the ionization state of PAHs (i.e., the 8.6μm emission is from the ionized shell and the 11.2μm emission is from the photo dissociation region), while the latter suggests different structure of PAHs (i.e., the 8.6μm emission is from PAH clusters and the 11.2μm emission is from chain- or sheet-like PAHs).

Careful selection/tailoring of filters will make spectral imaging more effective. Determining the spatial distribution of specific dust species will allow the development of models for the evolution of dust around evolved stars. Understanding dust evolution benefits other aspects of the physics involved such as grain nucleation and growth, the chemical evolution of the stellar atmospheres, and the nucleosynthesis of stars.

References

Allamandola et al. 1999, ApJ, 511, L115
Beintema, et al. 1996, A&A, 315, L369
Harrington et al. 1997, AJ, 113, 2147
Henning et al. 1995, A&AS, 112, 143
Molster et al. 1996, A&A, 315, L373
Speck, A.K. 1998, Ph.D Thesis, UCL
Waters et al. 1996, A&A, 315, L361
Waters et al. 1998, A&A, 331, L61

STRUCTURE AND PROPERTIES OF IRC+10216

Alexander B. Men'shchikov
Stockholm Observatory, SE–133 36 Saltsjöbaden, Sweden
sasha@astro.su.se

Yuri Yu. Balega
Special Astrophysical Observatory, Nizhnij Arkhyz, Karachaevo-Cherkesia, 357147, Russia

Thomas Blöcker, Roger Osterbart, Gerd Weigelt
Max-Planck-Insitut für Radioastronomie, Auf dem Hügel 69, D–53121 Bonn, Germany

Abstract We present selected results of the first detailed, two-dimensional radiative transfer modeling of the dusty envelope around the carbon star IRC+10216. The model reproduces very well the entire beam-matched spectral energy distribution of IRC+10216 from optical to centimeter wavelengths (at several phases of stellar luminosity), observed intensity profiles of the object at 1.25, 2.2, 10.5, 50, 100 μm, and 1.3 mm, a 10.5 μm lunar occultation intensity profile, our high-resolution J, H, K, and $H-K$ bispectrum speckle-interferometry images, and visibilities in J, H, K, L, M, and N bands. We present a new, accurate identification of the carriers of the well-known 11.3 μm and 30 μm dust features in the spectrum of IRC+10216. The dust model found in this study can also be successfully applied to some other carbon stars exhibiting broad emission features in the 10.3–12.6 μm and 25–37 μm wavelength regions. A firm result of our modeling is that the brightest compact near-infrared peak observed in IRC+10216 is not the direct light from the underlying star. In contrast to previous suggestions, it is the radiation emitted and scattered in the optically thinner southern outflow cavity of a bipolar dense shell moving away from the central star.

1. INTRODUCTION

Very soon after the initial studies three decades ago, IRC+10216 was identified as a longest-period pulsating carbon star surrounded by an opaque dust shell, perhaps in a state preceding that of a planetary nebula. Extremely large amounts of observational material have been acquired since then, making this object the best-observed carbon star. Dust continuum radiative transfer calcu-

R. Szczerba and S.K. Górny (eds.), Post-AGB Objects as a Phase of Stellar Evolution, 343–350.

lations in spherical geometry have also been repeatedly applied to IRC+10216 in order to understand in more detail physical, chemical, and evolutionary properties of the star and its molecular, dusty envelope. Recent highest-resolution observations have clearly demonstrated, however, that the inner dust shell of IRC+10216 is actually non-spherical and extremely clumpy, thus emphasizing the need of multidimensional radiative transfer calculations for proper understanding of the observations.

In our previous paper, the observations were interpreted in terms of a bipolar, dynamically evolving dusty envelope around the carbon star being at the very end of its AGB evolution (Osterbart et al. 2000). Here, we present some selected results of our detailed two-dimensional radiative transfer modeling of the dusty envelope of IRC+10216. For a more comprehensive introduction, review of the observational material, description of the modeling, and discussion of the model results, we refer to Men'shchikov et al. (2001).

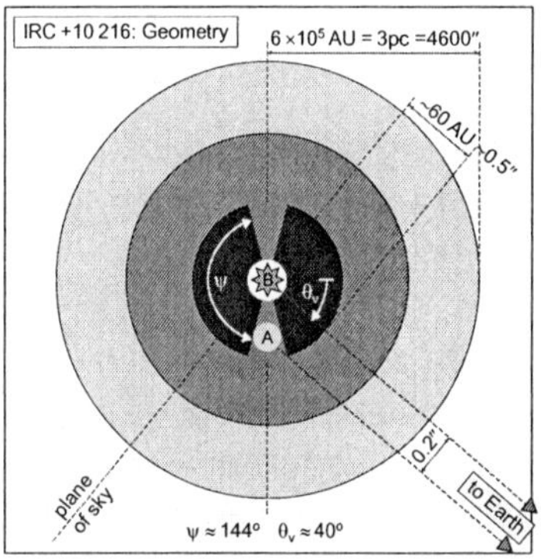

Figure 1 Geometry of the circumstellar envelope of IRC+10216. Schematically shown are three regions of the model envelope – the innermost dense core with bipolar cavities (dark color), the envelope where molecules are observed (medium color), and the outer extended envelope containing most of the mass lost by the central star (light color). The model is almost spherically-symmetric, except for the inner axially-symmetric region with bipolar outflow cavities. The geometry is defined by the opening angle of the cavities, $\omega = \pi - \psi \approx 36^\circ$ ($\psi \approx 144^\circ$) and the viewing angle, $\theta_v \approx 40^\circ$, between the equatorial plane and the line of sight. Bright components A and B labeled by Weigelt et al. (1998) correspond to those shown in Fig. 2.

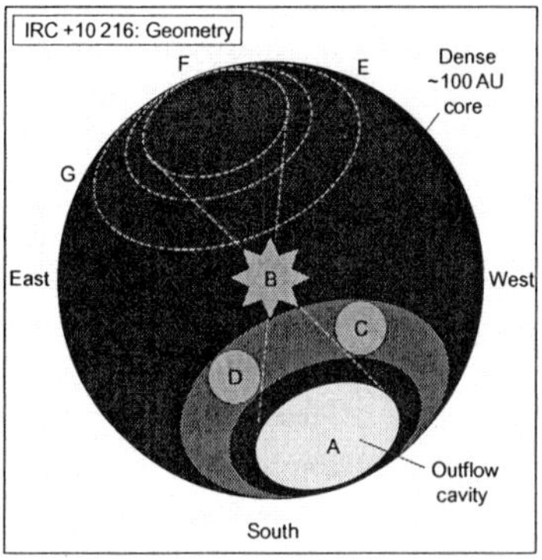

Figure 2 Geometry of the circumstellar envelope of IRC+10216 as it appears to an observer in projection onto the sky plane. Components A, B, C, D observed by Weigelt et al. (1998) are shown schematically, as well as much fainter components E, F, G found by Osterbart et al. (2000). Contrary to intuitive expectations, the star is located at the position of the fainter component (B), whereas the brightest peak (A) is produced by the light escaping from the southern cavity. The components C and D appear on a ring around the cavity most likely due to dust density (optical depth) fluctuations along the ring. Weak emission from the obscured northern cavity and inhomogeneous environment cause the appearance of the components E, F, and G.

2. MODEL RESULTS

We constructed the first two-dimensional radiative transfer model for IRC+ 10216, which describes in detail many aspects of its present-day structure (Figs. 1, 2). Reconstructing the density distribution throughout the huge, parsec-sized envelope, the model also gives insights into the past AGB evolution of the star. Quantitatively explaining all relevant manifestations of circumstellar dust grains in a self-consistent, extensive exploration of a very large parameter space, the model gives a reliable, coherent picture of the object.

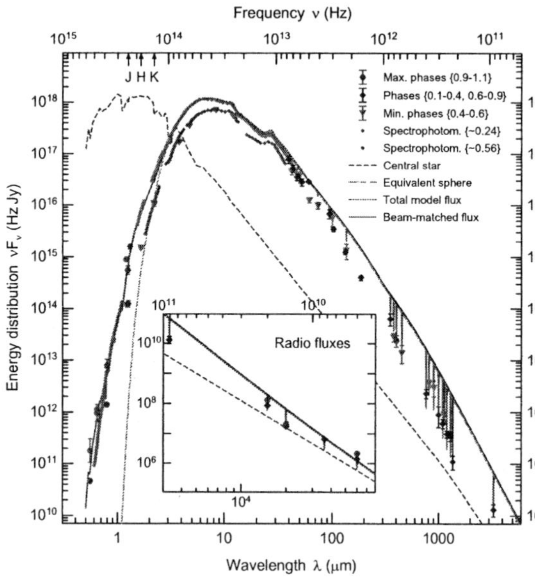

Figure 3 Observed SED of IRC+10216 compared to our model with $L_{\star} = 9500\, L_{\odot}$ ($\phi = 0.24$). Stellar continuum, total and beam-matched fluxes, and SED for an equivalent spherical envelope are plotted. The latter emphasizes effect of the bipolar outflow cavities. The observed broad-band fluxes are shown by different symbols to differentiate between those measured at around the maximum luminosity, intermediate phases, and minimum luminosity. If available, error bars for the fluxes are drawn, whenever they are larger than the symbols. Thick curves (dots) show all available spectrometry and spectrophotometry data; the upper and lower continua between $\sim 2\,\mu$m and $8.5\,\mu$m correspond to phases $\phi = 0.24$ and $\phi = 0.56$, respectively. The model assumes that we observe the envelope at $\theta_{\rm v} = 40°$ (relative to the midplane). Effect of beam sizes is shown by the vertical lines, clearly visible in the model SED at $\lambda > 50\,\mu$m. Whereas only the lower points of the vertical lines are relevant, we have connected them to the adjacent continuum by straight lines, to better visualize the effect. The insert compares the millimeter- and centimeter-wavelength portion of the model SED with available radio observations. Three arrows at the top axis indicate the central frequencies of the J, H, and K photometric bands.

STAR'S LOCATION. An important and firm result of the modeling is that the brightest component A, visible in the HST and our speckle interferometry

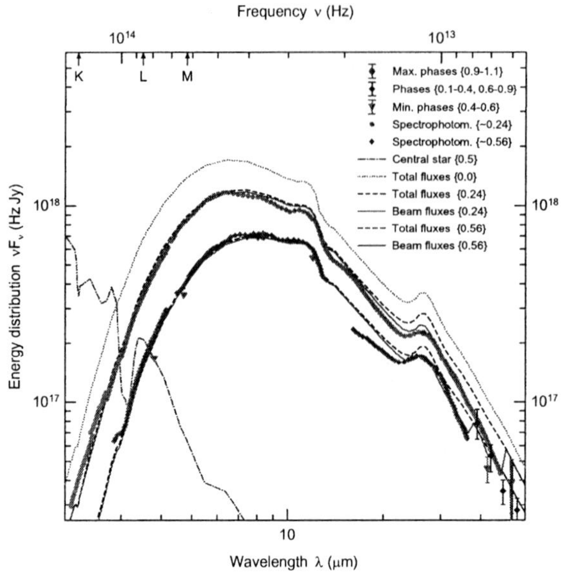

Figure 4 Top portion of the observed SEDs of IRC+10216 compared to our models with $L_\star = 9500\,L_\odot$ (ϕ=0.24) and $L_\star = 5500\,L_\odot$ (ϕ=0.56). Total model fluxes at maximum luminosity (ϕ=0.0) and the SED of a stellar atmosphere at minimum phase (ϕ=0.5) are also displayed. *Wavelength-dependent* effect of beam sizes is visible as the difference between the total and beam-matched fluxes.

images of IRC+10216, *does not* contain direct light from the central star. In contrast to an intuitive picture adopted in all previous works, the star is actually located at the projected position of the *fainter* component B. The brightest peak is only the light emitted and scattered by a narrow bipolar outflow cavity which is tilted toward us by $\theta_v = 40°$ (Fig. 1). Fainter components C and D appear most likely due to density deviations around the outflow cavities from axial symmetry on smaller scales (Fig. 2).

STELLAR PARAMETERS. Provided that the adopted distance of 130 pc is correct, stellar luminosity varies from 13000 to 5200 $L_\odot$ between the maximum and minimum brightness of IRC+10216 (Figs. 3, 4). We assumed that the effective temperature of the central star is changing from 2800 to 2500 K between the two phases. Being somewhat higher than the temperature usually adopted in radiative transfer calculations, it agrees better with stellar evolution models. Stellar radius varies in a relatively narrow range, from 500 to 390 $R_\odot$. During the last 1000 years, this star has experienced at least two episodes of very high mass loss with $\dot{M} \approx 9 \times 10^{-5}\,M_\odot\,\mathrm{yr}^{-1}$.

DENSITY DISTRIBUTION. The density distribution of dust *cannot* be described everywhere by a simple power law $\rho \propto r^{-2}$ characteristic for a stationary spherical mass loss. The density profile found in our modeling (Fig. 5) results

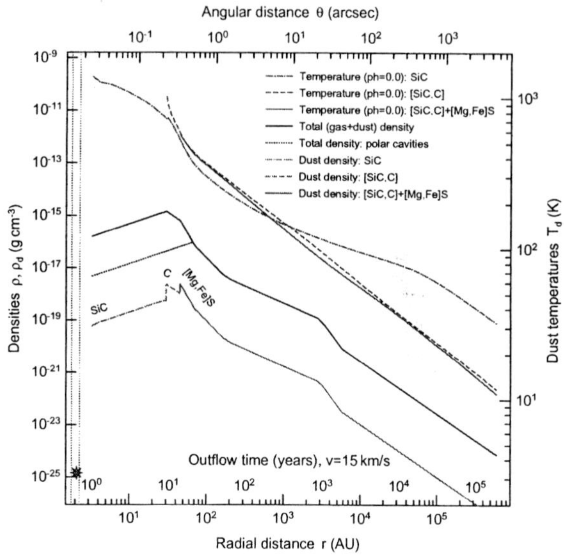

Figure 5 Temperatures and densities of dust grains in the envelope of IRC+10216 are displayed for the smallest grains only ($a = 0.01\ \mu$m) of all dust components; temperatures correspond to the luminosity maximum ($\phi = 0.0$). Also shown are the total (gas+dust) density profiles in the midplane and in bipolar outflow cavities along the symmetry axis.

most likely from a complex interplay between stellar pulsations, changing mass-loss rate, and radiative acceleration of dust and gas in the innermost parts of the envelope. All these processes are changing the structure of the dense subarc-second region of the envelope on a time scale of years. Polar outflow cavities in the axially-symmetric dense shell (Figs. 1, 2) cause the observed "clumpy" appearance of IRC+10216 in recent highest-resolution images. The huge envelope of 3 pc in radius has a total mass of 3 $M_\odot$, if we assume a dust-to-gas mass ratio of 0.004. The mass is consistent with an initial mass of the central star of $\sim 4\ M_\odot$ and the present-day core mass of 0.7 $M_\odot$. A lower initial mass of the star would imply a lower core and envelope masses and a higher $\rho_{\rm d}/\rho$.

DUST COMPOSITION. In agreement with theoretical expectations, but in contrast to all previous fits of the object's SED, our model predicts that both the radial distribution and composition of dust around IRC+10216 are strongly inhomogeneous. Our dust model consists of three dust components (Figs. 6, 7): Begemann et al. (1994): (1) silicon carbide SiC (α-SiC, Choyke & Palik 1985), (2) [SiC, C] aggregates made of SiC and carbon cel800 (Jäger et al. 1998), volume ratio 15:85, and (3) core-mantle [SiC, C]+[Mg, Fe]S grains, i.e. [SiC, C] cores covered by $[Mg_{0.5}Fe_{0.5}]S$ (Begemann et al. 1994) mantles of a thickness of $\Delta a/a = 0.15$. The inner dust boundary of the envelope, a subject of considerable debate over decades, is located very close to or even inside stellar photosphere, where the newly formed SiC grains have temperatures of $T_{\rm SiC} \approx 2000$ K. The most abundant carbon materials begin to nucleate in the wind when it reaches greater distances and lower temperatures ($T_{\rm C} \approx 1000$ K), leading eventually to the formation of [SiC, C] aggregates. When they enter the condensation zone of the [Mg, Fe] sulfides ($T_{\rm [Mg,Fe]S} \approx 600$ K), the latter form mantles on the existing inhomogeneous particles.

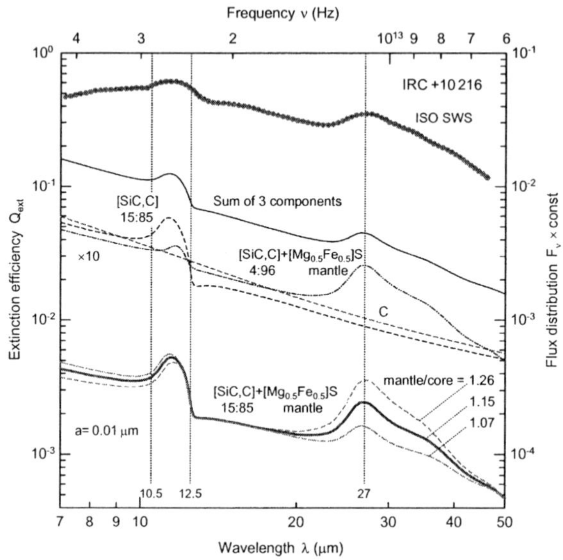

Figure 6 Inhomogeneous dust grains made of silicon carbide, carbon, and magnesium-iron sulfides explain both emission bands in the spectrum of IRC+10216 (Yamamura et al. 1998). The lower set of curves shows extinction efficiencies of small ($a = 0.01$ μm) spherical core-mantle grains composed of [SiC, C] (volume ratio SiC:C = 15:85) cores and [$Mg_{0.5}Fe_{0.5}$]S mantles with relative thicknesses of 7, 15, and 26%.

BROAD EMISSION BANDS. First detailed exploration of a large parameter space of various grain models enabled us to conclude that it is *inhomogeneous* conglomerate particles that are responsible for the well-known, yet poorly explained 11.3 μm and 27 μm (30 μm) emission features (Fig. 6). Our modeling strongly suggests that the 11.3 μm feature is produced by [SiC, C] grains, unorganized aggregates made of incompletely amorphous carbon with significant graphitic content and silicon carbide. The 27 μm emission band, also known as a 30 μm feature, is perfectly reproduced by the core-mantle [SiC, C]+[Mg, Fe]S grains. Although it is absolutely clear that the real situation in IRC+10216 is more complex, our dust model is the simplest approximation that consistently and quantitatively explains dust emission observed in this object. The same set of solid materials can be used to successfully model the whole *family* of broad emission bands observed in other carbon stars, with peaks in the 10.3–12.6 μm and 25–37 μm wavelength regions.

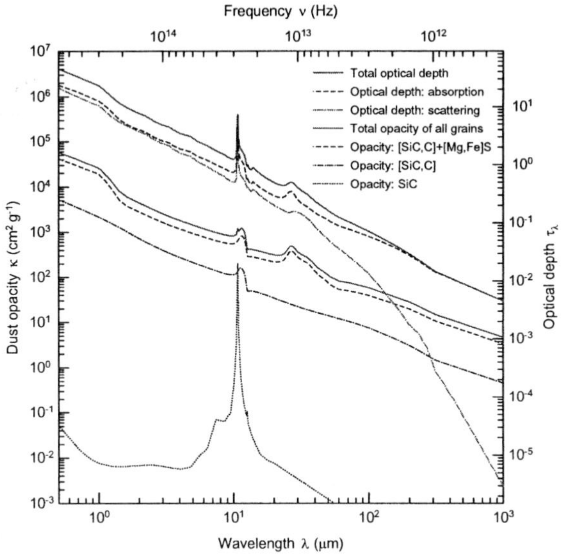

Figure 7 Properties of dust in our model of IRC+10216. Opacities for smallest grains ($a = 0.01\,\mu$m) of all three dust components as well as the total dust opacity are shown. Wavelength-dependent optical depth toward the star is also plotted for both absorption and scattering, along with the total extinction through the whole envelope produced by the entire population of dust grains.

IRC+10216 AND CRL 2688. We would like to draw several parallels between the object studied in this work and the proto-planetary nebula CRL 2688 , also known as the Egg Nebula (see, e.g., Sahai et al. 1998). Both objects have distinct dense, optically thick dusty circumstellar cores with biconical cavities presumably produced by high-velocity collimated outflows. The opening angle of the holes in CRL 2688 is $\omega \approx 30°$, almost the same as in our model of IRC+10216 (Fig. 1). Appearances of the two objects noticeably differ, however, primarily because they are inclined at different angles toward the observer (almost edge-on for CRL 2688 vs. $\theta_{\rm v} \approx 40°$ for IRC+10216). In both objects, shell-like density enhancements are observed in scattered light as incomplete arcs, which correspond to periodic increases of similarly high mass-loss rates ($\sim 10^{-4}\,M_{\odot}\,{\rm yr}^{-1}$) by a factor of 2–3. One can possibly identify the recent episode of extremely high $\dot{M}$ in IRC+10216 with a formation of one of such shells. CRL 2688 is in a significantly more advanced stage of a rapid transition into the planetary nebulae phase than IRC+10216. The above brief (and incomplete) comparison suggests that similar physical processes are operating within this evolutionary sequence.

SYMMETRY, CLOUDS, AND CAVITIES. We should stress that the widely used basic assumption of spherical symmetry appears to have a *very* limited applicability, as we "turn on" higher and higher angular resolutions. Furthermore, our model demonstrated that one should be very cautious while interpreting *asymmetric* optically thick dusty environments on the basis of common sense, instead of accurate quantitative modeling. Extending present results to other non-spherical objects, we believe that interpretations of observations in terms of *dust clouds* around mass-losing stars may be misleading. We expect that in many cases such "clouds" will turn out to actually be *outflow cavities* or other types of inhomogeneities in the density distributions of optically thick circumstellar shells.

3. CONCLUSIONS

IRC+10216 is the most difficult object we took for modeling, primarily because of its variability and enormous amounts of observational material that should be quantitatively explained by a realistic model. Having constructed the model that fits a wide variety of observational constraints, we are confident that the general structure and properties of IRC+10216 obtained in this study are indeed good approximations to the extremely complex reality. We should make it absolutely clear, however, that no model can provide ultimate answers. No matter how sophisticated, any model would ignore many important aspects of nature. Having explained all the observations of dust continuum *self-consistently*, our detailed model can serve as just a first step to a more complete and reliable understanding of IRC+10216 and carbon stars in general.

References

Begemann B., Dorschner J., Henning T., Mutschke H., Thamm E., 1994, ApJ 158, L133

Choyke W.J., Palik E.D., 1985, in: Palik E.D. (ed.), Handbook of Optical Constants of Solids, 587, Academic Press

Jäger C., Mutschke H., Henning T., 1998, A&A 332, 291

Men'shchikov A.B., Balega Y.Y., Blöcker T., Osterbart R., Weigelt G., 2001, A&A 368, 497

Osterbart R., Balega Y.Y., Blöcker T., Men'shchikov A.B., Weigelt G., 2000, A&A 357, 169

Sahai R., Trauger J.T., Watson A.M., et al., 1998, ApJ 493, 301

Weigelt G., Balega Y., Blöcker T., et al., 1998, A&A 333, L51

Yamamura I., de Jong T., Justtanont K., Cami J., Waters L.B.F.M., 1998, Ap&SS 255, 351

SPATIALLY RESOLVED SPECTRA OF THE RED RECTANGLE IN THE N-BAND

Takashi Miyata
SUBARU Telescope, National Astronomical Observatory,
650 North A'ohoku Place, Hilo, HI 96720, U.S.A.
takashi.miyata@nao.ac.jp

H.Kataza[1], Y.Okamoto[2], S.Sako[2], S.Takubo[2], T.Onaka[2], and T.Yamashita[3]
[1] *Institute of Space and Astronautical Science, Japan*
[2] *Department of Astronomy, University of Tokyo, Japan*
[3] *SUBARU Telescope, National Astronomical Observatory of Japan*

Abstract We have observed the Red Rectangle by our newly developed instrument COMICS attached on the SUBARU 8.2 m telescope, and obtained images and spectra in the N-band (8-13.2μm). Diffraction limited spatial resolution ($\sim$ 0.35 arcsec) was almost achieved, revealing differences of dust emission profiles with its distance from the central star. In the nebula region (distance $>$ 260 AU), UIR emissions at 7.7, 8.6, 11.27, and 12.7 μm were prominent, while spectra around the central star (distance $<$ 260 AU) are dominated by strong continuum emissions.

1. INTRODUCTION

The Red Rectngle is a remarkable bipolar nebula associated with the A0 type star HD 44179. It shows strong unidentified infrared (UIR) band emissions, and is one of the most studied sources. Especially, it is a suitable target for investigating formation and evolution processes of UIR carriers, because the Red Rectngle is near ($\sim$ 330 pc) from the earth. High spatial resolution achieved with large ground-based telescopes can reveal the differences of the UIR emissions with its distance from the central star.

2. OBSERVATIONS AND RESULTS

We observed the Red Rectangle with the COoled Mid-Infrared Camera and Spectrometer (COMICS) mounted at the Cassegrain focus of the 8.2m SUBARU Telescope. The COMICS is an istrument designed for ground-based

R. Szczerba and S.K. Górny (eds.), Post-AGB Objects as a Phase of Stellar Evolution, 351–354.

observations in mid-infrared wavelengths, and has imaging and spectroscopic capabilities (Kataza et al. 2000).

2.1. IMAGING

Imaging observations were carried out with three bandpass filters at 8.7, 9.8 and 11.6 μm. The pixel scale was 0.130 arcsec/pixel, and the field of view was 41×31 arcsec. We obtained diffraction limited images (FWHM $\sim$ 0.35 arcsec at 11 μm) with using offline shit-and-add reductions. On source integration time of the 8.7, 9.8, 11.6 μm filters were 0.37, 0.25, 0.37 sec, respectively.

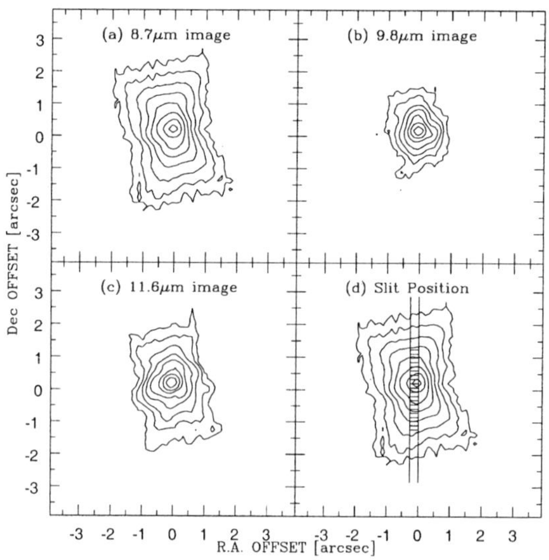

Figure 1 N-band image contours of the Red Rectangle. All contours are drawn in logarithmic scale. (a) 8.7 μm, (b) 9.8 μm, (c) 11.6 μm, and (d) slit position overlaying the 8.7 μm image.

Figure 1 shows obtained images of the Red Rectangle. All of the sources consist of a bright core and an extended nebula. The nebulae seem to be slightly different with their wavelengths; the 9.8 μm image is the most compact, while the nebula in the 8.7 μm filter is the most extended and looks like a rectangle. This agrees with results of previous observations in the N-band (e.g. Hora et al. 1996).

2.2. SPECTROSCOPY

The spectra of the Red Rectangle were obtained with long-slit low-resolution spectroscopic mode. The slit was adjusted on the center of the nebula along a north-south direction (See Figure 1(d)). The pixel filed of view was 0.165 arcsec, and the slit width was 0.33 arcsec which was close to the diffraction limited resolution. Spectral resolution was $\sim$ 250 at the wavelength of 12.8 μm. On source integration time was 5.0 sec.

We extracted a spectrum each 0.165 arcsec from the obtained 2 dimensional data and derived spectra at 25 positions of the Red Rectangle(11 of 25 are shown in Figure 2). At the central position (upper panel), continuum emission dominates the spectrum. This continuum becomes weaker as its distance from the central star, while unidentified infrared (UIR) emission features at 7.7, 8.6, and 11.3 μm enhanced. The spectra at a distance > 0.8 arcsec are almost dominated by the UIR features. This is consistent with the results of Sloan et al.(1993).

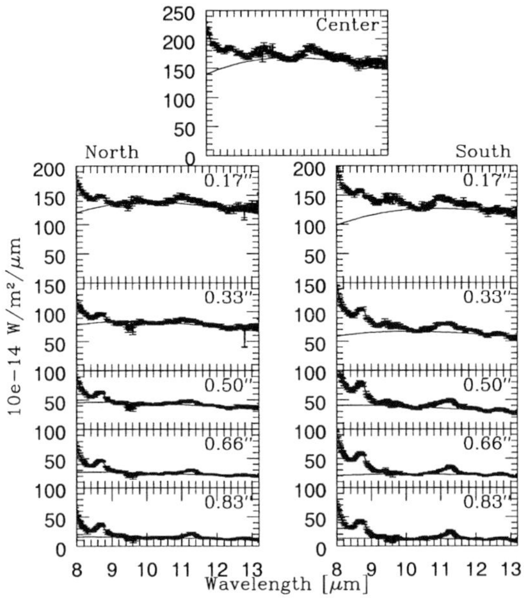

Figure 2 Spatially resolved spectra of the Red Rectangle in the N-band. Central 11 spectra are shown. A top panel shows a spectrum at the central position, and left and right panels show spectra of the northern and southern nebula, respectively. A line in each panel indicates 'pure' continuum fitted with a blackbody radiation.

The spectra are different in nothern and southern nebula. The UIR features are stronger at the southern nebula, especially the feature at 11.3 μm. On the other hand, continuum emissions show a little bit red at the northern nebula. These are remarkable just around the central region, while the spectra at the outer part are quite similar.

3. DUST FEATURES

In order to extract the dust features, we make a model fit of underlying continuum for each spectrum. We assumed that the observed spectra between 10.3-10.6 and 12.2-12.5 μm were the 'pure' continua and can be represented as blackbody radiations with a single temperature. The fitted continua are plotted as lines in the Figure 2, and the extracted dust features are shown in Figure 3.

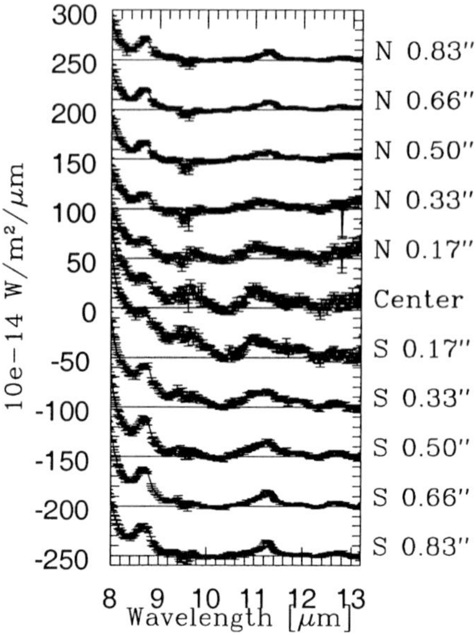

Figure 3 Extracted dust features. A solid line indicate zero level for each profile.

This clearly demonstrates that not only intensities but also profiles of the dust features differ as their positions. In the outer part of the nebula ($>$ 0.8 arcsec $\sim$ 260 AU), the feature around 11 μm shows a narrow profile. Its peak is located at 11.27 μm, and it is quite similar as the typical UIR emission. However, the plateau emission around 11 - 12 μm which is usually observed in UIR emission objects is not found. The profile gradually varies as the distance from the central region. A wing at the bule side of the 11.27 μm peak becomes prominent at a distance of 0.5 arcsec. At 0.33 arcsec, this wing getting prominent and another peak at 11.9 μm appears. At the central region, we cannot see the 11.27 μm peak, but two bumps at 11.0 μm and 11.9 μm. Similar tendency can be seen for the 8.6 μm feature. It is prominent and show the typical UIR profile in the outer region, while it disappers near the central region. These tendencies may be caused by ionization and/or dehydrogeneration of UIR carrier grains around the central star.

References

Hora J.L., Deutsch L.K., Hoffmann W.F., Fazio G.G., 1996, AJ., 112, 2064

Kataza H., Okamoto Y., Takubo S., Onaka T., Sako S. Naakamura K., Miyata T., and Yamashita T., 2000, Proc. of SPIE, 4008

Sloan G.C., Grasdalen G.L., LeVan P.D., 1993, ApJ, 409, 412

VI

MOLECULAR AND ATOMIC GAS AROUND PPNE

MOLECULES IN AGB (AND POST-AGB) CIRCUMSTELLAR ENVELOPES

Alain Omont
Institut d'Astrophysique de Paris – CNRS
omont@iap.fr

Abstract A number of recent developments about molecular abundances in AGB and post-AGB circumstellar envelopes and circumstellar chemistry are briefly summarized. Current problems and prospects with future facilities (especially ALMA) are stressed.

1. INTRODUCTION

Since the early detections of circumstellar molecules (see e.g. the review by Zuckerman 1980), circumstellar chemistry of envelopes of evolved stars is now more than twenty years old. Compared to its cousin, interstellar chemistry, it benefits from much better defined conditions, with steady spherical expansion and very short time constants. Because of such short time constants, it results a poorer chemistry in the case where the element abundances remain close to solar, with C/O $<<$ 1. However, this is partially compensated by the richness of carbon chemistry when C/O $>$ 1. Many excellent reviews have been devoted to the properties of AGB and post-AGB circumstellar envelopes (CSEs) and to their molecular observations (see e.g the last ones by Olofsson 1996, 1999a,b).

Most of the basic features of circumstellar chemistry, such as those revealed by the observed molecular abundances, are well understood (except those related to dust processes). Circumstellar chemistry has thus a well established status, nicely described by the review of A. Glassgold (1996). It relies on a vast number of observational measurements of molecular abundances, and on chemistry models by a few groups, including outstanding contributions from the Manchester group.

In this short review, I will try to stress and summarize recent developments and trends (especially from millimetre interferometers and from ISO), and foreseeable prospects, especially with the future Atacama large (sub-)millimetre array (ALMA).

R. Szczerba and S.K. Górny (eds.), Post-AGB Objects as a Phase of Stellar Evolution, 357–367.

2. CO AND MASS–LOSS

2.1. BEST TRACERS OF MASS-LOSS

Because of its stability, the circumstellar chemistry of the CO molecule is particularly simple. In the whole circumstellar envelope, from the molecular photosphere to the outer photodissociation region, its abundance practically equals that of the less abundant element among carbon and oxygen. It goes practically undamaged through the boisterous inner layers where shocks and dust form. Thanks to its self-shielding, it survives longer than most other molecules to dissociation by interstellar UV radiation [and by the inner UV radiation in planetary nebulae (PNe) and proto-planetary nebulae (PPNe)].

In addition to providing the values of the expansion and radial velocities, CO is rightly considered as the best tracer of the mass-loss rate $\dot{M}$, because its abundance is so well fixed and it is by far the easiest non-masering molecule to detect. The other most popular diagnosis of $\dot{M}$, from mid/far infrared dust emission (IRAS, ISO, etc.), suffers from the uncertainty in the gas-to-dust ratio and in the dust composition and emissity.

However, the accurate determination of $\dot{M}$ from the intensity of CO lines needs:

- a careful modelling of CO photodissociation and of its line formation: multi-line radiative transfer, and rotational excitation by infrared radiation and collisions which requires a modelling of the kinetic temperature.

- an accurate distance, which is presently provided by HIPPARCOS only for a few hundreds of optically bright AGB stars. The other distance determinations (kinematic, bolometric...) are rather uncertain, except for a few special locations, such as the Galactic bulge which is still however practically out of reach of easy millimetre CO detection.

2.2. SPACE AND TIME VARIATION OF MASS-LOSS

The gross features of CO emission in CSEs are generally close to spherial symmetry with an r^{-2} density dependence characteristic of constant mass-loss rate (see e.g. the IRAM interferometer survey by Neri et al. 1998, and also Sahai 1998, Kerschbaum and Olofsson 1999, Knapp et al. 1998, Olofsson 2000, etc.).

However, strong departures from such a perfect uniformity are relatively frequent, while weak departures seem the rule, reflecting fundamental properties of the mass-loss process.

First there are spectacular images of detached shells, possibly multiple, produced by millimetre interferometers, especially at IRAM. In the most striking case of TT Cyg (Olofsson et al. 2000) (see also S Sct, U Cam, etc.) one can see exquisite details in an amazingly thin perfectly spherical shell. Such shells are clearly characteristic of intermittent mass-loss, possibly related to thermal pulses. If such case studies are relatively rare with the present sensitivity and angular resolution, one can expect that ALMA will allow systematic surveys of such sharp variations of mass-loss in AGB stars.

One of the main advances in the field of the CSEs in the last years is the revelation of spectacular spherical "ripples" in their high quality optical images. They reflect a fine multi-shell structure, with period in the mass-loss process $\sim$100 yr. The case of IRC+10216 (CW Leo) is particularly nice (Mauron & Huggins 2000). The images of several PPNe and PNe are equally striking (Sahai et al. 1998). In IRC+10216, the multi-shell structure seems consistent with the ring structure observed in millimetre maps of various molecules (Lucas & Guélin 1999 and references therein, Mauron & Huggins 2000). Apparently, such a structure has not yet been detected in millimetre CO lines, probably because the CO lines presently available with interferometers are optically thick. Dedicated CO and ^{13}CO maps should be desirable. A parallel effort is needed to model photodissociation and millimetre line formation of CO (and ^{13}CO) with a multi-shell structure.

Spectacular asymmetrical structures, more or less bipolar with high-velocity flows, are common in PNe and in PPNe. This is probably a general feature of mass-loss at the final stages of evolution, whose origin is not yet completely clear. Asymmetries have also been detected in a number of AGB stars. In some cases, such as *o* Cet, it is clearly related to a binary or a multiple stellar system. In others, such as V Hya, it seems related to PPN bipolarity. Its presence in inner shells of nearby archetypes of massive AGB CSEs, such as CIT 6 and especially IRC+10216 (see Men'shchikov in this volume and references therein), could trace a systematic behviour. Understanding the physical origin(s) of mass-loss asymmetries, especially in the spectacular cases of PPNe, is clearly a major issue. Multi-line high angular resolution of CO and other molecules with ALMA will be a clue for such a purpose.

2.3. THE CASE OF VERY SMALL MASS-LOSS RATES

Although it is less spectacular, the case of very small mass-loss rates ($<<$ $10^{-7} M_{\odot}/yr$) deserves a particular attention. It probably corresponds to less luminous AGB stars, and it could be a different regime of mass-loss (Winters in this volume). It could have some relation with the mass-loss on the RGB which is responsible for a substantial fraction of the total mass loss of a star and which has never been systematically studied since the early work of Reimers (1975).

Weak AGB mass-loss has been relatively little studied in IRAS data and in CO lines, save for the comprehensive study by Kerschbaum and Olofsson (1999) and a few other works (Knapp et al. 1998, Groenewegen et al. 1999).

Such CO surveys should be pursued on nearby IRAS samples, updated with data from HIPPARCOS, MSX, 2MASS/DENIS and variability studies.

They should be completed by complementary studies on atomic lines and dust. Early works on atomic lines by Reimers (1995), Guilain & Mauron (1996), etc., should be extended with present sensitivities. New infrared surveys bring a lot of information on dust mass-loss rates. Systematic combination of 2MASS and IRAS data can give new insights for the solar neighbourhood. The limited survey ($\sim$20 deg^2) ISOGAL of the inner Galactic bulge and disk at 7 and 15 μm (Omont et al. 1999, 2000) has detected more than 10^5 stars, in majority AGB stars. Its combination with DENIS data, which also allows dereddening, provides a straightforward approximate estimate of $\dot{M}_{dust}$ from $K_o - [15]_o$. Approximate values of $\dot{M}$ are thus already available for thousands of stars in the Galactic bulge (Glass et al. 1999, Omont et al. 1999, ISOGAL & MACHO Collaborations 2001, Ojha et al. in preparation).

Correlation of these ISOGAL sources with long period variables of the micro-lensing MACHO survey, in two fields in the Baade's Windows, has already produced remarkable results (ISOGAL & MACHO Collaborations 2001). Most of these sources are M5-M9 weak semi-regular variables with short periods. Stars with relatively weak mass-loss (below $10^{-7} M_\odot/yr$) have periods between 70 and 150 days. However, not all the stars in this period range display an appreciable mass-loss.

2.4. CIRCUMSTELLAR DISKS

Circumstellar molecular and dust disks seem to naturally form in a number of binary AGB or post-AGB stars. A handful are presently known (see Jura & Kahane 1999, Lucas & Guélin 1999 and references therein: EU And, BM Gem, HD 44179 (Red Rectangle), AC Her, perhaps RV Boo and U Equ, etc.). They are especially important, in particular as long-lived reservoirs of circumstellar matter, allowing e.g. O-rich disks around carbon stars and special dust processing and segregation. They may host usual phenomena in disks, such as stellar accretion and even planet formation.

It is highly desirable to improve disk diagnosis from CO detections with very narrow profile, and to carry out more systematic searches for such disks around evolved stars in order to improve the statistics of their occurrence.

2.5. PROSPECTS

As in other fields of millimetre astronomy, ALMA will revolutionize the knowledge of molecular circumstellar matter of AGB and post-AGB stars from CO observations. The gain of almost two orders of magnitude in sensitivity will allow CO surveys and deep observations of particular objects through the whole Milky Way, in particular in the bulge. The parallel gain in angular resolution will much multiply the maps of CO shells of various classes of AGB and post-AGB stars and dedicated searches for multi-shells.

Parallel developments are expected in the infrared for dust studies of CSEs. SIRTF will allow systematic photometric and spectroscopic studies up to The Magellanic Clouds, and even the detection of most luminous CSEs in M31.

3. MOLECULAR ABUNDANCES AND CIRCUMSTELLAR CHEMISTRY

3.1. THE COLD OUTER ENVELOPE

About 60 circumstellar molecules, with up to 11 atoms, have been detected in CSEs, mostly in IRC+10216, mainly by millimetre lines, with a few notable IR detections of symmetrical molecules such as C_n. Most of them have been found in the outer layers of CSEs, with $r \geq 10^{16}$ cm and $T \leq 100$ K. It is well known that the gross features of the observable species much depend on the C/O ratio, with a much richer chemistry in the carbon rich case (C/O $>$ 1). The best direct molecular discriminator to check wether C/O is larger or smaller than one in CSEs, is the ratio of mm line intensity HCN/SiO (Bujarrabal et al. 1994, Olofsson et al. 1998).

In general, there is no serious problem to infer the molecular abundances from the intensities of optically thin millimetre lines. However, some care is needed to determine the rotational excitation by infrared radiation and collisions, either from modelling or from multi-line observations, as well as to model radiative transfer of optically thick lines (with some special problems for partially overlapping hyperfine components such as for HCN).

Modelling circumstellar chemistry in outer layers of CSEs is one of the best successes of cosmic chemistry (see e.g. the review by A. Glassgold 1996). It results from the simplicity of such objects with, in first aproximation, spherical symmetry, constant V_e and $\dot{M}$, uniform ambient UV interstellar radiation, well defined initial conditions, well defined and short time constants.... The initial upstream abundances are appreciable only for a limited number of stable molecules (H_2, CO, C_2H_2, HCN, CS, SiO, SiS, H_2O, H_2S, CH_4, NaCl, etc.) steaming out of the photosphere. They are then depleted by dust formation and possibly enriched by limited shock synthesis in the inner layers.

The rich outer chemistry well explains the extraordinary variety of the observed unstable species in IRC+10216 and in other C-rich CSEs. UV photodissociation, with dust shielding, directly accounts for observed simple radicals such as CN, C_2H, C_2, (as well as OH in O-rich shells). Such radicals and the corresponding ions created by photoionization ($C_2H_2{}^+$, C^+, etc.) initiate chains of fast reactions without important activation barrier, such as

$$C_2H_2^+ + HCN \rightarrow H_2C_3N^+ + H$$

$$H_2C_3N^+ + e^- \rightarrow HC_3N + H$$

$$CN + C_2H_2 \rightarrow HC_3N + H$$

Reasonably good agreement is thus reached for the abundances of most important species such as HNC, C_2, CN, HC_3N, HC_5N, HC_7N, C_3H, C_4H, etc.

The most striking confirmation of such a dominant photochemistry by interstellar UV radiation is probably the spectacular hollow shells observed with millimetre interferometers for the spatial distribution of such species (see e.g. Lucas & Guélin 1999 and references therein).

However, as mentioned, the details of the actual shells observed for various molecular species are strikingly coincident with the multi-shell structure observed in optical in IRC+10216 by Mauron & Huggins (2000). It is clear that the chemical models should be run again with multi-shell models for the gas density. One can expect that the enhanced density in such shells will accelerate the chemical reactions and will increase the shell contrast for the synthesized species.

The variety of species is much less spectacular in O-rich CSEs, where carbon is mostly locked in CO and silicon in SiO. The oxygen chemistry in gas phase is then dominated by the straightforward photodissociation of H_2O, responsible for OH masers and for the synthesis of SO and SO_2. However, non negligible amounts of C-bearing molecules, such as HCN, CS, etc., are commonly observed. Current models have some difficulties in explaining, with pure outer photochemistry, the observed abundances of such C-bearing molecules. It is more probable that such species could be formed by shocks in inner layers, as modelled e.g. by Duari et al. (1999).

The special case of the abundances of isotope varieties and of their ratios is very important to check nucleosynthesis in AGB stars (Kahane et al. 1992, 2000, Schoier & Olofsson 2000 and references therein). Most relevant species are ^{13}C, (^{14}C), ^{15}N, ^{18}O, ^{17}O, S, Si, ^{27}Al/^{26}Al, (^{35}Cl/^{37}Cl), etc. Isotope fractionation is believed to be negligible in most cases, except for ^{13}C/^{12}C where fractionation can be induced by selective photodissociation.

3.2. CHEMISTRY AND ABUNDANCES IN WARM INNER LAYERS

Inner layers of CSEs are much more complex than the regular flow of the outer ones, for many reasons: structure, homogeneity, dynamics, physics, etc. They are subject to the effects of pulsations which push up and levitate the upper layers of the photosphere, and produce shocks. The chemistry is also complex, although the abundances are mostly frozen from thermal equilibrium in dense layers of the photosphere. Shocks may synthesize molecules with abundances different from thermodynamical equilibrium, but only stable ones can survive. Dust formation is a complex process also out of equilibrium. It depletes the gas abundances of molecules entering in the dust composition, either totally for TiO, AlO, etc. (but see the peculiar case of U Equ, Barnbaum et al. 1996), or partially for C_2H_2, SiO, etc. Such chemical processes have a feedback on the physical conditions and the dynamics of these layers. The most important effect is radiation pressure on dust which plays a major role in mass-loss generation and in the acceleration of the expanding envelope.

Molecular observations are somewhat more difficult there than in the outer layers, mostly because of the small size of this region. However, near-infrared absorption FTS spectroscopy is very powerful in measuring column densities of abundant species (such as CO, C_2H_2, HCN, SiO and isotopes); multi-line observations and modelling allow to overcome confusion effects on the line of sight. ISO emission (and absorption) lines are providing a new insight in molecules such as H_2O, CO_2, SO_2 (and HCN, C_2H_2, SiO, etc.) (Barlow 1999 and references therein, Yamamura et al. 1999). Maser studies of lines forming there (SiO, H_2O) are still developing (see Section 3.3), as well as a few millimetre studies of non masering vibrationally excited rotational lines.

These layers will much benefit of the new possibilities opened by future facilities, both for the frequency range extended to submillimetre (FIRST, SOFIA, ALMA), and for the angular resolution and sensitivity (ALMA).

3.3. PROGRESS IN CIRCUMSTELLAR MASERS

By their intensity and their easy detection, circumstellar masers remain very important diagnosis tools of CSEs. They provide values of radial and expansion velocities and they trace conditions and fine spatial structures both in relatively outer (OH) and inner layers (SiO, H_2O). Recent years have seen a number of appreciable developments.

ISO observation of the 35μm OH rotational line has confirmed models showing that it is responsible for the pumping of 1612 MHz circumstellar OH masers, through rotation excitation followed by radiative cascade and differential radiative trapping of hyperfine components.

Various VLBI observations have produced spectacular high resolution maps of maser spots, especially of OH masers with MERLIN. They have also confirmed the potential of millimetre VLBI of SiO masers for astrometry and measurements of proper motions, especially to connect the radio and the optical reference frames.

Various particular studies have been pursued, especially of OH masers in PPNe and other sources (Lewis, Engels, Zijlstra in this volume). New surveys of SiO masers in the Galactic bulge from IRAS, ISOGAL and MSX sources are in progress (Deguchi et al. 2000, Messineo et al. in preparation).

4. PECULIARITIES OF MOLECULAR ABUNDANCES AND CHEMISTRY IN PPNE (AND IN PNE)

Although the studies are less developed because of their larger distances on average, there are no striking differences in molecules detected in well developed envelopes of proto-planetary nebulae with respect to those of AGB CSEs. This is well examplified for C-rich stars by the cases of AFGL 2688 where more than 20 molecules have been detected, and of the 21μm sources. The same photochemical processes are thus at work in the outer shells of PPNe as in AGB CSEs, and they still dominate the emission of most millimetre lines. The influence of the inner photodissociation and possible dust processing is not obvious from the properties of these lines. However, it probably intervenes in optical or near-infrared lines detected either in absorption (Bakker 1997 and references therein), or in emission in bipolar nebula.

More marked peculiarities appear when the objects depart more and more from the AGB regular envelopes, either by strong molecular flows or by the onset of photoionization. For instance, the peculiar O-rich source, OH 231.8+4.2 displays enhanced abundances of species synthesized in shocks in its high velocity molecular flow. AFGL 618 is probably the youngest planetary nebula known, with an ascent ionized region which rapidly evolves on a time scale of a few tens years. It displays an amazingly rich set of millimetre and infrared (ISO) molecular lines, especially of polyatomic C-bearing species. There is evidence of relatively high temperature excitation and possibly of shock chemistry related to its strong molecular flow (see Herpin in this volume). Very recent ISO results (Cernicharo et al. 2001a,b) have revealed the presence of methylpolyynes and of unique molecules such as C_6H_6 (benzene) and C_6H_2.

Of course, molecular abundances decrease in the next step of well developed young planetary nebulae, such as the massive molecular envelope of NGC 7027. Species like HCO^+, CN and HNC are enhanced with characteristics of strong photodissociation and of ion chemistry; the quasi-unique detection of CO^+ certainly reflects the exceptionally strong UV flux in NGC 7027. Some cases of spectacular spatial and velocity distribution of millimetre and near-infrared lines are discussed in detail by P. Cox and M. Meixner in this volume, as well as the surprizing evidence (especially from ISO) of molecular emission in late stages of evolution of planetary nebulae.

Such an amazingly rich field will certainly develop with the capabilities of future facilities, especially those with high angular resolution such as ALMA, 8-10m ground telescopes and NGST. It deserves a parallel effort of modelling, especially for the chemistry. Despite the complexity of such objects, the results of the first efforts, in particular by the Manchester group, are promising. Such transition objects are also central for understanding the occurrence and the processing of PAHs in the AGB and post-AGB stages (see next Section).

5. CONCLUSIONS

5.1. MAIN PROBLEMS IN CIRCUMSTELLAR CHEMISTRY

The most important and most difficult current problems are related to dust, especially to its formation. The latter is central, not only for the dust composition and its properties, but also for the the physics and the dynamics of the mass-loss, and for the depletion of gaseous species. Despite the complexity of the nucleation problem, the modelling of dust formation in CSEs has made much progress by the Berlin group in the carbon-rich case (see e.g. Winters in this volume and references therein), and even, more recently, for oxygen-rich CSEs (Jeong et al. 2000). The situation is thus ripe for a new effort for directly measuring, with the most powerful high angular resolution and spectral technics, the spatial variation of molecular depleted abundances in the region of dust formation. It is obviously important to vigourously pursue in parallel the improvement of modelling, encompassing all the complexity of the problem, in particular the various time scales of variation of stellar and mass-loss properties.

Other, more difficult, problems related to dust include surface reactions, mainly in inner regions close to dust formation, and grain sputtering and shattering. The latter could be particularly important in the violent conditions of the inner layers of PPNe and PNe. It is even perhaps not neglible in the milder conditions of grain drift through the gas in AGB CSEs.

The question of PAHs is related to dust. There is yet no evidence of PAH presence in AGB CSEs. However, they are possibly involved in early stages of C-rich dust formation, and the structure of circumstellar dust may have some reminiscence of PAHs. The apparition of infrared features of PAHs in C-rich PPNe, with peculiar aliphatic features, and their enhancement in young PNe are certainly made easier by the presence of more and more energetic visible and UV radiation which excites their infrared fluorescence. More important is probably PAH generation from dust desintegration in the violent conditions of UV radiation, shocks and grain shattering. However, the details of such processes are far from being clear.

Other unclear chemical processes include: shock chemistry in the inner layers with complex conditions, the influence of the fine multi-shell structure, chemistry in circumstellar disks around AGB or post-AGB stars (including dust processing and segregation).

5.2. PROSPECTS

In addition to the problems mentioned above, the capabilities of new facilities, especially ALMA, promise a lot of developments for circumstellar chemistry.

The increase in sensitivity and bandwith with ALMA (and GBT) will allow us to check the chemistry in AGB CSEs with various metallicities in the whole Milky Way, including the halo and the bulge. In addition, the high angular resolution of ALMA will provide detailed spatial variations of the abundances in distant CSEs, up to a few kpc. The gain in sensitivity and in observation rapidity will permit to focus on peculiar and rare objects. In particular, one will systematically address the cases of detached shells, multiple shells, very weak mass-loss, molecular flows, circumstellar disks, symbiotic stars and peculiar AGB stars, various classes of proto-planetary nebulae and planetary nebulae. New infrared surveys (ISOGAL, MSX, 2MASS/DENIS, SIRTF, etc.) will provide tens of thousands of potential sources and the way to select best ones for a given purpose.

As discussed, in Section 3.2, the breakthrough in angular resolution and in wavelength coverage, will be particularly important for the complex case of the inner warm layers. For the first time, detailed studies of the chemistry there will be thus possible, in the context of dust formation, mass-loss generation and shocks related to pulsations.

References

Bakker, E.J. 1997, Ap&SS 251, 73

Barlow, M.J. 1999, in *Asymptotic Giant Stars*, IAU Symp. 191, eds T. Le Bertre, A. Lebre & C. Waelkens (San Francisco:ASP), 353

Barnbaum C., Omont A., Morris M., 1996, A&A 310, 259

Bergman, P., Kerschbaum, F. & Olofsson, H. 2000, A&A 353, 257

Bujarrabal, V. 1999, in *Asymptotic Giant Stars*, IAU Symp. 191, eds T. Le Bertre, A. Lebre & C. Waelkens (San Francisco:ASP), 363

Bujarrabal, V., Fuente, A., Omont, A. 1994, A&A 285, 247

Cernicharo, J., Heras, A.M., Tielens, A.G.G.M., et al. 2001a, ApJ 546, L123

Cernicharo, J., Heras, A.M., Pardo J.R., et al. 2001b, ApJ 546, L127

Deguchi, S., Fujii, T., Izumiura, H., et al. 2000, ApJS 130, 351

Duari, D., Cherchneff, I., & Willacy, K. 1999, A&A 341, L47

Glass, I.S., Ganesh, S., Alard, C., et al. 1999, MNRAS 308, 127

Glassgold, A.E. 1996, ARA&A, 34, 241

Glassgold, A.E. 1999, in *Asymptotic Giant Stars*, IAU Symp. 191, eds T. Le Bertre, A. Lebre & C. Waelkens (San Francisco:ASP)

Groenewegen, M.A.T., Baas, F., Blommaert, J.A.D.L. et al. 1999, A&AS 140, 197

Guilain, C., Mauron, N. 1996, A&A, 314, 585
Habing, H.J. 1996, A&A Rev 7, 97
ISOGAL & MACHO Collaborations 2001, ApJ in press
Jeong, K.S., Sedlmayr, E., Winters, J.M. 2000, AGM 17, 30
Jura, M. & Kahane, C. 1999, ApJ 521, 302
Kahane, C., Audinos, P., Barnbaum, C., Morris, M. 1996 A&A 314, 871
Kahane, C., Cernicharo, J., Gomez-Gonzalez, J., Guélin, M. 1992, A&A 256, 235
Kahane, C., Dufour, E., Busso, M., et al. 2000, A&A, 357, 669
Kerschbaum, F., Olofsson, H. 1999, A&AS, 138, 299
Knapp, G.R., Young, K., Lee, E., Jorissen, A. 1998, ApJS 117, 209
Lucas, R. & Guélin, M. 1999, in *Asymptotic Giant Stars*, IAU Symp. 191, eds T. Le Bertre, A. Lebre & C. Waelkens (San Francisco:ASP), 305
Mauron & Huggins P.J. 1999, A&A 349, 203
Mauron & Huggins P.J. 2000, A&A 359, 70
Neri, R., Kahane, C., Lucas, R., Bujarrabal, V., Loup, C. 1998, A&AS 130, 1
Olofsson, H. 1996 Ap&SS 245, 169
Olofsson, H., Lindqvist, M., Nyman, L.-A., Winnberg, A. 1998, A&A 329, 1059
Olofsson, H. 1999a in *Asymptotic Giant Stars*, IAU Symp. 191, eds T. Le Bertre, A. Lebre & C. Waelkens (San Francisco:ASP), 3
Olofsson, H. 1999b in *Science with the Atacama Large Millimeter Array*, ed. A. Wootten (San Francisco: ASP)
Olofsson, H., Bergman, P., Lucas, R., et al. 2000, A&A 353, 583
Omont, A., Ganesh, S., Alard, C., et al. 1999, A&A 348, 755
Omont A. & The Isogal Collaboration, 2000, in Springer Lecture Notes of Physics Series, as part of the proceedings for *ISO Surveys of a Dusty Universe*, a workshop held at Ringberg Castle, Germany, November 8 - 12, 1999
Reimers, D., 1975, in *Problems in Stellar Atmospheres and Envelopes*, B. Baschek, W.H. Kegel, G. Traving (eds), Springer, Berlin, p 229
Sahai, R., Trauger, J.T., Watson A.M. et al 1998, ApJ 493, 301
Schoeier, F.L. & Olofsson, H. 2000, A&A 353, 583
Yamamura, I., de Jong, T., Onaka, T., Cami, J., Waters, L.B.F.M. 1999, A&A 341, L9
Zuckerman, B. 1980, ARA&A 18, 263

MOLECULAR GAS IMAGES OF PROTO-PLANETARY NEBULAE

M. Meixner, D. Fong & E.C. Sutton
University of Illinois, Dept. of Astronomy, MC-221, Urbana, IL, 61801, USA
meixner@astro.uiuc.edu

W.J. Welch
University of California, Radio Astronomy Lab, Berkeley, CA 94720-3411

Abstract We present preliminary results from a ^{12}CO J=1–0 imaging survey of evolved stars which includes the PPNe: IRAS 22272+5435, HD 161796 and AFGL 2343. This survey complements the low excitation atomic gas study presented here by Fong et al. and Castro-Carrizo et al. The CO emission images of these three PPNe reveal round and detached ring structures that are expanding. These simple structures contrast with the bipolar outflows embedded in the larger envelopes of the two best studied PPNe, AFGL 618 and AFGL 2688.

1. INTRODUCTION

Mass loss on the asymptotic giant branch (AGB) creates a dusty, molecular rich circumstellar envelope (see Omont, this volume) through two phases of mass loss: first via an AGB wind ($10^{-8} - 10^{-6}$ $M_\odot$ yr^{-1}) and finally via a superwind ($10^{-5} - 10^{-3}$ $M_\odot$ yr^{-1}). When the superwind mass loss stops, the object becomes a proto-planetary nebula (PPN) in which the circumstellar envelope has coasted away from the central star creating a gap between the star and envelope. The circumstellar envelope of a PPN is composed primarily of molecules and dust. However, as the central star evolves from cooler (2×10^3 K) to hotter ($< 3 \times 10^5$ K) temperatures, the inner edge of the nebula is photodissociated and then photoionized, hence enlarging the central hole in the molecular envelope. At some point during the PPN phase, a fast wind ($< 10^{-8}$ $M_\odot$ yr^{-1}, 1000 km s^{-1}) develops and plows into the slower moving AGB circumstellar envelope, creating shocked-gas regions, in which molecular species are dissociated and heated. The interstellar radiation field will photodissociate the circumstellar envelope from the outside and limit the outer radius of the molecular gas in a PPN.

R. Szczerba and S.K. Górny (eds.), Post-AGB Objects as a Phase of Stellar Evolution, 369–376.

The rotational transitions of carbon monoxide (CO) probe the kinematics and morphology of the molecular gas nebula in PPNe. Single dish measurements of ^{12}CO J=1–0 and 2–1 in PPNe reveal parabolic line profiles of 20–40 km s^{-1} in width indicating expanding shells (e.g., Bujarrabal, Alcolea & Planseas 1992). Millimeter interferometers, such as the Berkeley-Illinois-Maryland-Association (BIMA) array, simultaneously measure the spatial and velocity components of the ^{12}CO 1–0 emission, providing a 3-D view of the molecular gas (RA, Dec and velocity). In the case of a constant velocity, spherically symmetric, expanding shell, the channel maps, which are RA and DEC images at a certain velocities, can be interpreted as cross sectional area images of the shell along the line-of-sight dimension. The most blue-shifted and red-shifted velocity channel maps will appear as unresolved spots, which are the caps moving towards and away from the observer. At the systemic velocity, the CO emission will represent a cross section of the shell's center and the radius of the circular disk or ring will be maximum.

Figure 1 shows the central channel maps of ^{12}CO J=1–0 emission from an AGB star, CIT 6, (Meixner et al. 1998) and a young planetary nebula, NGC 7027, which serve as boundary examples of what we might expect for a PPN molecular envelope. The CO emission in CIT 6 is rather continuous and peaks at the center where the molecular gas is densest and warmest. On the other hand, the CO emission in NGC 7027 reveals a central ring of molecular gas. CIT 6 is still loosing mass. However, NGC 7027 stopped loosing mass $\sim$1000 yrs ago, and its central star is now hot enough to photodissociate and photoionize the inner edge of its molecular gas envelope creating a central depression in the CO emission (e.g. Latter et al. 2000). By imaging a sample of AGB stars, PPNe and PNe, in the ^{12}CO J=1–0 line, we can study the transformation of the molecular envelope as an object evolves from an AGB star to a PN. Moreover, the CO emission data cubes provide constraints on the history of mass loss, which is imprinted on the circumstellar nebula.

2. OBSERVATIONS

We have observed a small sample of 4 AGB stars, 5 PPNe and 3 PNe in the ^{12}CO J=1–0 line using the BIMA array (B, C and D configurations) and the NRAO 12 m telescope between 1997 and 2000. This project complements our ISO spectroscopy study of the low excitation atomic gas (see Fong et al. and Castro-Carrizo et al., this volume). The BIMA data reveal the small scale structure down to $\sim$2$''$ while the NRAO 12 m data provide the large scale structure information. The intent is to combine both data sets and create full synthesis images of the ^{12}CO emission. We then plan to model the ^{12}CO emission by adapting the radiative transfer models of Justtanont et al. (1996), in collaboration with Justtanont.

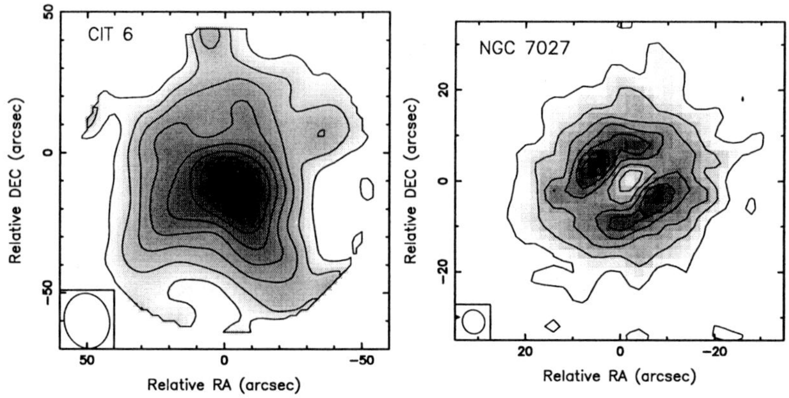

Figure 1 ^{12}CO J=1–0 emission from the channel map of the systemic velocity with the beam at the bottom left corner. Contours are 10% of peak emission. Left: The carbon rich AGB star, CIT 6, peak = 15 Jy beam^{-1} (from Meixner et al. 1998). Right: The young carbon rich PN, NGC 7027, peak = 5 Jy beam^{-1}.

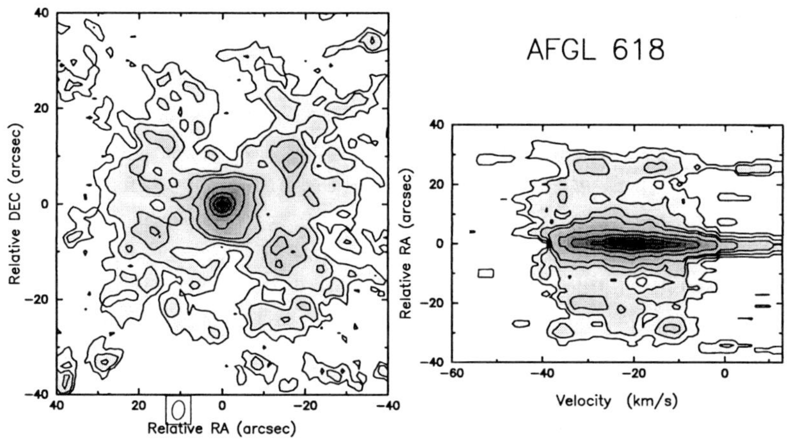

Figure 2 ^{12}CO J=1–0 emission from AFGL 618. Contours are 0.11, 0.29, 0.57, 0.86, 1.7, 2.8, 4.0, 5.1 Jy beam^{-1}. Left: The channel map of the systemic velocity with the beam in a box at the bottom. Right: A position-velocity image taken along the east-west bipolar axis. (From Meixner et al. 1998).

3. MOLECULAR GAS IMAGES OF PPN

Here we present some published images of full synthesis ^{12}CO J=1–0 images of AFGL 618 (Fig.2; Meixner et al. 1998) and some preliminary images derived from just BIMA data on IRAS 22272+5435 (Fig.3), HD 161796 (Fig.4) and AFGL 2343, which may be a supergiant and not a PPN, (Fig.5 and Fig.6). The preliminary results only show the compact structures because they are currently missing the information on large scales. In all four PPNe, the CO emission extends to larger radii than the mid-IR thermal dust emission, and

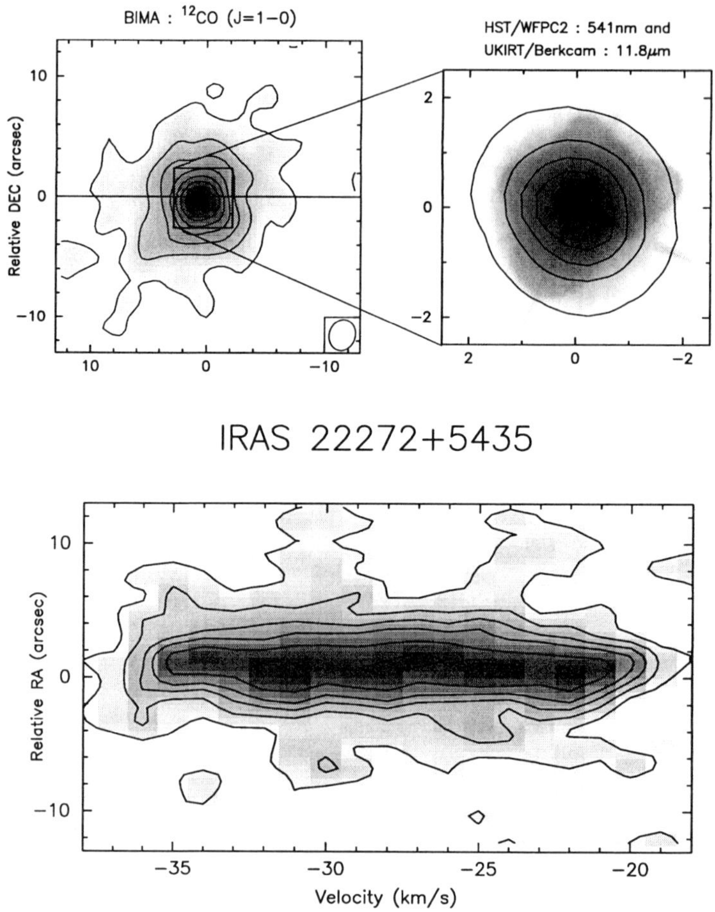

Figure 3 IRAS 22272+5435: Top left: Total integrated ^{12}CO J=1–0 intensity with the beam at the bottom right corner and with contour levels spaced at 12.5% of the peak, 33 Jy beam^{-1} km s^{-1}. The east-west line shows the axis of the position-velocity image at the bottom. The box shows the location of the combined mid-IR (contours) and HST images (greyscale) located at the top right (Meixner et al. 1997; Ueta, Meixner & Bobrowsky 2000).

the optical reflection nebula (see Ueta & Meixner, this volume; Hora et al. 1996). The structure is bipolar for AFGL 618; whereas, it is round/elliptical for IRAS 22272+5435, HD 161796 and AFGL 2343. The ring structure observed in AFGL 2343 is caused by expansion away from the central star, and not photodissociation (see Carrizo et al., this volume). The bulk of the CO emission in all four PPNe participates in a gentle expansion expected from an AGB

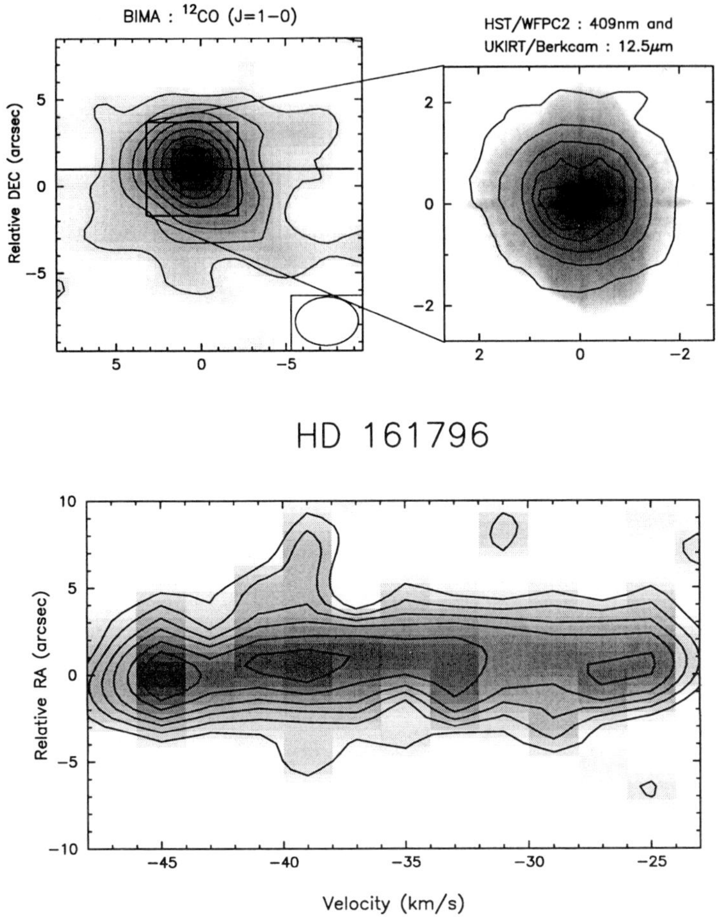

Figure 4 HD 161796: Same as Fig.3 except contour levels are spaced at 12.5% of the peak, 32 Jy beam^{-1} km s^{-1}, and mid-IR image from Skinner et al. (1994)

remnant envelope. However, AFGL 618 differs from the other three in that the core emission also includes a bipolar outflow in the east-west direction. AFGL 618 has evidence for shocks, resulting from the collision of this bipolar outflow with the surrounding molecular envelope (see Fong et al., Cox et al., and Herpin et al., this volume). AFGL 618 is quite similar to AFGL 2688, which is discussed by Cox et al. (this volume), in that both have massive molecular envelopes and central bipolar outflows.

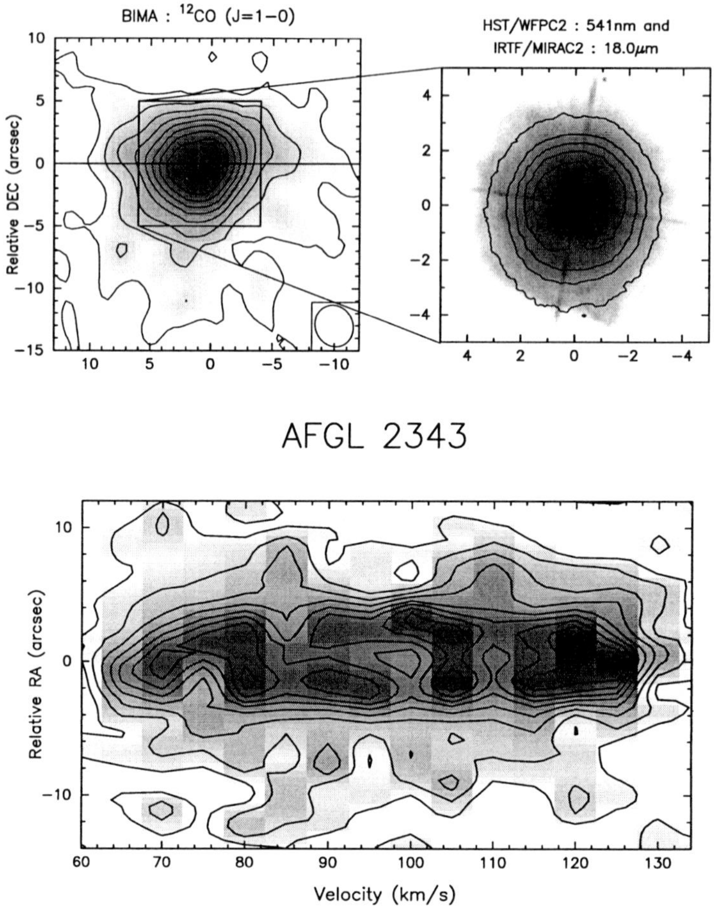

Figure 5 AFGL 2343: IRAS 19114+0002 Same as Fig.3 except contour levels are spaced at 11% of the peak, 87 Jy beam^{-1} km s^{-1}, and mid-IR image from Hawkins et al. (1995).

4. PRELIMINARY CONCLUSIONS

From the ^{12}CO emission, we observe structural and kinematic differences between AFGL 618, and AFGL 2688; and IRAS 22272+5435, HD 161796 and AFGL2343. AFGL 2688 has a similar spectral type star as HD 161796 suggesting that they are at a similar point in their evolution, yet they are quite different in their molecular gas properties. Therefore, we suggest that the PPNe, AFGL 618 and AFGL 2688, may be evolving in a different way than the PPNe, IRAS 22272+5435, HD 161796 and AFGL 2343.

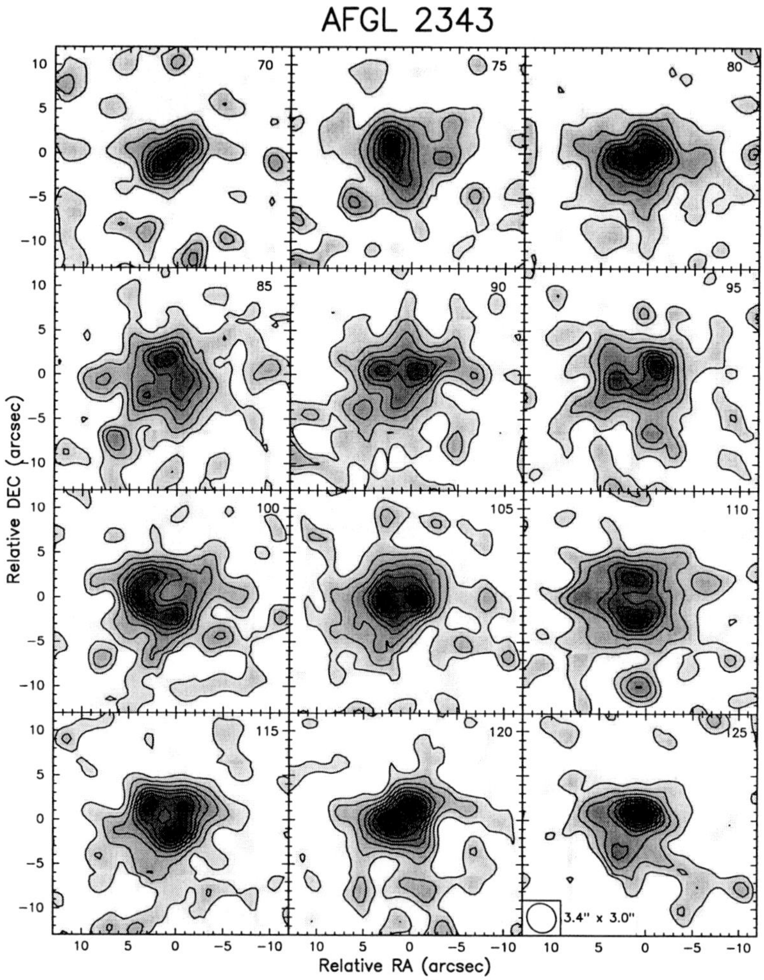

Figure 6 Channel maps of the ^{12}CO J=1–0 emission from AFGL 2343. The velocity in km s^{-1} is located at the top right corner. The beam is located in the bottom right frame. Contour levels are spaced at 11% of the peak, 2.0 Jy $beam^{-1}$. A ring of emission appears in many of the channel maps suggesting a detached shell of molecular gas.

Acknowledgments

Meixner, Fong and Sutton have been partially supported by NSF AST 99-81363. Meixner has also been supported by NSF AST 97-33697.

References

Bujarrabal V., Alcolea J., & Planesas P. 1992, A&A, 257, 701

Justtanont K., Skinner C.J., Tielens A.G.G. ., Meixner M., & Baas F., 1996, ApJ, 456, 337

Hawkins, G.W., Skinner, C.J., Meixner, M.M., Jernigan, J.G., Arens, J.F. Keto, E., and Graham, J. 1995, ApJ, 452, 314

Hora, J.L, Deutsch, L.K., Hoffmann, W.F., Fazio, G.G. 1996, AJ, 112, 2064

Latter W.B., Dayal, A., Bieging, J.H., Meakin, C., Hora, J.L., Kelly, D.M., & Tielens, A.G.G.M. 2000, ApJ, 539, 783

Meixner M., Campbell M.T., Welch W.J. & Likkel L. 1998, ApJ, 509, 392

Meixner M., Skinner C.J., Graham J.R., Keto E., Jernigan J.G., Arens J.F. 1997, ApJ, 482, 897

Skinner C.J., Meixner M., Hawkins G., Keto E., Jernigan J.G., and Arens J.F. 1994, ApJ, 423, L135

Ueta T., Meixner M., & Bobrowsky M. 2000, ApJ, 528, 861

H_2 SPECTROSCOPY OF PROTOPLANETARY NEBULAE

David A. Weintraub
Vanderbilt University
david.weintraub@vanderbilt.edu

Joel H. Kastner
Rochester Institute of Technology

Ian Gatley
Rochester Institute of Technology

Abstract Observations of planetary and protoplanetary nebulae in the spectral line of molecular hydrogen at 2.1218 μm are providing new insights into the processes of formation and evolution of post-AGB stars. In this paper, we present three significant results from the study of H_2 toward planetary and protoplanetary nebulae. First, the presence of H_2 emission is sufficient to determine that the source is *bipolar*; second, the formation of bipolar structure in protoplanetary nebulae *precedes* the event(s) that produces H_2 emission; and third, along the polar axes of PPNe, the velocity of molecular gas H_2 emission *decreases with distance from the central star*.

1. INTRODUCTION

Observers have known for two decades that some planetary and protoplanetary nebulae produce emission in the S(1) v=1$\rightarrow$0 ro-vibrational line of molecular hydrogen at 2.1218 μm; however, it is only with the introduction of very sensitive infrared imagers and spectrometers in recent years that such measurements have become useful for elucidating information about the structures of these sources and the astrophysics of their environments. One well established technique for studying sources in 2.1218 μm H_2 emission involves obtaining infrared images of sources through narrow band (full width half maximum = 1%) on-line (2.12 μm) and off-line (2.14 μm) filters. After appropriate calibrations, and assuming no significant continuum slope across the 2.12 - 2.14 μm

R. Szczerba and S.K. Górny (eds.), Post-AGB Objects as a Phase of Stellar Evolution, 377–382.

region and no contamination of the 2.12 μm image by He I line emission, the difference image $I_{2.12} - I_{2.14}$ yields an image of pure H_2 emission. In practice, the absence of He I line emission is a reasonable assumption for protoplanetary nebulae but not always so for planetary nebulae; on the other hand, most protoplanetaries show considerable continuum emission at 2 μm while many – but not all – planetaries do not. A newer technique, and one that is not dependent on the presence or absence of He I or continuum emission for an accurate assessment of the level of H_2 emission, takes advantage of a new generation of very high spectral resolution (R > 10,000) and moderate spatial resolution imaging spectrometers, which permit much more sensitive studies of emission in this line.

2. H_2 IS ASSOCIATED WITH BIPOLAR PNE AND PPNE

Zuckerman & Gatley (1988) and Kastner et al. (1994, 1996) demonstrated, and now Guerrero et al. (2000) have further confirmed, that the detection of H_2 emission from a planetary nebula is sufficient to confirm that the structure of that planetary nebula is bipolar (Gatley's rule). This is a remarkable result in that the identification of bipolar structure would appear to require a minimum spatial resoluton for the source while Gatley's rule indicates that one can determine the morphology of a source despite having no ability to obtain a spatially resolved image. These same studies reveal that bipolar nebulae are generally confined to low Galactic latitudes, suggesting that H_2-emitting, bipolar planetary nebulae are likely descended from massive progenitor stars.

While unresolved spectral data can inform us of the presence of bipolar structure, resolved imagery is necessary to determine whether the H_2 is in a disk or in the polar lobes and whether the H_2 is found throughout the disk and/or lobes or only at the inner or outer edges. Imagery also can help reveal the likely triggering mechanism(s) for producing H_2 emission.

Zuckerman & Gatley (1988) established that the brightest H_2-emitting regions of planetary nebulae are found just outside the ionized cores. Kastner et al. (1994) demonstrated that H_2 emission outlines the faint, remote, ionized halos of certain Ring-like planetary nebulae and argued that these faint H_2-emitting rings constitute a subset of bipolar nebulae whose polar axes are highly inclined out of the plane of the sky; Guerrero et al. (2000) show that the faint H_2 emission associated with the polar lobes lies just outside the optical [N II] emission zone. The brightest H_2 emission, however, is in the disk or ring of these systems, not in the lobes (Kastner et al. 1996; Guerrero et al. 2000)

With the advantage of a much more sensitive imaging system than was available for earlier studies, Kastner et al. (1994, 1996) completed a survey of H_2 emission from ~100 planetary nebulae. They found that 40 are detected in H_2

emission and, of these, 33 are classified as bipolar, and further confirmed that planetary nebulae detected in H_2 emission are confined to lower Galactic latitudes (Table 1). They concluded that the other seven, being classified as either rings or disks, were in fact bipolar sources observed at large inclinations. Thus, there is a very strong correlation between the progenitor mass and morphology of a planetary nebula and the presence or absence of H_2 emission at 2.1218 μm.

Table 1 Galactic Distribution of H_2-bright Planetary Nebula

PN Class	*Number of Sources*	*Height Above Galactic Midplane*
Detections:		
→ Bipolar	32	150 ± 25
→ Ring/Disk	6	150 ± 50
→ All Sources	38	150 ± 20
Nondetections:		
→ Bipolar	13	330 ± 100
→ Ring/Disk	14	1000 ± 320
→ All nonbipolar	33	730 ± 150
→ All Sources	54	660 ± 100

3. H_2 EMISSION POSTDATES BIPOLAR STRUCTURE

Having found that bipolar structure is closely correlated with the presence of emission from molecular hydrogen for planetary nebulae, it is natural to question how and why these two phenomena are related. Given that H_2 emission is sufficient to demonstrate the existence of bipolar structure among planetaries, can we extend this correlation back to protoplanetaries? And can we turn the correlation around and ask, do all bipolar protoplanetaries emit in the molecular hydrogen line?

Based on a study of seven sources, Weintraub et al. (1998) found that if a protoplanetary nebula is a source of H_2 emission (four sources) it is bipolar; thus, Gatley's Rule applies. Some bipolar protoplanetary nebulae (three sources), however, are not sources of H_2 emission. Specifically, the bipolar protoplanetary nebulae whose central stars have very late (M) spectral types are *not* sources of H_2 emission.

This result suggests that the event that triggers the production of emission from shocked H_2 occurs at an intermediate evolutionary epoch in the post-asymptotic giant branch evolution of transition objects, after the generation of bipolar structure and before the nebular envelopes are ionized. Kastner et al.

Table 2 Correlation of Spectral Type with H_2 Emission for Bipolar PPNe

Source	*Spectral Type of Central Star*	*H_2 Emission*	*References*
RAFGL 618	B0	yes	Burton et al. 1989
M 1–92	B0.5 IV	yes	Bujarrabal et al. 1998
IRAS 10178–5958	B1	yes	Sahai et al. 1999
IRAS 10197–5750	B4-A2	yes	Garcia-Hernandez et al. 2000
IRAS 16594–4656	B7	yes	Garcia-Hernandez et al. 2000; Van de Steene et al. 2000
RAFGL 2688	F5 Iae	yes	Sahai et al. 1998; Kastner et al. 2000
IRAS 17150–3224	G2 I	yes	Weintraub et al. 1998
IRAS 17119–5926	?	yes	Garcia-Hernandez et al. 2000
IRAS 17441–2411	?	yes	Weintraub et al. 1998
IRAS 09371+1212	M4	no	Weintraub et al. 1998
IRAS 07131–0147	M5 III	no	Weintraub et al. 1998
OH 231.8+4.2	M9 III	no	Weintraub et al. 1998

(2000) suggest that the unveiling of the central star, followed by the exposure of outflowing circumstellar material to a hotter and hotter photosphere, results in an acceleration of the un-ionized wind. Collisions of material from the faster, later epoch wind with the slower, earlier epoch wind results in the shocks seen in H_2.

In Table 2, we present an up-to-date list of 12 protoplanetary nebulae which now have been studied for H_2 emission. Of these sources, the only ones which are undetectable in H_2 emission remain the three sources with M spectral types originally reported by Weintraub et al. (1998). Nine protoplanetary nebulae now have been detected in H_2 emission; five of these have early (B) spectral types while two have intermediate (F, G) spectral types and two remain unclassified. Consequently, the correlation originally reported – that bipolar structure is generated very early in the late-AGB to post-AGB epoch, before any shocks heat the surrounding molecular gas – is strengthened by the new data.

4. H_2 VELOCITY GRADIENTS ALONG POLAR AXES OF PPNE

High resolution imaging spectroscopy of RAFGL 2688 and RAFGL 618 reveals that both of these protoplanetary nebulae have velocity gradients along their polar axes, as seen in molecular hydrogen emission (Kastner et al. 2000).

Velocity resolved maps of these two sources indicate that the highest velocity gas ($\sim$ 100 km sec^{-1}) lies *closest* to the central stars while the lowest velocity gas observed ($\sim$ 20 km sec^{-1}) lies furthest out along the polar axes (Figure 1).

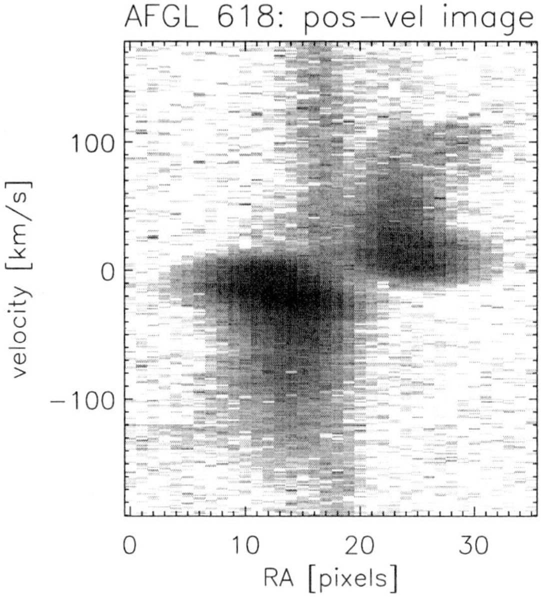

Figure 1 Calibrated Phoenix spectral image of RAFGL 618. The velocity scale of the image is centered on the systemic velocity of RAFGL 618. East is to the left; the spatial scale is 0.35″ per pixel, with the central star located very near pixel column 18 (continuum emission is present at this position in the form of vertical band across the image). The image is displayed in a logarithmic greyscale to bring out the faint line wing emission, which extends to $\sim \pm 120$ km s^{-1}.

These data are well explained if the outermost gas marks the outwardly moving loci of "slow," spherically symmetric winds from the AGB progenitor stars while the inner regions trace faster, collimated post-AGB winds. As the displacements from the central stars to the tips of the H_2 lobes are only about 6000 AU, the H_2 lobes have dynamical ages of only about 1500 yr. Thus, the transition from spherically symmetric AGB-star to axisymmetric, post-AGB star with a fully unveiled central star must be very rapid (Kastner et al. 2000; Cox et al. 2000). Kastner et al. (2000) attribute this quick transition to the stripping of the outer layers of the AGB star by an inwardly spiralling companion. The presence of H_2 emission then marks the "injestion" of the companion by the now post-AGB star and the beginning of the final, complete loss of the stellar envelope.

References

Burton, M.G., Brand, P.W.J.L., Geballe, T.R., Webster, A.S., 1989, MNRAS 236, 409

Bujarrabal, V., Alcolea, J., Sahai, R., Zamorano, J., Zijlstra, A.A., 1998, A&A 331, 361

Cox, P., Lucas, R., Huggings, P.J., Forveille, T., Bachiller, R., Guilloteau, S., Maillard, J.P. and Omont, A, 2000, A&A, 353, L25.

Garcia-Hernandez, D.A., Manchado, A., Garcia-Lario, P., Dominguez-Tagle, C., Conway, G., 2000, in "Post-AGB objects (protoplanetary nebulae) as a phase of stellar evolution."

Guerrero, M.A., Villaver, E., Manchado, A., Garcia-Lario, P., Prada, F., 2000, ApJS, 127, 125

Kastner, J.H., Gatley, I., Merrill, M., Probst, R., Weintraub, D.A., 1994, ApJ 421, 600

Kastner, J.H., Weintraub, D.A., Gatley, I., Merrill, M., Probst, R., 1996, ApJ 462, 777

Kastner, J.H., Weintraub, D.A., Gatley, I., Henn, L., 2000, ApJ, in press.

Sahai, R., Hines, D., Kastner, J.H., Weintraub, D.A., Trauger, J.T., Rieke, M.J., Thompson, R.I. & Schneider, G.S. 1998, ApJ 492, L163

Van de Steene, G.C., van Hoof, P.A.M., Wood, P.R., 2000, in "Post-AGB objects (protoplanetary nebulae) as a phase of stellar evolution."

Weintraub, D.A., Huard, T.L., Kastner, J.H, & Gatley, I., 1998, ApJ 509, 728

MOLECULAR HYDROGEN AROUND POST–AGB STARS

Domingo. A. García-Hernández
Instituto de Astrofísica de Canarias, E–38200 La Laguna, Spain
agarcia@ll.iac.es

Arturo Manchado
Instituto de Astrofísica de Canarias, E–38200 La Laguna, Spain

Pedro García–Lario
ISO Data Centre. Astrophysics Division, ESA. Aptdo. de Correos 50727, E-28080 Madrid, Spain

Carlos Dominguez–Tagle
Instituto de Astrofísica de Canarias, E–38200 La Laguna, Spain

Gail Conway
Dept. of Physics and Astronomy. University of Calgary, Calgary, Alberta, Canada T2N 1N4

Francisco Prada
Centro Astronómico Hispano–Alemán. Aptdo. de Correos 511, E–04080, Almeria, Spain

Abstract Near–infrared spectroscopy was conducted to search for molecular material around 40 post–AGB stars. We present the detection of molecular emission lines of H_2, 1–0S(1) at 2.122 μm and 2–1S(1) at 2.248 μm in 11 of the post–AGB observed. In addition, Br_γ lines at 2.166 μm also were detected in these objects. To investigate the excitation mechanism of molecular hydrogen, we have calculated the H_2 1–0S(1)/2–1S(1) ratio and vibrational temperature. Here we present the first results of a study to investigate the relation between the excitation mechanism of H_2 and the morphology / evolutionary stage for a sample of 11 post–AGB candidates that exhibit H_2 emission.

R. Szczerba and S.K. Górny (eds.), Post-AGB Objects as a Phase of Stellar Evolution, 383–386.

1. SAMPLE SELECTION

The objects considered in the present study were first identified as post–AGB stars through near–infrared photometry (García-Lario et al. 1997a). Five of these objects that exhibit H_2 emission have been observed with the Hubble Space Telescope (HST) using the Wide Field Planetary Camera (WFPC2), and they show bipolar morphology. Figure 1 show WFPC2 optical images for two of these objects.

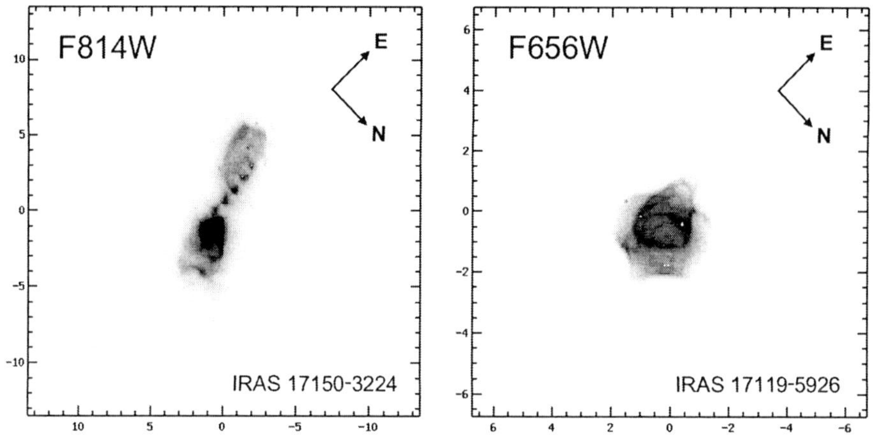

Figure 1 HST image of IRAS 17150, with WFPC2 F814W filter (left), and IRAS 17119 with WFPC2 F656W filter (right). The images are in logarithmic-scale and axes are in arcseconds.

Moreover, the bipolar structure present in the late stages of post–AGB stars is a well known but poorly understood phenomenon. Observations of molecular emission can help us to understand the origin of the bipolarity in these objects (Weintraub et al. 1998).

2. NEAR–INFRARED SPECTROSCOPY

Observations were conducted, March 1993 and January 1994, with IRSPEC at the 3.5 m NTT in La Silla.

The H_2 1–0S(1) emission line (2.122 μm) was detected in 11 objects ($\sim$28%) from the 40 IRAS sources that were observed (with eight new detections). All those that show H_2 1–0S(1) emission lines, except one (IRAS 17150), also present Br_γ emission lines (2.166 μm) and 6 of these ($\sim$60%) were detected in the H_2 2–1S(1) emission line (2.248 μm). Spectra in H_2 1–0S(1), Br_γ , H_2 2–1S(1), for two of the objects found in H_2 emission are shown in Figure 2.

We have calculated the flux lines (with an estimated uncertainty of 10% and they are not corrected for extinction), the H_2 1–0S(1)/ 2–1S(1) ratio and vibrational temperature for the objects detected in H_2 (see Table 1).

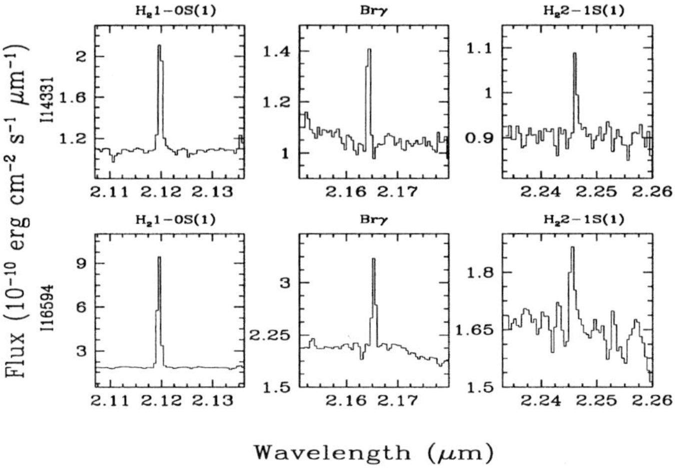

Figure 2 H_2 1–0S(1), Br_γ, and H_2 2–1S(1) spectra for IRAS 14331 and IRAS 16594.

3. H_2 EXCITATION MECHANISM

Since in our sample we expect to find low–mass stars ($\sim 1M_\odot$) with mainly B spectral-types for their central stars (García–Lario et al. 2001), it is reasonable to think that UV fields of radiation and gas densities will be moderate. In these conditions the study of the H_2 1–0S(1)/2–1S(1) ratio is a good test to identify the excitation mechanism of H_2. The shock–excited emission gives a value of the flux ratio of ~10 while fluorescent emission gives a value of ~2 (Aspin et al. 1993; Burton et al. 1998). We find that the shock–excited H_2 emission is the dominant excitation mechanism in our sample (see Table 1). This is consistent with the bipolar morphology that the sources that exhibit H_2 1–0S(1) emission generally show, where high velocity outflowing material is present.

However, we find that IRAS 17119 is one of the most evolved objects in the sample (it was identified as a very young PN with the earliest spectral type in our sample (O8V) by Parthasarathy et al. 1993), and its excitation mechanism of H_2 corresponds to pure fluorescence. This result may indicates that there is a relationship between the excitation mechanism of H_2 emission and different morphologies / evolutionary stages during the post–AGB phase.

The detection of H_2 1–0S(1) emission in 5 post-AGB candidates where HST images were available confirm the results obtained by Weintraub et al. (1998). They found that the detection of H_2 1–0S(1) emission is related with

Table 1 Flux and Ratio Lines for Post-AGB Stars Sample with H_2 emission

Object	$F[Br_\gamma]^a$	$F[1-0]^a$	$F[2-1]^a$	1−0/2−1	T_{vib}(K)
IRAS 06556+1623	65.9	3.3	$\leq$2.0	$\geq$1.7	$\leq$6841
IRAS 06562−0337	38.2	5.5	1.7	3.3	3724
IRAS 07027−7934	13.4	2.2	$\leq$1.9	$\geq$1.2	$\leq$12223
IRAS 10178−5958	26.0	36.5	1.5	24.8	1593
IRAS 10197−5750	21.6	8.9	$\leq$3.2	$\geq$2.7	$\leq$4273
IRAS 14331−6435	3.2	9.9	1.3	7.9	2366
IRAS 16594−4656	10.6	61.9	2.0	31.0	1498
IRAS 17119−5926	49.4	3.4	2.6	1.3	10098
IRAS 17150−3224	...	6.8	$\leq$0.7	$\geq$10.5	$\leq$2111
IRAS 17311−4924	1.7	34.3	3.3	10.2	2129
IRAS 18062+2410	0.6	2.7	...	...	...

[a] The line fluxes are in units of 10^{-14} erg cm^{-2} s−1. Note that $\leq$ symbol indicates an upper limit to the flux line.

the bipolar nature of the post−AGB stars. In addition, this work adds 3 new members (IRAS 10197, IRAS 16594 and IRAS 17119) to the short list of 5 bipolar proto−PNe with H_2 emission previously known (García-Lario, Riera, & Manchado. 1999).

References

Aspin et al., 1993, A&A, 278, 255

Burton M. G., Howe J. E., Geballe T. R. & Brand P. W. J. L., 1998, Publ. Astron. Soc. Aust., 15, 194

García-Lario P., Manchado A, Parthasarathy M., Pych W., & Pottasch S. R., 1997a, A&ASS, 126, 479

García-Lario P., Riera A., & Manchado A., 1999, ApJ, 526, 854

García-Lario P., Suárez O., Manchado A., & Manteiga M., 2001, (in preparation)

Parthasarathy M., Pottasch S. R., Clave J., García-Lario P., Manchado A., 1993, A&A, 267, L19

Weintraub D. A., Huard T., Kastner J., & Gatley I. 1998, ApJ, 509, 728

MORPHOLOGY AND KINEMATICS OF THE EGG NEBULA: CO AND H_2

Pierre Cox
Institut d'Astrophysique Spatiale
Université de Paris-Sud, F-91405 Orsay Cedex
Pierre.Cox@ias.u-psud.fr

Robert Lucas
300 rue de la Piscine, F-38406 Saint Martin d'Hères Cedex, France

Patrick J. Huggins
Physics Department, New York University, 4 Washington Place, New York, NY 10003, USA

Jean-Pierre Maillard
IAP, CNRS, 98b bd. Arago, F-75014 Paris, France

Abstract We report high angular and spectral resolution mapping results of the protoplanetary nebula RAFGL 2688 in the CO (J=2–1) line and the continuum at 230 GHz ($1.1'' \times 0.9''$), using the IRAM interferometer and in the H_2 1–0 S(1) transition at 4712.9 cm^{-1} (2.12 μm) and the nearby continuum ($0.5''$) obtained with *BEAR* at the CFHT. Both sets of observations reveal details of the structure and the kinematics of the gas ejected by the star over the past few hundred years. Two distinct, high-velocity outflow directions are detected emerging from a central core of gas which surrounds the star: one is oriented north-south along the optical axis, the other is oriented east-west, close to the equatorial plane. The north-south and east-west outflows are resolved into a striking series of collimated, bipolar outflows. The tips of the outflows correspond precisely to H_2 emission peaks, providing direct evidence for the impact and likely shaping effects of jets on the nearly spherical AGB molecular envelope. These outflows exemplify the mechanism by which point symmetries are imprinted on the structure of planetary nebulae at early stages of their formation.

R. Szczerba and S.K. Górny (eds.), Post-AGB Objects as a Phase of Stellar Evolution, 387–396.

1. INTRODUCTION

When a star leaves the AGB, it rapidly makes the transition to a planetary nebula through the so-called proto-planetary nebula (PPN) phase. The core of the central star is degenerate, and the envelope, which was ejected during the AGB phase, experiences a series of dramatic events. Strong stellar winds with velocities of several hundred km s^{-1} impact the inner regions of the envelope and, as the temperature of the star rises, the effects of ultraviolet radiation become apparent (e.g., Kwok 1993). The role played by high-velocity winds that interact with the slowly expanding AGB envelope has been recognized as essential in the shaping of planetary nebulae (Huggins et al. 1996; Sahai & Trauger 1998) and when the central star begins to emit ultraviolet radiation a new chemistry is triggered in the envelope which is radically different from that of the AGB phase (e.g., Cernicharo et al. 2001). However, the details of these physical mechanisms and the precise evolution from the symmetric AGB envelope to the asymmetries which characterize planetary nebulae are not well understood because relatively few objects are known to be in the rapid PPN transition phase.

The Egg Nebula in Cygnus (RAFGL 2688; Ney et al. 1975) is one of the prime examples of an object in the transition between the AGB and the planetary nebula phase. It is a bright infrared source with a bipolar optical and near–infrared nebula, scattering the light of the cool central star (F5 supergiant, $T_{eff} \sim 6500$ K). It is located at a distance of $\approx$ 1 kpc and is estimated to have evolved from the AGB about a hundred years ago (Jura & Kroto 1990). The cool star has not yet photoionized nor photodissociated the inner regions of the molecular envelope ejected during the AGB phase (e.g. Cox et al. 1996). However, the nearly circular, slowly expanding envelope is shocked by warm, optically thin, fast winds (e.g., Young et al. 1992; Jaminet et al. 1992). The optical (Crabtree & Rogers 1993; Sahai et al. 1998b) and near-infrared (Gatley et al. 1988; Smith et al. 1990; Latter et al. 1993; Cox et al. 1997; Skinner et al. 1997; Sahai et al. 1998a; Kastner et al. 1999, 2001) images show a remarkable juxtaposition of a series of concentric rings and axially symmetric structures which are centered on the obscured exciting star. Observations at near-infrared and millimeter wavelengths have revealed that high-velocity molecular gas is present both in the north-south and east-west direction suggesting multiple outflows (e.g., Cox et al. 1997; Sahai et al. 1998a; Cox et al. 2000) or a combination of an outflow and a rotating equatorial disk (Bieging & Nguyen-Q.-Rieu 1988, 1996; Kastner et al. 1999, 2001).

2. OBSERVATIONS AND RESULTS

Figure 1 summarizes high spatial resolution observations of the Egg nebula made in the CO J=2–1 transition at 1.3 mm using the Plateau de Bure interferometer (from Cox et al. 2000) compared to the *HST* image in the H_2 $1-0\,S(1)$ line emission and nearby 2.15 μm continuum from Sahai et al. (1998a). In the velocity integrated CO map, the distribution of the CO emission consists of a central core $\sim 4''$ in diameter, with extensions in both the north-south and east-west directions. At negative (approaching) velocities, the CO is detected in the northern and eastern extensions of the nebula. At positive (receding) velocities, the CO gas is extended to the south and west, with a similar velocity structure to the blue-shifted gas, reversed about the systemic velocity. The channel maps provide definite evidence of two, distinct, high-velocity outflow directions in AFGL 2688, one along a north-south axis at a P.A. of 17°, and the other in a roughly orthogonal direction east-west. Detailed examination of the CO data reveals that the two main outflows are resolved into a series of more collimated, bipolar outflows which are symmetric in direction and velocity about the center and which are identified in Fig. 1: four collimated outflows in the east-west direction, and three in the north-south direction. The identification of the outflows is supported by position-velocity diagrams along the main north-south axis and three east-west directions (see Cox et al. 2000 for further details). Note that the position of the central exciting star of AFGL 2688 derived from polarisation measurements (Weintraub et al. 2000) lies within the positional errors at the intersection of the outflows (Lucas et al. 2000 and Fig. 1).

The comparison between the *HST* image and CO data reveal that the tips of the CO outflows correspond precisely with the H_2 peaks seen in the near-infrared *HST* image. This striking correspondence is shown in greater detail in Figs. 2 and 3 where velocity resolved maps in the H_2 1–0 S(1) line emission are compared to the distribution of the CO J=2–1 line emission. Fig. 2 also shows the continuum in the near-infrared and in the millimeter.

The near-infrared observations were made in October 1999 using the *BEAR* Imaging FTS at the f/35 infrared focus of the 3.60-meter Canada-France-Hawaii Telescope (CFHT). The instrument *BEAR* – see Maillard (2000) for a detailed description – results from coupling the CFHT Fourier Transform Spectrometer with a 256×256 HgCdTe facility camera. The field of view is circular with a $24''$ diameter and the plate scale on the detector is $0.35''$/pixel. During the observations, the seeing at 2 μm was typically $0.5''$. In the camera, a narrow-band filter gave access to the spectral range 4705–4745 cm^{-1} which includes the H_2 1–0 S(1) transition at 4712.9 cm^{-1} (2.12 μm). The maximum path difference corresponds to a resolution limit (FWHM) of 0.137 cm^{-1}, i.e. 8.68 km s^{-1}, an improvement of a factor $\sim$ 4 as compared to the results presented in Cox et al. (1996) and close to the spectral resolution obtained by Kastner et al. (2001) with Phoenix.

The left panel of Fig. 2 shows the 2.1 μm continuum emission, which delineates the dust scattered light of the cool central star, superimposed on the 1.3 mm continuum which traces the thermal dust emission in the central $4''$.

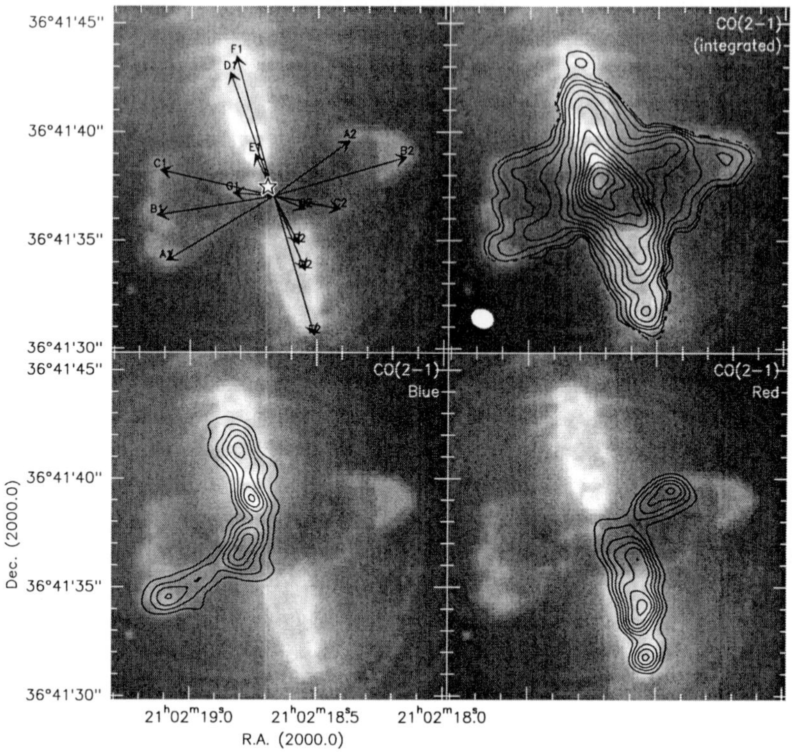

Figure 1 CO observations of AFGL 2688 superposed on the near infrared HST image from Sahai et al. (1998a). The HST image (in the background) shows reflected light from the remnant AGB envelope (suggesting spherical, episodic mass loss) and shocked molecular gas (seen in H_2 emission). The contours in the upper right panel show the inner regions of the AGB envelope traced by the velocity integrated CO(2–1) line emission. The lower panels show the blue-shifted (-80 to -60 km s^{-1}) and red-shifted (-22 to -2 km s^{-1}) CO emission. The cold, neutral gas seen in CO outlines a series of collimated, high-velocity outflows (identified in the left upper panel) that end in the shocked regions – from Cox et al. (2000). Note that the peaks A1, B1, C1 and B2 correspond to the peaks E2, E3, E4, and E1, respectively, in the nomenclature of Sahai et al. (1998a). The beam of the CO image (white ellipse) corresponds to a physical size of $\sim 1.5 \times 10^{16}$ cm. The position of the exciting star (determined from infrared polarimetric imaging by Weintraub et al. 2000) is indicated by the white star symbol in the upper left panel.

The 1.3 mm continuum lies at a P.A. of $\sim 54°$, comparable to the position angle of the centimeter continuum (Jura et al. 2000), and is not aligned with any of the outflow axes. The 1.3 mm continuum is structured: the peak emission is slightly offset to the west from the position of the central star (as determined by Weintraub et al. 2000) and, to the north, two weaker emission peaks encompass the near-infrared continuum. In the north part of the Egg nebula, there is thus a direct connection between the distribution of dust (and gas) in the central re-

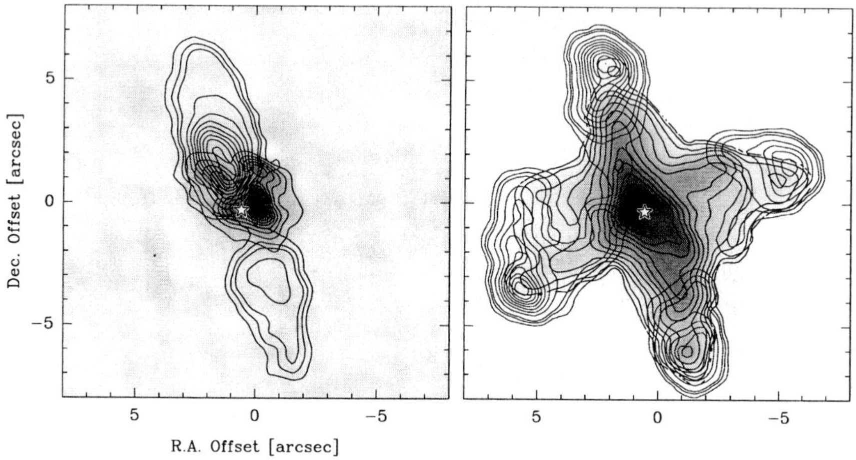

Figure 2 Left panel: Comparison of the continuuum at 1.3 mm (shaded area with contours) and at 2.1 μm (contours). Right panel: Comparison of the velocity integrated CO(2–1) (shaded area with contours) and 1–0 S(1) H_2 (contours) line emissions. The coordinates of the map center are $\alpha = 21^h02^m18^s.647$ $\delta = 36°41'37''.8$ (J2000.0). The position of the exciting star (determined from infrared polarimetric imaging by Weintraub et al. 2000) is indicated by the white star symbol.

gion, the scattered light and the 'searchlight beams' which have been described by Sahai et al. (1998b). Due to the inclination ($\sim$ 16° to the plane of the sky with the northern lobe towards the observer - Yusef-Zadeh et al. 1984), the southern side has a less favorable orientation than the northern side to discern the links between the scattered light and the structure of the dusty central core.

The distributions of the H_2 and CO emission are compared in the right panel of Fig. 2 where the striking correspondence between the H_2 peaks and the tips of the CO emission is further documented. The symmetry of the H_2 peaks with respect to the position of the central star is clearly seen. Note that the central CO core has no associated H_2 emission. Fig. 3 provides a detailed picture of the distribution and kinematics of the CO and H_2 in a series of channel maps. The distributions of the CO and H_2 emission are symmetric in both position and velocity with respect to the systemic velocity, $V_{sys} = -34\,\mathrm{km\,s^{-1}}$(LSR). Note that in the velocity channels -50 to $-55\,\mathrm{km\,s^{-1}}$, the blue-shifted CO emission is completely absorbed by the approaching side of the AGB wind that lies between us and the central regions. In the velocity channels -25 to $-55\,\mathrm{km\,s^{-1}}$, the H_2 emission lies at the tips of the CO emission. At the extreme red- and blue-shifted velocities the H_2 and CO emission are coincident. The velocity behaviour along the northern and eastern and southern and western extensions is complex. For instance, along the north extension, velocities increase away

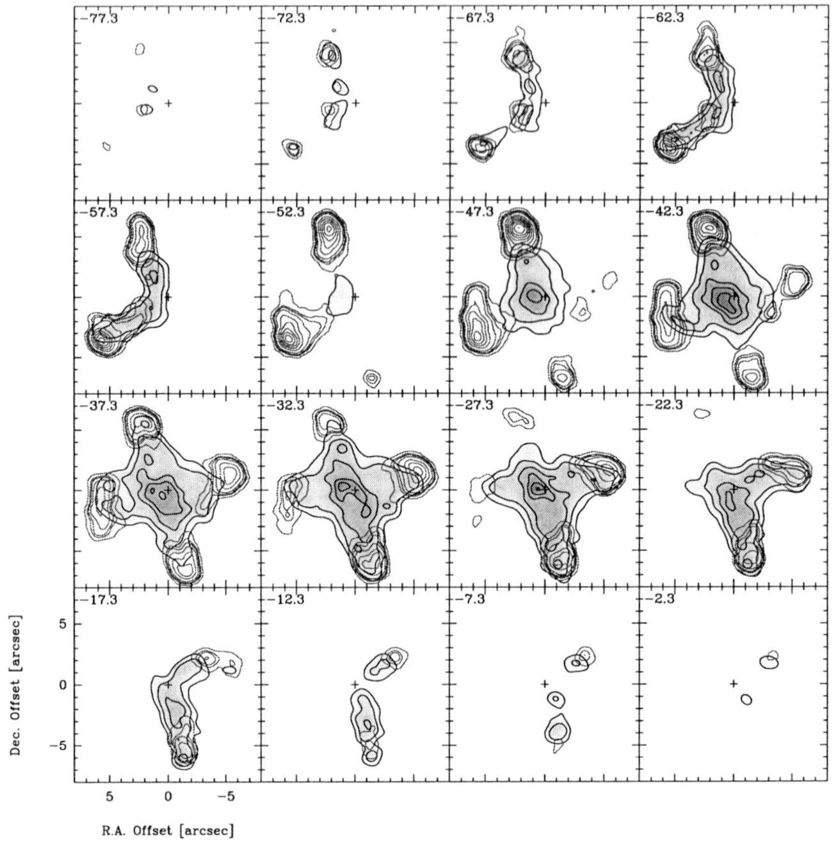

Figure 3 Comparison of the channel maps in the CO(2–1) and H_2 1–0 S(1) line emissions (shaded area with contours and grey contours, respectively). The channel separation is 5 km s^{-1}. The contours levels for the CO emission are 0.2, 0.5, and subsequent multiples of 0.5 Jy/beam. The contours of the H_2 are in steps of 10% of the peak value. The coordinates of the map center are $\alpha = 21^h02^m18^s.647$ $\delta = 36°41'37''.8$ (J2000.0) indicated by a cross.

from the center; the tip of the eastern arm is seen over a range of velocities from -60 to $-72\,\mathrm{km\,s^{-1}}$; and the highest velocity gas is found near the center of the nebula up to $\sim -80\,\mathrm{km\,s^{-1}}$. A similar behaviour is observed for the blue-shifted gas, reversed about the systemic velocity. Located behind the dense central gas, the high-velocity red-shifted CO gas has no clear H_2 counterpart, most likely as a result of extinction in the near-infrared.

Along all the outflow axes (A to G - see Fig. 1), the CO velocity increases with distance from the center. This implies that the observed CO gas is entrained by high-velocity jets which could be atomic or ionized. These jets shock the molecular gas of the AGB envelope and the points of impact are seen in the emission of molecular hydrogen (Fig. 2 and 3). Assuming the value $\sim 16^o$ for the inclination of the north-south axis to the plane of the sky, the de-projected flow velocity of F1 (at $6''$ from the center) is $22\,\mathrm{km\,s^{-1}}$, corresponding to a kinematic age of $\sim$ 1200 years. In comparison, the projected velocities of the jets D and E are $\sim 30\,\mathrm{km\,s^{-1}}$, implying ages of $\sim$250 and 125 years, if the same inclination is assumed.

As shown by Hora & Latter (1994) and Hora et al. (1999) from their study of H_2 emission lines in the near-infrared, the H_2 peaks in RAFGL 2688 do not show any significant deviation from a collisionally-excited spectrum. Further constraints on the shock conditions are provided by the far-infrared *ISO* LWS spectrum, which is dominated by strong rotational transitions of CO which are detected from J=14–13 up to J=27–26, but does not include the atomic fine structure lines [O I] and [C II] (Cox et al. 1996; Justtanont et al. 2000). Shock models provide reasonable agreement with the observations for pre-shock densities of $\leq 10^6\,\mathrm{cm^{-3}}$ and shock velocities of $20 - 30\,\mathrm{km\,s^{-1}}$, close to the velocities implied by the CO data. These shocks models also account for the lack of detection of [O I] and [C II] since the derived velocities are not strong enough to ionize the carbon or to produce atomic oxygen through the dissociation of the CO (Justtanont et al. 2000).

Finally, in addition to the collimated flows, the Plateau de Bure observations reveal in the central $2''$ a shell-like structure expanding with a velocity of $\sim 10\,\mathrm{km\,s^{-1}}$ (Fig. 3). This CO structure lies at a P.A. of $\sim 54^o$, which is comparable to that of the dust continuum emission at 1.3 mm (Fig. 2) or at 3.6 cm (Jura et al. 2000). The size of this structure ($1''$ or 1.5×10^{16} cm) implies that it was ejected $\sim$500 years ago and that it could trace the last episode of mass-loss on the AGB; the expansion velocity of this structure is much slower than that of the envelope ($\sim 20\,\mathrm{km\,s^{-1}}$). This inner core contains an estimated $\sim 0.01\,\mathrm{M_\odot}$ of dust (Jura et al. 2000). The dust opacity between 350 μm and 3.6 cm varies as $\nu^{0.6}$ which can be understood if grains at least as large as 0.5 cm have been produced in the very high density regions of the inner core of the Egg nebula. Apparently the Egg nebula experienced an intense mass ejection in which $\approx 1\,\mathrm{M_\odot}$ was expelled by the star in the past $\leq$ 1200 yr (Jura et al. 2000).

3. DISCUSSION

The CO and H_2 data presented in this paper provide detailed and complementary views of the distribution and the kinematics of the molecular gas in the inner regions of the Egg nebula. Neither the distribution of the dense gas at the center nor its kinematics indicate the presence of an equatorial disk in AFGL 2688. Instead, the CO kinematics reveal the presence of a central expanding shell-like structure (also traced in the dust continuum but not seen in H_2) and high-velocity gas, with a morphology comparable to the shocked H_2 gas tracing multiple, collimated outflows which are offset from the central core. The present data thus do not support the interpretation of an expanding, rotating equatorial disk or torus to explain the east-west kinematics in AFGL 2688 (Bieging & Rieu 1988, 1996). The recent model by Kastner et al. (2001), which combines expansion along the equator with rotation about the polar axis to account for the H_2 emission in the equatorial regions of RAFGL 2688, does not account for all the details of the kinematics revealed by the combination of the CO and H_2 observations.

In AFGL 2688, the series of young (a few 100 years), high velocity, collimated jets originate from the central star, and are directed along the north-south optical axis and in the east-west direction. The correlation of the CO outflows with H_2 emission provides details of their interaction with the nearly spherical envelope ejected on the AGB. The presence of multiple, collimated outflows in two, roughly orthogonal directions cannot be explained by the standard two-winds model (e.g., Balick 1987). In this model, a fast spherical wind interacts with the slowly expanding AGB envelope and develops a bipolar outflow due to the presence of an equatorial density enhancement, commonly assumed to be the result of a close binary companion affecting the mass expelled by the AGB star (Morris 1987). Although the similar linear extents of the north-south and east-west emission regions would not usually be expected in a precessing jet scenario, the fact is that proto-planetary nebulae with morphologies similar to AFGL 2688 have recently been found (e.g., Kwok et al. 1998) and multipolar jets appear to be a common phenomenon in young planetary nebulae (Forveille et al. 1998; Huggins et al. 2000). Whatever detailed processes are involved, the high-velocity winds must be intimately linked to the abrupt transition in the evolution of the star after the AGB phase. One important consequence of the jets is their shaping effect on the molecular envelope ejected on the AGB. They generate complex point symmetries in the envelope that later emerge in the ionized gas during the planetary nebula phase.

Acknowledgments

The financial support for one of us (P.C.) by the 'Jumelage Franco-Polonais' is gratefully acknowledged. P.J.H. acknowledges support of NSF grant AST-9986159.

References

Balick B. 1987, AJ 94, 671

Bieging, J.H., & Nguyen-Q.-Rieu 1996, ApJ, 324, 516

Bieging, J.H., & Nguyen-Q.-Rieu 1996, AJ, 112, 706

Cernicharo, J., Heras, A.M., Tielens, A.G.G.M., Pardo, J.R., Herpin, F., Guélin, M., & Waters, L.B.F.M. 2001, ApJ, 546,

Cox, P., González-Alfonso, E., Barlow, M.J. et al. 1996, A&A, 315, L265

Cox P., Maillard J.-P., Huggins P.J., Forveille T., Simons D. Guilloteau S., Rigaut F., Bachiller R., & Omont A. 1996, A&A, 293, 871

Cox P., Lucas R., Huggins P.J., Forveille, T., Bachiller, R., Guilloteau, S., Maillard, J.-P., & Omont, A. 2000, A&A, 353, L25

Crabtree, D.R. & Rogers, C. 1993, in Mass Loss on the AGB and Beyond, ed. H.E. Schwarz (Garching: ESO), p. 255

Gatley, I., DePoy, D.L., Fowler, A.M. 1988, Science, 242, 1264

Forveille T., Huggins P.J., Bachiller R., & Cox P. 1998, ApJ, 495, L111

Hora, J.L., & Latter, W.B. 1994, ApJ 437, 281

Hora, J.L., Latter, W.B., & Deutsch, L. 1999, ApJS, 124, 195

Huggins P.J., Bachiller R., Cox P., & Forveille T. 1996, A&A, 315, 284

Huggins, P.J., Forveille, T., Bachiller, R., & Cox, P. 2000, ApJ, 544, 889

Jaminet, P.A., Danchi, W.C., Sandell, G., & Sutton, E.C. 1992, ApJ 400, 535

Jura M., & Kroto H. 1990, ApJ 351, 222

Jura M., Turner J.L., Van Dyk S., & Knapp G.R. 2000, ApJ, 528, L105

Justtanont, K., Barlow, M.J., Tielens, A.G.G.M., Hollenbach, D., Latter, W., Liu, X.-W., Sylvester, R.J., Cox, P., Nguyen.-Q.-Rieu, & Skinner, C.J. 2000, A&A, 360, 1117

Kastner J.H, Henn L., Weintraub D.A., Gatley I. 1999, in IAU Symp. 191, Asymptotic Giant Branch Stars, ed. T. Le Bertre, A. Lèbre, C. Waelkens (San Francisco: PASP), 431

Kastner, J.H., Weintraub, D.A., Gatley, I., & Henn, L., 2001, ApJ, 546, 287

Kwok S. 1993, ARA&A, 31, 63

Kwok S., Su Y.L.S., & Hrivnak B.J. 1998, ApJ, 501, L117

Latter, W.B., Hora, J.L., Kelly, D.M., Deutsch, L.K., Maloney, P.R. 1993, AJ, 106, 260

Lucas R., Cox P., & Huggins P.J. 2000, in ASP Conference Series 199, Asymmetrical Planetary nebulae II: From Origins to Microstructures, ed. J.H. Kastner, N. Soker & S. Rappaport (San Francisco: ASP), p. 285

Maillard, J.P., 2000, in ASP Conference Series 195, Imaging the Universe in 3 Dimensions, ed. E. van Breughel & J. Bland-Hawthorn, p. 185

Morris M. 1987, PASP, 99, 1115

Ney, E.P., Merrill, K., Becklin, E.E., Neugebauer, G., & Wynn-Williams, C.G. 1975, ApJ, 198, L129

Sahai R., Hines D.C., Kastner J.H., Weintraub, D.A., Trauger, J.T., Rieke, M.J., Thompson, R.I., & Schneider 1998a, ApJ, 492, L163

Sahai, R., Trauger, J.T., Watson, A.M., Stapelfeldt, K.R., Hester, J.J., Burrows, C.J., Ballister, G.E., Clarke, J.T., Crisp, D., Evans, R.W., Gallagher, J.S. III,

Griffiths, R.E., Hoessel, J.G., Holtzman, J.A., Mould, J.R., Scowen, P.A., & Westphal, J.A. 1998b, ApJ, 493, 301

Sahai R., & Trauger J.T. 1998, AJ, 116, 1357

Skinner, C.J., Meixner, M., Barlow, M.J., Collison, A.J., Justtanont, K., Blanco, P., Pina, R., Ball, J., Keto, E. Arens, J.F., & Jernigan, J.G. 1997, A&A, 328, 290

Smith, M.G., Geballe, T.R., Sandell, G., Aspin, C. 1990, in Submillimeter Astronomy, ed. G.D. Watt & A.S. Webster (Kluwer, Dordrecht), p. 29

Weintraub D.A., Kastner J.H., Hines D.C., & Sahai R. 2000, ApJ 531, 401

Young K., Serabyn G., Phillips T.G., Knapp G.R., Güsten R., & Schultz A. 1992, ApJ, 385, 265

Yusef-Zadeh F., Morris M., & White R.L. 1984, ApJ, 278, 186

O-BEARING MOLECULES IN CARBON-RICH PROTO-PLANETARY OBJECTS[1]: STUDY OF CRL 618

F. Herpin

Dept. Fisica Molecular, I.E.M., C.S.I.C, Serrano 121, E-28006 Madrid, Spain

herpin@isis.iem.csic.es

J. Cernicharo

Dept. Fisica Molecular, I.E.M., C.S.I.C, Serrano 121, E-28006 Madrid, Spain

Abstract We present ISO LWS observations of the proto-planetary nebula CRL 618, a star evolving very fast to the planetary nebula stage. In addition to the lines of ^{12}CO , ^{13}CO , HCN and HNC, we report on the detection of H_2O and OH emission together with the fine structure lines of [OI] at 63 and 145 μm. We suggest that O-bearing species other than CO are produced in the innermost region of the circumstellar envelope. The UV photons from the central star photodissociate most of the molecular species produced in the AGB phase and allow a chemistry dominated by standard ion-neutral reactions. Not only allow these reactions the formation of O-bearing species, but they also modify the abundances of C-rich molecules like HCN and HNC for which we found an abundance ratio of $\simeq 1$, much lower than in AGB stars. The molecular abundances in the different regions of the circumstellar envelope have been derived from radiative transfer models and our knowledge of its physical structure.

Furthermore, we compare ISO LWS observations of three C-rich objects typical of each step of the fast transition of an AGB star to the Planetay Nebula stage: CRL 2688, a very young Proto-Planetary Nebula, CRL 618 a Proto-Planetary Nebula, and NGC 7027 a young Planetary Nebula.

[1] Based on observations with ISO, an ESA project with instruments funded by ESA Member States (especially the PI countries: France, Germany, the Netherlands and the United Kingdom) and with participation of ISAS and NASA.

R. Szczerba and S.K. Górny (eds.), Post-AGB Objects as a Phase of Stellar Evolution, 397–402.

1. INTRODUCTION

CRL 618 is one of the few clear examples of an AGB star in the transition phase to the Planetary Nebula stage. It has a compact HII region created by a hot central star (Kwok & Feldman 1981). This object is observed as a bipolar nebula at optical, radio and infrared wavelengths (Carsenty & Solf 1982, Bujarrabal et al. 1988, and Hora et al. 1996). The expansion velocity of the envelope is around 20 km s^{-1}, with also a high-velocity outflow with velocities up to 300 km s^{-1}(Cernicharo et al. 1989). The high velocity wind and the UV photons from the star perturb the circumstellar envelope (CSE) producing shocks and photodissociation regions (PDRs) which modify the physical and chemical conditions of the gas (Cernicharo et al. 1989, Neri et al. 1992). We present here (see Herpin & Cernicharo 2000) an ISO Long Wavelength Spectrometer (LWS) spectrum between 43 and 197 μm.

2. O-BEARING MOLECULES IN CRL 618

2.1. OBSERVATIONS AND RESULTS

The rotational lines of ^{12}CO (J=14-13 to J=41-40), ^{13}CO (J=14-13 to J=19-18), HCN, HNC, [OI] (at 63.170 and 145.526 μm) and C^+ (at 157.74 μm) are observed. Several lines of OH and H_2O are also detected for the first time in a C-rich star (see Fig. 1 and 2). The far-infrared spectrum is essentially dominated by the CO lines.

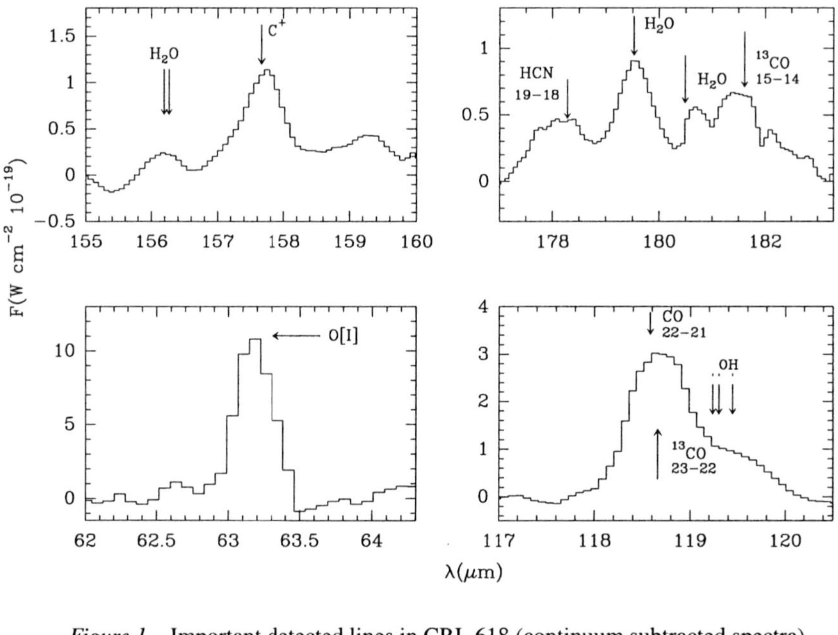

Figure 1 Important detected lines in CRL 618 (continuum subtracted spectra).

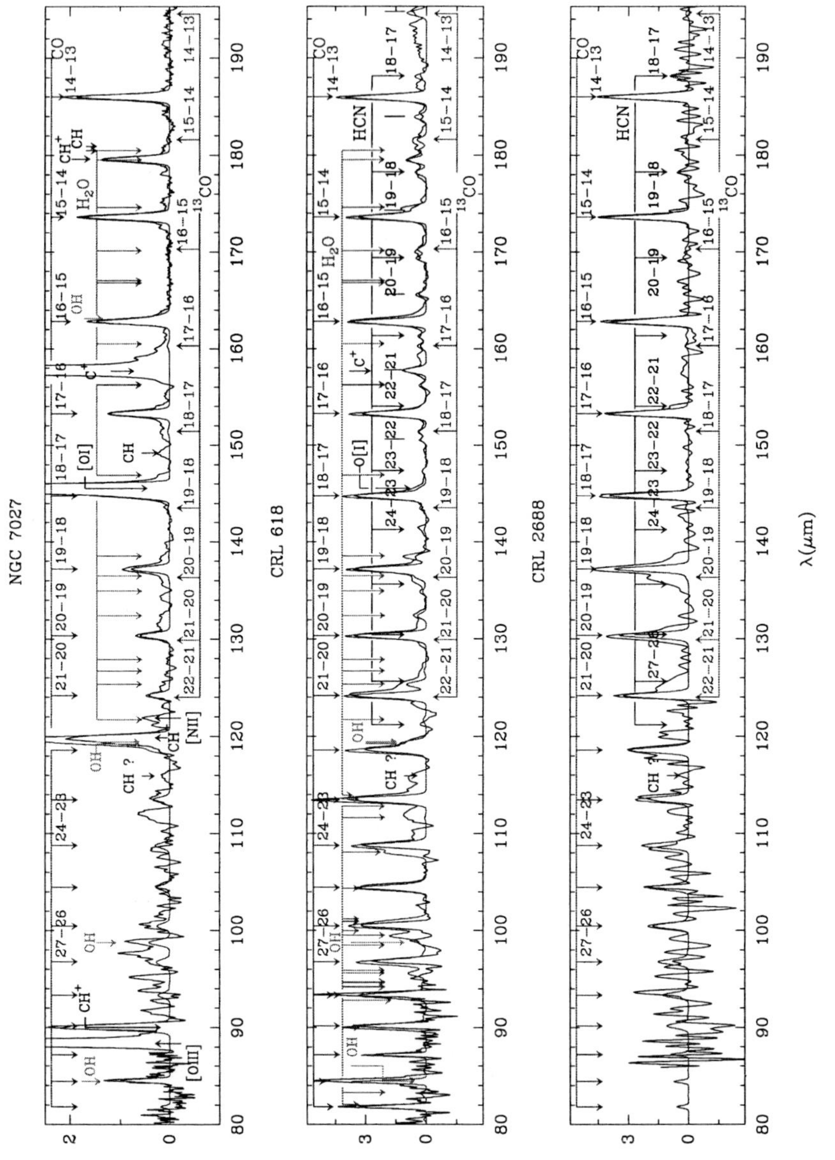

Figure 2 Continuum subtracted spectra of CRL 618 (LW detectors). The result of our model is shown by the continuous line. The lines of CO, ^{13}CO , HCN, H_2O and OH are indicated by arrows while those of HNC are indicated by vertical lines (from J=22-21 at 150.627 μm to J=17-16 at 194.759 μm). The C^+ transition is not included in our models; the plot indicates a gaussian fit to this feature.

2.2. DISCUSSION

Previous works on CRL 618 underlined the presence of a central torus associated with a Photo-Dissociated Region, an extended AGB remnant envelope and high velocity wind regions (HVW) as bipolar lobes (Weintraub et al. 1998, Neri et al. 1992, Carsenty & Solf 1982), whose physical parameters, like the velocity of the gas or the angular size, were investigated (Martin-Pintado et al. 1995, Meixner et al. 1998, Hajian et al. 1996). In order to reproduce the emission, we have to introduce several spatial components, in terms of kinematics, temperature and density based on the knowledge we have of this object. The adopted geometry is (see parameters in Table 1): (i) a central torus with the PDR; (ii) the extended AGB remnant envelope; (iii) the lobes (bipolar jets, HVW). We tried to reproduce the LWS observations by simulating (LVG-model) a spectrum (^{12}CO +^{13}CO +HCN+H_2O +OH+[OI], see Fig. 2).

In the torus the ^{12}CO column density (*N*) is 10^{19} cm^{-2} and the ^{13}CO , H_2O , HCN, HNC and OH abundances relative to ^{12}CO are $1/20$, $1/25$, $1/1000$, $1/1000$ and $1/1250$ respectively. As we reach the inner and warmer edge of the PDR, only the atomic oxygen seems to be present (abundance of 4.5 relative to CO), whereas all molecular species are photodissociated. The contribution from the bipolar lobes corresponds to high velocity and dense warm shock-heated molecular gas (Cernicharo et al. 1989, Herpin et al. in preparation), which could dominate the CO high rotational transitions (e.g., Hollenbach & McKee 1989). A contribution from this HVW region for H_2O , OH and [OI] line emission is likely because the molecules in the AGB remnant must be dissociated in the shocks, and in the molecular recombination some O-bearing species could be found like in the inner part of the torus. However, due to the lack of velocity resolution, we can not discriminate both components. The contribution of the extended AGB remnant envelope is as a colder zone for ^{12}CO and ^{13}CO .

A [^{12}CO /^{13}CO] ratio of 20 is derived, lower than the common interstellar value of 40-45, but equal to the detected ratio of Kahane et al. (1992) in this object. Our high-velocity HCN component comes from the lobes where we find a ratio [HCN/CO] of 10^{-1} equal to the ratio obtained by Neri et al. (1992). An abundance of CO of a few 10^{-4} leads to a HCN abundance of a few 10^{-5} for our data ([HCN/HNC] $\simeq$ 10). This is the expected abundance ratio in a C-rich star in the AGB phase (Cernicharo et al. 1996), and proves that this HCN material has not been processed by the UV photons from the central hot star. Furthermore, Deguchi et al. (1986) found in the disk surrounding the source an abundance of HCN of order of 2 10^{-8} per H_2 molecule, much less than Neri's value. For this region we have found a ratio [HCN/CO] of 10^{-3} which leads to an abundance of a few 10^{-7}. We can conclude that in this region HCN molecules are already efficiently dissociated. The HCN and HNC abundances are similar, which implies a high degree of reprocessing of the molecular gas (χ(HCN) >> χ(HNC) in standard C-rich AGB stars).

The detection of the fine structure lines of [OI] at 63 and 145 μm indicates that the reservoir of CO molecules formed in the AGB phase of CRL 618 has started to be reprocessed through UV photons and shocks. In CRL 618 Cernicharo et al. (1989) also found H_2CO, and Bujarrabal et al. (1988) found HCO^+. These molecules are not formed efficiently in the AGB phase of a C-rich star. How are they formed in CRL 618 ? As soon as some atomic oxygen is released into the gas phase from the photodissociation of CO, standard ion-neutral reactions can form quickly H_3O^+ and HCO^+. From the former, and through electronic recombination, H_2O and OH are efficiently produced. We expect these molecules to arise from the innermost region of the envelope where the UV flux is largest. In our models, the abundances of H_2O and OH relative to CO are 4 10^{-2} and 8 10^{-4} respectively. Obviously, in the PDR itself all molecules could be destroyed and most of the [OI] dominant coolant in high density PDRs emission is probably coming from that region.

Table 1 Table of the column densities (in cm^{-2}) for all molecules present in each region of the object as seen by the model. N is the density of H_2 in cm^{-3}.

Molecules	^{12}CO	^{13}CO	HCN	HNC	o-H_2O	OH	O[I]
TORUS (included PDR)							
T=250 K, $\theta = 1.5''$ $V = 20$ km s^{-1}, $N = 5\ 10^7$	10^{19}	5 10^{17}	10^{16}	10^{16}	3 10^{17}	8 10^{15}	
T=800 K, $\theta = 1.1''$ $V = 20$ km s^{-1}, $N = 5\ 10^7$	6 10^{17}	3 10^{16}					
T=1000 K, $\theta = 0.6''$ $V = 20$ km s^{-1}, $N = 7\ 10^7$	1 10^{19}	5 10^{17}					4.5 10^{19}
HVW							
T=200 K, $\theta = 1.7''$ $V = 50$ km s^{-1}, $N = 10^7$	5 10^{18}	2.5 10^{17}	3 10^{17}	2 10^{17}			
T=1000 K, $\theta = 1.5''$ $V = 200$ km s^{-1}, $N = 10^7$	2 10^{16}	1 10^{15}	2 10^{15}	10^{14}			
AGB remnant							
T=100 K, $\theta = 5''$ $V = 20$ km s^{-1}, $N = 5\ 10^5$	7 10^{18}	3.5 10^{17}	7 10^{17}	7 10^{15}			

3. PERSPECTIVES: A GLOBAL VIEW

To identify more precisely the changes occurring during the transition from the AGB to the PN stage, we compare ISO LWS observations of three C-rich objects typical of each step of the fast transition of an AGB star to the Planetay Nebula stage (Herpin, Goicoechea & Cernicharo 2000): CRL 2688, a very young Proto-Planetary Nebula, CRL 618, and NGC 7027 a young Planetary Nebula. We underline the violent changes that occur in the chemical composition of these objects due to the increasing UV radiation field and the strong shocks generated by fast stellar winds. The importance of these mechanisms depends on the degree of evolution of the star. This is clearly indicated by the atomic and ionic lines appearing in CRL 618, and predominant in the spectra of NGC 7027. In contrary, the shocks are less important as the evolution goes on. We confirm that O-bearing species other than CO are produced in the innermost region of the circumstellar envelope (see previous section): the UV photons from the central star photodissociate most of the molecular species produced in the AGB phase and allow a chemistry dominated by standard ion-neutral reactions; CO molecules are reprocessed, and the ^{13}CO abundance increases. HCN molecules are also reprocessed, leading to strong HNC emission close to the PDR (same abundance of HCN and HNC). At this point, CO lines and [OI] atomic lines are the dominant coolants. As the star reaches the PN stage, all the *old* AGB material has been reprocessed: the CO and outher molecules are constantly produced and destroyed. The spectra is now dominated by atomic and ionic lines. New species appear like CH^+. There is only weak HCN, or HNC emission. The disappearance of H_2O in NGC 7027 may be due to the fact that H_2O is more sensible to the UV than CO and can be fastly dissociated; it can also be incorporated in grains.

References

Bujarrabal V. et al., 1988, A&A 204, 242
Carsenty U., Solf J., 1982, A&A 106, 307
Cernicharo J. et al., 1989, A&A 222, L1
Cernicharo J. et al., 1996, A&A 315, L201
Deguchi S., Claussen M.J., Goldsmith P.F., 1986, ApJ 303, 810
Hajian A.R., Phillips J. A., Terzian Y., 1996, ApJ 467, 341
Herpin F., Cernicharo J., 2000, ApJ 530, L129
Herpin F., Goicoechea J., Cernicharo J., in preparation
Herpin F., et al., in preparation
Hollenbach D., McKee C.F., 1989, ApJ 342, 306
Hora J.L., Deutsch L.K., Hoffmann W.F., Fazio G.G., 1996, AJ 112, 2064
Kahane C. et al., 1992, A&A 256, 235
Kwok S., Feldman P.A., 1981, ApJ 247, L67
Martin-Pintado J. et al., 1995, ApJ 446, 687
Meixner M., Campbell M. T., Welch W. J., Likkel L., 1998, ApJ 509, 392
Neri R. et al.1992, A&A 262, 544
Weintraub, D. A., Huard, T., Kastner, J.H., Gatley, I., 1998, ApJ 509, 728
Yamamura I., Shibata K.M., Kasuga T., Deguchi S., 1994, ApJ 427, 406

LOW-EXCITATION ATOMIC GAS AROUND EVOLVED STARS: ISO OBSERVATIONS OF C-RICH NEBULAE

D. Fong, M. Meixner & E.C. Sutton
University of Illinois, Dept. of Astronomy, MC-221, Urbana, IL 61801, USA
d-fong@astro.uiuc.edu

A. Castro-Carrizo & V. Bujarrabal
Observatorio Astronómico Nacional, Apartado 1143, E-28800, Alcalá de Henares, Spain

W.B. Latter
Caltech, SIRTF Science Center, MS 314-6, Pasadena, CA 91125, USA

A.G.G.M. Tielens
Kapteyn Astronomical Institute, P.O. Box 800, 9700 AV Groningen, The Netherlands

D.M. Kelly
Univ. of Arizona, Steward Obs., Tucson, AZ, 85721, USA

Abstract We present ISO LWS and SWS observations of far-infrared atomic fine structure lines in 12 carbon-rich evolved stars from the AGB phase to the formation of the PN. Our analysis shows that photodissociation and not shocks dominates the evolution of the circumstellar envelope by transforming the initially molecular AGB envelopes into the atomic gas found in PPNe and PNe. Atomic mass estimates are also derived.

1. INTRODUCTION

Past studies of PNe circumstellar envelopes (CSE) mainly focused on the molecular and the ionized gas, but relatively little work have gone into studying the low excitation atomic component before ISO. In our investigation we are trying to understand the chemical evolution of the atomic gas from the AGB

R. Szczerba and S.K. Górny (eds.), Post-AGB Objects as a Phase of Stellar Evolution, 403–408.

phase to the formation of the PNe. Three questions will be addressed: 1) when does the neutral atomic component start to develop, 2) what process dominates the conversion of the molecular gas into the atomic gas and 3) how significant is the atomic mass?

The circumstellar material of an AGB star is mostly molecular, but the circumstellar shell in a PN is increasing being ionized, so somewhere between this transition, the atomic gas must emerge. The PPNe circumstellar shell is primarily molecular, but two processes can transform the molecular gas into atomic gas: 1) FUV photons emitted by the hardening radiation field of the central star which creates a photodissociation region (PDR) or 2) shocks as the fast wind ($\sim$ 1000 km s^{-1}) interacts with the slower AGB wind ($\sim$ 10 km s^{-1}).

Theoretical studies of PDRs (e.g. Tielens & Hollenbach 1985, Latter & Tielens in preparation) and shocked regions (e.g. Hollenbach & McKee 1989) show that they both contain regions of warm ($\sim$500 K) atomic gas. PDRs and shocked gas regions are cooled by far-infrared (FIR) atomic fine structure transitions of [OI] 63 and 146 μm, [CII] 158 μm, [SiII] 35 μm, [SiI] 68 and 130 μm, [SI] 25 μm, [FeII] 26 and 35 μm, and [FeI] 24 and 35 μm. We observed these lines using the ISO LWS and SWS in both the grating and Fabry-Perot (FP) modes. Our sample includes 24 objects which span the AGB, PPNe and PNe phases of evolution. Half of the objects are carbon-rich (C-rich) and the other half are oxygen-rich (O-rich), which will be discussed by Castro-Carrizo et al. (in these proceedings).

2. HR DIAGRAM

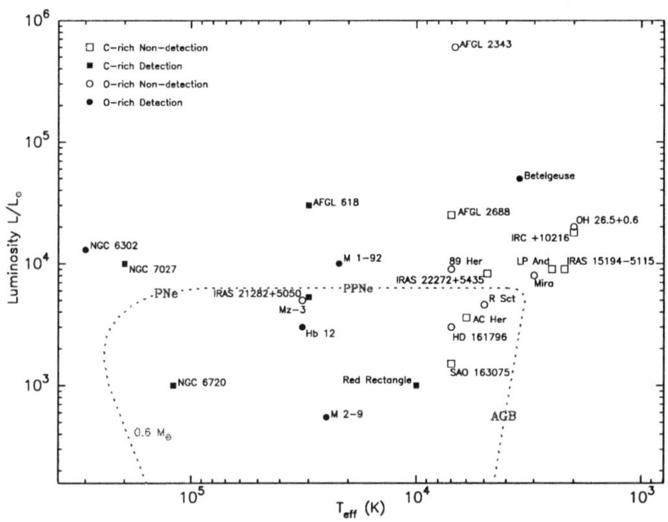

Figure 1 The observed source distribution plotted in the HR diagram.

The HR diagram (Fig. 1) graphically displays the evolutionary distribution of our evolved star sample for both the C-rich and O-rich sources. The dashed line represents the evolutionary track for a star with a core mass of $0.6\ M_{\odot}$. Only 10 out of the 24 objects had FIR line detections: all 3 PNe, 3 young PNe/PPNe, 3 PPNe, 1 red supergiant star and no AGB stars. The detections were evenly split between C-rich and O-rich sources, so there appears to be no preferential chemical type where FIR lines were seen. Except for Betelgeuse, that has excess UV emission from an active chromosphere, the HR diagram shows a clear temperature boundary between where atomic cooling lines were detected and not detected. FIR line emission was not seen until the sources attained T_{eff} above $\sim$ 10000 K. The Red Rectangle apparently defines the lowest T_{eff} where FIR was observed and only the [OI] 63 μm line was seen. With higher T_{eff} the number of detected lines increased as well as the intensity of the line emission. This trend is expected for PDRs because with increasing T_{eff}, more UV photons are available to photodissociate and heat the gas, which increases the strength of the atomic cooling lines. Below 10000 K, there are probably too few photons with high enough energies to photodissociate CO. The HR diagram is consistent with photodissociation acting as the main driving agent of chemical evolution, but this analysis alone does not preclude shocks.

3. KINEMATICS AND THE ISO FP

Our study is distinct from previous ISO studies because the spectral resolution on our FP observations was high enough to resolve the lines. For NGC 7027, the FP spectrum of the [OI] 63 μm line is shown in Fig. 2 (Left). A Gaussian fit can falsely suggest the presence of line wings, so we developed a program which convolves model line profiles with the instrumental profile. The convolved model fit reveals that the line wings are instrumental in nature. The HWHM velocity of 23 km s^{-1}, derived from the convolved fits, agree with the molecular expansion velocity of 20 km s^{-1} observed in CO and the ionized gas velocity of 27 km s^{-1} that is traced by [NeII] 12.8 μm. The ionized, atomic, and molecular gas layers are all flowing out at roughly at the same speeds – a kinematic behavior expected for PDRs. Post-shocked gas, on the other hand, would be accelerated with respect to the ambient gas and hence the line profile would be broadened. From this analysis, we find that the kinematics of the atomic gas for most of the sources agrees with PDRs.

The only object with an unusual line profile is AFGL 618. In Fig. 2 (Right), we show the fits from 3 convolved models: a single component Gaussian with a V_{HWHM} of 20 km s^{-1} to approximate the main CO line profile, the best-fit single component model for the line wings with a V_{HWHM} of 68 km s^{-1}, and a double component model. The best overall fit was provided by the double component model. The narrow component suggest PDR emission, while the broad

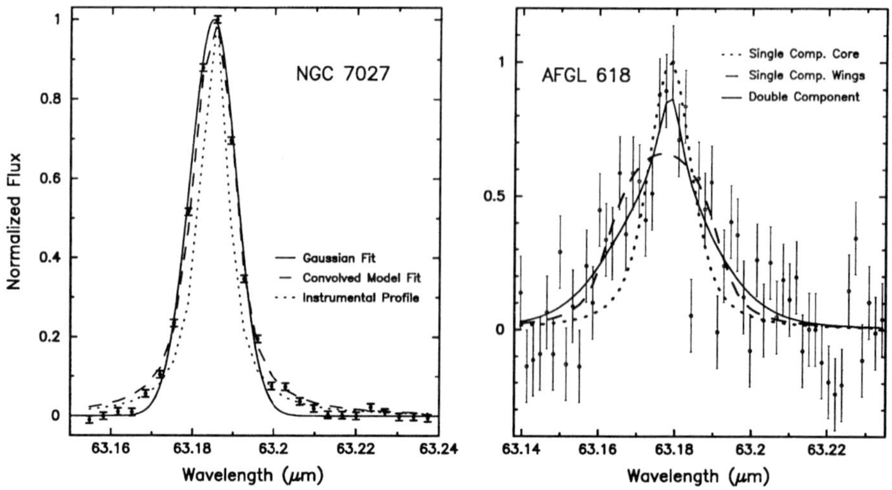

Figure 2 (Left) NGC 7027: Comparison of the Gaussian fit (solid), convolved model fit (dashed) and the instrumental profile (dotted). (Right) AFGL 618: Comparison of the best fit single component core (dotted) and line wing (dashed) models with the double component model (solid).

velocity component implies post-shocked, accelerated gas. The H_2 spectrum of AFGL 618 show that the emission is dominated by shocks, but excitation from PDRs is also present (Hora, Latter & Deutsch 1999).

4. PDR AND SHOCK MODELS

Our remaining two tests to discern the relative importance of PDRs and shocks is to compare the observed line intensities with the line intensities predicted from theoretical PDR and shock models. The results of the PDR model for the [OI] 63 μm line is shown in Figure 3. A new PDR code that properly treats enhanced carbon abundances (Latter & Tielens in preparation) was used to model the observations. The incident FUV radiation field, G, depends on the stellar parameters such as luminosity, nebular size and the fraction of far-UV photons emitted by the central star. The PDR model for [OI] 63 μm as well as the other FIR cooling lines (see Fong et al. 2000) agree reasonably well with the observed data.

Shock models, on the other hand, do not explain the observations. The results of J Shock model for the [CII] 158 μm line is shown in Figure 3 (adapted from van den Ancker 1999). The observed [CII] emission generally lies about 2 orders of magnitude higher than the J shock model predictions. J shock and C shock model results of the remaining lines (see Fong et al. 2000) show similar inconsistencies with the observations.

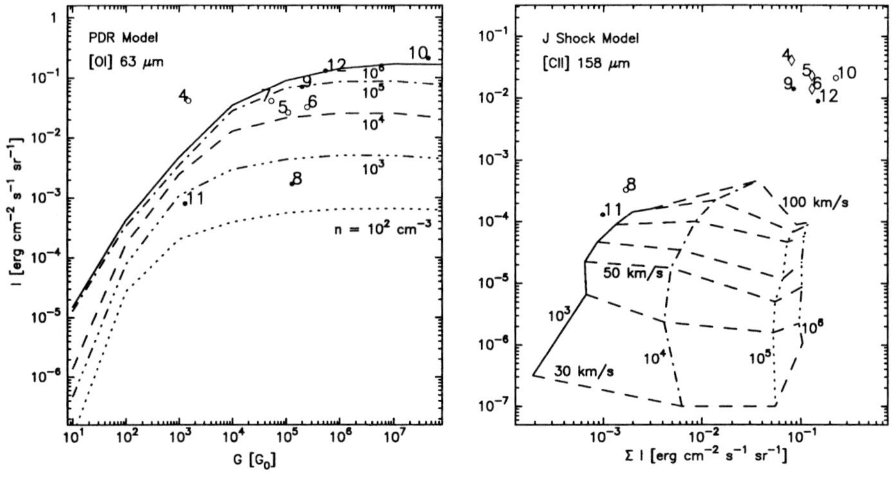

Figure 3 (Left) Comparison of the observed FIR lines with the PDR model. The curves represent the predicted intensities for models with densities from 10^2 to 10^6 cm^{-3}. (Right) Comparison of the observed FIR lines with J shock models. Σ I is the summation of all the detected FIR line intensities. The curves show predicted intensities for models with shock velocities of 40 to 100 km s^{-1} and pre-shock gas densities of 10^3 to 10^6 cm^{-3}. (Both Plots) Filled circles signify detections, open circles represent upper limits in just the y-axis and diamonds indicate upper limits in both axes. The numeric symbols represent: 4–IRAS 22272+5435, 5–AC Her, 6–AFGL 2688, 7–SAO 163075, 8–Red Rectangle, 9–IRAS 21282+5050, 10–AFGL 618, 11–NGC 6720 and 12–NGC 7027.

5. ATOMIC MASSES

One of the highlights of our work is the estimation of the atomic masses, especially for the PPNe since this is the first time that atomic masses have been calculated for these transition objects. The [CII] 158 μm line serves as a good tracer for the low excitation atomic regions. In these warm regions, the excitation temperature of [CII] is $\gg$ 92 K, and [CII] is optically thin. These conditions allow the observed line strength to be independent of the density and temperature, and dependent only on the emitting mass. The calculated atomic masses are listed in Table 1 along with the ionized and molecular masses gathered from the literature. For the young PNe IRAS 21282+5050 and NGC 7027, the molecular component harbors most of the nebular mass, while the atomic component becomes more significant for the older PN NGC 6720. These results are consistent with the growing importance of the PDR as the PN evolves.

6. SUMMARY

Our analysis supports photodissociation as the main process in the chemical evolution of the atomic gas. Photodissociation becomes significant once the

Table 1 Low-Excitation Atomic Gas Masses

Source	$M_{ionized}$	M_{atomic}	$M_{molecular}$
IRAS 22272+5435	–	< 0.01	0.56
AC Her	–	< 0.01	$1.1\ 10^{-4}$
AFGL 2688	–	< 0.02	0.6
Red Rectangle	–	< 0.002	$2.5\ 10^{-5}$
IRAS 21282+5050	0.008	0.3	2.7
AFGL 618	$4\ 10^{-4}$	0.01*	1.6
NGC 7027	0.04	0.2	1.4
NGC 6720	0.2	0.2**	0.34

* questionable [CII] detection
** mass of the observed portion of NGC 6720 ($\sim \frac{1}{3}$ of source)

source attains T_{eff} above $\sim$ 10000 K. The kinematics of our sample agrees with motions expected in PDRs, but AFGL 618 also shows signs of post-shocked, accelerated gas. The atomic gas occupies only a small fraction of the total circumstellar mass for PPNe, but grows appreciably as the PN evolves.

References

Fong D., Meixner M., Castro-Carrizo A., Bujarrabal V., Latter W.B., Tielens A.G.G.M., Kelly D.M., & Sutton E.C. 2000, A&A, submitted

Hora J.L., Latter W.B. & Deutsch L.K., 1999, ApJS, 124, 195

Hollenbach D. & McKee C.F., 1989, ApJ, 342, 306

Latter W.B. & Tielens A.G.G.M., in preparation

Tielens A.G.G.M. & Hollenbach D.; 1985, ApJ, 291, 722

van den Ancker M., 1999, *PhD Thesis*, University of Amsterdam

LOW-EXCITATION ATOMIC GAS IN PPNE: ISO OBSERVATIONS OF FIR LINE EMISSION IN O-RICH OBJECTS

Arancha Castro-Carrizo, Valentín Bujarrabal
Observatorio Astronómico Nacional, Apdo. 1143, E-28800, Alcalá de Henares, Spain
carrizo@oan.es

Dave Fong, Margaret Meixner
University of Illinois, Dept. of Astronomy, MC-221, Urbana, IL61801, USA

Alexander G.G.M. Tielens
Kapteyn Astronomical Institute, P.O.Box 800, 9700 AV Groningen, The Netherlands

William B. Latter
Calthec, SIRFT Science Center, MS 314-6, Pasadena, CA 91125, USA

Michael J. Barlow
Dept. Physics and Astronomy, Un. College London, Gower Street, London WC1E6BT, UK

Abstract We present ISO observations of fine-structure atomic lines in the far-infrared (FIR) from a sample of protoplanetary nebulae (PPNe). Only nebulae surrounding stars hotter than 10000 K were detected. We have compared our data with emission models of photodissociation regions (PDRs) and of shocked regions. The dependence of the emission on the stellar temperature, and the good agreement with the PDR models suggest that emission comes from PDRs instead of shocked regions. The [C II] 158 μm line intensity has been used to measure the mass of the low-excitation atomic gas.

1. INTRODUCTION

PPNe evolve from a molecule-rich envelope that surrounds an AGB star into an ionized planetary nebula. Many PPNe are intense emitters in molecular

R. Szczerba and S.K. Górny (eds.), Post-AGB Objects as a Phase of Stellar Evolution, 409–412.

lines, but some evolved PPNe or young PNe show emission from ionized gas. The transition from molecular gas toward ionized gas must occur through a phase where the region is photodissociated (PDR, e.g. Tielens & Hollenbach 1985). Also, the presence of shocks in PPNe could play an important role in the molecular dissociation, leading to the formation of relatively wide regions of atomic-rich gas (Hollenbach & McKee 1989).

2. ISO OBSERVATIONS AND ANALYSIS

We have observed with the ISO satellite (grating and FP spectrometers) a sample of 24 evolved objects, mostly PPNe but also including a few PNe and AGB stars for comparison. We also observed off source points to avoid confusion with interstellar cloud emission. The carbon-rich sources are presented by Fong et al. (in these proceedings). Here, we mainly discuss the case of the oxygen-rich sources. Our purpose was to detect the fine-structure atomic lines [O I] (63.2 μm and 145.5 μm), [C II] (157.7 μm), [N II] (121.9 μm), [Si I] (68.5 μm and 129.7 μm), [Si II] (34.8 μm), [S I] (25.2 μm), [Fe I] (24.0 μm and 34.7 μm), and [Fe II] (26.0 μm and 35.3 μm). An H-R diagram with all our observed sources can be found in Fong et al. (in these proceedings), where detected sources are indicated.

Some of the observed lines in M 2−9, Mz 3, Hb 12, and NGC 6302 were spectrally resolved. After deconvolution of the instrumental profile, we have checked that those FIR profiles are comparable to their low-velocity CO emission. This suggests that the atomic line emission arises from PDRs, instead of shocked regions, where higher velocities would be expected. Only the detected FIR emission from M 2−9 could partially come from shocked gas.

3. INTERPRETATION OF THE DATA

Essentially, nebulae surrounding stars with effective surface temperature (T_{eff}) $\gtrsim$ 10000 K are detected (see H-R diagram in Fong et al., in these proceedings). Moreover, line intensities are higher for hotter sources. Furthermore, we note that: 1) Betelgeuse is a supergiant with an active chromosphere, 2) the non-detection for Mz 3 is uncertain because of strong galactic contamination, 3) although AFGL 618 and AFGL 2688 have very similar envelopes, atomic FIR emission was detected in AFGL 618, whereas no lines were found in AFGL 2688, and 4) there is no atomic emission from AFGL 2343 nor HD 161796, in spite of their very massive molecular envelopes. This dependence on T_{eff} argues that the emitting region is a PDR, that is, molecular dissociation is mainly due to stellar UV photons, and not to shocks nor interstellar radiation.

To compare with the PDR models (Tielens & Hollenbach 1985, and van den Ancker 1999), we calculate the incident far-ultraviolet radiation in the PDR,

G. Note that photoelectric heating, that is the main gas heater, depends on G, which is distance independent. PDR models calculate the FIR line intensity I from G, and from the density of the emitting region n. The comparison of our observations with the predictions allows us to estimate n for the PDR of each source. Such a comparison is quite satisfactory, since densities required

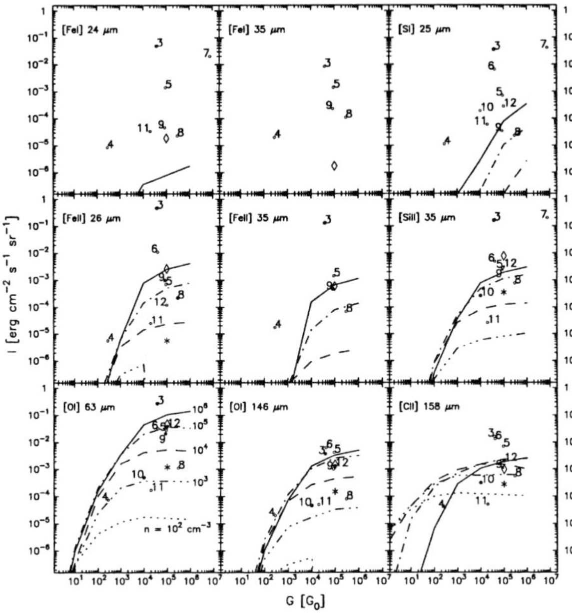

Figure 1 Comparison of the O-rich source observations with the PDR model predictions. The compared sources are: (3) Betelgeuse, (4) R Sct, (5) AFGL 2343, (6) HD 161796, IRAS 17436+5003 (7) 89 Her, (8) M 1−92, (9) M 2−9, (10) Hb 12, (11) Mz 3 and (12) NGC 6302. Detections are represented by filled circles and upper limits by open circles.

to explain the data are compatible enough for the different lines (except for Si^+, whose abundance may be too low in the model). However, models do not seem to account for the strong contrast found between the atomic emission of nebulae around stars with more or less than ~ 10000 K. In some of the cooler PDRs, the comparison of the observations with models predict much smaller densities than expected. The reason could be that the initial assumptions of the models only account for the paramenters G and n. The fraction of UV photons able to dissociate CO strongly decreases when T_{eff} is below 10000 K, and this fact is not accounted for in the calculations of G, which involves a large range of frequencies.

We have also compared our data with theoretical shock model predictions for both J- and C-type shocks (from van den Ancker 1999). Those comparisons are much less satisfactory. We cannot find definite shock characteristics that can explain the detected intensities and upper limits consistently for every source. Particularly, the [C II] 158 μm line intensity cannot be reproduced by any kind of shock.

Therefore, comparisons with PDR and shock models also suggest that the detected FIR emission mainly arises from PDRs.

4. CALCULATION OF THE LOW-EXCITATION ATOMIC MASS IN EVOLVED STARS

The best tracer of the low-excitation atomic regions may be C^+, because C^0 is ionized almost at the same time that CO is dissociated. Moreover, we have checked that the excitation temperature of the transition [C II] 158 μm is $\gg$91.2 K, and its opacity is < 1 for our sources. Such conditions allow us to calculate the total emitting mass (if C abundance is known) indenpendently of the n and the temperature, depending only on the [C II] flux and the distance to the source:

$$M(g) = 4.15\,10^{43}\,F(erg\,cm^{-2}s^{-1})\,D(kpc)^2 \qquad (64.1)$$

Table 1 Low-excitation atomic gas masses derived from the eq.1.1. Upper limits are shown when the [C II] line was not detected. Mass estimates of the molecular and ionized regions are given by comparison, when they have been found in general literature.

Source	$M(M_\odot)$	M_{mol}	*Source*	$M(M_\odot)$	M_{mol}	M_{ion}
R Sct	$< 10^{-3}$	0.002	M 2–9	< 0.03	0.005	0.004
AFGL 2343	< 2	4.8	Hb 12	0.6	<0.001	0.015
HD 161796	< 0.05	0.68	Mz 3	< 0.3	0.5	0.2
M 1–92	< 0.2	0.9	NGC 6302	1.5	0.1	0.2

5. SUMMARY

We have observed atomic fine-structure lines in the FIR from PPNe. Only nebulae surrounding stars with $T_{eff} \gtrsim 10000$ K have been detected. This seems to imply that low-excitation atomic emission comes from PDRs. Comparison with PDR line emission models is quite satisfactory, and allows us to estimate the density of the emitting gas. On the other hand, comparison with models of shocked regions cannot reproduce the detected intensities. Furthermore, the spectral analysis also point out that the emission does not arise from high-velocity regions. We can calculate the total mass of the low-excitation atomic gas from the detected [C II] flux. The well determined masses for NGC 6302 and for Hb 12 are very high, $\sim 1\ M_\odot$. This work is described more in detailed by Castro-Carrizo et al. (2000).

References

Castro-Carrizo A., Bujarrabal V., Fong D., Meixner M. et al., 2000, A&A submitted.

Fong D., Meixner M., Sutton E.C., Castro-Carrizo A. et al., 2000, in these proceedings.

Hollenbach D. & McKee C.F., 1989, ApJ 342, 306

Tielens A.G.G.M. & Hollenbach D., 1985, ApJ 291, 722

Van den Ancker M.,1999, *PhD Thesis*, University of Amsterdam.

PROBING THE POST-AGB NATURE OF HD 179821

E. Josselin, A. Lèbre

GRAAL, CC72, UMR 5024–ISTEEM, CNRS/Univ. Montpellier II, Place E. Bataillon, F-34095 Montpellier Cedex, France

josselin@graal.univ-montp2.fr

Abstract In spite of recent detailed studies, the nature of HD 179821 is still puzzling. It is classified either as an intermediate-mass post-AGB star or as a massive supergiant.

We analyze here new high sensitivity radio observations of its circumstellar envelope in the ^{12}CO and ^{13}CO lines. We also report the new detection of HCO$^+$ in this object. The ^{12}CO line intensities allow estimations of the mass-loss rate of HD 179821. The comparison between ^{12}CO and ^{13}CO lines indicates an isotopic ratio of at most 5. Such a low value is characteristic of (post-)AGB stars with low-mass progenitors. Furthermore, the HCO$^+$ molecular ion indicates that an active photochemistry occurs in the envelope of HD 179821, probably induced by UV photons emitted as the central post-AGB star warms up.

1. INTRODUCTION

The distinction between post-AGB stars and supergiants is a difficult task. Both types of objects are of luminosity class I. Their circumstellar envelopes can present large expansion velocities ($\gtrsim$ 15 km s^{-1}). The nature of low-latitude stars of spectral types B-G and luminosity class I is then difficult to assess.

HD 179821 (spectral type G5 Ia) is one of these puzzling objects. It is oxygen-rich and has a double-peaked spectral energy distribution, similar to other post-AGB stars which have experienced interruption of mass loss (Kwok 1993). We present here an analysis of new millimeter observations of this source, which clearly favour the post-AGB hypothesis. More details can be found in Josselin & Lèbre (2001, submitted to A&A).

2. SOURCE PROPERTIES

In their detailed analysis of high-resolution optical spectra of this source, Reddy & Hrivnak (1999) found high s-process element abundances, which support the idea that HD 179821 is a post-AGB object. However, interstellar absorption components in the NaI doublet suggest a large distance for HD 179821

R. Szczerba and S.K. Górny (eds.), Post-AGB Objects as a Phase of Stellar Evolution, 413–416.

($\gtrsim$ 4 kpc) implying a high luminosity (M_{bol} $\sim$ -9 mag) and a massive supergiant nature (15–20 $M_\odot$).

On the contrary, Thévenin et al. (2000) have based their composition analysis on medium-resolution spectra, arguing that the strong macroturbulence in HD 179821 dominates the spectral line widths and implies strong line blending. They did not find the high s-process elements abundances and they pointed out that moderate metal deficiency is inconsistent with the massive population I supergiant hypothesis.

Based on mid-infrared observations, Jura & Werner (1999) studied the detached dust shell around this star. All their results were interpreted assuming a 30 $M_\odot$ supergiant. They derived a total mass-loss rate for HD 179821 of $\sim$4 10^{-4} $M_\odot$ yr^{-1} which occured about 1800 yrs ago, for a distance of 6 kpc and a gas-to-dust ratio of 100.

3. THE PROBLEM OF THE DISTANCE

One of the main issue in determining the nature of HD 179821 is the determination of its distance. The kinematic distance of 6 kpc (Zuckerman & Dyck 1986) is widely used to favour the massive star hypothesis. But one has to keep in mind that this value is very uncertain. Indeed, this method is valid only if the star lies within the galactic disk. For a distance of 6 kpc, the height of HD 179821 above the galactic plane would be 524 pc, well above the scale height for 30 $M_\odot$ stars ($\sim$ 90 pc, Miller & Scalo 1979). Then, the high radial velocity may rather reflect a past runaway phenomenon.

The hypothesis that HD 179821 is a low-mass post-AGB star thus seems to be more robust. We consistently adopt a distance of 1 kpc for HD 179821 in the analysis of the observations exposed in the next section and we will show that it is fully consistent with the properties of its circumstellar envelope.

4. THE MOLECULAR ENVELOPE

4.1. OBSERVATIONS AND RESULTS

In August 1999 we performed new observations at the IRAM 30m telescope, on Pico Veleta (near Granada, Spain). Four lines were observed simultaneously: ^{12}CO(1–0) and (2–1), ^{13}CO(2–1) and HCO^+(1–0).

The parameters of the observed lines are given in table 1. Although it is weak, the detection of the HCO^+(1–0) is reliable as it has the same kinematic properties as the strong CO lines.

For an optically thick CO(1–0) line the mass-loss rate is estimated from the formula given by Loup et al. (1993) :

$$\dot{M} = 2.7\,10^{-11} \frac{T_{\rm mb}\, V_{\rm exp}^2\, D^2}{f_{\rm co}^{0.85}\, F(R_{\rm co})}$$

For a distance of 1 kpc we obtain $\dot{M} = 2\,10^{-4}$ $M_\odot$ yr^{-1}. This value is in good agreement with the mass-loss rates inferred for post-AGB stars (Bujarrabal 1999). Assuming a stellar luminosity of $\sim 10^4$ $L_\odot$ at 1 kpc and this mass-loss rate, the ratio between the momentum of the envelope ($\dot{M} V_{exp}$) and the stellar momentum ($L_\star$/c) is $\sim$ 34. This value is inconsistent with models of radiation-driven mass loss, but it is still consistent with the idea of a quasi-explosive ejection of matter during the post-AGB phase (Bujarrabal 1999).

4.2. THE $^{12}C/^{13}C$ ISOTOPIC RATIO

The observed intensity ratio I(^{12}CO(2–1))/I(^{13}CO(2–1)) is 3.2. To convert this value into the $^{12}C/^{13}C$ isotopic ratio, we have to assume that the filling factor is the same for both lines, observed with the same beam size. It is also reasonable to assume that the rotational levels are thermalized, since the dipole moment of CO is quite small ($\sim$ 0.1 Debye). Then, the main effect which has to be taken into account is the ^{12}CO(2–1) line saturation, due to optical thickness. This effect should not be higher than 30% (Knapp & Morris 1985). We thus obtain an upper limit of $\sim$ 5 for the $^{12}C/^{13}C$ abundance ratio.

This value is remarkably low, but not unusual for a low-mass post-AGB star. Indeed, for PNe of initial mass $\sim$ 1–2.5 $M_\odot$, Palla et al. (2000) found isotopic ratio similar to the one of HD 179821. Interestingly, for such a low initial mass, the star probably did not experience a 3rd dredge-up as precisely found by Thévenin et al. (2000).

4.3. CIRCUMSTELLAR CHEMISTRY

HCO^+ may result from an active photochemistry. In particular, it is probably essentially formed through the reaction

$$CO^+ + H_2 \rightarrow HCO^+ + H$$

Table 1 Results of the observations.

line	V_{lsr} (km s^{-1})	V_{exp} (km s^{-1})	I (K km s^{-1})	T_{mb} (K)	rms (K)
^{12}CO(1–0)	96	36	64.5	1.17	0.02
^{12}CO(2–1)	97	37	172.4	3.45	0.08
^{13}CO(2–1)	98	34	53.8	1.24	0.09
HCO^+(1–0)	95	35	0.3	0.01	0.003

Then, HCO^+ is associated with CO^+, which should be formed through the propagation of an ionization front into neutral gas rich in CO. The initiation of such ionization front is one of the characteristics of PPNe (Kwok 1993) and HCO^+ seems indeed to be formed during this stage (Bachiller et al. 1997).

5. CONCLUSION

The analysis of our new high-sensitivity radio observations favour the hypothesis that HD 179821 is a low-mass post-AGB star. First, the $^{12}C/^{13}C$ isotopic ratio is rather low, but comparable to observations towards PPNe and PNe. Furthermore, we detect for the first time towards HD 179821 the $HCO^+(1–0)$ line, and thus put in evidence an active photochemistry, which is characteristic of post-AGB stars.

References

Bachiller R., Forveille T., Huggins P.J., Cox P., 1997, A&A 324, 1123.
Bujarrabal V., 1999, in "Asymptotic Giant Branch Stars", IAU Symp. 191, Eds. T. Le Bertre, A. Lèbre, C. Waelkens, p. 363.
Jura M., Werner M.W., 1999, ApJ 525, L113.
Knapp G.R., Morris M., 1985, ApJ 292, 640.
Kwok S., 1993, ARA&A 31, 63.
Loup C., Forveille T., Omont A., Paul J.F., 1993, A&AS 99, 291.
Miller G.E., Scalo J.M., 1979, ApJS 41, 513.
Palla F., Bachiller R., Stanghellini L., et al., 2000, A&A 355, 69.
Reddy B.E., Hrivnak B.J., 1999, AJ 117, 1834.
Thévenin F., Parthasarathy M., Jasniewicz G., 2000, A&A 359, 138.
Zuckerman B., Dyck H.M., 1986, ApJ 311, 345.

MASER DIAGNOSTICS TO IDENTIFY STARS DEPARTING FROM THE AGB

Dieter Engels

Hamburger Sternwarte, Gojenbergsweg 112, D-21029 Hamburg, Germany

dengels@hs.uni-hamburg.de

Abstract In a number of Proto-Planetary Nebulae (PPN) candidates maser emission has been found, most often from OH. Their properties are distinctively different from those in AGB stars, reflecting the change from a radially symmetric mass-loss process on the Asymptotic Giant Branch to other mostly axisymmetric geometries. Interferometric maps indicate that during the PPN stage, when a bipolar morphology develops, the OH masers are located in radial outflows associated with the equatorial disc. H_2O masers can be associated with the dying superwind or with the newly developing bipolar outflows. SiO masers are rarely detected. Peculiar maser properties of IRAS sources with infrared colors similar to Mira variables and OH/IR stars can be used to pinpoint PPN in the making.

1. INTRODUCTION

Sometimes during the transition from the AGB to the PN stage stars change their outer appearance from an apparently spherical symmetric shape to a variety of geometrical (often axisymmetric) shapes. This change reflects changes of the mass-loss process, as the outer appearance is determined by the circumstellar dust and gas shell on the AGB and by the ionized nebula as a PN. Proto-Planetary Nebulae (PPN) are objects thought to be in the rather short transition phase ($\approx 10^3 - 10^5$ years) and their study might unveil the physical processes making different mass-loss mechanisms to work.

In the past the term "PPN" has been assigned to a large number of objects, which were considered as evolved but showed observationally deviating, sometimes unique properties. To introduce some systematics into the classification of PPN, the reappearence of the central star in the optical or near-infrared is introduced as a criterium (cf. Kwok, these proceedings), which is completely obscured during the high mass-loss "superwind" phase on the late AGB. Also in theoretical models the end of the AGB is arbitrarily defined as the end of the short phase, in which the mass-loss rates rapidly decreases from $\approx 10^{-4}$ down to $\approx 10^{-8}$ $M_\odot$ yr^{-1} (Steffen et al. 1998).

R. Szczerba and S.K. Górny (eds.), Post-AGB Objects as a Phase of Stellar Evolution, 417–422.

There is however increasing evidence accumulating that non-spherical mass-loss processes already develop on the AGB, while the stars are still deeply buried in their circumstellar shell. For example, in the center of the prototype carbon-rich AGB variable IRC+10216, a bipolar feature was discovered in the infrared, which was interpreted by Skinner et al. 1998 as the "birth of a Planetary Nebula". The onset of the axisysmmetric winds, which are so prominent during the late PPN and PN stages, might therefore be difficult to observe, and even the identification of stars currently developing such structures is a challenge. Part of them can be found by selecting objects in very red regions of the IRAS color-color diagram, but others might have escaped discovery, because the IRAS colors are similar to those of ordinary AGB stars. In the case of oxygen-rich stars the SiO, H_2O, and OH masers, which probe different parts of the circumstellar shell and which sensibly react on the properties of the mass-loss process, can help to find them. Many studies during the past two decades were able to define the systematics of the maser properties on the AGB, allowing to use deviations from these systematics to identify peculiar objects, to which possibly many PPN belong. In the following, the term "PPN" will not be restricted to objects with optically visible counterparts.

2. OH MASERS

OH masers in the main lines at 1665 and 1667 MHz and in the satellite line at 1612 MHz have been detected with a broad range of properties among a number of objects considered to be PPN. They were searched for among known PNs and PPNs or among cool IRAS sources. It was found that often mainline emission is strong or even prevails, in contrast to what is seen in OH/IR stars. The profiles have widely different shapes, also contrasting the regular double-peaked profiles of the 1612 MHz masers of OH/IR stars. In a number of cases the outflow velocities are considerably higher than is usual on the AGB. Although the OH masers often are thought to reside in the remnant AGB shell, new imaging studies usually associate the masers with low-latitude outflows in the regions, where an equatorial disk has been developed.

First evidence that maser emission might be present also in advanced stages of the transition process came from the detection of OH maser emission in the compact PN Vy 2–2 (Davis et al. 1979). A subsequent search among PNs yielded only one further object: NGC 6302 (Payne et al. 1988). Later searches by Zijlstra et al. 1989, Likkel 1989, Te Lintel Hekkert 1991, Hu et al. 1994, and Te Lintel Hekkert & Chapman 1996 focused on samples of cold IRAS sources suspected to contain PPN candidates and got a considerably higher detection rate, although the presence of masers is still the exception rather than the rule. Among PN, OH masers were detected in four cases at 1612 MHz and in one case at 1667 MHz. The masers generally are weak and consist of a single feature

(Ziljstra et al. 1989, 1991; Te Lintel Hekkert & Chapman 1996). Shepherd et al. 1990 observed the 1612 MHz maser of Vy 2−2 interferometrically and concluded that the emission originates from the remnant AGB shell.

Mainline OH masers dominate in a number of objects with mostly bipolar optical reflection nebulae. The velocity ranges are typically $\Delta v = 40$–$60\ \mathrm{km\,s^{-1}}$, which is almost twice as large as on the AGB. Different profiles are observed ranging from double-peaked to broad plateau-like shapes. Objects of this category are Roberts 22, IRAS 16559−2957 (Hu et al. 1994), IRAS 17150−3224 (Hu et al. 1993), He 3−1475 (Bobrowsky et al. 1995), and M 1−92. Roberts 22 (Sahai et al. 1999b) and M 1−92 (Seaquist et al. 1991) were observed interferometrically and in both cases the OH emission is attributed to radially expanding gas associated with the dark equatorial region. Related to these objects is probably IRAS 16342−3814 which first attracted attention due to its spectacular high velocity outflows $\Delta v > 100\ \mathrm{km\,s^{-1}}$ traced by OH and H_2O masers (Sahai et al. 1999a).

One of the best studied PPN candidates OH 231.8+4.2 resembles in many properties the bipolar reflection nebulae discussed above. The 1667 MHz emission is dominating and extending over a large velocity range of $\approx 100\ \mathrm{km\,s^{-1}}$ (Morris et al. 1984). However, its unique feature is a pulsating Mira variable as central star (Kastner et al. 1998), which indicates that this object is probably in a different evolutionary stage.

Also among high-latitude Post-AGB stars with F–G-type optical spectra OH maser emission was detected, but with widely varying properties. IRAS 07331+0021 (AI CMi), IRAS 08005−2356 (1612 MHz masers dominating), and IRAS 17436+5003 (HD 161796), IRAS 18095+2704 (1667 MHz maser dominating) have small outflow velocities $v_e = \Delta v/2 < 20\ \mathrm{km\,s^{-1}}$, while the 1667 MHz masers of IRAS 11385−5517 (HD 101584), and IRAS 18025−3906, indicate $v_e \geq 40\ \mathrm{km\,s^{-1}}$ (Lewis et al. 1990; Te Lintel Hekkert et al. 1992; Hu et al. (1994). Interferometric observations by Te Lintel Hekkert et al. of HD 101584 show that the masers are located in a bipolar outflow. The membership of IRAS 19114+0002 (HD 179821) within this PPN group is questionable (Reddy & Hrivnak, 1999), beeing instead a massive supergiant comparable to the "hypergiant" IRC+10420 (Nedoluha & Bowers 1992; Kastner & Weintraub 1995). Both also have wide OH maser ranges.

Since many years "non-variable" OH/IR stars are considered as PPN because the large amplitude pulsation has stopped and evidence was gathered from infrared measurements that they have detached circumstellar shells (Bedijn 1987, van der Veen et al. 1989). In these stars profiles, intensities and velocity ranges of the OH maser emission are indistinguishable from those of variable OH/IR stars, indicating that the masers are coming from the remnant AGB shell. Confirmation to their PPN nature might provide IRAS 17245−3951, which has an ordinary double-peaked OH 1612 MHz spectrum (Sevenster et al. 1997), but has a faint (V=15.5 mag) bipolar nebula as counterpart (Hrivnak et al. 1999). The very red IRAS colors, and its non-variability in the near infrared and IRAS bands are in accordance with the "non-variable" OH/IR stars, although the OH velocity range is relatively small (Δv=23 km s^{-1}).

During surveys for OH maser emission a number of obscured sources with large outflow velocities and sometimes prominent mainline emission were discovered. The high velocities indicate that these masers are not located in the AGB remnant shell, but may trace the emerging bipolar outflow, as seen in the optical reflection nebulae. Examples are IRAS IRAS 15405−4945 (Δv=155 km s^{-1}, Te Lintel Hekkert & Chapman 1996), IRAS 17347−3139 (Δv=125 km s^{-1}, Zijlstra et al. 1989), and IRAS 18491−0207 (Δv=200 km s^{-1}, Te Lintel Hekkert 1991). Also IRAS IRAS 16342−3814, in which the bipolar nebula was detected recently (Sahai et al. 1999a), belong to this group.

In a number of cool IRAS sources OH masers and radio continuum emission were detected, indicating the emergence of ionizing radiation. The object best studied is OH 0.9+1.3 (Zijlstra et al. 1989; Shepherd et al. 1990).

3. H_2O AND SIO MASERS

Almost all PPN candidates detected in OH do not show H_2O masers. However, among the "non-variable" OH/IR stars a few objects with remarkable H_2O maser properties are known. OH 15.7+0.8 and OH 17.7−2.0, show Type A maser spectra, e.g. the emission is concentrated at the radial velocity of the star. This is unusual, because their inferred mass-loss rates are $>10^{-5}\,M_\odot\,yr^{-1}$, while Type A maser profiles occur on the AGB only in stars with lower mass-loss rates ($<10^{-6}\,M_\odot\,yr^{-1}$). At higher mass-loss rates the maser consists of two line complexes close to the velocities of the OH maser peaks (Type B spectra) (Engels et al. 1986). This suggests that the H_2O masers in these two stars trace the Post-AGB wind with a mass-loss rate of $\approx 10^{-7}\,M_\odot\,yr^{-1}$, which is in this case simply the final phase of the decaying AGB wind. Given typical shell sizes of OH masers, the start of the decay of the mass-loss rate occured less than 200-2000 years ago (Engels 1997). Furthermore, the H_2O maser of OH 17.7-2.0 has undergone a dramatic decrease in intensity since its discovery in 1983 and disappeared after 1990. Engels attributed this decrease of the H_2O maser luminosity to a rapid decrease of the mass-loss rate in OH 17.7–2.0 by a factor $\approx$10 well below a level of $10^{-7}\,M_\odot\,yr^{-1}$, where masers are not supported any longer.

The presence of Type A H_2O maser spectra among red IRAS sources can serve therefore as pointer to objects, in which the superwind has ceased not long ago. This is corroborated by the cases of the Type A H_2O masers in

IRAS 19566+3423 and AI CMi, two stars for which independent evidence of their PPN nature exists. IRAS 19296+2227 (Engels 1996) and IRAS 19312+1950 (Nakashima & Deguchi 2000) were classified as PPN using their H_2O maser spectra.

Two other "non-variable" OH/IR stars OH 12.8−0.9 and OH 37.1−0.8 have water masers outside the OH maser velocity range. Their inferred H_2O outflow velocities are ≈11 and ≈12 km s^{-1} higher than their OH outflow velocities, which is incompatible with the standard model of AGB circumstellar shells, in which the expansion velocity increases outwards. These stars might be linked to the "water fountain" sources IRAS IRAS 16342−3814, W43A, and IRAS 19134+2131 with H_2O and OH (if present) outflow velocities of 40–130 km s^{-1} (Likkel et al. 1992). The H_2O velocities of IRAS IRAS 16342−3814 are considerably higher than those of OH, leading Likkel et al. and Sahai et al. 1999a to the suggestion that in contrast to OH the H_2O masers trace the fast bipolar wind directly.

SiO masers are rarely detected in PPN (Nyman et al. 1998). Secure detections were made only in OH 231.8+4.2 at 43 GHz by Jewell et al. 1991 and in IRAS 19312+1950 at 43 and 86 GHz by Nakashima & Deguchi 2000.

4. CONCLUSIONS

The broad variety of maser properties observed in PPN candidates as are the profiles, the relative intensities of the different OH transitions, and their velocities, make it difficult to assign particular properties to particular stages during the transition process. A first attempt was made by Lewis 1989 with his "chronological sequence", in which the restrengthening of the OH mainline emission was incorporated as a characteristic feature of the Post-AGB phase, reflecting the dilution of the remnant AGB shell. It seems however, that in parallel a new mass-loss process develops, giving rise to non-radial symmetric geometries. In a number of sources imaging observations associate the OH masers with outflowing regions related to the newly developed equatorial disc (Seaquist et al. 1991; Sahai et al. 1999a; Sahai et al. 1999b). It is therefore questionable, if any of the OH masers detected in objects, in which the central object is already visible directly or in reflected light, emerge in regions belonging to the remnant AGB shell, as it was argued i.e. in the case of the compact PN Vy 2−2 by Shepherd et al. 1990. This is also true for part of the obscured PPN candidates, namely those showing large maser velocity ranges in either H_2O or OH, which indubitably are associated with the fast new emerging wind. Possibly, the observed variety of maser properties is a result of the variety of developing geometries of the mass outflow, its density, and its inclination with respect to our line of sight. This is fundamentally different to the situation on the AGB, where radial symmetry prevails, and maser properties can be used to establish a chronogical sequence.

Although masers are of little help to serve as chronometer during the transition process, their peculiar properties have the advantage to provide a simple tool to identify new PPN candidates. In particular they are useful in the very early stages, when the infrared colors (e.g. in the IRAS bands) may differ only marginally from those of ordinary AGB stars.

References

Bedijn P.J., 1987, A&A 186, 136
Bobrowsky M., Zijlstra A.A., Grebel E.K., et al., 1995, ApJ 446, L89
Davis L.E., Seaquist E.R., Purton C.R., 1979, ApJ 230, 434
Engels D., Schmid-Burgk J., Walmsley C.M., 1986, A&A 167, 129
Engels D., 1996, A&A 315, 521
Engels D., 1997, in IAU Symp. 180, Eds. H.J. Habing & J.G.L.M. Lamers, p. 348
Hrivnak B.J., Kwok S., Su K.Y.L., 1999, ApJ 524, 849
Hu J.Y., Slijkhuis S., Nguyen-Q-Rieu, de Jong T., 1993, A&A 273, 185
Hu J.Y., Te Lintel Hekkert P., Slijkhuis F., et al., 1994, A&AS 103, 301
Jewell P.R., Snyder L.E., Walmsley C.M., et al., 1991, A&A 242, 211
Kastner J.H., Weintraub D.A., 1995, ApJ 452, 833
Kastner J.H., Weintraub D.A., Merrill K.M., Gatley I., 1998, AJ 116, 1412
Lewis B.M., 1989, ApJ 338, 234
Lewis B.M., Eder J., Terzian Y., 1990, ApJ 362, 634
Likkel L., 1989, ApJ 344, 350
Likkel L., Morris M., Maddalena R., 1992, A&A 256, 581
Morris M., Bowers P.F., Turner B.E., 1984, ApJ 259, 625
Nakashima J., Deguchi S., 2000, PASJ, in press
Nedoluha G.E., Bowers P.F., 1992, ApJ 392, 249
Nyman L-Å., Hall P.J., Olofsson H., 1998, A&AS 127, 185
Payne H.E., Phillips J.A., Terzian Y., 1988, ApJ 326, 368
Reddy B.E., Hrivnak B.J., 1999, AJ 117, 1834
Sahai R., Te Lintel Hekkert P., Morris M., et al., 1999a, ApJ 514, L115
Sahai R., Zijlstra A., Bujarrabal V., Te Lintel Hekkert P., 1999b, AJ 117, 1408
Seaquist E.R., Plume R., Davis L.E., 1991, ApJ 367, 200
Sevenster M.N., Chapman J.M., Habing H.J., et al., 1997, A&AS 124, 509
Shepherd M.C., Cohen R.J., Gaylard M.J., West M.E., 1990, Nature 344, 522
Skinner C.J., Meixner M., Bobrowsky M., 1998, MNRAS 300, L29
Steffen M., Szczerba R., Schönberner D., 1998, A&A 337, 149
Te Lintel Hekkert P., 1991 A&A 248, 209
Te Lintel Hekkert P., Chapman J.M., Zijlstra A.A., 1992, ApJ 390, L23
Te Lintel Hekkert P., Chapman J.M., 1996, A&AS 119, 459
van der Veen W.E.C.J., Habing H.J., Geballe T.R., 1989, A&A 226, 108
Zijlstra A.A., Te Lintel Hekkert P., Pottasch S.R., et al., 1989, A&A 217, 157
Zijlstra A.A., Gaylard M.J., Te Lintel Hekkert P., et al., 1991, A&A 243, L9

IRAS 18455+0448: THE BIRTH AND DEATH OF OH/IR STARS

B. M. Lewis
Arecibo Observatory
HC3 Box 53995
Arecibo PR00613
blewis@naic.edu

Abstract The exponential decay over the last 706 days of the 1612 MHz maser in IRAS 18455+0448 points to a PPN status for the star, and heralds its imminent death as an OH/IR star. PPN also show us that the immediate precursors of OH/IR stars are mimics: the first emergence of 1612 MHz masers from a mimic's shell then flags its birth as an OH/IR star. Prototype examples of recent births are IRAS 19566+3423 and V1511 Cyg.

1. INTRODUCTION

The immediate precursors of oxygen-rich proto planetary nebulae (PPN) are usually OH/IR stars. Indeed the prototype PPN, IRAS 18095+2704, still has OH masers (Hrivnak, Kwok and Volk 1988), as does IRAS 19114+0002 and a recently identified member of this subset, IRAS 19386+0155 (Lewis 2000). Since PPN are actively evolving into PN, which characteristically have ionised rather than neutral gas shells, the decay and loss of circumstellar masers is expected during the early PPN phase of evolution, as Engels (this volume) has discussed. This report concerns (i) results from monitoring 1612 MHz masers in IRAS 18455+0448, which suggest it is a PPN; (ii) pointers that masers in PPN provide about the onset and duration of the 1612 MHz emission phase. PPN inform us about both the birth and death of OH/IR stars.

2. IRAS 18455+0448

This OH/IR star had a classic 1612 MHz spectrum at its discovery in 1988, with 2.07 & 0.64 Jy 1612 MHz masers. When reobserved in 1998, the stronger maser had weakened by a factor of 20 to 0.11 Jy, while the second had disappeared. Figure 1 shows the result of subsequent observations (Lewis, Oppen-

R. Szczerba and S.K. Górny (eds.), Post-AGB Objects as a Phase of Stellar Evolution, 423–428.

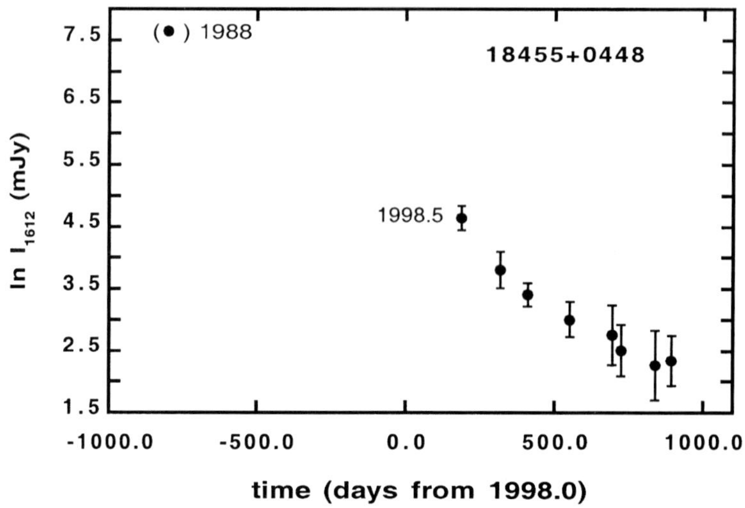

Figure 1 Decay of the last 1612 MHz maser in IRAS 18455+0448

heimer & Daubar 2000). When the peak intensity is assessed at a consistent 0.57 km s^{-1} resolution, the maser exhibits a steady exponential decline over the 706 day span of our observations, to give an e-folding time of 319 d. It is now 200 times weaker than it was at discovery, and is near the threshold for easy detection at Arecibo.

There are two scenarios for IRAS 18455+0448: it is either (i) a PPN, so the loss of masers is permanent; or (ii) it is heading into a deep local minimum in its luminosity evolution at the conclusion of a thermal pulse, whereon it will eventually be reborn as an OH/IR star again. The first scenario fits the presently known facts best, as the IR colors and expansion velocity are similar to the candidate PPN IRAS 20531+2909. And there is no hint of periodicity in either the mainline masers or its mid-IR fluxes. We thus appear to have caught a PPN in the very act of losing its 1612 MHz maser.

3. IRAS 18095+2704

Is IRAS 18095+2704 near this stage too? This thought is prompted by its small (70-240 mJy) OH peak intensities in 1987-8 relative to its 126 Jy 25μm flux (Lewis *et al.* 1990). Moreover its SED, as fitted by Hrivnak *et al.* provides an estimate for the size of the central hole in its circumstellar shell, which, with the measured OH velocities, gives it a 103 yr expansion age, as well as its age as a PPN (Lewis 2000). Its OH masers were therefore reobserved in June 2000: Figure 2 shows the resulting 1612 MHz spectrum. While the peak intensities of both mainline masers are weaker than in 1987, the reverse is true for its 1612

MHz emission. Indeed the -11.4 km s^{-1} feature at 234 mJy is three times as strong, and still 80% stronger when averaged down to the resolution used in 1987, while the integrated 1612 MHz flux is now 64% larger. This warns us not to presume that 1612 MHz masers always decline monotonically as a PPN evolves.

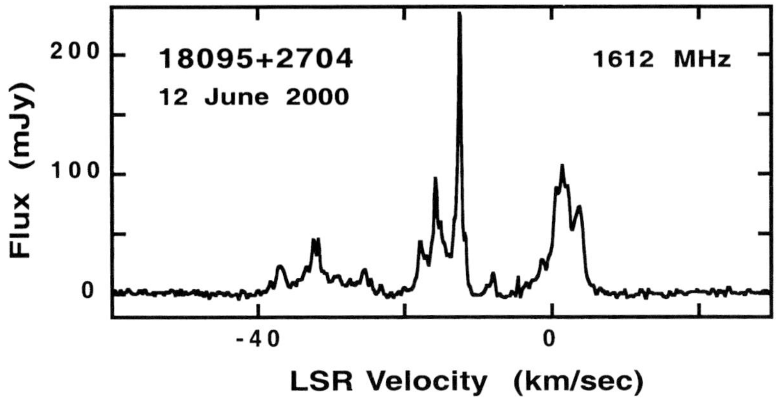

Figure 2 The current 1612 MHz masers from 18095+2704

IRAS 18095+2704 has a b = 20°. When all IRAS sources in the Arecibo ($0° < \delta < 38°$) sky with $|b| > 10°$ are investigated, those with 1612 MHz masers predominently have $-0.7 < (25-12)\ \mu m < -0.38$, as Figure 3 shows. The only 4 redder shells with OH masers are PPN, so the relative duration of the OH/IR star and/or IR emission phase is given by the ratio between their numbers and the associated PPN. Since there are 62 Arecibo OH/IR stars with $|l| < 90°$ and S(25) > 2 Jy, the OH/IR star phase in these stars is $\sim$ 16 times longer than the 1612 MHz emission phase of a PPN. Setting that at the 103 yr of IRAS 18095+2704 suggests that the entire 1612 MHz emission phase of a high latitude star is of order 103 ([62 + 4]/4) = 1700 yr (Lewis 2000). Nevertheless 118 objects in our $|l| < 90°$ sample do not exhibit 1612 MHz masers; these are dubbed "OH/IR star color mimics", or mimics for short. But there is only one PPN candidate (25-12) μm > -0.38 without an OH maser to match to them, where $\sim$ 8 are expected if mimics usually evolve directly into PPN. Inferentially, therefore, most mimics evolve into OH/IR stars before becoming PPN. So the birth of an OH/IR star is flagged by the rise of a 1612 MHz maser from an erstwhile mimic.

4. NEWLY BORN OH/IR STARS

IRAS 19566+3423 is in part the inverse case to IRAS 18455+0448. At discovery in 1986 it had just two, narrow, 1 Jy features: Galt (1993) documents

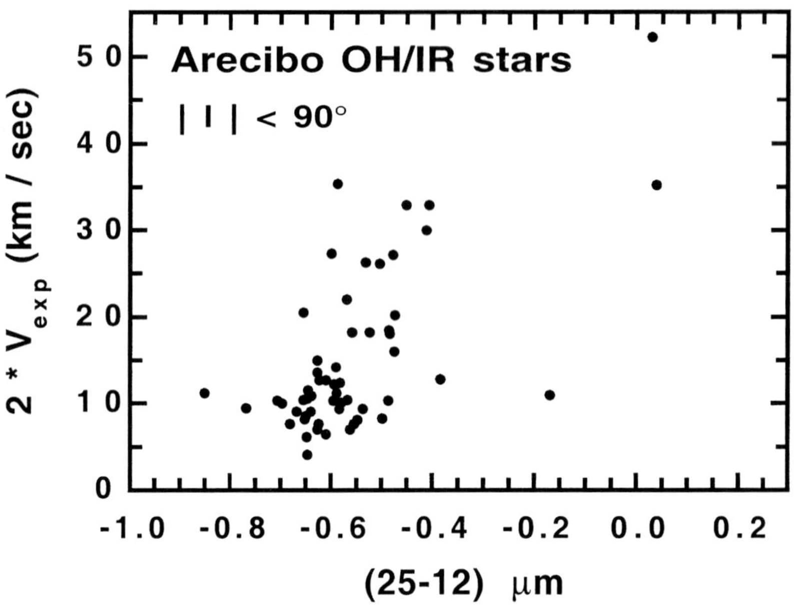

Figure 3 The expansion velocity - color plot for Arecibo OH/IR stars with $|l| < 90°$. Three PPN plot onto the right half of the figure, the last (missing) being detected in only one 1667 MHz feature

the subsequent rise in the peak intensity of several 1612 MHz features. But in 1999 its 1612 and 1665 MHz spectra both had intensities > 3 Jy and factor of 2 larger velocity ranges (Lewis 1999): Figure 4 shows the changes that have occurred in its 1612 MHz spectrum. Strong masers can thus emerge from nothing in less than 10 yr. Moreover the (1986 - 1992) masers are from local clumps in its shell that probably would not have been detectable 10 yr earlier. IRAS 19566+3423 is clearly a recently born OH/IR star, although it is also a $\sim 180,000 L_{\odot}$ supergiant.

V1511 Cyg (alias IRAS 19586+3637) is a second example of a newly born OH/IR star. It was found to exhibit the classic 1612 MHz profile of a normal OH/IR star in 1999. This is shown in Figure 5, where the peak intensities are 72 & 332 mJy, though it had not exhibited 1612 MHz masers in four published searches from 1986-1991. We are currently monitoring the changes in its masers with pulsation period.

Acknowledgments

This work is supported by the National Astronony and Ionosphere Center, which is operated by Cornell University under a management agreement with the National Science Foundation.

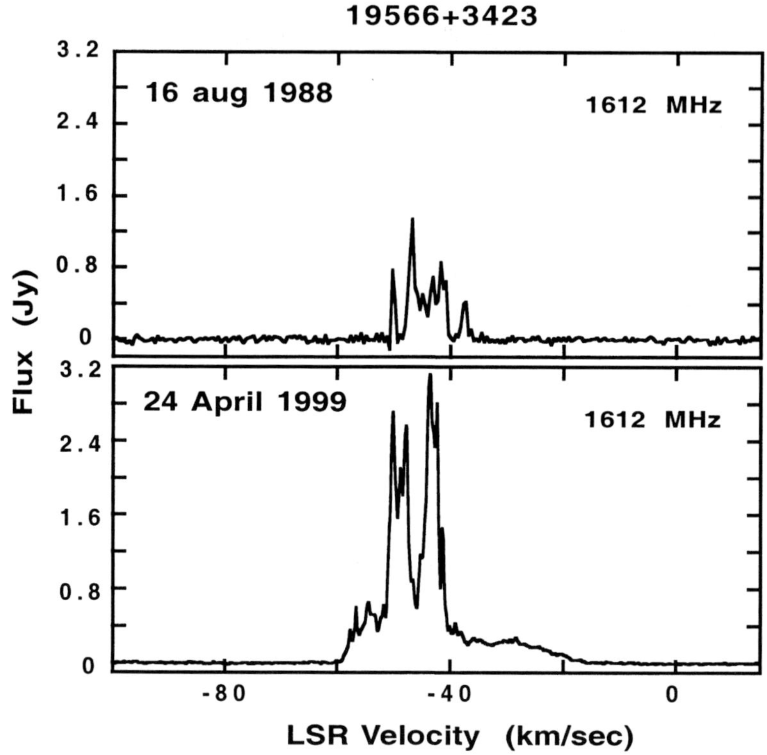

Figure 4 The change over 11 years in the 1612 MHz masers of IRAS 19566+3423

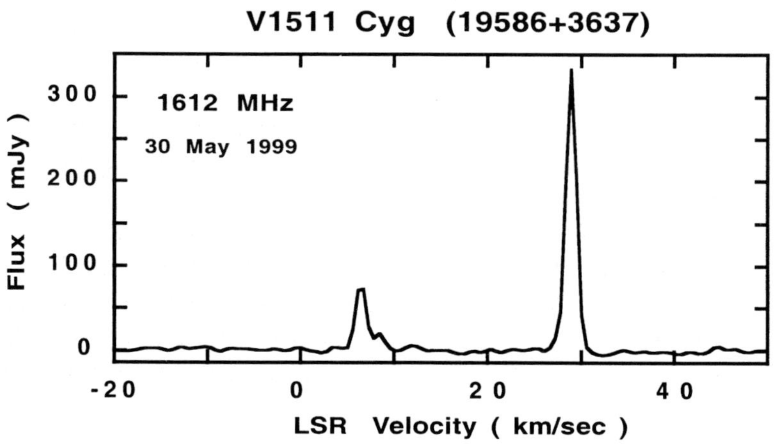

Figure 5 The new 1612 MHz masers of V1511 Cyg

References

Galt, J. 1993, in “Astrophysical Masers", Lecture Notes in Physics v412, ed. A. W. Clegg and G. E. Nedoluha, (Springer Verlag: Heidelburg) p 283

Hrivnak, B. J., Kwok, S., & Volk, K. M. 1988, ApJ, 331, 832

Lewis, B. M. 1999, BAAS, 31, 1437

Lewis, B. M. 2000, ApJ, 533, 959

Lewis, B. M., Eder, J., & Terzian, Y. 1990, ApJ, 362, 634

Lewis, B. M., Oppenheimer, B. D., & Daubar, I. J. 2000, ApJL (in press), or ftp://ftp.naic.edu/pub/publications/bml/18455.ps.Z

OH MASER MAPPING OF POST-AGB STARS

M. Szymczak
Toruń Centre for Astronomy, Nicolaus Copernicus University, Gagarina 11, PL-87100 Toruń, Poland

A.M.S. Richards
Jodrell Bank Observatory, University of Manchester, Macclesfield, Cheshire, SK11 9DL, UK

Abstract The OH 1612 MHz maser emission from four objects which are probably in transition between the AGB and the PN stage has been mapped with MERLIN. The basic parameters of their circumstellar envelopes are determined. There are indications of episodic variations in mass loss rates in two objects. The structure of linearly polarized emission is well resolved in two envelopes indicating highly ordered circumstellar magnetic fields.

1. INTRODUCTION

Some OH/IR stars are characterized by an infrared colour temperature below 200 K and bimodal distribution of energy (Bedijn 1987). In a few cases possible optical counterparts, of spectral type earlier than M, have been identified (Le Bertre 1987). The OH 1612 MHz maser emission still has a double-peaked profile but the flux density is non-variable or exhibits only small (< 0.3 mag) amplitude variations (Herman 1983). The H_2O 22 GHz masers in some objects have completely disappeared during last 10-20 years, or individual maser features have appeared and disappeared on time scale of a few years (Engels 1997). These properties suggest that such objects are in a stage of transition from AGB to PN. In transition objects one can expect to observe the first signs of changes in the geometry. We present the first results of observations of a sample of suspected transition objects. Our purposes are to determine the basic parameters of the outermost envelopes and to verify a possibility that the outflows are shaped by the stellar magnetic fields.

R. Szczerba and S.K. Górny (eds.), Post-AGB Objects as a Phase of Stellar Evolution, 429–432.

Table 1 Estimated stellar positions.

star	$\alpha(1950)$ (h m s)	$\delta(1950)$ (o ′ ″)
OH 15.7+0.8	18 13 33.797±0.006	−14 56 10.95±0.12
OH 17.7−2.0	18 27 39.778±0.006	−14 31 03.83±0.12
OH 77.9+0.2	20 26 40.115±0.004	38 57 00.90±0.10
OH 85.4+0.1	20 51 50.010±0.004	44 45 59.91±0.10

Table 2 OH 1612 MHz maser envelope parameters.

star	V_s ($\mathrm{km\,s^{-1}}$)	V_e ($\mathrm{km\,s^{-1}}$)	R_0 (″)
OH 15.7+0.8	−0.8 ± 0.2	14.5±0.4	1.12/0.50±0.06
OH 17.7−2.0	60.8±0.3	12.5/15.5±0.6	0.67/0.25±0.07
OH 77.9+0.2	−39.8 ± 0.2	11.4±0.2	0.29±0.04
OH 85.4+0.1	−21.9 ± 0.3	13.6±0.3	–

2. OBSERVATIONS AND RESULTS

Four sources OH 15.7+0.8, OH 17.7−2.0, OH 77.9+0.2 and OH 85.4+0.1 were observed on 1999 May 27 with MERLIN (6 antennae) at 1612 MHz transition in full polarization spectral mode. Each target was observed for 15 min at 4-5 different hour angles to optimize the u-v plane coverage. The data were processed in several steps (see Szymczak et al. 1998 for details) taking special care to calibrate the polarization. 0200+359 was used to correct for instrumental feed leakage and parallactic angle rotation; 3C286 was used to establish the primary flux scale and the polarization position angle and 3C84 was used as a bandpass calibrator. The strongest emission of point like structure was used as phase reference in a few cycles of self-calibration. Map cubes in all Stokes parameters were made and cleaned with 1000 cycles per channel. The angular resolution was 0.15 arcsec and the spectral resolution was 0.36 $\mathrm{km\,s^{-1}}$. A typical rms level in the line-free channels of I Stokes was about 30 mJy.

To estimate the stellar position and shell parameters, the equation for a thin uniformly expanding shell, $R = R_0[1 - (V - V_s)^2/V_e^2]^{0.5}$, was fitted to the data. Here, R is the projected radius at radial velocity V, R_0 is the radius of the shell expanding with velocity V_e and V_s is the stellar radial velocity. Parameters for the best-fitting model are given in Tables 1 and 2. For OH 85.4+0.1 the OH 1612 MHz emission was unresolved and the position of the strongest blueshifted

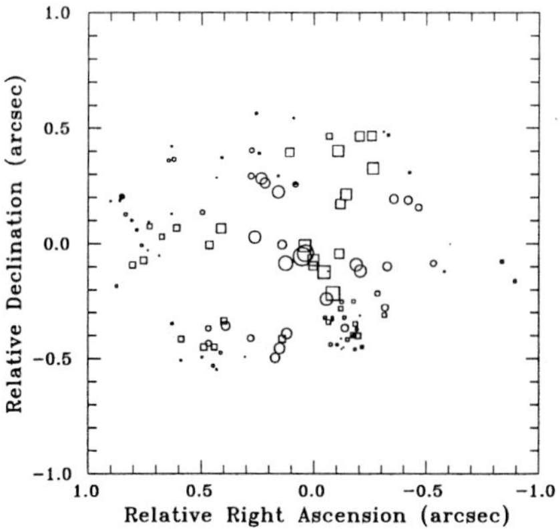

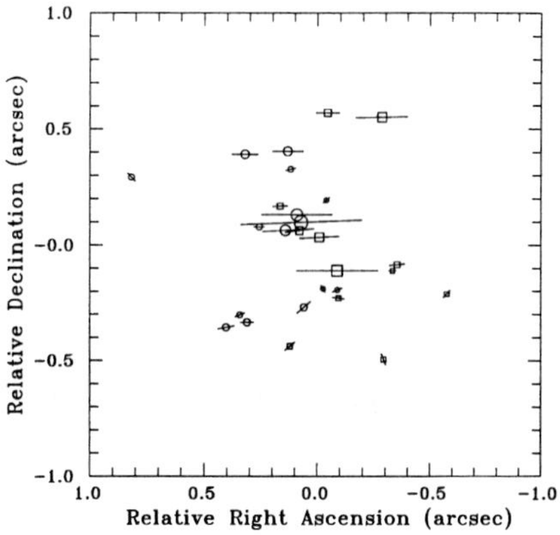

Figure 1 Distributions of the OH 1612 MHz masers of OH 15.7+0.8 relative to the estimated stellar position. *Left.* I Stokes maser components. *Right.* $\sqrt{Q^2 + U^2}$ Stokes maser components. Circles and squares represent blueshifted and redshifted components respectively, relative to the stellar velocity. The symbol sizes are proportional to the logarithm of the peak brightness of components. The bars indicate the plane of the electric field vector and have length proportional to the linearly polarized intensity, where 1 arcsec=1.1 Jy beam^{-1}.

components is given in Table 1, while the expansion velocity and the systemic velocity was inferred from the maser spectrum.

The maser emission from OH 15.7+0.8 is patchy and the overall distribution of maser spots slightly deviates from a spherical symmetry (Fig. 1). For a thin-shell model the data can be described by two shells of radii 0.5 and 1.12 arcsec expanding with the same velocity (Fig. 2). For the assumed distance 2.3 kpc the expansion times are 465 and 841 yrs respectively. This suggests that the mass loss rate has varied episodically over the last 1000 yrs. However, the clumpy nature of the masers may mask the underlying mass density distribution. Distribution of the electric field vectors of linearly polarized emission is evidence for a highly aligned magnetic field. The mean position angle of the projected axis of the magnetic field is $5 \pm 3°$ for dominant components. There are also localized structures in the magnetic field, particularly at intermediate velocities. The magnetic field direction is nearly perpendicular to the major axis of ellipsoidally distributed I Stokes components.

In OH 17.7−2.0 the extreme low- and high-velocity emission is not positionally coincident. The separation between the strongest components is 0.21 arcsec along the position angle of about 45°. If the central object is located just between these two clumps a model of two shells of radii 0.25 and 0.67 arcsec expanding with velocities of 15.5 and 12.5 km s^{-1} respectively, can be fitted to the 1612 MHz data. For the assumed distance 5.4 kpc the expansion times are

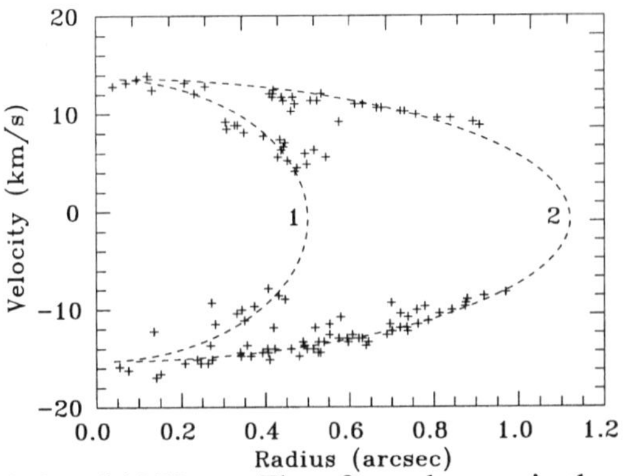

Figure 2 Angular separations of maser spots of OH 15.7+0.8 from the estimated stellar position versus the LSR velocity. The symbol sizes are proportional to the logarithm of intensity. The dashed lines 1 and 2 show the thin-shell model fits to the emission from "young" and "old" envelopes, respectively.

412 and 1370 yrs. Therefore, changes in the mass loss rate during last 1500 yrs are possible.

In OH 77.9+0.2 we found a weak linearly polarized feature at extreme blueshifted velocity only. No linear and circular polarization was found in OH 85.4+0.1.

OH 15.7+0.8 and OH 17.7−2.0 exhibit some degrees of asymmetry in the 1612 MHz maser distribution. A thin shell model can be fitted reasonably well to the data, but more complicated models which incorporate anisotropic or episodic mass loss cannot be excluded. Polarization morphology of both sources evidence a regular magnetic field with some local structures in the outer parts of envelopes. It is likely that the circumstellar magnetic fields channel asymmetric structures in the outflows.

References

Bedijn, P.J., 1987, A&A 186, 136

Engels D., 1997, In: Habing H.J., Lamers H.J.G.L.M.(eds.) Planetary nebulae. IAU Symp.180, Kluwer, Dordrecht, p.348

Herman J., 1983, PhD thesis, University of Leiden

Le Bertre T., 1987, A&A 180, 160

Szymczak M., Cohen R.J., Richards A.M.S., 1998, MNRAS 297, 1151

THE TIME VARIATION IN THE WATER VAPOR BANDS IN OXYGEN-RICH MIRA VARIABLES *

Mikako Matsuura, Issei Yamamura, Hiroshi Murakami

The Institute of Space and Astronautical Science (ISAS), Yoshino-dai 3-1-1, Sagamihara, Kanagawa 229-8510, Japan †

(M.M.):mikako@ir.isas.ac.jp

Takashi Onaka

Department of Astronomy, School of Science, University of Tokyo, Hongo 7-3-1, Bunkyo, Tokyo, 113-0033, Japan

Abstract We examine the time variation in the near-infrared H_2O bands in oxygen-rich Mira variables. We analyzed four Mira variables, which were monitored for longer than one period of their optical light curves with the ISO/SWS. In the wavelength range from ~3.5 to 4.0 μm, all of the sample stars show absorption features of H_2O bands around minima and emission features around maxima. The model analysis shows that the H_2O layer with an excitation temperature of ~2000 K is responsible for these features and that the layer is located at about 1 stellar radius around minima, and expands to about 2 stellar radii around maxima. These results are explained by either the dynamical motion of the matter in the extended atmosphere, or the variation of the temperature structure in the extended atmosphere through the shocks caused by the pulsation.

1. INTRODUCTION

Strong pulsations in AGB stars, especially in the Mira variables, extend their atmospheres. This extended atmosphere is enriched with various kinds of molecules. Its temperature ranges from ~2000 to ~1000 K (Woitke et al. 1999). Recent infrared spectroscopic observations from the space enable us to carry

* Based on observations with ISO, an ESA project with instruments funded by ESA Member States (especially the PI countries: France, Germany, the Netherlands and the United Kingdom) with the participation of ISAS and NASA. The SWS is a joint project of SRON and MPE.

† M.M. is also at Department of Astronomy, School of Science, University of Tokyo, Hongo 7-3-1, Bunkyo, Tokyo 113-0033, Japan

R. Szczerba and S.K. Górny (eds.), Post-AGB Objects as a Phase of Stellar Evolution, 433–438.

out detailed studies of the molecules in the extended atmosphere, based on the data unaffected by the terrestrial atmosphere.

Using the Short-Wavelength Spectrometer (SWS; de Graauw et al. 1996) onboard the Infrared Space Observatory (ISO; Kessler et al. 1996), Yamamura et al. 1999 studied the near-infrared H_2O bands in the Mira variables. They found that H_2O bands in the $\sim$3.5–4.0 μm region were observed as emission in *o* Cet, which was observed at maximum in the visual light curve. In contrast, another star Z Cas observed at minimum showed the bands in absorption at the same wavelengths. They analyzed the spectra of these stars in the 2.5–5.0 μm region with a simple 'slab' model. This model consists of two H_2O layers ('hot' and 'cool' layers) with independent excitation temperatures, column densities, and radii. The excitation temperature of the H_2O molecules responsible for the spectra in the $\sim$3.5–4.0 μm region was estimated as 2000 K in both stars. Yamamura et al. 1999 showed that the 'hot' layer is as large as $\sim$2 R_* in *o* Cet, while it was only $\sim$1 R_* in Z Cas, where R_* is the radius of the central star. They suggested that the difference between *o* Cet and Z Cas is mainly caused by the different radius of the 'hot' layer, and that the 'hot' water layer is more extended at maximum than at minimum.

2. THE TIME VARIATION IN THE H_2O BANDS

We examine the time variation in the H_2O bands in the 2.5–4.0 μm region in oxygen-rich Mira variables. We selected four stars (R Aql, R Cas, T Cep, and Z Cyg) from the ISO Data Archive. They were observed by the SWS several times over the period longer than one variable cycle (Onaka et al. 1999; Loidl et al. 1999). Fig. 1 shows an example of the time variation of the spectra between 2.5 and 4.0 μm.

We apply the same model as described in Yamamura et al. 1999. The plane-parallel configuration and local thermodynamic equilibrium (LTE) were assumed in the model calculation. A disk-shaped background source and molecular layers were placed one over the other. The background source, which represents the stellar continuum, is assumed to be a 3000 K blackbody. In most cases only H_2O molecules are included in the model except for two spectra in R Aql and four spectra in T Cep, for which OH molecules are also taken into account to reproduce sharp absorption features. We calculated more than ten thousands of spectra, over a large parameter space of radius and column density of two layers, and the excitation temperature of the cool layer with sufficiently fine grid points. According to Yamamura et al. 1999, we fix the excitation temperature of the 'hot' layer as 2000 K. We apply the χ^2 test to find the best-fit spectra. Fig. 2 indicates the best fitted results for Z Cyg corresponding to the observed spectra in Fig 1. Our simple 'slab' model satisfactorily reproduces the global shape as well as the small features in the observed spectra.

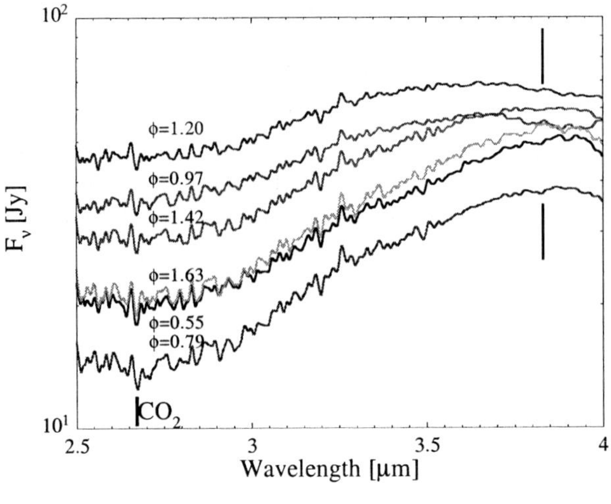

Figure 1 The time variation in the Z Cyg spectra obtained by the ISO/SWS. The optical variability phase (ϕ) at each observation, determined by the AAVSO light curve, is indicated in the figure by the labels. The '3.83 μm feature' (see text and Fig. 3) is marked by the vertical lines. The sharp absorption feature at 2.68 μm is CO_2.

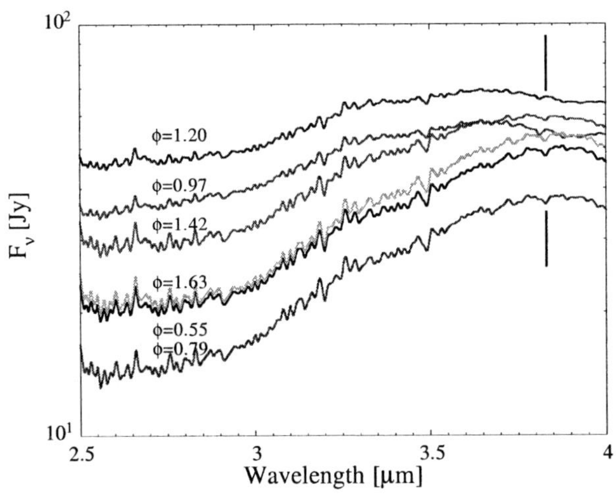

Figure 2 The synthesized spectra corresponding to Fig. 1 are indicated. Global shape and most of the small features in this wavelength range are reproduced by the model including only H_2O.

Here we focus on the most interesting parameter, the radius of the 'hot' layer (R_H). Other parameters are discussed in Matsuura et al. 2000. The 'hot' layer has an excitation temperature of 2000 K. The radius of the 'hot' layer

(R_H) affects the features in the wavelengths longer than $\sim$3.5 μm (Fig. 3). The overlaying 'cool' layer causes absorption features in the wavelengths shorter than $\sim$3.5 μm and masks features arising from the 'hot' layer. The features from the hot layer become in emission in the 3.8 μm region when R_H is larger than $\sim$1.6 R_*, while they are seen in absorption when R_H is smaller than that. This is clearly demonstrated by a sharp peak at 3.83 μm in Fig. 3. In the observed spectra in Fig. 1, this feature, marked by the lines, are seen in emission around maximum, and in absorption around minima. Thus, from Fig. 1, we estimate that R_H is larger than $\sim$1.6 R_* around maximum and smaller than that around minima.

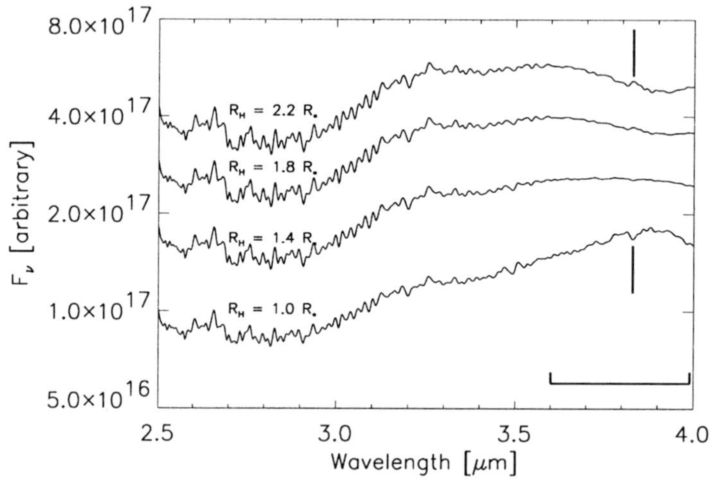

Figure 3 The effects of R_H on the synthesized spectra. From top to bottom, R_H is 2.2, 1.8, 1.4, 1.0 R_*. The features longer than $\sim$3.5 μm vary with R_H. The features become emission when R_H is larger than $\sim$1.6 R_*. The effects are prominently seen in the 3.83 μm feature which is either seen in emission or in absorption depending on R_H. The wavelengths shorter than $\sim$ 3.5 μm are subjected to the absorption feature from the overlaying 'cool' layer, and the features are insensitive to the parameters of the hot layer in this figure. The absolute flux varies owing to the variation of radius of the emitting region. Other parameters are fixed as T_H = 2000 K, T_C = 1200 K, $N_H = 1 \times 10^{22}$ cm^{-2}, $N_C = 1 \times 10^{21}$ cm^{-2}. In these calculations, R_C is given as $R_C^2 = R_H^2 + 1^2$ $[R_*^2]$.

The variation of R_H derived from the model fitting is plotted in Fig. 4. In all of the four stars, R_H reaches about 2 R_* around maxima (ϕ=0.0, 1.0, and 2.0) and becomes $\sim$1 R_* around minima. These results are in agreement with Yamamura et al. 1999. The variation in R_H is calculated as the ratio to the radius of the stellar continuum. According to the interferometric observations by Tuthill, Haniff & Baldwin 1995, the variation in the 'stellar radius' of Mira is smaller than 7 %. Theoretical work (Bessell, Scholz, Wood 1996) indicates the range of about 20 %. The amplitude of variation in the stellar radius is smaller than that of R_H. Thus the location of the 'hot' H_2O layer actually varies.

The periodical variations of the H_2O bands suggest that they are related to the pulsation. One possible explanation for the motion of the 'hot' H_2O layer is that pulsations drive the matter, including the H_2O molecules, into the outer region of the atmosphere. Another possible mechanism is the heating by the pulsation shocks. When a shock runs through the extended atmosphere, it heats up the matter locally. The location of the heated shell moves outwards as the shock moves, and the effective radius of the hot H_2O layer thus expands. To confirm which is more dominant, comparisons with theoretical works are necessary. More detailed discussion is given in Matsuura et al. 2000.

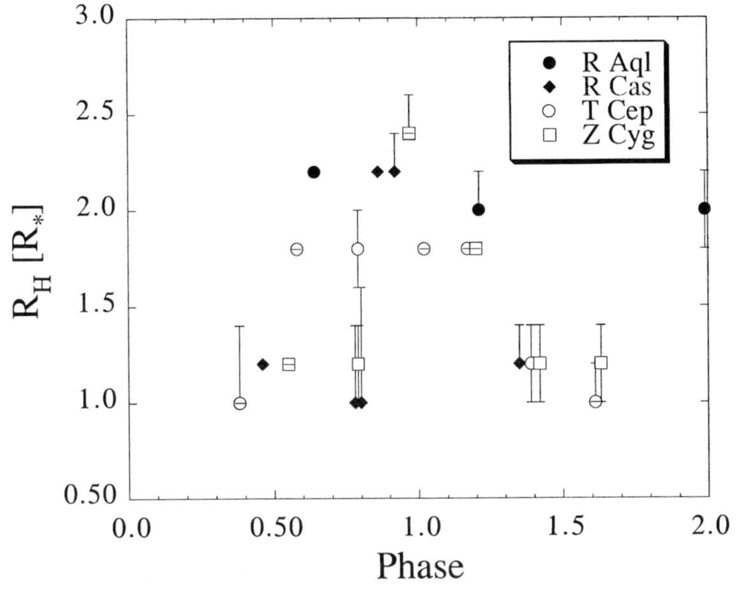

Figure 4 The derived R_H for four Mira variables are indicated against the optical phase. The radius of the 'hot' layer is about ~1 R_* around minimum, and about ~2 R_* around maximum.

Acknowledgments

The authors acknowledge the American Association of Variable Star Observers (AAVSO) to provide us the light curves. We are grateful to Prof. de Jong for the discussions. M.M. thanks the hospitality of the Astronomical Institute, University of Amsterdam during her stay. M.M. is a Research Fellow of the Japan Society for the Promotion of Science for the Young Scientists.

References

Bessell M.S., Scholz M., Wood P.R., 1996, A&A 307, 481
de Graauw Th., Haser L.N., Beintema D.A., et al., 1996, A&A 315, L49
Kessler M., Steinz J.A., Anderegg M.E., et al., 1996, A&A 315, L27
Loidl R., Aringer B., Hron J., Kerschbaum F., Höfner S., Jørgensen U.G., Windsteig W., 1999, ESA-SP 427, 365
Matsuura M., Yamamura I., de Jong T., et al., 2000, A&A, in preparation
Onaka T., de Jong T., Yamamura I., Cami J., Tanabé T., 1999, ESA-SP 427, 281
Tuthill P.G., Haniff C.A., Baldwin J.E., 1995, MNRAS 277, 1541
Woitke P., Helling Ch., Winters J.M., Jeong K.S., 1999, A&A 348, L17
Yamamura I., de Jong T., Cami J., 1999, A&A 348, L55

FURTHER DETECTIONS OF OH MASERS IN CARBON STARS WITH SILICATE FEATURES

M. Szymczak
Toruń Centre for Astronomy, Nicolaus Copernicus University, Toruń, Poland

R. Szczerba
Nicolaus Copernicus Astronomical Center, PAS, Toruń, Poland

P.S. Chen
Yunnan Astronomical Observatory, CAS, Kunming, P.R. China

Abstract A sample of J–type carbon stars was searched for OH maser emission. The new detection of three OH lines towards two silicate carbon stars is reported. In V778 Cyg, previously known as the main–lines (1665 and 1667 MHz) maser source, the satellite 1612 MHz emission was discovered while in NSV 2814 the main OH lines were detected. The presence of OH maser lines confirms the former suggestion that oxygen–rich material is located in the vicinity ($\approx 10^{15-16}$ cm) of silicate carbon stars.

1. INTRODUCTION

The silicate emission features at about 10 and 18 μm are characteristic for oxygen–rich dust envelopes. However, these features were also discovered in some optically classified carbon stars (Little-Marenin 1986; Willems & de Jong 1986), named later silicate carbon stars. Furthermore, many silicate carbon stars are recognized as ^{13}C-rich (J-type) carbon stars (Lambert et al. 1990; Lloyd Evans 1990 and references therein). The most recent compilation of 22 known and suspected silicate carbon stars is presented by Chen et al. (1999). Note also the discovery of the first extragalactic silicate carbon star (IRAS 04496−6958) in the Large Magellanic Cloud (Trams et al. 1999). The detection of silicate emission from some (silicate) carbon stars suggests that their relatively close surroundings contain oxygen–based dust, in spite of their photospheric chemical composition which shows C/O > 1. An additional argument for the existence

R. Szczerba and S.K. Górny (eds.), Post-AGB Objects as a Phase of Stellar Evolution, 439–442.

of oxygen–rich material in the vicinity of these stars comes from the detection of water and hydroxyl maser lines towards some of them (see Little-Marenin et al. 1994 and Engels 1994 and references therein). To date H_2O masers were found in four (J–type) silicate carbon stars and only one of them (V778 Cyg) exibits OH maser emission in the main–lines. In addition 1612 MHz maser emission was detected in FJF 270 (te Lintel Hekkert 1991). Here we report the results of a high sensitivity search for OH emission towards 14 J–type carbon stars including 7 known silicate carbon stars.

2. OBSERVATIONS

The observations were performed with the Nançay radio telescope on April 6–13, 1999. The HPBW was 3.5′ in right ascension by 18′ in declination. A dual channel receiver was used. The system temperature was about 50 K. The ratio of flux density to antenna temperature was 1.1 Jy K^{-1} at 0° declination. The 1024–channel autocorrelation spectrometer was split into 4 banks, each covering a bandwidth of 0.2 MHz, for the three OH lines. The 1612 MHz satellite line was observed in both circular polarizations while the 1665 and 1667 MHz main–lines were observed in left and right circular polarization, respectively. The spectra were taken in frequency switching mode with channel spacings of 0.28 and 0.29 km s^{-1} at the main–lines and the satellite line, respectively. Upper flux density limit for the non–detections was 80 mJy (3σ). W 12 was observed to provide the flux density calibration which was accurate to about 10%.

3. RESULTS AND DISCUSSION

Our program stars are listed in Table 1 with the designation from the General Catalogue of Cool Galactic Carbon Stars (CCGCS, Stephenson 1989) given in the first column. Known silicate carbon stars are marked in boldface. Previously detected H_2O and/or OH maser lines are marked by a '+' sign while our new detections are depicted by a '++' sign. OH masers were found in NSV 2814 (main–lines) and in V778 Cyg (satellite line). Note that the 1612 MHz maser line towards V778 Cyg has not been detected either by Barnbaum et al. (1991) nor by Little-Marenin et al. (1994).

The spectra are shown in Fig. 1 where, for the sake of comparison, re–observed OH main–lines from V778 Cyg are presented. In V778 Cyg the 1667 MHz feature at -16.0 km s^{-1} has a peak flux density of 0.30 Jy, which is a factor of 1.5 higher than the 1665 MHz peak of 0.19 Jy at -16.9 km s^{-1}. Within the measurement errors the intensities and radial velocities of these lines are the same as observed 12 years ago (Little–Marenin et al. 1988). The 1612 MHz feature of 0.12 Jy peaks at -14.4 km s^{-1}. This indicates that the 1612 MHz emission is slightly redshifted relatively to the OH main–lines but coincides with a -14 km s^{-1} component of the H_2O maser (Engels & Leinert 1994).

Table 1 List of observed stars.

CCGCS	other name	OH 1612	OH 1665	OH 1667	H_2O
1158	**NSV 2814**	–	++	++	+
1653	**BM Gem**	–	–	–	–
1682		–	–	–	+
1698		–	–	–	+
1921		–	–	–	–
2028		–	–	–	–
2156	MT Hya	–	–	–	–
3066	HD 100764	–	–	–	–
3935	**FJF 270**	+	–	–	–
4222	**NC 83**	–	–	–	–
4923	**V778 Cyg**	++	+	+	+
5526	**MQ Cyg**	–	–	–	–
5548	V1415 Cyg	–	–	–	+
5848	**EU And**	–	–	–	+

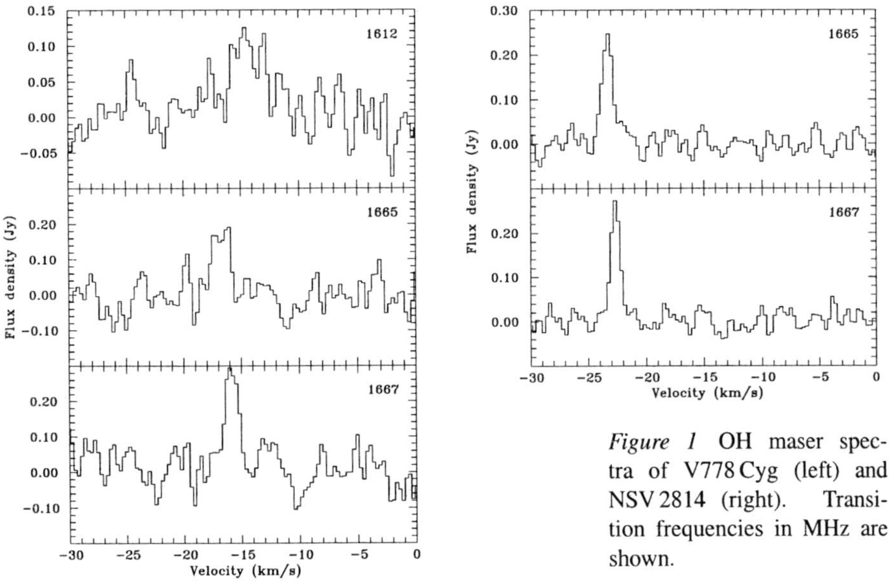

Figure 1 OH maser spectra of V778 Cyg (left) and NSV 2814 (right). Transition frequencies in MHz are shown.

In NSV 2814 the peak intensities of the 1665 and 1667 MHz are 0.24 and 0.27 Jy, respectively. The velocities of OH features at -23.5 and -22.8 km s^{-1} are very different from the H_2O maser velocity of -13 km s^{-1} detected about 7 years ago (Engels 1994). Although temporal variations may affect the maser profiles, the difference in velocities of the OH and H_2O maser features of about

$10\,\mathrm{km\,s^{-1}}$ is quite high for an AGB object. This may suggest that each maser is associated with a different object. Little-Marenin et al. (1994) noted that CCGCS 1158 may be incorrectly associated with the IRAS source. Furthermore, the intensity ratio of 1667/1665 MHz transitions of about 1.1 suggests optically thin LTE conditions, so that the masers may arise in the intervening interstellar medium. The case of NSV 2814 needs further careful studies.

In spite of almost 15 years of debate, the formation of silicate carbon stars still remains controversial. A stability of the silicate feature (Yamamura et al. 2000) and OH main–lines (this work) in V778 Cyg makes the scenario of *fast* transition from M–type AGB stars into carbon stars difficult to reconcile with the lifetime of silicate emission from an expanding detached shell (see e.g. Yamamura et al. 2000). A plausible scenario may be some kind of long–lived reservoirs of oxygen–rich material in a binary system (e.g. Lloyd Evans 1990 and Jura & Kahane 1999) with (possibly) a main sequence secondary (e.g. Lambert et al. 1990 and Yamamura et al. 2000). Silicate carbon stars seem to be related to the *normal* J–type carbon stars in the sense that both groups show H_2O maser emission (Engels 1994) and that the mechanism responsible for their low $^{12}C/^{13}C$ ratios might be the same (Ohnaka & Tsuji 1999).

Acknowledgments

The authors acknowledge support from grant No. 2.P03D.002.13 and No. 2.P03D.024.18p01 of the Polish State Committee for Scientific Research.

References

Barnbaum C., Morris M., Likkel L, et al., 1991, A&A 251, 79
Chen P.S., Wang X.H., Wang F., 1999, Chin. Astron. Astrophys. 23, 371
Engels D., 1994, A&A 285, 497
Engels D., Leinert Ch., 1994, A&A 282, 858
Jura M., Kahane C., 1999, ApJ 521, 302
Lambert D.L., Hinkle K.H., Smith V.V., 1990, AJ 99, 1612
Little–Marenin I.R., 1986, ApJ 307, L15
Little–Marenin I.R., Benson P.J., Dickinson D.F., 1988, ApJ 330, 828
Little–Marenin I.R., Sahai R., Wannier P.G., et al., 1994, A&A 281, 451
Lloyd Evans T., 1990, MNRAS 243, 336
Ohnaka K., Tsuji T., 1999, A&A 345, 233
Stephenson C.B., 1989, Pub. Warner and Swasey Obs., Vol. 3, No.2 (CCGCS)
te Lintel Hekkert P., Caswell J.L., Habing H.J., et al., 1991, A&AS 90, 327
Trams N.R., van Loon J.Th., Zijlstra A.A., et al., 1999, A&A 344, L17
Willems F.J., de Jong T., 1986, ApJ 309, L39
Yamamura I., Dominik C., de Jong T., et al., 2000, A&A 363, 629

Object Index

γ Peg, *278*
ω Cen, *80*–81
o Cet, *173*, *215*–219, *359*, *434*
19W32, *194*
89 Her, *8*–9, *273*, *411*
AC Her, *9*, *262*, *267*, *273*, *360*, *407*–408
AD Aql, *273*
AFGL 2343 – *see* IRAS 19114+0002
AFGL 2688, *5*, *8*, *13*, *55*, *57*, *89*, *115*–119, *197*–198, *266*–268, *273*, *291*, *294*, *324*, *333*–337, *349*, *364*, *369*, *373*–374, *380*, *387*–395, *397*, *401*, *407*–408, *410*
AFGL 3068, *3*, *15*
AFGL 618, *8*, *13*, *197*, *333*–337, *364*, *369*, *371*–374, *380*–381, *397*–401, *405*–408, *410*
AG Peg, *222*
AI CMi – *see* IRAS 07331+0021
AR Pup, *272*
AS 201, *222*
BD+30°3639, *56*, *194*, *339*–341
BD+33°2642, *31*, *38*, *272*, *277*–282
BI Cru, *222*
BM Gem, *360*, *441*
BPS CS 22946−005, *272*
BPS CS 29493−046, *273*
BT Lib, *272*
Barnard 29, *37*–38, *272*, *315*–316
Betelgeuse, *405*, *410*–411
CCGCS 1682, *441*
CCGCS 1698, *441*
CCGCS 1921, *441*
CCGCS 2028, *441*
CCGCS 2156, *441*
CCGCS 3066, *441*
CE Vir, *272*
CH Cyg, *221*–222, *225*–226
CIT 6, *289*–294, *359*, *370*–371
CPD−56°8032 – *see* Hen 3−1333
CPD−61°455, *272*
CT Ori, *272*
CW Leo – *see* IRC+10216
Cn 3−1, *61*
DDDM 1, *315*
DS Aqr, *273*
DY Aql, *273*
DY Ori, *272*
EN TrA, *272*
EP Lyr, *273*
EU And, *360*, *441*
Egg Nebula – *see* AFGL 2688
FJF 270, *440*–441
Frosty Leo, *9*, *15*, *55*, *57*, *380*
GP Ori, *291*–293
HBV 475, *222*
HD 101584 – *see* IRAS 11385−5517
HD 107369, *260*, *272*
HD 112374, *260*
HD 119608, *272*
HD 137569, *31*
HD 161796 – *see* IRAS 17436+5003
HD 168625, *309*–312
HD 172481, *283*–288
HD 177566, *273*
HD 179821 – *see* IRAS 19114+0002
HD 187885 – *see* IRAS 19500−1709
HD 326971, *31*
HD 331319, *301*–303
HD 44179 – *see* Red Rectangle
HD 46703 – *see* IRAS 06338+5333
HD 52961, *272*, *279*, *314*
HD 56126 – *see* IRAS 07134+1005
HD 95767, *272*
HD 97534, *296*
HH 30, *61*
HM Sge, *222*, *225*
HR 1865, *284*
HR 4049, *279*, *314*
HR Car, *312*
H 1−36, *222*
H 2−2, *222*
Hb 12, *59*, *410*–412
Hb 4, *194*
Hb 5, *115*–119
Hb 7, *33*
Hen 1−1, *155*
Hen 2−104, *221*–224
Hen 2−113, *194*

Hen 2−115, *59*
Hen 2−138 – *see* IRAS 15513−6600
Hen 2−142, *59*, *163*, *165*–166
Hen 2−147, *222*
Hen 2−161 – *see* IRAS 16206−5956
Hen 2−180, *60*
Hen 2−320, *60*
Hen 2−351, *61*
Hen 2−436, *160*
Hen 2−437, *194*
Hen 2−47, *60*
Hen 2−90, *58*, *61*
Hen 3−1333, *163*, *165*–166, *194*
Hen 3−1357 – *see* IRAS 17119−5926
Hen 3−1383, *222*
Hen 3−1475 – *see* IRAS 17423−1755
Hen 3−401 – *see* IRAS 10178−5958
Hu 2−1, *156*
IC 2553, *153*–155
IC 418, *58*, *61*
IC 4593, *156*
IRAS 01005+7910, *317*–320
IRAS 02143+5852, *46*–47
IRAS 04296+3429, *5*, *46*–47, *55*, *57*, *112*, *267*–268, *272*
IRAS 04440+2605, *267*, *272*
IRAS 04496−6958, *439*
IRAS 05040+4820, *46*–47
IRAS 05089+0459, *46*–47
IRAS 05113+1347, *46*–47, *112*
IRAS 05238−0626, *46*–47
IRAS 05341+0852, *46*–47, *272*, *287*, *314*
IRAS 05381+1012, *46*–47, *268*
IRAS 06176−1036, *66*
IRAS 06338+5333, *46*–47, *272*
IRAS 06530−0213, *46*–47, *55*
IRAS 06556+1623, *31*, *386*
IRAS 06562−0337, *386*
IRAS 07027−7934, *159*, *194*, *386*
IRAS 07131−0147, *46*–47, *380*
IRAS 07134+1005, *5*, *66*–68, *105*–110, *112*, *266*–267, *272*, *314*
IRAS 07171+1823, *31*, *46*–47
IRAS 07253−2001, *46*–47
IRAS 07284−0940, *267*
IRAS 07331+0021, *267*, *272*, *419*, *421*
IRAS 07399−1435, *66*, *190*–193, *364*, *380*, *419*, *421*
IRAS 07430+1115, *46*–47, *103*, *112*, *261*, *272*, *287*
IRAS 08005−2356, *193*–194, *419*
IRAS 08187−1905, *46*–47
IRAS 09276+4454, *267*
IRAS 09371+1212 – *see* Frosty Leo
IRAS 09470−4617, *31*
IRAS 10082−5647, *31*
IRAS 10178−5958, *31*, *34*, *36*, *55*, *57*, *380*, *386*
IRAS 10197−5750, *380*, *386*
IRAS 10215−5916, *305*–306, *308*
IRAS 10301+0933, *272*
IRAS 11065−6026, *31*
IRAS 11385−5517, *193*, *272*, *295*–298, *306*, *419*
IRAS 12419−5414, *9*
IRAS 12584−4837, *31*, *35*
IRAS 13110−5425, *25*–26
IRAS 13185−6922, *31*
IRAS 13266−5551, *31*, *35*
IRAS 14072−5446, *31*
IRAS 14331−6435, *31*, *34*–35, *385*–386
IRAS 14488−5405, *31*
IRAS 15154−5258, *31*
IRAS 15405−4945, *193*, *420*
IRAS 15513−6600, *31*, *59*
IRAS 16206−5956, *31*, *61*, *272*
IRAS 16342−3814, *55*, *57*, *66*–68, *193*–194, *419*–421
IRAS 16559−2957, *419*
IRAS 16594−4656, *55*, *198*, *200*, *380*, *385*–386
IRAS 17041−2709, *31*
IRAS 17074−1845, *31*, *35*
IRAS 17106−3046, *7*–8, *200*–202
IRAS 17119−5926, *29*–34, *36*–37, *39*, *380*, *384*, *386*
IRAS 17150−3224, *8*, *55*, *66*–68, *115*–119, *198*–199, *298*, *324*, *380*, *384*, *386*, *419*
IRAS 17203−1534, *31*, *34*–35
IRAS 17245−3951, *55*, *198*, *419*
IRAS 17253−2831, *193*
IRAS 17279−1119, *273*
IRAS 17308+0822, *51*
IRAS 17311−4924, *31*, *34*–35, *386*
IRAS 17347−3139, *420*
IRAS 17381−1616, *31*
IRAS 17418−4604, *31*
IRAS 17423−1755, *31*, *34*, *36*, *55*, *67*–68, *193*, *209*–211, *213*–214, *419*
IRAS 17427−3010, *31*, *59*
IRAS 17436+5003, *8*, *66*–68, *103*, *260*, *267*, *273*, *313*–314, *324*, *333*–337, *369*, *371*, *373*–374, *410*–412, *419*
IRAS 17441−2411, *67*–68, *115*, *118*–119, *198*–199, *298*, *380*
IRAS 17460−3114, *31*
IRAS 17476−4446, *25*
IRAS 18023−3409, *31*
IRAS 18025−3906, *260*, *419*
IRAS 18062+2410, *8*, *29*–31, *33*–34, *39*, *267*, *273*, *386*
IRAS 18095+2704, *5*, *102*, *104*, *267*, *273*, *324*, *419*, *423*–425
IRAS 18131−3008, *31*
IRAS 18237−0715, *31*
IRAS 18281+2149 – *see* AC Her
IRAS 18313−1738, *31*
IRAS 18371−3159, *31*

IRAS 18379−1707, *31*, *35*
IRAS 18442−1144, *31*
IRAS 18455+0448, *423*–425, *427*
IRAS 18491−0207, *193*, *420*
IRAS 18548+0823, *52*
IRAS 19035+0801, *52*
IRAS 19102+0329, *267*
IRAS 19114+0002, *66*–68, *103*, *267*, *313*–316, *339*–341, *369*, *371*, *374*, *410*–416, *419*, *423*
IRAS 19134+2131, *421*
IRAS 19157−0247, *31*
IRAS 19296+2227, *421*
IRAS 19312+1950, *421*
IRAS 19336−0400, *31*
IRAS 19386+0155, *423*
IRAS 19399+2312, *31*
IRAS 19477+2401, *200*
IRAS 19500−1709, *5*, *259*, *273*, *314*, *325*–326, *407*
IRAS 19566+3423, *421*, *423*, *425*–427
IRAS 19590−1249, *31*, *34*, *273*
IRAS 20000+3239, *103*
IRAS 20004+2955, *103*, *267*, *273*, *324*
IRAS 20028+3910, *200*
IRAS 20160+2734, *52*
IRAS 20462+3416, *31*, *33*–34
IRAS 20531+2909, *424*
IRAS 21282+5050, *407*–408
IRAS 22023+5249, *31*
IRAS 22036+5306, *193*
IRAS 22223+4327, *102*, *104*, *267*–268, *273*
IRAS 22272+5435, *5*, *7*, *55*, *66*–68, *102*, *104*, *112*, *267*, *273*, *287*, *291*–292, *294*, *326*, *369*, *371*–372, *374*, *407*–408
IRAS 22495+5134, *31*
IRAS 22574+6609, *200*
IRAS 23304+6147, *5*, *46*–47, *103*, *267*–268, *273*
IRAS Z02229+6208, *272*, *287*
IRC+10216, *ix*, *96*–98, *100*, *115*, *119*–121, *123*, *125*, *241*–244, *289*, *343*–350, *359*, *361*–362, *418*
IRC+10420, *3*, *266*–267, *269*, *419*
IW Car, *272*
K 2−16, *163*, *165*
K 3−35, *194*
K 3−66, *59*
K 648, *77*–80, *82*
L^2 Pup, *98*
LB 3116, *273*
LB 3193, *272*
LB 3219, *272*
LS II +34 26 – *see* IRAS 20462+3416
LS IV −12 111 – *see* IRAS 19590−1249
LS IV −4 01, *273*
LS 1021, *31*
LS 1339, *31*
LS 1392, *31*
LS 3591 – *see* IRAS 16206−5956
LS 4039, *273*
LS 4825, *273*
MQ Cyg, *441*
MSX050825.4−685359, *75*
MSX052048.5−701210, *75*
MSX052148.5−700955, *75*
MSX053010.3−690932, *75*
MSX053307.1−703033, *75*
MSX053349.0−701325, *75*
M 1−26 – *see* IRAS 17427−3010
M 1−37, *60*
M 1−92, *9*, *56*, *380*, *411*–412, *419*
M 15, *77*–78
M 2−43, *163*, *165*
M 2−9, *3*, *36*, *194*, *410*–412
M 22, *78*
M 31, *77*–78, *81*–82
M 4−18, *60*
Me 2−2, *60*
MyCn 18, *56*
Mz 3, *9*, *410*–412
NC 83, *441*
NGC 205, *77*, *82*
NGC 5882, *153*, *155*–156
NGC 5986, *77*–78, *80*–81
NGC 6302, *194*, *410*–412, *418*
NGC 6441, *78*
NGC 6543, *115*–116, *118*–119
NGC 6572, *155*
NGC 6720, *407*–408
NGC 6826, *137*
NGC 6891, *137*
NGC 7027, *22*, *58*, *115*–116, *118*–119, *327*–328, *364*, *370*–371, *397*, *401*, *405*–408
NGC 7662, *155*–156
NSV 2814, *439*–442
OH 0.9+1.3, *420*
OH 12.8−0.9, *421*
OH 15.7+0.8, *420*, *430*–432
OH 17.7−2.0, *420*, *430*–432
OH 231.8+4.2 – *see* IRAS 07399−1435
OH 37.1−0.8, *421*
OH 77.9+0.2, *430*, *432*
OH 85.4+0.1, *430*, *432*
PG 1159 stars, *163*, *249*, *251*, *253*
PG 1323−086, *34*, *272*
PG 1704+222, *34*, *273*
PHL 1580, *273*
PHL 174, *273*
Pal 6, *78*
ROA 24, *80*, *272*
RR Lyr variables, *81*–82
RS Oph, *222*
RV Boo, *360*
RV Tau – see IRAS 04440+2605
RV Tau stars, *8*, *17*, *101*, *107*–108, *110*, *113*–114, *262*–263, *271*

RX Pup, *221*–222, *225*, *227*–233
R Aql, *434*
R Aqr, *222*, *225*, *232*
R Cas, *434*
R Scl, *97*, *99*, *127*–130
R Sct, *273*, *411*–412
R Sge, *273*
Red Rectangle, *9*, *262*–263, *314*, *351*–353, *360*, *405*, *407*–408
Roberts 22, *9*, *55*, *193*–194, *419*
SAO 163075 – *see* IRAS 19500−1709
SAO 239853, *259*–260, *272*
SAO 243754, *272*
SAO 34504 – *see* IRAS 22272+5435
SAO 85766 – *see* IRAS 18062+2410
SMP 58, *160*
SNP 61, *160*
SS 96, *222*
SS Gem, *272*
S Sct, *99*, *127*, *358*
Sakurai's object, *249*, *251*, *255*–256, *267*
Sp 3, *33*
SwSt 1, *163*–165
TT Cyg, *99*, *127*, *135*, *138*, *358*
TT Oph, *272*
TW Cam, *272*
T Ari, *98*
T Cep, *434*
UU Her, *267*, *272*
UU Her stars, *8*, *101*
UY Ara, *273*
U Ant, *127*, *129*–130
U Cam, *127*, *358*
U Equ, *360*, *363*
U Hya, *127*
U Mon, *9*, *272*
V1016 Cyg, *3*, *222*
V1415 Cyg, *441*
V1511 Cyg, *423*, *426*–427
V348 Sgr, *252*
V360 Cyg, *273*
V417 Cen, *222*
V453 Oph, *273*
V605 Aql, *252*
V778 Cyg, *439*–442
VV Cep, *235*–238
V Hya, *359*
Vy 2−2, *418*–419, *421*
W43A, *421*
Y CVn, *127*
Y Tau, *291*–293
Z Cas, *434*
Z Cyg, *434* –435

Author Index

Acker A., *163*
Alksnis A., *289*
Altmann M., *277*
Alves, *77*
Bachiller R., *215*
Balega Y.Yu., *241*, *343*
Barlow M.J., *409*
Barthès D., *105*
Blöcker T., ***241***, *343*
Bobrowsky M., *209*
Bond H.E., ***77***
Brandi E., *227*
Bujarrabal V., *403*, *409*
Castro-Carrizo A., *403*, ***409***
Cernicharo J., *397*
Chen P.S., *439*
Cohen M., *71*
Conway G., *383*
Corradi R.L.M., *153*, ***221***
Cox P., ***387***
Dayal A., *339*
Delgado D.G., ***127***
Deutsch L.K., *339*
Dominguez–Tagle C., *383*
Dominik C., *121*
Engels D., *49*, ***417***
Fazio G., *339*
Ferrer O.E., *227*
Fokin A., *105*
Fong D., *369*, ***403***, ***409***
Frankowski A., ***173***
Fujii T., *29*, ***45***, ***111***
García L.G., *227*
García-Lario P., *21*, *49*, *209*, *295*, *301*, *305*, ***309***, ***329***, *383*
García-Segura G., *147*
García-Hernández D.A., ***383***
Gatley I., *377*
Gauba G., *29*
Gawryszczak A.J., ***205***
Gesicki K., ***163***, ***167***
Gillet D., *105*
Gonçalves D.R., ***153***
Graczyk D., ***235***
Górny S.K., *13*
Heber U., *277*
Herpin F., ***397***
Herrmann M., *277*
Herwig F., ***249***
Hoffmann W.F., *339*
Hora J.L., *339*
Hrivnak B.J., ***101***, ***197***, *323*
Hu J., ***317***
Huggins P.J., *387*
Icke V., *121*
Jasniewicz G., *313*
Jeske K., *139*
Jiménez-Esteban F., ***49***
Josselin E., ***215***, ***413***
Kastner J.H., *377*
Kataza H., *351*
Kelly D.M., *403*
Klochkova V.G., ***265***
Kwok S., ***3***, ***115***, *197*, *323*
Lèbre A., ***105***, *413*
Latter W.B., *403*, *409*
Le Bertre T., *93*
Lewis B.M., ***423***
Lucas R., *387*
Maillard J.-P., *387*
Mampaso A., *153*
Manchado A., ***21***, *147*, *209*, *295*, *301*, *305*, *309*, *383*
Manteiga M., *21*
Marang F., *227*
Matsuura M., ***433***
Mauron N., *105*, *215*
Meixner M., *65*, *333*, *339*, ***369***, *403*, *409*
Men'shchikov A.B., *241*, ***343***
Mikołajewska J., *205*, ***227***
Mikołajewski M., *235*
Miyata T., ***351***
Murakami H., *433*
Nakada Y., *29*, *45*, *111*
Napiwotzki R., ***277***
Okamoto Y., *351*
Olofsson H., *127*
Omont A., ***357***
Onaka T., *351*, *433*

Osterbart R., *241*, *343*
Panchuk V.E., *265*
Parthasarathy M., ***29***, *45*, *111*, *295*, *301*, *305*, *309*, *313*
Perea Calderøn J.V., *329*
Perinotto M., *153*
Planesas P., *215*
Pottasch S.R., *21*, *295*
Prada F., *383*
Rózyczka M., *205*
Reyniers M., *257*, ***283***
Richards A.M.S., *429*
Riera A., ***209***
Sahai R., ***53***
Sako S., *351*
Schönberner D., ***85***, *131*
Schmidt M.R., ***271***
Siódmiak N., *139*
Simis Y., ***121***
Sivarani T., ***295***, ***301***, ***305***, *309*
Soker N., ***181***
Spēlmanis R., *289*
Speck A.K., ***333***, *339*
Steffen M., *85*, ***131***
Stoesz J.A., *115*
Su K.Y.L., *115*, *197*
Suárez O., *21*
Sutton E.C., *369*, *403*
Szczerba R., ***13***, *139*, *265*, *323*, *439*
Szymczak M., ***429***, ***439***
Takubo S., *351*
Thévenin F., ***313***
Tielens A.G.G.M., *403*, *409*
Tylenda R., ***139***
Ueta T., ***65***, *333*, ***339***
van Hoof P.A.M., *41*
Van Winckel H., ***257***, *283*
Van de Steene G.C., ***41***
Villaver E., ***147***, *153*
Volk K., ***323***
Weigelt G., *241*, *343*
Weintraub D.A., ***377***
Welch W.J., *369*
Whitelock P.A., *227*
Winters J.M., ***93***
Wood P.R., *41*, ***71***
Yamamura I., *433*
Yamashita T., *351*
Začs L., ***289***
Zalfresso–Jundziłło M., *13*
Zijlstra A.A., ***157***, *167*, ***189***